# WILEY BLACKWELL STUDENT DICTIONARY OF HUMAN EVOLUTION

# WILEY BLACKWELL STUDENT DICTIONARY OF HUMAN EVOLUTION

Edited by
**BERNARD WOOD**
The George Washington University

Executive Editor
**AMANDA HENRY**
Max Planck Institute for Evolutionary Anthropology

Editorial Assistant
**KEVIN HATALA**
Max Planck Institute for Evolutionary Anthropology

**WILEY Blackwell**

This edition first published 2015 © 2015 by John Wiley & Sons, Ltd

*Registered Office*
John Wiley & Sons, Ltd, The Atrium, Southern Gate, Chichester, West Sussex, PO19 8SQ, UK

*Editorial Offices*
9600 Garsington Road, Oxford, OX4 2DQ, UK
The Atrium, Southern Gate, Chichester, West Sussex, PO19 8SQ, UK
111 River Street, Hoboken, NJ 07030-5774, USA

For details of our global editorial offices, for customer services and for information about how to apply for permission to reuse the copyright material in this book please see our website at www.wiley.com/wiley-blackwell.

The right of the author to be identified as the author of this work has been asserted in accordance with the UK Copyright, Designs and Patents Act 1988.

All rights reserved. No part of this publication may be reproduced, stored in a retrieval system, or transmitted, in any form or by any means, electronic, mechanical, photocopying, recording or otherwise, except as permitted by the UK Copyright, Designs and Patents Act 1988, without the prior permission of the publisher.

Designations used by companies to distinguish their products are often claimed as trademarks. All brand names and product names used in this book are trade names, service marks, trademarks or registered trademarks of their respective owners. The publisher is not associated with any product or vendor mentioned in this book.

Limit of Liability/Disclaimer of Warranty: While the publisher and author(s) have used their best efforts in preparing this book, they make no representations or warranties with respect to the accuracy or completeness of the contents of this book and specifically disclaim any implied warranties of merchantability or fitness for a particular purpose. It is sold on the understanding that the publisher is not engaged in rendering professional services and neither the publisher nor the author shall be liable for damages arising herefrom. If professional advice or other expert assistance is required, the services of a competent professional should be sought.

*Library of Congress Cataloging-in-Publication Data applied for*

A catalogue record for this book is available from the British Library.

Wiley also publishes its books in a variety of electronic formats. Some content that appears in print may not be available in electronic books.

Set in 10/13pt Minion by SPi Publisher Services, Pondicherry, India
Printed and bound in Malaysia by Vivar Printing Sdn Bhd

1   2015

# CONTENTS

Preface and Acknowledgments ................................................................................ vi
*Wiley-Blackwell Encyclopedia of Human Evolution* .............................................. viii
Hominin Fossil Abbreviations ................................................................................. ix

| | | | |
|---|---|---|---|
| A.................................................. 1 | | N.................................................. 288 | |
| B.................................................. 27 | | O.................................................. 300 | |
| C.................................................. 47 | | P.................................................. 318 | |
| D.................................................. 86 | | Q.................................................. 364 | |
| E.................................................. 103 | | R.................................................. 366 | |
| F.................................................. 124 | | S.................................................. 380 | |
| G.................................................. 144 | | T.................................................. 419 | |
| H.................................................. 166 | | U.................................................. 442 | |
| I.................................................. 200 | | V.................................................. 447 | |
| J.................................................. 209 | | W.................................................. 453 | |
| K.................................................. 212 | | X.................................................. 458 | |
| L.................................................. 227 | | Y.................................................. 459 | |
| M.................................................. 249 | | Z.................................................. 460 | |

# PREFACE AND ACKNOWLEDGMENTS

Not long ago the only information a student needed in order to do well in a course about human evolution was an appreciation of general evolutionary principles, a familiarity with a relatively sparse fossil record and its context, and knowledge of a few simple analytical methods. But times have changed. The fossil record has grown exponentially, imaging techniques allow researchers to capture previously unavailable gross morphological and microstructural evidence in previously unimaginable quantities, analytical methods have burgeoned in scope and complexity, phylogeny reconstruction is more sophisticated, molecular biology has revolutionized our understanding of genetics, evolutionary history, modern human variation, and development, and a host of different advances in biology, chemistry, earth sciences, and physics have enriched evidence about the biotic, climatic, and temporal context of the hominin fossil record. In short, the fossil evidence and the range of methods used to study human evolution have grown by several orders of magnitude in the past six decades. Yet there is no single reference source where students can go to find out about topics as diverse as sagittal crest, *Sahelanthropus tchadensis*, Saint-Césaire, sampling with replacement, the Sangiran Dome, sapropel, savanna, and satellite imagery.

The *Wiley Blackwell Student Dictionary of Human Evolution* is based on the principles that were used to determine the content of the *Wiley Blackwell Encyclopedia of Human Evolution*, but the layout and content are deliberately different and new. We used our combined student and teaching experience to cull the entries in the *Encyclopedia*, select the ones most relevant to students, and then rewrite them with an emphasis on explaining the relevance of each entry to studies of human evolution. We are indebted to all the editors and contributors who were involved in the assembly of the *Encyclopedia*, for without that as a template, our task would have been much more difficult.

Kelvin Matthews at Wiley Blackwell, and Nik Prowse, our freelance copy editor and project manager, made substantial and important contributions to any success this student dictionary enjoys. We are also grateful to those who helped us improve the text. Laurel Poolman, a George Washington University undergraduate archeology major, read through an early draft and alerted us to topics we needed to explain more clearly or where we needed to do a better job of explaining why they were included in the *Dictionary*. After BW and AH responded to these suggestions the revised text was read in its entirety by two students in George Washington University's hominid paleobiology graduate program, Kevin Hatala and Laura Reyes. Their comments were invaluable, in terms of both catching errors and making many constructive suggestions for improvement. Charlotte Krohn's help with the final stages of preparing the manuscript is greatly appreciated. However, despite the best efforts of Laurel, Kevin, and Laura, in a project like this errors will have been made. If you see one, please contact us (bernardawood@gmail.com, amanda_henry@eva.mpg.de) and we will make sure it is corrected in later editions.

Bernard Wood<br>
Amanda Henry<br>
July 2014

# WILEY-BLACKWELL ENCYCLOPEDIA OF HUMAN EVOLUTION

This comprehensive A to Z encyclopedia provides extensive coverage of important scientific terms related to improving our understanding of how we evolved. Specifically, the 5,000 entries cover evidence and methods used to investigate the relationships among the living great apes, evidence about what makes the behavior of modern humans distinctive, and evidence about the evolutionary history of that distinctiveness, as well as information about modern methods used to trace the recent evolutionary history of modern human populations. This text provides a resource for everyone involved in the study of human evolution.

**Visit the companion site www.woodhumanevolution.com to browse additional references and updates from this comprehensive encyclopedia.**

# HOMININ FOSSIL ABBREVIATIONS

| | |
|---|---|
| A.L. (or AL) | Lower Awash River, Hadar, Ethiopia |
| ALA-VP | Alayla – Vertebrate Paleontology, Western Margin, Middle Awash, Ethiopia |
| ARA-VP | Aramis – Vertebrate Paleontology, Middle Awash, Ethiopia |
| ATD | Atapuerca – Gran Dolina, Sierra de Atapuerca, northern Spain |
| ATE | Atapuerca – Sima del Elefante, Sierra de Atapuerca, northern Spain |
| BAR | prefix for the fossils recovered at Lukeino, Tugen Hills, Baringo District, Kenya from 2000 onwards (e.g., BAR 1000'00, the holotype of *Orrorin tugenensis*) |
| BBC | Blombos Cave, South Africa |
| BC | Baringo Chemeron Formation, Kenya |
| BEL-VP | Belohdelie – Vertebrate Paleontology, Ethiopia |
| BK | Baringo – Kapthurin, Kenya, and Blimbingkulon, Indonesia |
| BOD | Bodo, Ethiopia |
| BOU | Bouri, Middle Awash, Ethiopia |
| BOU-VP | Bouri – Vertebrate Paleontology, Middle Awash, Ethiopia |
| BS | Busidima Formation, Gona, Ethiopia |
| DIK | Dikika, Ethiopia |
| DNH | Drimolen hominid, Drimolen, South Africa |
| HCRP UR | Hominid Corridor Research Project, Uraha, Malawi |
| KBS | Kay Behrensmeyer site, Koobi Fora, Kenya |
| KHS | Kamoya's Hominid Site, Kibish, Omo Basin, Ethiopia |
| KNM-ER | Kenya National Museum followed by the site code for East Rudolf (now referred to as Koobi Fora), Kenya |
| KNM-KP | Kenya National Museum followed by the site code for Kanapoi, Kenya |
| KNM-LT | Kenya National Museum followed by the site code for Lothagam, Kenya |
| KNM-LU | Kenya National Museum followed by the site code for Lukeino, Tugen Hills, Baringo region, Kenya |
| KNM-TH | Kenya National Museum followed by the site code for the Tugen Hills, Baringo region, Kenya |

| | |
|---|---|
| KNM-WT | Kenya National Museum followed by the site code for **W**est **T**urkana, Kenya |
| L. | prefix used by the American contingent of the International Omo Research Expedition for localities in the Shungura Formation (e.g., L. 396). The prefix is also used for fossils found within one of the L. localities (e.g., L. 40-19). |
| LB | Liang **B**ua, Flores, Indonesia |
| LH (or L.H.) | Laetoli **h**ominin, Tanzania |
| MH | **M**alapa **h**ominin |
| MLD | **M**akapansgat **L**imeworks **D**eposit, South Africa |
| MKM | **M**a**k**ha **M**era collection area, Woranso-Mille study area, Central Afar, Ethiopia |
| OH (or O.H.) | **O**lduvai **h**ominid, Tanzania |
| Omo | prefix for fossils collected by the French-led group from the Shungura Formation, **Omo** Basin, Ethiopia |
| SK | **S**wart**k**rans hominin, South Africa |
| Sts | fossil hominins recovered from the **St**erkfontein type **s**ite between 1947 and 1949, South Africa |
| Stw or StW | **St**erkfontein **W**itwatersrand, South Africa |
| TD | **T**rinchera **D**olina, Gran Dolina, Sierra de Atapuerca, Spain |
| TM | **T**ransvaal **M**useum, South Africa |
| UA | **U**adi **A**alad, Ethiopia |
| UR | **Ur**aha, Malawi |
| WLH | **W**illandra **L**akes **h**ominid, Australia |

### a
The abbreviated form of annum (from L. *annus* = year). In 2006 the joint IUPAC-IUGS Task Group urged that the Systéme International (SI) unit "a" be used for both ages and time spans (e.g., 36 ka for 36 thousand years and 2.3 Ma for 2.3 million years). The same report discouraged the use of y, yr, and yrs in combination with k, K, m, M, etc.

### abductor
A muscle that moves a limb away from the midline reference plane (e.g., deltoid, gluteus medius) or a digit away from the reference digit (e.g., dorsal interossei of the hand).

### abiotic
All "nonliving" factors (e.g., climate, physical catastrophies such as massive volcanic eruptions or tsunamis, etc.) that might have influenced the outcome of human evolution. (Gk *a* = not and *bios* = life.) *Compare to* **biotic**.

### absolute dating
Dating methods (e.g., potassium-argon, radiocarbon, thermoluminescence, and uranium-series) that are based on physical or chemical systems with predictable dynamics. Geochronologists are moving away from the old categories of absolute and relative dating methods. Instead, they refer to absolute dating methods as methods that provide a "numerical age estimate." (L. *absolutus* = free or unrestrained.) *See also* **geochronology**.

### acceleration
A technical term used in growth and development studies. It refers to a change in the relative timing of events that results in the acceleration of shape change without any corresponding increase in the rate of change in the size of the individual. Adult size and the duration of growth are unchanged. (L. *accelerationem* = a hastening.) *See also* **heterochrony**; **neoteny**; **pedomorphosis**; **peramorphosis**.

---

*Wiley Blackwell Student Dictionary of Human Evolution*, First Edition. Edited by Bernard Wood.
© 2015 John Wiley & Sons, Ltd. Published 2015 by John Wiley & Sons, Ltd.

### accessory cusp
A cusp on a maxillary (upper) or mandibular (lower) molar tooth that is not one of the main cusps. Examples of accessory cusps are the metaconule, which is between the metacone and protocone on a maxillary molar, and the tuberculum sextum, which is between the entoconid and the hypoconulid on a mandibular molar. *See also* **cusp**.

### accretion model
A model that suggests that the distinctive morphology of *Homo neanderthalensis* emerged gradually (i.e., accreted) over a period of several hundred thousand years. Fossils with different levels of expression of these features are divided into stages, with Stage 1 being the most primitive and Stage 4 being the most derived. (L. *accrescere* = to grow.) *See also* **Homo heidelbergensis**.

### Aché
A group of modern human foragers (also known as the Guayaki) that lives in eastern Paraguay. Behavioral ecological studies of their diet suggest that both plant and animal foods are dietary staples and the different foraging goals of men and women reflect this. Among the Aché hunting may be a social signal as well as a method of provisioning.

### Acheulean
A stone tool industry proposed in the 19thC by Gabriel de Mortillet that takes its name from the French village of Saint-Acheul in the Somme river valley. It is defined by handaxes and similar implements (e.g., cleavers and picks), but these tool types are not confined to the Acheulean. However, when they occur in other contexts they are rare and are typically outnumbered by flakes, cores, and other smaller modified tools such as scrapers. The Oldowan industry preceded the Acheulean and many of the nonhandaxe Acheulean tools are Oldowan-like. Similarly, some later Acheulean sites (e.g., the Somme river valley and in the Kapthurin Formation of Kenya) show evidence of the use of Levallois technology for the production of large flakes. The Acheulean, which is also known as the Acheulean industrial complex, is unique in the sense that neither before nor since has such a distinctive technology dominated the activities of hominins for so long over so much of the planet. Currently the earliest evidence of the Acheulean is found at the 1.76 Ma site of Kokiselei, in West Turkana, Kenya; the most recent sites date to *c*.0.16 Ma.

### Acheulian
*See* **Acheulean**.

### actualistic studies
Studies in which researchers try to recreate objects and circumstances encountered by archeologists and paleontologists. For example, in archeology, researchers use replicas of ancient tools in controlled circumstances to help determine what the ancient tools were used for. In paleontology, actualistic research involves studying the factors that determine the formation and nature of present-day bone assemblages and then applying that knowledge to the paleontological record. Actualistic studies are explicitly uniformitarian in that they assume that the objects and processes used in the past and the present have the same functions, products, or outcomes. (L. *actus* = an act.) *See also* **uniformitarianism**.

## adaptation

A useful feature or trait that (a) promotes survival and reproduction and (b) is shaped by natural selection. Adaptations must be heritable *and* perform a function. Adaptation can also be used as an adjective in connection with a taxon, as in "the dentition of *Paranthropus boisei* is better adapted for chewing than for slicing food." In such usage, adaptation is being used in an informal sense (i.e., "better adapted" can be read to mean "functions better"). Adaptations will tend to be under-recognized (i.e., the process of recognizing adaptations is prone to Type I error). It can be difficult to establish whether a given trait has been subjected to natural selection in fossil hominins, so the identification of adaptations in paleoanthropology is inevitably conjectural. Formulating and testing hypotheses of adaptation is a major focus of paleoanthropological research. (L. *adaptare* = to fit.) *See also* **structure–function relationship**.

## adaptive radiation

The rapid diversification of a lineage into species that evolve a range of new adaptive strategies that enable them to occupy new adaptive zones. An example is the simultaneous appearance of multiple pig species at the same sites at several times during the Pliocene and Pleistocene. During the last 4 Ma approximately six pig species have coexisted at the same locality (i.e., they are sympatric) at the same time (i.e., they are synchronic). A number of these species exhibit tall (hypsodont) third molar crowns and craniofacial elaboration (bosses, crests, tusks). Such changes are usually interpreted as an adaptive response to more open grassland environments. Archaic hominins may be an example of an adaptive radiation. (L. *adaptare* = to fit and *radius* = ray.)

## adaptive strategy

Any set of traits that enables the members of a species to survive and reproduce. When different species have similar adaptive strategies, it could be because (a) they were inherited from a recent common ancestor or (b) they may be the result of independent adaptations to similar environmental conditions (i.e., parallel evolution or convergent evolution). In the latter case the morphology involved would be a homoplasy. (L. *adaptare* = to fit and *strategos* = general.) *See also* **grade**.

## adductor

Any muscle that moves a limb towards the midline reference plane (e.g., teres major, adductor longus), or a digit towards the reference digit (e.g., palmar interossei of the hand).

## aDNA

*See* **ancient DNA**.

## adolescence

The period of life history in modern humans between puberty and maturity. Adolescence ends when skeletal lengths and dental development reach their adult state, and when sexual maturation is attained. This usually occurs between 17 and 25 years in modern human populations. The defining characteristic of human adolescence is a rapid height increase; nonhuman primate species undergo growth spurts in craniofacial dimensions and overall body mass, but not height. The intensity and duration of this height spurt varies among and within modern human populations. It has been suggested that the adolescent stage evolved in either premodern *Homo* or anatomically modern *Homo sapiens*. It is difficult to determine whether extinct hominin taxa underwent a growth spurt and, if so, whether it was more similar to that of modern humans or to that of nonhuman primates. (L. *adolescentia* = youth.)

## aeolian
Sediments deposited primarily by wind action. The large volumes of glacially derived silts subject to aeolian transport are responsible for the loess deposits that are common in central China. Some of the sediments in the Laetolil Beds at Laetoli, Tanzania, are formed from airfall tuffs that have been reworked by aeolian processes. (Gk *Aeolus* = god of the winds.)

## Afar Rift System
The part of the East African Rift System comprising a series of river valleys and basins that are mainly in modern-day Ethiopia and Eritrea. The Dikika, Gona, and Middle Awash study areas are all in the Afar Rift System.

## aff.
Abbreviation of **affinity** (*which see*).

## affinity
A term used in taxonomy (usually abbreviated as "aff.") to suggest that a specimen belongs to a hypodigm that is closely related to, but not necessarily synonymous with, a taxon. Thus, a small piece of thick cranial vault might be assigned to "*Homo* aff. *H. erectus*."

## Afro-European hypothesis
See *Homo heidelbergensis*; **out-of-Africa hypothesis**; **replacement with hybridization**.

## age estimate
The number of years that are estimated to have elapsed between an event (e.g., the deposition of a bone or artifact) and the present day. Paleoanthropological age estimates are expressed in thousands (ka) and millions (Ma) of years. *See also* **geochronology**.

## agenesis
Absence or lack of development of an anatomical structure (e.g., third molars). Examples of agenesis in fossil hominins include at least one individual of *Homo floresiensis*, LB1, that shows agenesis of the lower right second premolar ($RP_4$) and upper right third molar ($RM^3$). The *Homo ergaster* associated skeleton, KNM-WT 15000, has agenesis of both lower third molars ($M_3$s). (Gk *a* = absence or without, and *genesis* = birth or origin.)

## *Ailuropoda-Stegodon* fauna
A cave fauna named after two consistent components: *Ailuropoda*, the only genus in the subfamily Ailuropodinae of ursids (i.e., bears), and *Stegodon*, a genus of proboscideans (i.e., elephant precursors) within the extinct subfamily Stegodontinae. The *Ailuropoda-Stegodon* fauna is found in caves in southern China, Vietnam, and Laos. The presence of these fossils has been used as a means of dating several East Asian sites.

## Ain Hanech
This $c.1.8$ Ma Algerian site (also known as Aïn Hanech) contains some of the oldest stone artifacts in North Africa. The artifacts found in the early layers are considered a North African variant of the Oldowan industrial complex. Artifacts recovered from overlying sediments have been attributed to the Acheulean industrial complex. (Location 36°16′39″N, 08°19′00″E, Algeria.)

## A.L. 288-1
Also known as "Lucy," this specimen was the first relatively complete hominin associated skeleton of great antiquity and it remains the best preserved associated skeleton of *Australopithecus afarensis*. It components were found on the surface at Hadar, Ethiopia, in 1974 by Donald Johanson and his team; it is dated to $c.3.2$ Ma. The cranial vault remains include portions of the parietals, occipital, left zygomatic, and frontal bones. The mandible includes the right $P_3$–$M_3$, and the left $P_3$, $M_3$, and two $M_1$ fragments. The postcranial skeleton is represented by the right scapula, humerus, ulna, radius, a portion of the clavicle, the left ulna, radius, and capitate, and the axial skeleton by lumbar and thoracic vertebrae and ribs. The left pelvic bone, sacrum, and left femur are well preserved. Remains of the right leg include fragments of the tibia, fibula, talus, and some foot and hand phalanges. This individual has an endocranial volume of 375–400 cm$^3$.

## A.L. 444-2a–h
The first well-preserved (75–80% complete) skull of *Australopithecus afarensis* that was found at Hadar in 1992 by Yoel Rak; it is dated to $c.3.0$ Ma. The skull includes parts of the cranial vault, the face, maxillary dentition, the right side of the mandibular corpus, the symphyseal region and part of the mandibular dentition.

## A.L. 666-1
The first example of a *Homo habilis*-like morphotype in the middle Pliocene. It was found at Hadar in 1994 in a layer with Oldowan tools; it is dated to $c.2.35$ Ma. It comprises a maxilla, broken along the intermaxillary suture, with the left $P^3$–$M^3$, the right $P^3$–$M^1$ with $M^2$ and $M^3$ roots, plus other isolated dental fragments.

## ALA-VP-2/10
The holotype of *Ardipithecus kadabba* was found at Alayla, Ethiopia, in 1999; its age is $c.5.8$–$5.2$ Ma. It comprises the right side of the mandibular corpus with $M_3$, together with associated teeth (left $I_2$, C, $P_4$, $M_2$, and part of the $M_3$ root).

## albumin
This protein, which is present in plasma serum, was given its name because it turns white when it is heated or coagulated. It was one of the first molecules that was used to precisely measure the closeness of the relationships between the extant great apes. When fresh albumins from the extant great apes other than *Pan* meet and react with modern human antiserum they coagulate and form a white spur. In contrast, modern human albumin does not create a spur in the presence of modern human antiserum, because the albumin is not recognized as foreign and thus does not prompt a reaction. There is also no spur when modern human antiserum meets and reacts with *Pan* albumin. This suggests that this test, or assay, cannot discriminate between modern human and chimpanzee albumin. (L. *alba* = white.) *See also* **immunochemistry**.

## Alcelaphini

A tribe of the family Bovidae that includes wildebeest, hartebeest, bonteboks, and their allies. Alcelaphine bovids are grazers with a preference for open grassland habitats that are characterized by tall (hypsodont) tooth crowns and cursorial (running) limb adaptations. When researchers attempt paleoenvironmental reconstructions of African fossil assemblages, examining the frequencies of alcelaphine bovids is one way to track the presence of open grasslands.

## allele

The form of a gene at a specified site, or locus, in the genome, or the form of a particular DNA sequence. If the locus in the genome is a "street address," the allele at that locus is analogous to the type of house present at that address. All houses share the same basic attributes, although one may be a luxury mansion while another may be a modest single-story residence. The genome is arranged into units called chromosomes, and with one exception (the genes on the X chromosome in males of many types of animal) every chromosome is present as a pair in the cell. Therefore, for each gene there is a pair of alleles. The particular combinations of alleles at a locus can have significant effects on function. For example, in modern humans the S allele at the beta-globin locus is protective against malaria if present with a wild-type allele (A) in the heterozygous form (i.e., AS). However, if both copies of the beta-globin allele are the S type (i.e., the homozygous form, SS) that individual will suffer from sickle cell anemia. (Gk *allos* = another.)

## Allen's Rule

Attributed to Joel Allen in the late 19thC, it states that animals living in locations with lower average temperatures tend to have smaller appendages (i.e., shorter limbs or tails). *Homo neanderthalensis* and some other high-latitude archaic *Homo* specimens have the type of body proportions (i.e., relatively shorter distal limb lengths and larger bi-iliac breadths) that would be predicted from Allen's Rule.

## allometry

The study of the growth, or size, of one part of an organism with respect to the growth, or size, of the whole (or another part that is taken as a proxy for the whole) of the same organism. The term allometry is used in two senses. It is often used to refer generally to the study of the "consequences of differences in size." In this sense, allometry is equivalent to the term scaling. However, allometry can also be used in a more specific sense to refer to changes in shape of a part or the whole of an organism that are associated with changes in the overall size of the organism. When a variable increases in size more slowly than overall body size, this is called negative allometry (i.e., the variable becomes proportionally smaller as overall body size increases). The term used when a variable increases in size more quickly than overall body size is positive allometry (i.e., the variable becomes proportionally larger as overall body size increases). In both negative and positive allometry any change in size will result in a change in shape. When used in this sense, the opposite of allometry is isometry, which is when shape is maintained as size increases. In other words, an isometric variable increases in size at the same rate as body size. (Gk *allos* = other and *metron* = measure.) *See also* **scaling**.

## allopatric speciation
A mode of speciation in which new species evolve as a consequence of the original species population being subdivided by a geographic barrier. The resulting physical isolation leads to loss of gene flow, and the accumulation of genetic differences in the new populations is due to genetic drift, natural selection, and mutation. Allopatric speciation is thought to be the most common cause of speciation in mammals, including hominins. See also **parapatric speciation**; **sympatric speciation**.

## allopatry
When two organisms have geographic ranges that are entirely separate and distinct (i.e., there is no overlap). Given the nature of the fossil record it is difficult to be certain whether hominin species were truly allopatric, but, for example, *Australopithecus africanus* (known only from southern Africa) and *Australopithecus afarensis* (known only from East Africa) were probably allopatric. (Gk *allos* = other and *patris* = fatherland.) See also **speciation**; **vicariance biogeography**.

## alluvial
Nonmarine sediments deposited by water that is flowing. If there is evidence to attribute the sediments to a more specific depositional mechanism (e.g., fluvial, lacustrine, etc.) then the term alluvial should be avoided. (L. *alluere* = to wash against.) See also **riverine**.

## alpha taxonomy
According to Ernst Mayr alpha taxonomy is the process of "characterizing and naming" species. Beta taxonomy involves arranging species in "a natural system of lesser and higher categories," and gamma taxonomy involves the "analysis of intraspecific variation." (Gk *alpha* = first and *taxis* = to arrange or "put in order.") See also **systematics**; **taxonomy**.

## altricial
Taxa with newborn offspring that are still at a relatively early stage of development at the time of birth. Altricial offspring possess little to no ability to move independently and are reliant on parents or relatives for varying lengths of time after birth for temperature regulation, food, and transport (e.g., newborn kittens rely on the mother to clean them, transport them, and direct them to the nipple). Compared to nonhuman primates, most of which are relatively precocial at birth, modern human babies are altricial and require intensive parental care. (L. *alere* = to nourish.) See also **precocial**.

## Alu repeat elements
A family of short interspersed nucleotide elements (or SINEs) of DNA that are common in all primates including the great apes and modern humans. Each Alu repeat element is approximately 300 base pairs (bp) in length. Alu repeat elements, which are a class of retrotransposons (i.e., sequences that are transcribed from DNA to messenger RNA (mRNA) and then the mRNA is copied back into DNA, which is inserted elsewhere in the genome), were originally named for the Alu restriction enzyme cut site that is typically found within each element. Alu elements, which account for as much as 10% of the modern human genome,

are useful for phylogenetic analyses and for studies of population history. This is because (a) the insertion of an Alu element has an unequivocal ancestral state (no Alu insertion), (b) each Alu insertion is almost certainly homologous, as the probability of two insertions at the same location within the genome is very small, (c) they are stable, and (d) they are relatively easy to analyze.

## alveolar process
The inferior part of the upper jaw (i.e., the maxilla) and the superior part of the body, or corpus, of the lower jaw (i.e., the mandible) into which the roots of the upper and lower teeth, respectively, are embedded. (L. *alveolus* = small hollow, dim. of *alveus* = hollow, and *processus* = to go forward or advance.)

## alveolus
The name for the socket in the alveolar process of the maxilla or mandible into which the root of a tooth is embedded. (L. *alveolus* = small hollow, dim. of *alveus* = hollow; pl. alveoli.)

## ameloblast
The name given to secretory and maturational (i.e., functional) enamel-forming cells. During enamel formation, secretory ameloblasts move away from the enamel–dentine junction, secreting enamel matrix as they go. The secreted matrix forms elongated enamel prisms approximately 5 µm in diameter. Secretory ameloblasts cease to lay down enamel matrix when the final thickness of enamel is completed. The subsequent mineralization of the matrix is a separate process. Short-period and long-period incremental lines produced by ameloblasts represent interruptions in the secretion or mineralization of the matrix. (Gk *amel* = pertaining to enamel and *blastos* = germ.) See also **enamel development**.

## amelogenesis
The process of enamel formation by ameloblasts. (Gk *amel* = pertaining to enamel and *genesis* = birth or origin.) See also **enamel development**.

## amino acid
A relatively small molecule that is the building block of proteins. There are 20 different standard amino acids. Amino acids are transported by specific transfer RNA (tRNA) and then they are joined together in a sequence encoded by messenger RNA (mRNA) to form a polypeptide chain. The latter process, which is catalyzed by ribosomes, is referred to as translation. Proteins consist of one or more polypeptide chains. (Gk *ammoniacos* = the pungent resin that is the source of ammonia, $NH_3$, which was first collected from near the temple of Amen in Libya.) See also **protein**.

## amino acid racemization
Amino acids exist in two forms called antimeres: a "right-handed" or D form and a "left-handed" or L form. When proteins are assembled in cells the component amino acids are all in the L form, but they convert at a predictable rate by a process called racemization to the D form. Racemization is also known as epimerization. See also **amino acid racemization dating**.

## amino acid racemization dating

The apparently regular and predictable process of amino acid racemization has been used as a molecular clock for dating biological specimens, but because the process proved to be temperature-dependent, the dates were found to be unreliable and the method fell into disuse. Recently, the principle has been revived and applied to the epimerization of isoleucine, an amino acid preserved within the calcite crystals of ostrich eggshell, to estimate the age of those shells. However, the problem of temperature-dependency persists.

## AMS radiocarbon dating

Accelerator mass spectrometry (or AMS) dating enables the direct measurement of individual $^{14}C$ atoms; AMS can routinely date samples of 1 mg of carbon. This means that smaller and previously undatable samples, like single hominin teeth and individual grains of domesticated cereals, can now be dated. The AMS method also allows for more thorough chemical pretreatment of samples. This is particularly important for older samples (>25 ka BP) where small amounts of modern carbon contamination may have a large effect on the measured $^{14}C$ fraction and hence the date. *See also* **radiocarbon dating**.

## Amud

A cave approximately 5 km/3 miles northwest of the Sea of Galilee, in Israel. Excavations recovered Amud 1, a fairly complete but poorly preserved, presumed male, adult of *Homo neanderthalensis*. Also recovered were fragments of at least three other individuals including Amud 7, an associated skeleton of a *H. neanderthalensis* neonate that may have been intentionally buried. Recent thermoluminescence dating on a number of burned lithic artifacts for the various stratigraphic horizons indicates two occupation events, one c.70 ka and the other c.55 ka; the hominin remains are associated with the younger age. Archeological evidence includes Middle and Upper Paleolithic lithics and signs of fire-related behavior. (Location 32°52′N, 35°30′E, Israel.)

## anagenesis

An evolutionary pattern (or mode) in which an ancestral species evolves into a descendant species without lineage splitting. For example, it has been claimed that *Australopithecus anamensis* and *Australopithecus afarensis* are time-successive species in the same lineage and are therefore an example of an anagenetic relationship. Anagenesis is the alternative to cladogenesis. (Gk *ana* = up and *genesis* = birth or origin.) *See also* **cladogenesis**.

## analogous

A trait (structure, gene, or developmental pathway) in two or more taxa that was *not* inherited from their most recent common ancestor. Analogous morphology is the cause of homoplasy. (Gk *analogos* = resembling, from *ana* = according to and *logos* = ratio.) *See also* **analogue**; **homoplasy**.

## analogue

An organism that is a functional proxy for another organism without being closely related to it. For example, the differences between the masticatory systems of bears and pandas, animals that are only distantly related to hominins, have been compared to the differences between the masticatory system of *Australopithecus africanus* or *Australopithecus afarensis*, on the one hand, and that of *Paranthropus robustus* or *Paranthropus boisei*, on the other. In this case, pandas serve as an analogue for *P. robustus* and *P. boisei*. (Gk *analogos* = resembling, from *ana* = according to and *logos* = ratio.) *See also* **homology**.

### analysis of covariance
(or ANCOVA) A variant of multiple regression in which a continuous variable is dependent on continuous and categorical variables (where the categorical variables are converted to binary dummy variables). It is typically used to determine whether the slopes and/or intercepts of scaling relationships between continuous variables differ between groups.

### analysis of variance
(or ANOVA) A statistical test commonly used to determine whether there is a significant difference in the mean of a continuous variable between two or more groups. For example, if cranial capacity is known for samples of crania belonging to three different species, ANOVA can be used to identify whether a significant difference exists between the three species in mean cranial capacity. Results from an ANOVA performed for two groups (as opposed to three or more) are equivalent to the results of a *t* test. ANOVA is a parametric statistical test; the equivalent nonparametric statistical test is the Kruskal–Wallis test.

### anatomical position
The position of the body used as a reference when describing the surfaces of the body, the spatial relationships of the body parts, or the movements of the axial and postcranial skeleton. In modern human anatomy, the anatomical position assumes an individual is upright, looking forward, with their legs and feet together, their arms by their side, and with the palms facing forward. All the surfaces that face towards the front are called anterior or ventral. All of the surfaces that face towards the back are called posterior or dorsal. Superior is nearer to the crown of the head; inferior is nearer to the soles of the feet. Medial is nearer to the midline; lateral is further from the midline. With respect to the limbs, proximal is in the direction of the root of the limb, where it is attached to the torso; distal is in the direction of the tips of the fingers or toes. Moving a whole limb forward is to flex it; moving it backwards is to extend it. Moving a limb away from the body is to abduct it; moving it back towards the midline is to adduct it. These latter terms also apply to movements of the fingers and toes, except that the movements are described relative to one of the digits rather than to the whole body (NB: the reference digit of the hand is the middle finger and the reference digit of the foot is the second toe).

### anatomical terminology
Many anatomical terms were based on the everyday Latin (and sometimes Greek) vocabulary. Thus, the "cup-like" articular surface of the hip joint on the pelvis is called the acetabulum because Pliny thought it resembled a Roman vinegar (*acetum*) receptacle (*abrum*) and the condylar process of the mandible takes its name from the Greek word for a knuckle. The latest version of official modern human anatomical terminology is the *Terminologia Anatomica* (1998). *See also* **paleoanthropological terminology**.

## ancient DNA
(or aDNA) Deoxyribonucleic acid (DNA) that is extracted from old bone, teeth, hair, tissue, or coprolites. Current problems addressed by ancient DNA research include the relationships among *Homo neanderthalensis*, the Denisovans, and modern humans, the initial colonization of the Americas, regional population history, social organization at a particular site, diet, the sex of individuals, and relationships among individuals within a cemetery. Ancient DNA research initially targeted mitochondrial DNA (mtDNA) because of its high copy number in cells, but as methods have improved ancient nuclear DNA has become a tractable source of evidence. (OF *ancien* from the L. *ante* = before and DNA = deoxyribonucleic acid.)

## Andresen lines
Long-period (greater than circadian) incremental features in dentine that correspond to striae of Retzius in enamel. *See also* **incremental features**.

## Anglian
*See* **glacial cycles**.

## anisotropy
Materials (e.g., bone or enamel) are anisotropic when their material properties (e.g., stiffness) are sensitive to direction. Isotropy is when material properties are the same in all directions. (Gk *an* = not, *iso* = equal, and *tropus* = direction.) *See also* **dental microwear**.

## ANOVA
Acronym for **analysis of variance** (*which see*).

## antelope
The informal name for a member of any of the taxa within the Antilopini, the tribe of the family Bovidae that includes the gazelles and their allies. (Gk *antholops* = a fabulous beast from the orient.)

## anterior teeth
The two incisors and canine in each quadrant of the jaws. The rest of the teeth in each quadrant are called postcanine teeth.

## anthropogenic bone modification
Any alteration of a bone resulting from hominin activity. It includes bone surface modifications (e.g., cutmarks and hammerstone percussion marks), fracture/breakage patterns, heating, burning, and use wear on bone tools. Recognition of anthropogenic bone modifications is central to demonstrating that a fossil bone assemblage has been accumulated and/or altered by hominins, as opposed to other taphonomic agents such as carnivores, porcupines, or fluvial processes. *See also* **bone breakage patterns**.

### anthropoid
Primates that are relatively modern human-like. This term is usually used in one of two senses: to refer either to the nonhuman higher primates (i.e., chimpanzee, gorilla, and orangutan and their immediate ancestors), as in "anthropoid apes," or to all the members of the Anthropoidea (i.e., living anthropoids include all the extant New World monkeys, Old World monkeys, and apes, plus modern humans). Strictly speaking the latter use is the correct one. (Gk *anthropos* = human being.)

### antibody
Antibodies (also known as immunoglobulins) are proteins produced by lymphocytes (a type of white blood cell) when the latter react with foreign particles collectively called antigens (e.g., bacteria, pollen, and viruses). Each antibody reacts to a specific antigen, binds with it, and then tags it for destruction by other parts of the immune system. In the case of an organism the antibody prevents it from growing or causing damage. Antibodies raised against foreign albumins were the basis of one of the experiments undertaken to investigate the relationships among the great apes. (Gk *anti* = opposite and ME *body* = container.) *See also* **albumin**.

### anticline
A type of fold in structural geology in which the oldest rocks occupy the center and rocks become progressively younger towards the margins. (Gk *anti* = against and *klinein* = to slope.)

### anticodon
A sequence of three nucleotides in a transfer RNA (tRNA) molecule that is complementary to a codon (i.e., a sequence of three nucleotides) in a messenger RNA (mRNA) molecule. *See also* **genetic code**.

### antigen
Any foreign molecule capable of stimulating the production of an antibody or of provoking other responses by the immune system. (From *anti*body *gen*eration.) *See also* **antibody**.

### Antilopini
A tribe of the family Bovidae that includes the gazelles and their allies. In paleoenvironmental reconstructions of fossil assemblages, high frequencies of antilopine bovids are generally interpreted as indicating open habitats. (Gk *antholops* = a fabulous beast from the orient.)

### antimere
Refers to the version of a bilateral structure that belongs to the opposite side of the body. For example "the crown area of the right $P_3$ of KNM-ER 992 is larger than its antimere" (i.e., the crown of the left $P_3$). (Gk *anti* = opposite and *meros* = a part.)

### anvil
A stationary object against which another object (e.g., a bone or core) can be struck to fracture it. Anvils are generally made of stone although materials such as wood may be used when stone is not available (e.g., Taï forest nut-cracking chimpanzees). Stone cores flaked using hammer-and-anvil or bipolar techniques usually have flakes removed from both ends. *See also* **bipolar percussion**.

## apatite

Apatite is one of the common names (hydroxyapatite and bioapatite are others) for the mineral phase of bone, cementum, dentine, and enamel. Apatite makes up approximately 96% of the mineral phase of mature (i.e., fully mineralized) enamel, and this high proportion is responsible for enamel's extreme hardness and resilience to diagenesis. (Gk *apate* = deceit, because of apatite's reputation for being confused with other minerals.)

## ape

An informal taxonomic category that is coincident with the superfamily Hominoidea. The extant taxa in this superfamily are chimpanzees, bonobos, gorillas, orangutans, gibbons, and siamangs. The fossil taxa are all the extinct forms that are more closely related to chimpanzees, bonobos, gorillas, orangutans, gibbons, and siamangs than to any other living taxon. (OE *apa* = ill-bred and clumsy; before apes had been investigated scientifically and appreciated on their own terms they were regarded as being "clumsy" because they lacked dexterity; syn. hominoid.)

## apical tuft

The distal part of the distal phalanx of a digit (finger or toe). The apical tuft provides bony support for the nail and the soft tissue (pulp) that lies beneath the nail. [L. *apex* = point and OF *tof(f)e* = projection; syn. ungual process, tuberosity, or tuft.]

## apomorphic

A "catch-all" word that refers to any derived character state. Apomorphic is used in cladistic analysis to refer to the state of a character that is different from the ancestral or primitive condition of that character. Apomorphy is one of several terms used in cladistics that is relative. The same morphology can be derived, or apomorphic, in one context and primitive, or symplesiomorphic, in another; it depends on the taxa used as comparators. (Gk *apo* = different from and *morphe* = form.) See also **autapomorphy**; **synapomorphy**.

## appendicular skeleton

The hard-tissue (bone and cartilage) components of the upper and lower limbs. In the upper limb it comprises the pectoral or shoulder girdle (scapula and clavicle), the bone of the arm (humerus), and the bones of the forearm (radius and ulna) and hand (carpals, metacarpals, and phalanges). In the lower limb it comprises the pelvic girdle (pelvic bone made up of the ilium, ischium, and pubic bones, but not the sacrum), the bone of the thigh (femur), the patella, and the bones of the lower leg (tibia and fibula) and foot (tarsals, metatarsals, and phalanges). (L. *appendere* = to hang upon.)

## appositional enamel

Although all enamel is technically appositional because it is deposited in layers, this term usually refers to the cuspal enamel formed during the initial phase of enamel formation (i.e., it excludes imbricational enamel). Striae of Retzius do not reach the surface of appositional enamel. (L. *appositus* = to put near.) See also **cuspal enamel**; **enamel development**.

## appositional growth

See **ossification**.

## approximal wear
See **interproximal wear**.

## aptation
There are two main categories of aptation. If a functional trait was fixed in a population by natural selection and it still performs that function, then it is referred to as an *ad*aptation. But if there is evidence the trait now performs a *different* function, or if a functional trait was non-functional prior to being co-opted for its current function, then the trait is referred to as an *ex*aptation. (L. *adaptare* = to fit.) *See also* **adaptation**; **exaptation**.

## Arago
See **Caune de l'Arago**.

## aragonite
See **calcium carbonate**.

## Aramis
The type site of *Ardipithecus ramidus*. It is situated between the headwaters of the Aramis and Adgantoli drainages on the west side of the Awash River in the Middle Awash study area in the Afar Depression in the Afar Rift System in Ethiopia. All of the localities (ARA-VP) are in the Sagantole Formation. Specimens recovered from the c.4.4 Ma Aramis Member include the holotype of *Ar. ramidus* ARA-VP-1/1 and a remarkably complete associated skeleton, ARA-VP-6/500. A left maxilla from the site, ARA-VP-14/1, has been attributed to *Australopithecus anamensis*. No archeological evidence has been found. (Location 10°28′N, 40°26′E, Ethiopia.)

## $^{40}Ar/^{39}Ar$
See **argon-argon dating**.

## ARA-VP-6/1
The holotype of *Ardipithecus ramidus*. It was found at Aramis, Ethiopia, in 1993 and it is dated to c.4.4 Ma. It comprises several associated teeth, including the left $I^1$, C, $P^3$, and $P^4$, and right $I^1$, C, $P^4$, $M^2$, $P_3$, and $P_4$.

## ARA-VP-6/500
This exceptionally complete c.4.4 Ma associated skeleton is the centerpiece of the fossil evidence for *Ardipithecus ramidus*. Functional interpretations of ARA-VP-6/500 form the main evidence for the proposal that *Ar. ramidus* is a basal hominin. The first fragments were recognized at Aramis, Ethiopia, in 1994, but more evidence of it was recovered in subsequent years. One hundred and thirty recognizable fragments belonging to a single individual were recovered, but the fragments are so fragile that the cranial morphology had to be recovered from micro-computed tomography scans of cranial fragments still embedded in matrix. The endocranial volume is estimated at $c.300\,cm^3$. The crowns and roots of all of the upper teeth on the right side, and the left lower canine through to the $M_3$, are preserved. The right forearm is intact apart from the distal end of the ulna; the partial right hand includes carpal bones and a complete ray. The only evidence of the left forearm is part of the radial shaft, but much of the

skeleton of the left hand is preserved. Evidence of the thorax includes a few vertebrae and the left first rib. Much of the left pelvic bone is preserved but it is crushed and distorted, as is a piece of the lower part of the body of the sacrum and part of the right ilium. All that remains of the lower limb is a substantial length of the shaft of the right femur, most of the right tibia, and all but the proximal end of the right fibula. Between them the two preserved foot skeletons provide most of the bones of the tarsus and the toes.

## ARA-VP-7/2

Fragments of the long bones of an arm, including the proximal end of the humerus, found at Aramis, Ethiopia, in 1993 and dated to c.4.4 Ma. It has been assigned to *Ardipithecus ramidus*. Its discoverers used the size of the humeral head to generate the c.40 kg estimate for the body mass of this individual. They also concluded that the arm of *Ar. ramidus* had some characters usually associated with great apes.

## arboreal

A term used to describe animals that live in trees. Some of the defining features of primates are important for life in the trees (e.g., binocular vision helps animals to judge distances). The vast majority of primates are dependent on trees, with platyrrhines being exclusively arboreal. Nonetheless, many living and extinct primates have successfully radiated into terrestrial niches or have combined life in the trees with life on the ground. The early hominins are a good example of a combination of arboreality and terrestriality and some researchers have suggested that at least one form of hominin bipedalism may have emerged as a way of moving or foraging in trees. (L. *arbor* = tree.) See also **locomotion**.

## arboreality

The tendency to live partially, or wholly, in the trees. (L. *arbor* = tree.) See also **arboreal**.

## Arcy-sur-Cure

A series of caves in the limestone cliffs above the Cure river in central France, including the Grande Grotte, the Grotte du Renne, the Grotte du Hyène, the Grotte des Fées, and the Grotte du Loup. The Grande Grotte is best known for its c.28–33 ka cave paintings, which are the second oldest such paintings in France. The Grotte du Renne is best known for several beads and objects of personal ornamentation found in the Châtelperronian level, alongside a juvenile *Homo neanderthalensis* and several Neanderthal teeth. Researchers debate whether these finds reflect independent Neanderthal invention of so-called modern behaviors, or evidence of acculturation from interaction with modern humans, or if the finds were the product of Neanderthals at all. In the Grotte du Hyène, several hominin remains including a nearly complete mandible with dentition were recovered from the lower Mousterian levels. The other caves also contain Mousterian and some transitional and Upper Paleolithic layers. (Location 47°35′N, 03°45′E, France.)

## *Ardipithecus* White et al., 1995

A genus established in 1995 by White et al. to accommodate the species *Ardipithecus ramidus*. Subsequently a second, more primitive, and temporally older species, *Ardipithecus kadabba*, was recognized and included in the same genus. The type species is *Ardipithecus ramidus* (White et al., 1994) White et al., 1995. (*ardi* = ground or floor in the Afar language and Gk *pithekos* = a postfix that means ape or "ape-like.") See also **Ardipithecus kadabba**; **Ardipithecus ramidus**.

### Ardipithecus kadabba Haile-Selassie, 2001

A hominin subspecies with this name was established in 2001, and it was subsequently elevated to species rank in 2004. All of the hypodigm was recovered from five c.5.8–5.2 Ma localities in the Middle Awash study area, Ethiopia. Four of the localities are in a region called the Western Margin, and one is in the Central Awash Complex. The main differences between *Ardipithecus kadabba* and *Ardipithecus ramidus* involve the upper canine and the $P_3$. The postcranial evidence is generally ape-like. Researchers have suggested that there is a morphocline in upper canine morphology, with *Ar. kadabba* exhibiting the most ape-like morphology, and *Ar. ramidus*, *Australopithecus anamensis*, and *Australopithecus afarensis* interpreted as becoming progressively more like the lower and more asymmetric crowns of later hominins. The first discovery was ALA-VP-2/10 (1997). The holotype is ALA-VP-2/10. (*ardi* = ground or floor in the Afar language, Gk *pithekos* = a postfix that means ape or "ape-like," and *kadabba* = a "family ancestor" in the Afar language.)

### Ardipithecus ramidus (White et al., 1994) White et al., 1995

Hominin species established in 1994 to accommodate cranial and postcranial fossils recovered from c.4.5–4.4 Ma localities at Aramis on the northeastern flank of the Central Awash Complex in the Middle Awash study area, Ethiopia. The taxon was initially included within the genus *Australopithecus*, but in 1995 it was transferred to a new genus, *Ardipithecus*. Additions to the *Ardipithecus ramidus* hypodigm have come from the Gona study area, the Aramis locality, and Kuseralee Dora and Sagantole, two other localities in the Central Awash Complex. Initial estimates based on the size of the shoulder joint suggested that *Ar. ramidus* weighed approximately 40 kg, but researchers claim the enlarged hypodigm indicates an estimated body mass of approximately 50 kg. The position of the foramen magnum, the form of the reconstructed pelvis, and the morphology of the lateral side of the foot have all been cited as evidence that the posture and habitual gait of *Ar. ramidus* were respectively more upright and bipedal than is the case in the living apes. Timothy White and his colleagues claim that *Ar. ramidus* is a basal hominin, yet the inclusion of *Ar. ramidus* in the hominin clade necessitates substantial amounts of convergent evolution in the closely related great ape clades. The hypothesis that *Ar. ramidus* is not a hominin, but instead is a member of an extinct ape clade, would, in many respects, be more parsimonious than assuming it is a basal hominin. The first discovery, ARA-VP-1/1 was made in 1993, but if either the mandible KNM-LT 329 from Lothagam, Kenya, or the mandible KNM-TH 13150 from Tabarin, Kenya, prove to belong to the *Ar. ramidus* hypodigm, then they would be the initial discovery. The holotype is ARA-VP-6/1 and the main sites are localities in the Gona and Middle Awash study areas, Ethiopia. (*ardi* = ground or floor in the Afar language, Gk *pithekos* = a postfix that means ape or "ape-like," and *ramid* = "root" in the Afar language.)

### argon-argon dating

An isotopic dating method based upon the potassium-argon (K/Ar) system, in which radioactive $^{40}K$ is driven to $^{40}Ar$ in a reactor, and used as a proxy for the K content. Subsequent analyses can be done in a single experiment, using the same sample, by measuring isotopes of Ar in a mass spectrometer. This approach avoids the necessity of measuring K and Ar in different aliquots of a sample, thus reducing potential error. The current analytical methods are so sensitive they can be applied to single crystals of feldspar or volcanic glass; this more precise version is referred to as single-crystal laser fusion $^{40}$Ar/$^{39}$Ar dating.

## armature
A term used to refer to any body part or implement used for the offense or defense of an organism. In archeology the term armature refers to most points and to any other obvious hunting equipment (e.g., the wooden spears from Schöningen). (L. *armatura* = armor or equipment.)

## art
The use of non-utilitarian images for symbolism or self-expression. Prehistoric art is divided into mobile (L. *mobilis* = to move) or portable art (e.g., small figurines), and parietal (L. *paries* = wall) or fixed art (e.g., wall paintings or engravings). (L. *art* = art.)

## artifact
Any portable object made, modified, or used by hominins. The earliest artifacts presently known are stone tools and their manufacturing debris from the site of Gona, Ethiopia, dating to 2.55 Ma, although indirect traces of stone tool use may be preserved as cutmarks on bones from the surface at Dikika, Ethiopia dating to 3.39 Ma. Because they preserve well, stone artifacts form the largest part of the early archeological record. Artifacts are one of the fundamental units of data used by archeologists when they reconstruct the behavior of extinct hominins and prehistoric modern humans.

## Artiodactyla
The mammalian order that includes all of the taxa with an even number of hoofed toes. Artiodactyls are terrestrial and largely herbivorous, although some artiodactyl taxa (e.g., the Suidae) are more omnivorous. The most diverse family of artiodactyls is the Bovidae (antelopes and their allies). Other artiodactyl families include the Suidae (pigs), the Hippopotamidae (hippopotami), and the Giraffidae (giraffes). Other, less common artiodactyls in African hominin sites are the Camelidae (camels) and the Tragulidae (chevrotains). In Eurasia, the Cervidae (deer) are common at hominin sites, as are the Moschidae (musk deer). (Gk *artios* = even and *daktulos* = toe; literally, the "even-toed.")

## Asa Issie
An area of fossiliferous sediments, which is 10 km/6 miles west of Aramis, Ethiopia, containing hominin fossils assigned to *Australopithecus anamensis*. The fossils, dating to between 4.2 and 4.1 Ma, are claimed to be transitional between *Ardipithecus ramidus* and *Au. anamensis*. No archeological evidence has been found. (Location 11°10′N, 40°20′E, Ethiopia.)

## ascertainment bias
Synonymous with experimental bias, ascertainment bias refers to a systemic distortion of results attributable to nonrandom sampling. Such biases can lead to incorrect inferences about an entire population either because of distorted or nontypical sampling of the population or

because the data (i.e., a specific marker) used for the analysis were identified in a biased way. [L. *ad* = near and *certus* (the root of "certain") = to determine.]

## As Duma
Site located on the west side of the Awash River in the Gona Western Margin sector of the Gona Paleoanthropological study area in the Afar Depression, in Ethiopia. It contains several *Ardipithecus ramidus* fossils and dates to c.4.4 Ma. No archeological material was found. (Location 11°10′N, 40°20′E, Ethiopia.)

## assemblage
An archeological assemblage is a stratigraphically bounded, spatially associated set of artifacts. For example, a single archeological site may contain several artifact assemblages (e.g., many of the excavations at Olduvai Gorge). These may derive from different strata at the site, or from different facies within the same stratum (e.g., from channel and floodplain deposits of the same river system). Assemblages form one of the basic comparative units above the level of single artifacts or artifact types, and variations in the range of tool types found within an assemblage have been used to infer past site function(s). Assemblage is also used as an inclusive term to describe the paleontological evidence from a site.

## assimilation model
A model for the origin of modern humans that accepts an African origin but rejects the total replacement of local archaic populations (e.g., *Homo neanderthalensis*) by modern humans as they spread into Eurasia. This model relies on fossil evidence that suggests minor morphological traits show within-region continuity through the transition between the archaic and modern forms. In contrast, the replacement with hybridization model accepts the theoretical possibility of admixture, but rejects any morphological evidence for it. The assimilation model is consistent with the mostly out-of-Africa genetic model of modern human origins. It is also supported by recent evidence from the Neanderthal genome project indicating that Neanderthals made low-level contributions to the genomes of modern Eurasian populations, and evidence that the genomes of modern Melanesian, Oceanian, and Southeast Asian populations contain contributions from Denisovan hominins. *See also* **candelabra model**; **multiregional hypothesis**; **out-of-Africa hypothesis**; **replacement with hybridization**.

## associated skeleton
Refers to a fossil specimen that includes more than one skeletal element from the same individual. Most fossil hominin taxa are diagnosed and identified on the basis of skull morphology, so associated skeletons that include skull and postcranial elements can help determine which limb bones go with which skulls. For example, for a long time the lack of a securely associated skeleton of *Paranthropus boisei* that preserves both taxonomically distinctive skull evidence *and* evidence of the postcranial skeleton has hampered attempts to sort into taxa hominin postcranial fossils from East Africa. Well-preserved associated skeletons allow researchers to compare the size of the teeth with the rest of the body, or the relative lengths of the limbs and/or limb segments, or the relative sizes of joint surfaces.

Examples of associated skeletons include A.L. 288-1 (*Australopithecus afarensis*), KNM-WT 15000 (*Homo ergaster*), and Dederiyeh 1 (*Homo neanderthalensis*). (L. *associare* = to join with.)

### astronomical theory

A theory espoused by Milutin Milanković, a Serbian astronomer, suggesting that cyclic changes in three important aspects of the Earth's orbital geometry (precession, obliquity, and eccentricity) largely determine long-term climate changes. *See also* **astronomical time scale**. (Gk *astron* = star and *kronos* = time.)

### astronomical time scale

A geological time scale based on regular changes involving three aspects of the Earth's orbital geometry (the way it rotates about its axis and the shape of its orbit around the sun), namely precession, obliquity, and eccentricity. Precession (the "wobble" of the Earth's axis of rotation, which has a 19–23 ka cycle) controls seasonal changes in the intensity of the sun's rays on the Earth's surface (insolation). Obliquity (the tilt of the Earth's axis, which has a dominant periodicity of *c*.41 ka) controls the length of the winter polar darkness. Eccentricity (the elliptical nature of the Earth's orbit, which has approximately 100 and 400 ka cycles) is thought to determine the timing (pacing) of the northern hemisphere glacial cycles. Only eccentricity changes the global magnitude of insolation, and even then by a small amount; the other orbital cycles change only seasonality or the latitudinal distribution of insolation. Precession has had a long-term influence on the strength of the monsoons ("23 ka world"). Only at times of global cooling and northern hemisphere glaciation, as has been the case for the last 3 Ma, is there evidence of a strong signal of obliquity ("41 ka world"). It is obliquity that determines the timing of glacial/interglacial cycles. For the last 1 Ma, *c*.100 ka-long cycles have dominated ("100 ka world"). The regularity of these various cycles is so predictable that astrochronology is used to calibrate ("tune") other methods of age estimation. (Gk *astron* = star and *kronos* = time; syn. astrochronology). *See also* **eccentricity**; **obliquity**; **orbital tuning**; **precession**:

### Atapuerca

The Sierra de Atapuerca is a series of eroded limestone hills 14 km/9 miles east of Burgos in northern Spain. It is permeated by several sediment-filled cave systems, one of which is the Cueva Mayor-Cueva del Silo, and within this system are several cave/fissure complexes including the Sima de los Huesos. The nearby Trinchera del Ferrocarril cave system includes several sites (e.g., Galería, Gran Dolina, and Sima del Elefante) that have yielded hominin fossils and archeological evidence. The Sima del Elefante site, which may be 1.2 Ma, has provided the oldest hominin remains from the Atapuerca hills. (Location 42°21′N, 03°31′W, Spain.) *See also* **Gran Dolina**; *Homo antecessor*; *Homo heidelbergensis*; *Homo neanderthalensis*; **Sima del Elefante**; **Sima de los Huesos**.

### atavism

A morphological variant or anomaly in a current specimen that is more closely associated with a presumed ancestor (recent or distant). For example, the occasional appearance of contrahentes muscles in the palm of the hand of modern humans is an atavism. The presence of an independent os centrale in the carpal bones of the hand of a modern human is also atavistic because in modern humans the os centrale normally fuses with the main part of the scaphoid to form the scaphoid tubercle. (L. *atavus* = ancestor, from *atta* = father plus *avus* = grandfather.)

## ATD6-1–12

A mandible and several isolated teeth from the same individual from the site of Gran Dolina that comprise the holotype of *Homo antecessor*. They were found in the Aurora stratum of TD6 in 1994 by the team led by Eudald Carbonell and Jóse Maria Bermúdez de Castro. Geochronological dating suggests these remains are more than 780 ka (TD6 is thought to be in the Matuyama chron), and uranium-series dating and electron spin resonance spectroscopy dating suggest an age of 731 ± 63 ka.

## ATE9-1

A fragment of the mandibular corpus that extends from the alveolus of the left $P_4$ to the alveolus of the right $M_1$. Provisionally assigned to *Homo antecessor*, it was the earliest reliably dated hominin from Europe at the time of its discovery in 2007. It was found at the Sima del Elefante by a team led by Eudald Carbonell and Jóse Maria Bermúdez de Castro. A combination of biostratigraphy, the observed reversed magnetic polarity (consistent with the TE16 and older lithostratigraphic layers being in the Matuyama chron), and cosmic radionuclide dating suggest the ATE9-1 mandible is 1.2–1.1 Ma.

## Aterian

An industrial complex of the African Middle Stone Age, likely dating to *c*.35–90 ka. Aterian sites occur in northern Africa, from the Maghreb (the northern parts of Morocco, Algeria, and Tunisia), south to Niger, throughout the Sahara, and east to Egypt. Tanged pieces (especially points and scrapers) as well as bifacial points characterize Aterian sites. The Aterian is one of the best examples of regional diversity in the archeological record of the Middle Stone Age. Aterian populations occupied a range of habitats, including desert areas in the present Sahara. A perforated *Nassarius gibbosulus* shell from Oued Djebbana and the green silicified tuff used to make artifacts at Adrar Bous, Algeria, are evidence for long-distance transport. Aterian artifacts are associated with hominin fossils generally considered to be robust examples of *Homo sapiens*. (The name is based on Bir el Ater, a community near the type site of this industrial complex.)

## attrition

*See* **tooth wear**.

## auditory ossicles

Three small bones (malleus, incus, and stapes) in the middle ear, or tympanic cavity, of mammals. They connect the tympanic membrane (at the medial end of the external ear) with the oval window (behind which is the outer of the two fluid-filled cavities of the inner ear). The ossicles occasionally survive as fossils (e.g., a *Paranthropus robustus* incus, SK 848, and an *Australopithecus africanus* stapes, Stw 151). Approximately 25 ear ossicles have been recovered from the Sima de los Huesos at Atapuerca, Spain. (L. *audire* = to hear and *ossiculum* = small bone.) *See also* **bony labyrinth**.

## auditory tube
The tube connecting the middle ear and nasopharynx. The lateral one-third is bone (this is the only part that fossilizes) and the medial two-thirds consists of fibrocartilage. The auditory tube is inclined close to the sagittal plane in extant apes and in archaic hominins, and more horizontally in modern humans. Some specimens of *Paranthropus robustus* and *Australopithecus africanus* have club-like processes (or Eustachian processes) at the medial ends of the bony auditory tube. (L. *audire* = to hear and *tubus* = hollow cylinder; syn. Eustachian tube, pharyngotympanic tube, tympanic tube.)

## Aurignacian
A material culture named after the site of Aurignac in France. Generally thought to be the oldest modern human (or Cro-Magnon) culture in Europe, its hallmarks are the use of blades, bone tools, beads, and other objects of personal decoration, as well as figurines and other figurative art. Its appearance in Eastern Europe *c*.43 ka and in Western Europe between 40 and 36 ka is consistent with a migration of anatomically modern humans from the Near East. In many regions the Aurignacian is replaced by the Gravettian culture between 28 and 26 ka.

## Australasian strewn-field tektites
From time to time, the Earth collides with showers of small meteorites called tektites. The *c*.800 ka Australasian strewn-field, the largest and youngest strewn-field, extends across most of Southeast Asia (Vietnam, Thailand, southern China, Laos, and Cambodia), as well as the Philippines, Indonesia (including Java), and Malaysia. The identification of Australasian strewn-field tektites in sediments is one of the many ways researchers have tried to date hominins from China and Southeast Asia.

## australopith
Informal name for some, or all, of the fossil hominins *not* included in the genus *Homo*. This term is not used consistently in paleoanthropology. Some workers use it to categorize *all* non-*Homo* hominins, while others include only non-*Homo* hominins that exhibit postcanine megadontia or hyper-megadontia (i.e., species typically attributed to *Australopithecus* and/or *Paranthropus*). The term is increasingly used in place of australopithecine, because the latter should *only* be used if the writer believes that all *Australopithecus* and *Paranthropus* taxa belong in their own hominin subfamily, the Australopithecinae. Australopith taxa include the type species of *Australopithecus*, *Australopithecus africanus*, and *Paranthropus boisei*. See also **australopithecine**.

## australopithecine
Informal name for the subfamily Australopithecinae. Strictly speaking, this term should only be employed if the user supports elevating the archaic hominin taxa included in genera such as *Australopithecus* and *Paranthropus* to the level of a subfamily. More generally, the term has been used to refer to all or most of the early hominins that do not belong to the genus *Homo*, but researchers are increasingly adopting the term australopith in place of australopithecine. *See also* **australopith**.

## *Australopithecus* Dart, 1925

Hominin genus established by Raymond Dart in 1925 to accommodate the type species *Australopithecus africanus*. The list of species assigned to the genus *Australopithecus* has varied over time. From its discovery up until the seminal publications of John Robinson, the genus subsumed three species, *Au. africanus* from the site of Taung, *Australopithecus transvaalensis* (later, *Plesianthropus transvaalensis*) from Sterkfontein, and *Australopithecus prometheus* from Makapansgat. Robinson sank *Au. prometheus* and *Pl. transvaalensis* into *Au. africanus*, but he did not include the fossil hominins from Swartkrans and Kromdraai, which he interpreted as belonging to the genus *Paranthropus*. In subsequent decades it became conventional to assign all of the above hominins as well as other species such as *Zinjanthropus boisei* and *Meganthropus africanus* to the genus *Australopithecus*. In the 1980s, researchers who favored the hypothesis that the "robust" species are a monophyletic group revived the genus *Paranthropus* for *Paranthropus robustus*, *Paranthropus boisei*, and, for some, *Paranthropus aethiopicus*, but many researchers still retain these species in *Australopithecus*. Since 1996, four new species of *Australopithecus* (*Australopithecus anamensis*, *Australopithecus bahrelghazali*, *Australopithecus garhi*, and *Australopithecus sediba*) have been described. In recent years, the evident paraphyly of *Australopithecus* has led some researchers to advocate removing species from the genus (e.g., transferring the hypodigm of *Australopithecus afarensis* to the genus *Praeanthropus*) until there is sound evidence that it is clearly monophyletic. (L. *australis* = southern and Gk *pithekos* = ape.) See also **Praeanthropus**.

## *Australopithecus aethiopicus* (Arambourg and Coppens, 1968)

A taxon used by researchers who recognize neither *Paraustralopithecus* nor *Paranthropus* as separate genera but who do recognize the pre-2.3 Ma hyper-megadont hominins from the Omo region as a species separate from *Australopithecus boisei*. (L. *australis* = southern, Gk *pithekos* = ape, and *aethiopicus* = Ethiopia.) See also **Paraustralopithecus aethiopicus**.

## *Australopithecus afarensis* Johanson, 1978

A hominin species established in 1978 by Donald Johanson to accommodate the c.3.7–3.0 Ma cranial and postcranial remains recovered from Laetoli, Tanzania, and Hadar, Ethiopia. In 1981, Tim White and colleagues made a compelling case for recognizing *Australopithecus afarensis* as a distinct species with a generally more primitive craniodental anatomy than *Australopithecus africanus*. It remains to be seen whether the dental, facial, and mandibular similarities between the Laetoli remains and those of *Australopithecus anamensis* sustain the hypothesis that *Au. anamensis* evolved via anagenesis into *Au. afarensis*. Most body mass estimates for *Au. afarensis* range from approximately 30 to 45 kg, and known endocranial volumes range between 385 and 550 cm$^3$. This is larger than the average endocranial volume of a chimpanzee, but the brain of *Au. afarensis* is not substantially larger than that of *Pan* relative to estimated body mass. The incisors of *Au. afarensis* are smaller than those of extant chimps/bonobos, but its premolars and molars are larger. The appearance of the pelvis and the relatively short lower limb suggests that although *Au. afarensis* was capable of bipedal walking, it was not adapted for long-range bipedalism. The discovery at Laetoli of several trails of fossil footprints provide direct evidence that a contemporary hominin, presumably *Au. afarensis*, was capable of bipedal locomotion. The upper limb, especially the hand and the shoulder girdle, retains morphology that some workers suggest reflects a significant element of arboreal locomotion. The size of the Laetoli footprints, plus the length of the limb bones, suggest that the standing height of adult *Au. afarensis* individuals was between 1.0 and 1.5 m. Most researchers interpret the fossil evidence for *Au. afarensis* as consistent with a substantial level of sexual dimorphism. The holotype is LH 4 (1974) and the main sites are Belohdelie (tentative), Dikika, Hadar, Maka, and White Sands, Ethiopia; Koobi Fora, Kenya; and Laetoli, Tanzania. Some researchers prefer to exclude *Au. afarensis* from *Australopithecus*, in which case it reverts to the genus *Praeanthropus*, as *Praeanthropus afarensis*. (L. *australis* = southern, Gk *pithekos* = ape, and *afarensis* recognizes the contributions of the local Afar people).

## *Australopithecus africanus* Dart, 1925

A hominin species established in 1925 by Raymond Dart for an immature skull recovered from the limeworks at Taungs (now called Taung, South Africa) in 1924. The taxon includes, in addition to the type specimen from Taung, fossils from Member 4 at Sterkfontein, fossils initially assigned to *Australopithecus prometheus* from Members 3 and 4 at Makapansgat, and fossils recovered from limeworks dumps and extracted *in situ* from the breccia exposed at Gladysvale. The hypodigm of *Australopithecus africanus* spans the period between *c*.3.0 and 2.5 Ma; it remains to be seen whether the associated skeleton StW 573 from Sterkfontein Member 2, and 12 hominin fossils recovered from the Jakovec Cavern since 1995, belong to *Au. africanus*. The cranium, mandible, and dentition of *Au. africanus* are well sampled, but the postcranium, particularly the axial skeleton, is not well represented. There is at least one specimen of each of the long bones, but many of the fossils have been crushed and deformed. Morphological and functional analyses suggest that although *Au. africanus* was capable of bipedalism, it was probably more arboreal than most other archaic hominin taxa. *Au. africanus* had relatively large chewing teeth and, apart from the reduced canines, the skull is relatively ape-like. Its mean endocranial volume is approximately 460 cm$^3$. The Sterkfontein evidence suggests that *Au. africanus* exhibited sexual dimorphism, but probably not to the degree exhibited in *Au. afarensis*. For a long time *Au. africanus* was regarded as the common ancestor of all later hominins, but in most cladistic analyses *Au. africanus* is either the sister taxon of *Homo* or *Paranthropus*, or the sister taxon of the common ancestor of the *Homo* and *Paranthropus* clades. The holotype is Taung 1 (1924), and the main sites are Gladysvale, Makapansgat (Members 3 and 4), Sterkfontein (Member 4), and Taung. (L. *australis* = southern, Gk *pithekos* = ape, and L. *africanus* = pertaining to Africa.)

## *Australopithecus anamensis* Leakey et al., 1995

A hominin species established in 1995 by Meave Leakey and colleagues to accommodate a left distal humeral fragment recovered by Bryan Patterson and cranial remains from *c*.3.9–4.2 Ma localities at Allia Bay and Kanapoi, Kenya. Additional fossils from the two sites were added to the hypodigm 3 years later. Leakey et al. (1995) claimed that the dentition (e.g., the mandibular canine morphology, the asymmetry of the premolar crowns, and the relatively simple crowns of the deciduous first mandibular molars) of *Australopithecus anamensis* is more primitive than that of *Australopithecus afarensis*, but in other respects (e.g., the low cross-sectional profiles and bulging sides of the molar crowns) the teeth of *Au. anamensis* show similarities to *Paranthropus*. Upper limb fossils assigned to *Au. anamensis* were described as being australopith-like, but a tibia has features associated with obligate bipedalism. Fossils from the Middle Awash study area, dating to c.4.2–4.1, were added to the hypodigm in 2006. Bill Kimbel and others have made a compelling case that *Au. anamensis* and *Au. afarensis* are parts of the same anagenetically evolving lineage. The holotype is KNM-KP 29281 (1994); the main sites are Allia Bay and Kanapoi in Kenya, and Aramis and Asa Issie in Ethiopia. (L. *australis* = southern, Gk *pithekos* = ape, and *anam* = means "lake" in the Turkana language.)

### *Australopithecus bahrelghazali* Brunet et al., 1996

A hominin species established by Michel Brunet and colleagues in 1996 to accommodate a midline mandible fragment and an upper premolar tooth recovered from c.3.5–3.0 Ma sediments in the Bahr el Ghazal region, Koro Toro, Chad. The researchers claim that it has thicker enamel than *Ardipithecus ramidus*, a more vertically orientated and more gracile mandibular symphysis than *Australopithecus anamensis*, more complex premolar roots than *Australopithecus afarensis* and *Australopithecus africanus*, and larger incisors and canines than *Au. africanus*. However, most researchers interpret these differences as geographical variation within *Au. afarensis*. The holotype is KT 12/H1 (1995) and the main site is Bahr el Ghazal, Chad. (L. *australis* = southern, Gk *pithekos* = ape, and *bahr el ghazali* = place of discovery.)

### *Australopithecus boisei* (Leakey, 1959)

The genus *Zinjanthropus* and the species *Zinjanthropus boisei* were established in 1959 to accommodate fossil hominins recovered in 1955 and 1959 from Bed I, Olduvai Gorge, Tanzania, but Louis Leakey subsequently suggested that *Zinjanthropus* be subsumed into *Australopithecus*. Many now refer to this taxon as *Paranthropus boisei*. (L. *australis* = southern, Gk *pithekos* = ape, plus *boisei* to recognize the substantial help provided to Louis Leakey and Mary Leakey by Charles Boise.)

### *Australopithecus garhi* Asfaw et al., 1999

A hominin species established by Berhane Asfaw and colleagues to accommodate a fragmented cranium and two partial mandibles recovered from the c.2.5 Ma Bouri Formation in the Middle Awash study area, Ethiopia. *Australopithecus garhi* combines a primitive cranium with very large-crowned, hyper-megadont, postcanine teeth, especially the premolars. But unlike *Paranthropus boisei* the incisors and canines of *Au. garhi* are large and the enamel apparently lacks the extreme thickness seen in *P. boisei*. A partial skeleton exhibiting a long femur with a long forearm was found nearby, but it is not associated with the type cranium and these fossils have not been formally assigned to *Au. garhi*. Its discoverers claim that despite its large postcanine tooth crowns, the cranium of *Au. garhi* lacks the derived features of *Paranthropus*. They suggest *Au. garhi* may be ancestral to *Homo*, but what little evidence there is from phylogenetic analyses does not support a close link with *Homo*. The holotype is BOU-VP-12/130 (1997) and the taxon is presently only known from one site, Bouri, in Ethiopia. (L. *australis* = southern, Gk *pithekos* = ape, and *garhi* = "surprise" in the Afar language.)

### *Australopithecus ramidus* White et al., 1994

A species of *Australopithecus* proposed by Tim White and colleagues to accommodate 17 fossils recovered from localities at Aramis in the Central Awash Complex of the Middle Awash study area, Ethiopia. A year later, in a corrigendum to the original paper, the same authors made a new genus name, *Ardipithecus*, available along with a brief diagnosis. (L. *australis* = southern, Gk *pithekos* = ape, and *ramid* = "root" in the Afar language.) See **Ardipithecus ramidus**.

## *Australopithecus sediba* Berger et al., 2010

A hominin taxon established by Lee Berger and colleagues to accommodate two hominin associated skeletons, MH1 and MH2, recovered from Malapa in southern Africa. These researchers suggest that although the lower limb of *Australopithecus sediba* is like those of other archaic hominins, it has cranial (more globular neurocranium, gracile face), mandibular (more vertical symphyseal profile, a weak chin), dental (simple canine crown, small anterior and postcanine tooth crowns), and pelvic (acetabulocristal buttress, expanded ilium, and short ischium) morphologies that are otherwise only seen in *Homo*. The holotype is MH1 (2008) and presently this taxon is only known from Malapa in southern Africa. (L. *australis* = southern, Gk *pithekos* = ape, and Se Sotho *sediba* = fountain or wellspring.)

## autapomorphy

A term used in cladistic analysis for a derived character state that is confined to one taxon, or operational taxonomic unit, used in that analysis. Autapomorphies can be used for taxonomic identification in light of the results of a cladistic analysis, but character states cannot be assumed to be autapomorphic without first performing a cladistic analysis. Because autapomorphic morphology is by definition confined to a single taxon, it cannot be used to explore how closely one taxon is related to another. Examples of probable autapomorphies within the hominin clade are the enlarged talonid of the mandibular postcanine teeth of *Paranthropus boisei* and the large globular brain case of *Homo sapiens*. (Gk *autos* = self and *morphe* = form.)

## autecology

The branch of ecology that focuses on the interactions between an individual (or a single species) and its environment. Many ecological studies of hominins are essentially autecological, as some species are represented by a single fossil and in many cases research focuses on a single species. (Gk *auto* = self.)

## available

To be "available" a taxonomic name must have been generated according to the rules and recommendations of the International Code of Zoological Nomenclature. For example, the rules stipulate that an available name must not have been used in that context before, must be formed from the 26 letters of the alphabet, and must not have any commercial connotation. However, just because a taxonomic name is available does not mean it is valid. (L. *valere* = to be worthy.) See also **International Code of Zoological Nomenclature**; **valid**.

## Awash River Basin

The part of the Awash River that passes through the Afar Triangle in Ethiopia. It was not until geologist Maurice Taieb undertook reconnaissance work there in the 1960s that its potential as a source of fossils was realized. In the early 1970s, Taieb, together with Yves Coppens, John Kalb, and Don Johanson, formed the International Afar Research Expedition (or IARE) to explore the northern part of the basin.

### awl
A small pointed tool commonly made of bone or metal and used for engraving or perforating. Bone awls, manufactured by longitudinal splitting of long bones and subsequent sharpening of the pointed tip, occur as components of early bone industries (e.g., Blombos Cave, southern Africa). (ME *aul* = pointed tool.)

### axial digit
Term used for the reference digit of the hand (the middle finger) or the reference digit of the foot (the second toe). (L. *axis* = a straight line around which a body or object rotates, or a line around which something is symmetric.)

### axial skeleton
Comprises the skull, the vertebral column including the sacrum, the ribs, and the sternum (i.e., it is what is left of the skeleton after the limbs and limb girdles have been excluded).

### axis
Four meanings of axis are relevant to human evolution. First, it refers to the midline, or axis, of the body, as in axial skeleton. Second, it refers to the straight line, or axis, around which a part of the body rotates. Third, it refers to the reference digit of the hand (the middle finger) and foot (the second toe). The fourth meaning refers to the second (C2) of the seven cervical vertebrae. (L. *axis* = a straight line around which a body or object rotates, or a line around which something is symmetric.)

# B

## baboon
The common name for a member of the Old World monkey genus *Papio*. The term is sometimes used more inclusively for a grade that includes the large-bodied papionin monkeys (including *Papio*, *Theropithecus*, and *Mandrillus* plus a range of extinct genera). Aspects of baboon behavior and ecology provide a popular framework for comparative models to develop hypotheses regarding early hominins. *See also* **Theropithecus**.

## Ba/Ca ratios
*See* **strontium/calcium ratios**.

## backed
Term used to describe the presence of a blunted or abraded edge on a microlith. The blunting, or backing, strengthens the edge so the microlith can be mounted to form a composite tool. The opposing edge is usually left unmodified.

## Bahr el Ghazal
A collection area approximately 45 km/28 miles east of Koro Toro, Chad. More than 17 numbered localities have been identified in exposures of Pliocene fossiliferous sediments. At one of the sites, KT 12, researchers recovered the remains of two hominins, one of which is the type specimen of *Australopithecus bahrelghazali*. (Ar. *Bahr el Ghazal* = river of the gazelles.)

## balanced polymorphism
When two (or more) distinct forms of a phenotypic character (e.g., S and A hemoglobin) are present in a population in greater proportions than would be possible solely from recurrent mutation. The determining alleles (in this case the S-type and wild-type alleles of the beta-globin gene) are maintained at elevated frequencies by balancing selection or frequency-dependent selection.

## balancing selection
*See* **natural selection**.

---

*Wiley Blackwell Student Dictionary of Human Evolution*, First Edition. Edited by Bernard Wood.
© 2015 John Wiley & Sons, Ltd. Published 2015 by John Wiley & Sons, Ltd.

## balancing side
*See* **chewing**.

## Bapang Formation
Formerly known as the Kabuh Formation, it is the younger of the two fossil-hominin-bearing geological formations in Central and East Java, Indonesia. Most of the hominin fossil specimens from the Sangiran Dome come from the Bapang Formation. Argon-argon ages of the lowermost Bapang hominins are 1.51–1.47 Ma, whereas the later specimens are 1.33–1.24 Ma. (Named after a local village near the type section.)

## BAR
Abbreviation for Baringo. From 2000 onwards it is the prefix for fossils recovered at Lukeino, Tugen Hills, Baringo District, Kenya (e.g., BAR 1000′00, the holotype of *Orrorin tugenensis*).

## BAR 1000′00
The holotype of *Orrorin tugenensis*, which consists of fragments of an adult mandible, plus the left $M_2$ and left and right $M_3$. Found at Lukeino, Tugen Hills, Baringo District, Kenya, in 2000 in *c*.6 Ma deposits. *See also* ***Orrorin tugenensis***.

## bare area
The area on the squamous part of the occipital bone between lambda and the superior nuchal lines that was not covered by either the temporalis or the nuchal muscles. Extant apes seldom have a significant bare area, but megadont and hyper-megadont archaic hominins (e.g., MLD 1, OH 5) may have a bare area.

## barium/calcium ratios
*See* **strontium/calcium ratios**.

## Barnfield Pit
*See* **Swanscombe**.

## basalt
A hard, dense, basic (between 45 and 52% $SiO_2$) igneous rock, rich in the minerals plagioclase and pyroxene, that is formed when liquid magma cools either on the Earth's surface as a lava flow or at shallow depths as an intrusion. In rift valleys, basalts often underlie sedimentary rocks, or are visible as dykes or sills. During the early stages of dating the sediments at Olduvai Gorge, a layer of basalt beneath Bed I provided a maximum potassium-argon age for the fossil-bearing sediments above it.

## base
*See* **base pair**; **basicranium**; **cranium**; **mandible**.

## base pair
(or bp) The fundamental unit of a double-stranded DNA molecule. A single-stranded DNA molecule consists of nucleotides that each contain one of four bases: adenine, guanine, cytosine, or thymine. The bases are joined in pairs by hydrogen bonds to form the "rungs" that make up

the double-helix ladder-like structure of double-stranded DNA. Bases are divided into two classes: purines (adenine and guanine) and pyrimidines (cytosine and thymine). In the DNA double helix, guanine always binds with cytosine and adenine always binds with thymine. The length of a DNA molecule is expressed as the number of base pairs it contains (e.g., 1789 bp).

## basicranium

Term for the part of the bony cranium beneath the brain. Its inner or endocranial surface is divided into three hollowed areas called cranial fossae. The anterior and middle cranial fossae lie, respectively, beneath the frontal and temporal lobes of the cerebral cortex, and the posterior cranial fossa lies beneath the cerebellum. The unpaired bones that contribute to the cranial base are, from front to back, a small part of the frontal, the ethmoid, sphenoid, and the occipital, except for the occipital squama; only one paired bone, the temporal, contributes to the cranial base. Studies of the cranial base or basicranium usually consider either its midline morphology (e.g., cranial base angle) or its parasagittal morphology (i.e., the morphology that lies on either side of the sagittal plane). The cranial base completes its growth and development before the face and the neurocranium, and therefore it constrains their development. Basicranium is synonymous with the more modern term cranial base. (Gk *kranion* = brain case and L. *basis* = base.)

## basin

Teeth A depressed area on the surface of the crown of a postcanine tooth bounded by enamel ridges or cusps (e.g., talonid basin). (syn. fovea, fossa.) Geology Basin is used to describe large-scale structural formations that formed after the deposition of the strata (structural basins), or contemporaneously or prior to the deposition of the strata, in response to tectonic processes (e.g., crustal stretching and rifting during rift valley formation). Sedimentary basins (e.g., Turkana Basin) provide the major locations for the deposition and preservation of terrestrial sediments, and thus favor the preservation of fossils.

## bauplan

Term used to describe the general organization, or body plan, of a taxon. It is usually used for large taxonomic groups (e.g., the four limbs of tetrapods), but the term can also be applied to a species (e.g., a flat face, robust mandible, and large postcanine tooth crowns would be part of the bauplan of *Paranthropus boisei*). (Ge. *bau* = architectural and *plan* = sketch or drawing.)

## Bayesian methods

Bayesian statistical methods trace their roots to the Reverend Thomas Bayes (1701–61), who formulated a rule that described how to calculate the posterior probability of a hypothesis after observing data and after considering one's prior belief in the hypothesis (i.e., at a time before the data could be observed). Bayes' theorem is difficult to solve exactly for most real-world problems, but posterior probabilities can be approximated using Markov chain Monte Carlo (MCMC) procedures. Hypotheses of adaptive trait evolution can also be tested in a Bayesian context by comparing phylogenetic generalized least squares and related models through a MCMC process. Bayesian methods are gaining in popularity for phylogenetic tree inference and for examining correlated trait evolution in a phylogenetic context. *See also* **maximum likelihood; Monte Carlo; phylogenetically independent contrasts; phylogenetic generalized least squares**.

## BBC
Abbreviation for **Blombos Cave** (*which see*).

## BC
An abbreviation for both Baringo Chemeron and Border Cave, and a prefix used for fossils recovered from the Baringo Chemeron Formation, Tugen Hills, Baringo District, Kenya (e.g., KNM-BC 1), and for fossils from Border Cave (e.g., BC 1). *See also* **Border Cave; Chemeron Formation**.

## bed
A layer, or stratum, of rock that can be distinguished from the layers, or strata, that lie above it and below it. The bed is the smallest stratigraphic unit that is routinely used in the formal geological nomenclature (e.g., Bed I at Olduvai Gorge can be distinguished from the basalt below it and the base of Bed II above it). (ME *bedd* = bed.)

## bedrock
A general term for the deep rock layers beneath a more recent, usually Quaternary or later Cenozoic, stratigraphic sequence. For example, at Olduvai Gorge the sedimentary rocks that make up Beds I–IV, etc., lie superficial to basalt bedrock.

## Beeches Pit
A Marine Isotope Stage 11 and 12 Acheulean site in England that preserves some of the earliest evidence of controlled fire in Western Europe. The evidence for repeated episodes of burning at Beeches Pit is consistent with evidence from other early sites such as Gesher Benot Ya'akov in Israel and Schöningen in Germany. (Location 52°18′56″N, 00°38′20″E, England.)

## behavioral ecology
The study of the evolutionary and ecological factors that influence behavior. Behavioral ecologists might consider, for example, whether the costs of an activity or behavior outweigh its benefits. Applying behavioral ecological principles to modern humans and to their evolution has contributed to debates over the sexual division of labor, foraging, provisioning, and life history strategies. *See* **optimal foraging theory**.

## Belohdelie
One of the named drainages/subdivisions within the Bodo-Maka fossiliferous subregion of the Middle Awash study area in Ethiopia. It was first explored for its paleoanthropological potential in 1981 by a multidisciplinary team organized by the Ethiopian Ministry of Culture and Sports Affairs and led by Desmond Clark. A hominin frontal bone, BEL-VP-1/1, was recovered during the initial survey.

## BEL-VP
Prefix used for fossils recovered from Belohdelie, Ethiopia (e.g., BEL-VP-1/1).

## BEL-VP-1/1
Seven cranial vault fragments, three of which comprise most of the frontal bone, found at Belohdelie, Ethiopia, in 1981. If its proposed age of c.3.89–3.86 Ma is confirmed, the Belohdelie frontal would be among the earliest, if not the earliest, fossil evidence for *Australopithecus afarensis*.

## bending energy
*See* **transformation grid**.

## benthic
Organisms that live in, on, or close to the sea floor. Cores extracted from the floor of the deep ocean are used to sample the shells of ancient benthic foraminifera. The $^{18}O/^{16}O$ ratios of these shells contain a temporal record of deep sea temperatures and glacial ice volume. Conversely, planktonic (i.e., subsurface) foraminifera provide a record of sea surface temperature. (Gk *benthos* = deep in the sea.)

## bent-hip bent-knee walking
A form of bipedal walking with flexed hips and knees that is used by chimpanzees when they walk bipedally. John Napier suggested archaic hominins might have employed this technique, and given the skeletal morphology and inferred muscular arrangements seen in A.L. 288, Jack Stern and Randall Susman suggested this was the form of gait most likely used by *Australopithecus afarensis*.

## Berg Aukas
A breccia deposit in Namibia that yielded a right proximal half of a massive femur in 1965 (Berg Aukas 1). Its large head, low neck-shaft angle, and thick cortical bone resemble the presumed *Homo erectus* specimen KNM-ER 736 from Koobi Fora, Kenya, and the femoral fragment from Castel di Guido, Italy, attributed to *Homo heidelbergensis*. (Location 19°30′58″S, 18°15′10″E, Namibia.)

## Bergmann's Rule
The observation that animals living at lower average temperatures tend to have larger body sizes. Bergmann's Rule has not been extensively tested in nonhuman primates, and the studies that have been undertaken provide limited support. However, it does seem to apply to modern humans with respect to mean body mass. *Compare to* **Allen's Rule**.

## beta taxonomy
According to Ernst Mayr, beta taxonomy involves arranging species in "a natural system of lesser and higher categories" (i.e., phylogeny reconstruction). (Gk *beta* = the second letter in the Greek alphabet and *taxis* = to arrange or "put in order.") *See also* **systematics**; **taxonomy**.

### Biache Saint-Vaast
A Marine Isotope Stage 7 open-air site located in the Pas-de-Calais, France, that preserves evidence of several episodes of hominin occupation during warm interglacial periods, and a large faunal assemblage that includes the fragmentary remains of two of the oldest known *Homo neanderthalensis* crania. (Location 50°18′54″N, 02°56′55″E, France.)

### bias
A term that refers to error in the estimation of a population parameter. Statistical bias refers to a difference between the estimate of a population parameter (such as the mean) in a sample of a population and the value of that parameter in the population itself. Ascertainment bias occurs when the population sample is a nonrandom one. Consider a fossil assemblage that only preserves the largest individuals in a population; in that case, the mean value of any measurement for that sample would be an overestimate of the actual value of the mean for the population from which the sample was drawn. Systematic bias is when a systematic measurement error results in all of the measurements being off by the same amount (e.g., if a caliper is not correctly zeroed). (Fr. *biais* = slant, from the Gk *epikarsios* = slanted.)

### biface
A stone tool that has been worked on opposed surfaces, or faces, as opposed to tools that are worked on only one side. Handaxes are a type of biface. (L. *bi* = two and *facies* = surface.)

### Bilzingsleben
An open-air Marine Isotope Stage 11 site in central Germany which contains many flint artifacts and controversial evidence of dwelling structures, hearths, and worked bone. (Location 51°16′52″N, 11°04′07″E, Germany.)

### bimaturism
When two organisms, or structures, differ in size because the larger one has grown for a longer period of time (e.g., the larger male canines in extant hominoids). (L. *bi* = two and *maturus* = ripe.)

### binominal
See **Linnaean binominal**.

### biochronology
A correlated-age method of relative dating that uses fossils preserved in sedimentary rocks. Many fossil sites lack the raw materials (e.g., igneous rocks) needed for absolute dating. Thus, these sites have to be dated using any fossils they contain. Fossils are matched with those found at other sites that can be dated by numerical-age and calibrated-age methods (e.g., isotopic or radiogenic) of absolute dating. If the first appearance datum (or FAD) or last appearance datum (or LAD) of a taxon is the same across several absolutely dated fossil sites, then that taxon can be used alone, or preferably in combination with other time-delineated taxa, to generate a biochronometric date for a stratum. Data from one taxon, or a series of taxa, can be combined to define one, or more, biozones. An example of a single-taxon system is the use of the extinct pig *Metridiochoerus andrewsi* as a marker taxon in East African strata;

an example of a multi-taxon system is the European mammal neogene (or MN) system for dating Miocene and Pliocene sites whose ages range from 24 to 2 Ma. (Gk *bios* = life, *chronos* = time, and *ology* = the study of.) *See also* **biostratigraphy**; **European mammal neogene**; **geochronology**.

## biogenetic law
Summarized as "ontogeny recapitulates phylogeny," this law proposes that during embryonic development animals pass through the adult stages of their ancestors. The strictest interpretation of the biogenetic law is now understood to be false; that is, evolution does not occur in a linear manner such that the developmental stages of ancestors are literally compressed into the ontogeny of descendant species. There are broad similarities in embryonic development among all mammals, among mammals and reptiles, and among mammals and fish, but these mostly occur early in embryogenesis. (Gk *bios* = life and *genesis* = birth or origin.)

## biogeography
The scientific investigation of how and why organisms came to be distributed on the Earth in the way they are. Charles Darwin is often considered the father of evolutionary biogeography. Some of the most important hypotheses in paleoanthropology concern biogeographic events (e.g., the dispersals of various australopith and early *Homo* species within Africa during the Plio-Pleistocene, and the later dispersal of hominins out of Africa). Key biogeographical questions related to paleoanthropology include whether hominins dispersed in the same direction and at the same time as other mammals, whether hominin dispersals track changes in range of vegetation zones, and how, if at all, climate change has affected hominin biogeographic patterns. (Gk *bios* = life, *geo* = earth, and *graphein* = to write.)

## biological species
A biological species is an extant species that is diagnosed on the basis of evidence of interbreeding among individuals belonging to that species, and the lack of any evidence of interbreeding with individuals belonging to another species. *See also* **species**.

## biological species concept
A "process" species definition developed by Ernst Mayr in which species are defined as "groups of interbreeding natural populations reproductively isolated from other such groups." This definition requires there to be at least one additional species; it is a "relational" and not a "freestanding" species definition. The biological species concept is of limited value in considering extinct hominins because it is presently impossible to collect direct evidence for interbreeding from the fossil record. *See also* **species**.

## biome
A distinctive regional ecological community of plants and animals adapted to the prevailing climate and soil conditions. Biomes are collections of ecosystems that have similar biotic and abiotic characteristics. It is conventional to recognize six major terrestrial biomes: forest, grassland, woodland, shrubland, semidesert shrub, and desert. Each of these biomes can be further subdivided on the basis of climatic conditions and elevation. (Gk *bios* = life.)

## biospecies
A contraction of **biological species** (*which see*).

## biostratigraphy
The process of linking rock layers within and among stratigraphic sequences by comparing the fossil assemblages preserved within them. This can be via either the matching of individual taxa (e.g., the rodent *Mimomys* in Pleistocene European sites), or by using a system such as the European mammal neogene (or MN), which uses a combination of taxa. The use of rock layers linked in this way to infer the age of the strata is also called biochronology. (Gk *bios* = life, *stratos* = to cover or spread, and *graphein* = to write.) *See also* **biochronology**.

## biotic
Describes all the living factors (e.g., competition for food, predator pressure) that affect the evolutionary history of a taxon, or of a clade. (Gk *bios* = life.) *Compare to* **abiotic**.

## bioturbation
The displacement and mixing of sediment or soil by biotic agents or processes. In Africa, for example, both burrowing termites and rodents can move artifacts and ecofacts from their original context. Searching for evidence of bioturbation is an important part of considering the temporal and contextual integrity of a paleontological or archeological site, especially when evidence from the site is used to either develop or test hypotheses about hominin activity patterns or reconstruct the paleoenvironment. (Gk *bios* = life and *turbe* = turmoil.)

## biozone
A series of strata at one or more sites, all of which sample the same fossil taxon, or suite of fossil taxa. An example would be the *Metridiochoerus andrewsi* biozone in East Africa. *See also* **biochronology; biostratigraphy**.

## biped
A creature whose resting posture is achieved by standing on the feet of the two hind, or lower, limbs, and whose locomotor repertoire involves a significant amount of bipedal travel. (L. *biped* = two-footed.) *See also* **bipedal**.

## bipedal
Form of posture or locomotion in which the body is supported exclusively on the feet of the two hind, or lower, limbs. In 1980, J. H. Prost usefully distinguished between taxa in which adults are *able* to travel bipedally, a locomotor mode which he called "facultative bipedalism," and taxa such as modern human adults for which a bipedal gait is the *only* efficient way for them to move any distance. Prost referred to the latter mode as "preeminent bipedalism," but others prefer the term "obligate bipedalism." The earliest hominins were most likely facultative bipeds, whereas *Homo neanderthalensis* was almost certainly an obligate biped. The possession of some of the morphological traits seen in known bipeds (i.e., bipedal traits) does not mean the taxon that possesses them was necessarily a biped. (L. *biped* = two-footed.)

## bipedalism
The act of walking or running using only the feet of the two hind, or lower, limbs for support. *See also* **bipedal**.

## bipedalism, facultative
*See* **bipedal**.

## bipedalism, obligate
*See* **bipedal**.

## bipedal locomotion
*See* **bipedal**; **walking cycle**.

## bipedal trait
A term for a morphological feature found in known bipeds that is functionally related to either an upright posture, or bipedal locomotion, or both. For example, an adducted hallux is a bipedal trait established prenatally that is present in all known hominin bipeds and that plays an important role in propulsion at the end of the stance phase of bipedal gait. *See also* **bipedal**.

## bipolar percussion
A term used to refer to a specific technological approach for making stone flakes in which the core is rested on an anvil and then struck to split the core, or to remove a flake. Chimpanzee populations that crack nuts mostly use bipolar technology. (L. *bi* = two and *polus* = axis.)

## birds
True bird fossils are relatively rare in the fossil record, but trace fossils of birds (e.g., prints and tracks) and fossilized egg shell are more common. Traces of birds in the fossil record can be useful paleoenvironmental indicators (e.g., wading birds indicate the nearby presence of water; ostrich egg shells suggest dry conditions).

## bite force
*See* **chewing**.

## BK
Prefix used for fossils recovered from the Kapthurin Formation, Tugen Hills, Baringo District, Kenya (e.g., KNM-BK 67) and from the site of Blimbingkulon in Java, Indonesia (e.g., BK 7905). It is also the acronym for Bell's Korongo at Olduvai Gorge.

## Blaauwbank valley
A shallow valley in the vicinity of Krugersdorp in Gauteng Province, South Africa. There are many substantial breccia-filled solution cavities in the limestone hills that form the high veldt and several (e.g., Cooper's Cave, Drimolen, Kromdraai, Malapa, Sterkfontein, and Swartkrans) in the Blaauwbank valley contain fossil hominins.

## blade

A type of flake that is at least twice as long as it is wide, or one that has parallel edges or multiple flake scars on its outer surface. Blade production was most likely invented multiple times, and perhaps by more than one hominin taxon. Blades appear intermittently in stone-tool industries from the Middle Pleistocene onwards, and they are particularly common in Upper Paleolithic assemblages. Particularly small (<10 mm wide) blades are known as bladelets. (OE *blaed* = sharp-edged tool.)

## Blimbingkulon

A hominin site within the Sangiran Dome, in Java, Indonesia. Two hominin mandibular fragments found at the site most likely come from the *c*.1.5 Ma Grenzbank zone, between the older Sangiran Formation and the younger Bapang Formation. (Location 07°20′S, 110°58′E, Indonesia.)

## Blombos Cave

A cave on the coast approximately 300 km/185 miles east of Cape Town, South Africa, that contains approximately 50 m² of sediments. The Middle Stone Age levels date to between 80 and 70 ka. Blombos is mainly important because of the recovery of likely evidence of symbolic behaviors (e.g., cross-hatch patterns on pieces of ochre and deliberately pierced "Dunker" shells), bone tools, pressure flaking, pyrotechnology, and the exploitation of shallow-water fish. All of these behaviors are more commonly associated with the Later rather than the Middle Stone Age. (Location 34°25′S, 21°13′E, South Africa.)

## BOD

Prefix for the paleontological evidence collected in the Bodo area of the Middle Awash study area, Ethiopia.

## Bodo

Locality within the Bodo-Maka fossiliferous subregion of the Middle Awash study area, Ethiopia, that has yielded abundant vertebrate fossils, Acheulean artifacts, and three hominin fossils, including a cranium (Bodo 1). The Bodo fauna suggests the site dates from the Middle Pleistocene and argon-argon dating and stratigraphic correlation suggest an age for the site of *c*.600 ka. (Location 10°37.5′N, 40°32.5′E, Ethiopia.)

## Bodo 1

A *c*.600 ka partial cranium reconstructed from many fragments recovered from Bodo, in the Middle Awash study area of Ethiopia. Its face and supraorbital torus are large and its nose is wide. Many aspects of the Bodo cranium resemble the morphology of *Homo erectus* (e.g., projecting brow ridges, low frontal, angular torus, approximately 1250 cm³ endocranial volume), but other morphology suggests affinities with later *Homo* (e.g., the arched parieto-temporal suture, frontal torus separated into medial and lateral elements). Bodo 1 resembles the hominin fossils Petralona, Arago, Kabwe, and Saldanha, and many regard it as consistent with the hypothesis that these specimens all belong to *Homo heidelbergensis*. If Bodo 1 is as old as claimed, then, along with the type specimen from Mauer, it is among the earliest widely accepted evidence for that taxon.

## Bodo-Maka
The only major fossiliferous subregion within the Middle Awash study area on the east side of the Awash River, Ethiopia.

## body mass estimation
The body mass of fossil hominins is often estimated from fragmentary fossil remains using relationships between the size of the bone or tooth and body mass derived from living animals. The body mass of a fossil hominin might be estimated from a general mammalian, primate, or ape data set, but it is not valid to estimate an australopith body mass from a regression equation calculated using only modern humans, since australopiths are not modern humans. All body mass estimates for fossil species should be reported with explicit recognition of the statistical uncertainty involved (e.g., a confidence interval).

## body size
A distance, area, volume, or mass measurement designed to reflect the overall dimensions of an organism in a single value. Most applications in paleoanthropology refer to body mass (although the terms "mass" and "weight" are sometimes used interchangeably, the correct measurement is mass). Scaling seeks to explain much of the functional variation among organisms in relationship to variation in body size. It is necessary to take body size into account when using brain size to judge cognitive evolution, tooth size to infer diet, or sexual dimorphism to infer social organization.

## Boker Tachtit
One of several sites found in the late 1970s in the Negev, Israel. Abundant stone tools from four main occupation levels apparently document the Middle to Upper Paleolithic transition. (Location 30°50'N, 34°46'E, Israel.)

## Bølling-Allerød
A warm interval (c.13–11 ka) that followed the abrupt termination of the Last Glacial Maximum.

## bone
Bone is one of the two connective tissues (cartilage is the other) that form the skeleton. All connective tissues consist of cells and fibers embedded in a matrix. Bone (and cartilage) are hard and stiff because they are mineralized; that is, their matrix and fibers contain crystals of hydroxyapatite [$Ca_{10}(PO_4)_6(OH)_2$]. Osteoblasts make, deposit, and mineralize the matrix, osteoclasts remove it, and osteocytes are what osteoblasts turn into once the matrix has matured (they are quiescent osteoblasts). The collagen fibers in bone are arranged in concentric bundles in lamellar (layered) bone. In immature woven (nonlamellar) bone, they are arranged in irregular bundles. Mature bone is either compact or cancellous; the latter has trabeculae, the former does not. Long bone shafts have an outer layer (cortex) of compact bone and an inner cavity (medulla) of cancellous bone. The spaces in cancellous bone are filled either by fat (yellow bone marrow) or blood-forming tissues (red bone marrow), or a mixture of the two. The combination of the collagen fibers and the mineralized matrix gives bone its unique material properties. (ME *bon* = bone.) *See also* **bone strength**.

## bone breakage patterns

Bones can be fractured by numerous taphonomic agents and processes, including hominins, carnivores (i.e., by chewing), large mammals (i.e., by trampling), by falling rocks within cave sites, and by sediment compaction. Identifying the taphonomic agent responsible for the fractured bones involves a detailed study of bone breakage patterns. Various types of fracture or cracking are known to result from different modes of loading on different kinds of bone (long bones, vertebrae) depending on the state of the bone (fresh, wet, dry, weathered, fossilized). Bone breakage patterns reveal whether bones were broken while fresh (spiral fractures characterized by curved outlines and oblique fracture edges) possibly during carcass processing, or while dry (transverse fractures and perpendicular fracture angles) long after discard. The best way of determining the agent responsible for bone fracture is the detailed study of bone surface modifications (e.g., percussion marks, carnivore tooth marks).

## bone remodeling

Refers to any internal reorganization within a bone that does *not* result in a net change in mass or in any change of external shape. It involves osteoclasts resorbing existing bone, and osteoblasts laying down an equivalent mass of new secondary osteonal or Haversian bone. Note that "modeling" is when bone is modified so that its mass and/or its external shape *are* changed (e.g., during growth or as the result of excessive use). (L. *re* = in a different way and *modus* = a measure; syn. Haversian remodeling.) *See also* **ossification**.

## bone strength

In material science strength is the ability to withstand an applied stress without failure. The strength of bone tissue is a material, and not a structural, property. You can measure the strength of a bone by loading it to failure, but that is a measurement of its structural strength, and depends on the shape of the bone in addition to the inherent material strength of the bone tissue. The material strength of different types of bone (i.e., compact, cancellous) is measured by recording their ability to resist being indented by a fine point.

## bone tool

Any bone object used during the performance of a task. For example, bone tools found in Members 1–3 at Swartkrans, South Africa, have been interpreted as implements used to extract termites, whereas bones from Olduvai Gorge, Tanzania, were deliberately modified and shaped through flaking (the technique is similar to that used for stone tool production) to make handaxe-like pieces. Bone tools are common in Upper Paleolithic and Later Stone Age sites.

## Bonferroni correction

A method that addresses the problem of identifying the appropriate significance level when multiple significance tests are performed. Suppose that in the course of an analysis one were to conduct 20 significance tests. By random chance we would expect that 5% of the tests, or 1 out of 20, would yield a statistically significant result even though no real relationship exists. The Bonferroni correction recognizes this problem and adjusts the value at which a result is deemed statistically significant downwards to compensate for conducting multiple tests.

## bonobo
See *Pan paniscus*.

## bony labyrinth
The complex of cavities in the petrous part of the temporal bone of the cranium that contains the membranous labyrinth. It includes the cochlea (anterior), the vestibule (middle), and the semicircular canals (posterior) and it can be visualized using computed tomography (or CT). (Gk *labyrinthos* = complex of cavities.) *See also* **inner ear**; **semicircular canals**.

## bootstrap
A resampling procedure that can be used to calculate a confidence interval and/or the degree of bias for sample statistics (e.g., mean, variance, median, correlation coefficient, regression slope, etc.). The bootstrap is used to calculate the confidence interval of interest (e.g., for the median) when data are not normally distributed and/or when sample sizes are small. The resampling can be with, or without, replacement.

## Border Cave
A site in South Africa that includes Middle and Later Stone Age artifacts. Two hominins were found early on at the site but their provenance is uncertain. Later discoveries are probably from the Iron Age. (Location 27°01′13″S, 31°59′08″E, KwaZulu-Natal, South Africa.)

## Bose Basin
A large basin north of Nanning in southern China. Seven river terraces have been identified in the basin with Paleolithic and Neolithic archeological remains surface collected and excavated from the first four terraces (T1–T4). To date, about 80 Paleolithic open-air sites have been identified, the best known of which are Damei, Fengshudao, and Gaolingpo. Australasian strewn-field tektites excavated from T4 suggest an age of between 800 and 700 ka. The evidence of handaxes at sites in the Bose Basin has been used to refute the idea of Movius' Line. (Location approximately 23°66′N, 106°37′E, Guangxi Zhuang Autonomous Region, China.)

## Boskop
An open-air site in present-day Gauteng Province, South Africa. The remains recovered include a calvaria that attracted intense interest at the time because of its large endocranial volume. Initially it was thought to be old, but later authors emphasize its similarity to a variety of other large-brained crania from the Holocene of the Cape. (Location 26°34′S, 27°07′E, Gauteng Province, South Africa.)

## bottleneck
An evolutionary event during which genetic diversity is reduced due to a substantial decrease in population size. A population bottleneck increases genetic drift (since the latter is a stronger force of evolution in smaller populations), and it also increases inbreeding. Researchers suggest that modern humans have undergone one or more population bottlenecks, possibly in connection with natural disasters, disease, or migrations.

## BOU
Prefix used for fossils from localities in the Bouri peninsula, Middle Awash study area, Ethiopia.

## Bouri
The name given to the Bouri Fault Block, a peninsula in the Middle Awash study area, Ethiopia. Three members of the Bouri Formation are exposed on the peninsula. From oldest to youngest these are: the Hata Member (an abbreviation of Hatayae), the Daka Member (an abbreviation of Dakanihylo), and the Herto Member. Research at Bouri has resulted in the recovery of the type specimen of *Australopithecus garhi*, hominin postcranial remains, isolated Mode 1 artifacts, and cutmarked and percussed mammal bones from the c.2.5 Ma Hata Member, Acheulean artifacts and hominins including the BOU-VP-2/66 calvaria from the c.1.0 Ma Daka Member, and the Herto cranium from the c.150–200 ka Herto Member. (Location approximately 10°15–18'N, 40°30–34'E, Ethiopia.)

## BOU-VP-2/66
A well-preserved calvaria, retaining much of the basicranium except for the sphenoid and ethmoid, recovered from the surface of the Daka Member at Daka, Ethiopia, in 1997. It is younger than a pumiceous unit at the base of the Daka Member dated to $c.1.04 \pm 0.01$ Ma. The Daka calvaria shows a mix of features associated with *Homo ergaster* and *Homo erectus*, supporting the conclusion that the two are conspecific. The specimen possesses some morphological features seen in *Homo heidelbergensis* (e.g., more vertical parietal walls).

## BOU-VP-12/130
Associated cranial fragments together with a complete and well-preserved dentition recovered from the surface of the c.2.5 Ma Hata Member at Bouri, Ethiopia, in 1997. It is the holotype of *Australopithecus garhi*.

## BOU-VP-16/1
A well-preserved cranium, lacking the left side of the face and part of the left side of the cranial vault, recovered *in situ* in the Upper Herto Member of the Bouri Formation at Herto, Ethiopia, in 1997. It is dated to between $160 \pm 2$ and $154 \pm 7$ ka, and its large endocranial volume (approximately 1450 cm$^3$) and *Homo sapiens*-like morphology led it to be designated the holotype of a subspecies, *Homo sapiens idaltu*. Until the redating of Omo I and Omo II from Omo-Kibish, Ethiopia, the BOU-VP-16/1 cranium was the earliest evidence of *H. sapiens* in Africa.

## bovid
The informal name for the artiodactyl family, the Bovidae (i.e., antelopes and their allies).

## Bovidae
A family that includes eight subfamilies: Bovinae (cattle and buffalo); Antilopinae (dik-diks and gazelles); Cephalophinae (duikers); Reduncinae (kob); Aepycerotinae (impala); Caprinae (ibex and musk ox); Hippotraginae (antelopes and oryx); and the Alcelaphinae (wildebeest, topi, etc.). They are recognized on the basis of their cranial morphology and particularly the

size and shape of the bony horn cores (the horns themselves are made of keratin and do not survive, but horns have a bony core that does fossilize). (L. *bos* = cow.)

## bovid size classes
Subdivisions of bovids ordered by estimated weight. Class I: less than 23 kg (e.g., duikers and dwarf antelopes); class II: 23–84 kg (e.g., springbok and Grant's gazelle); class III: 84–296 kg (e.g., wildebeest and hartebeest); class IV: more than 296 kg (e.g., eland and buffalo); and class V: more than 900 kg (e.g., extinct giant buffalo *Pelorovis antiquus*). Ethnographic and experimental observations suggest that the body size of a bovid can influence a number of behavioral and/or taphonomic factors, including butchery patterns, carcass transport strategies, and carnivore ravaging.

## bovine
Informal name for the artiodactyl subfamily, the Bovinae, comprising cattle and buffalo. (L. *bos* = cow.) *See also* **Bovidae**.

## Boxgrove
A quarry near the southern coast of England that dates to Marine Isotope Stages 12 and 13. It has yielded Acheulean artifacts, a hominin tibia, and two hominin incisors attributed to *Homo heidelbergensis*. (Location 50°51′31″N, 00°42′48″W, England.)

## bp
Acronym for **base pair** (*which see*).

## BP
An abbreviation of "before the present" that is often used in radiocarbon dating. To standardize radiocarbon dates the "present" is defined as January 1, 1950. Thus "5000 years BP" refers to a date 5000 years before 1950.

## brachial index
Ratio of the lengths of the radius and humerus, calculated as (radius length/humerus length × 100). It measures the relative length of the forearm. The brachial index is often used to discuss the locomotor regime and climate preferences of primates and hominins. Animals that regularly engage in below-branch swinging or climbing (brachiators) tend to have higher brachial indices than arboreal or terrestrial quadrupeds or bipeds. Because the ability to lose or retain heat is proportional to relative surface area, primates (including modern humans) in hot, dry environments have elongated distal limb elements (high brachial indices), while those in cold, humid habitats have relatively short distal elements (low brachial indices). The lower-limb equivalent is the crural index. (L. *brachium* = arm.) *See also* **Allen's Rule**.

## brachium
The most proximal segment of the upper limb, from the shoulder to the elbow. (L. *brachium* = arm.)

## brain

The brain is the part of the central nervous system that lies within the endocranial cavity. The caudal (i.e., towards the tail) part of the neural tube, which is the majority of its length, becomes the spinal cord. The cranial part of the neural tube has three swellings, or vesicles, that give rise to the brain. The vesicle that connects with the spinal cord at the level of the foramen magnum is the hindbrain (rhombencephalon), the middle swelling is the midbrain (mesencephalon), and the rostral-most (i.e., toward the mouth) of the three swellings is called the forebrain (prosencephalon). The forebrain is comprised of two divisions, the diencephalon (including the thalamus and hypothalamus), and the telencephalon (including the cerebral hemispheres and basal ganglia). The cerebral hemispheres are composed of gray matter (the cell bodies of neurons) externally in the cerebral cortex, and white matter (neuron cell processes) internally. In primates, but especially in the great apes and particularly in modern humans, it is the cerebral hemispheres that undergo the greatest expansion. (OE *braegen* = brain, most likely from the Gk *bregma* = the front of the head.) *See also* **brain size**; **brain volume**.

## brain size

The size of the brain can be expressed in terms of either volume or weight (or mass). Brain size is equal to the endocranial volume (also called cranial capacity) minus the volume (or weight) of the meninges, the extracerebral cerebrospinal fluid (cerebrospinal fluid outside the ventricles), the intracranial (but extracerebral) blood vessels, and the cranial nerves within the cranial cavity. *See also* **endocranial volume**.

## brain size evolution, molecular basis of

The genes involved in regulating brain size may also be among the molecular mechanisms underlying brain evolution within the hominin clade. The microcephaly-related genes *ASPM* and *MCPH1* are involved in regulating brain size. All primary microcephaly genes (*MCPH1* to *MCPH6*) influence brain size by regulating the proliferation of neural precursor cells. Microcephalin shows evidence of positive selection primarily at the base of the hominid clade, and *CDK5RAP2* shows especially high rates in the hominin and panin clades. Another gene involved in neural precursor proliferation, *ADCYAP1*, which helps regulate the transition from proliferative to differentiated states during neurogenesis, has also shown evidence of accelerated evolution in the hominin clade. Examples of genes with critical roles in determining the relative size of areas within the cerebral cortex include *Emx2*, *Pax6*, and *COUP-TFI* (*Nr2f1*). Variation in some of these genes among modern humans leads to cognitive deficits.

## brain volume

Usually defined as the sum of the volume of the brain tissue, the cerebrospinal fluid within the ventricles of the brain (intracerebral cerebrospinal fluid), and any cranial nerves and meninges that may adhere to the brain. In practice, adult human brain volume is approximately 85% of the endocranial volume or cranial capacity.

## branch length

The lengths of the branches of a phylogenetic tree, which are sometimes scaled according to the numbers of character-state changes or time intervals that separate the tree's nodes. The length of the branch that separates two nodes may indicate either the amount of evolutionary change that has taken place since the ancestral taxon gave rise to the descendant taxon, or the amount of time that has elapsed since the ancestral taxon gave rise to the descendant taxon.

## breccia
A rock containing angular fragments that have been cemented together in a fine-grained matrix (e.g., soil or cave earth), which is then hardened with a mineral cement such as calcium carbonate. Nearly all of the early hominins from southern Africa have been extracted from breccia-filled limestone caves (e.g., Drimolen, Sterkfontein). The breccia and speleothem that adheres to many of these fossils can be removed mechanically, with precision drills, or chemically, by immersion in dilute acetic acid, under strictly controlled conditions. (Ge. *brehhan* = to break, or It. *breccia* = rubble.)

## bregma
The name for the midline landmark where the sagittal suture between the two parietal bones meets the coronal suture, which separates the frontal from the parietal bones. (Gk *brechein* = moist.)

## bristle-cone pine
See **radiocarbon dating**.

## broad-sense heritability
($H$ or $H^2$) Defined as the genetic variance divided by the phenotypic variance. It is called broad-sense (as opposed to narrow-sense) heritability since it reflects *all* of the possible genetic contributions to the phenotypic variance. *See also* **heritability**.

## Broca's area
The region of the inferior frontal gyrus of the cerebral cortex that is important for the production of speech in modern humans (e.g., syntax, lexical retrieval, and the processing of rules relating to recursion). Broca's area is generally thought to include Brodmann's cytoarchitectonic areas 44 and 45. Homologues of Brodmann's areas 44 and 45 have been identified in macaque monkeys and great apes. When compared to the other cortical areas, area 44 and area 45 on the left side are among the most enlarged in the modern human brain cerebral hemispheres. (Named after Paul Broca.)

## Brodmann's areas
Numbered regions of the cerebral cortex. They are named after Korbinian Brodmann who, after studying sections of the cerebral cortex, subdivided it according to what he interpreted as differences in the neuronal morphology, cell packing density, and relative thickness of the six main cortical layers. For example, the primary motor speech areas of the cerebral cortex are given the numbers 44 and 45, and the primary visual cortex is numbered 17.

## Broken Hill
See **Kabwe**.

### brown striae
Another term for striae of Retzius, the long-period incremental lines in enamel. They are called that because striae of Retzius appear as brown lines when viewed using a regular transmitted light microscope.

### brow ridge
An informal inclusive term that applies to either discrete thickening of the bone above each superior orbital margin (e.g., as seen in *Homo neanderthalensis*) or to a continuous bar of thickened bone (e.g., as seen in Trinil 2) that extends from the lateral end of one superior orbital margin across the midline to the lateral end of the superior orbital margin on the other side. The term brow ridge is usually not applied to the compound structure seen above each orbital opening in *Homo sapiens*, which consists of the superciliary eminence medially and the supraorbital trigon laterally. (ME *brow* = eyelash.)

### browser
An animal (e.g., extant giraffes and elephants) that feeds on above-ground vegetation (e.g., leaves, fruits, and shoots in bushes and trees). *See also* **browsing**.

### browsing
The consumption of leaves, fruits, and shoots of usually dicotyledonous plants (flowering plants with two embryonic seed leaves or cotyledons) that grow above ground level. This term is often used in paleoanthropology to refer to the inferred diets of fossil herbivores. In Africa, browsers feed on plants, such as shrubs, trees, and nongrass herbs, that use the $C_3$ photosynthetic system and are depleted in $^{13}C$ (i.e., have less $^{13}C$) relative to grasses and other plants which use the $C_4$ and CAM photosynthetic systems. Thus, African browsing animals have bones and enamel that are depleted in $^{13}C$ relative to the bones of grazing animals that consume plants which use the $C_4$ and CAM systems. (OF *broust* = shoot or twig.) *See also* **$C_3$ and $C_4$**; **$^{13}C/^{12}C$**; **stable isotopes**.

### Brunhes
*See* **geomagnetic polarity time scale**.

### BSN49/P27
Known as the Gona pelvis, this fossil comprises most of the right side and a substantial part of the left side of a pelvis, plus the associated sacrum and the last lumbar vertebra. The pelvic brim is preserved except for the medial end of the right pubic body. It was found on the surface and during excavation of the Upper Busidima Formation at Gona in 2001. This exceptionally well-preserved c.0.9–1.4 Ma pelvis has been interpreted as providing evidence about the size of the birth canal of a *Homo erectus* adult female, but some researchers have suggested that the pelvis could belong to an archaic hominin, possibly *Paranthropus boisei*.

### Bubing Basin
Bubing is southwest of the Bose Basin in China. Among the caves in the basin are Chuifeng and Mohui (c.2 Ma), Wuyun and Upper Pubu (c.280–76 ka), Luna Cave (70–20 ka), and Zhongshan (12–6 ka). Two hominin teeth excavated from Mohui have been allocated to *Homo*

cf. *erectus*. Eight core and flake tools were excavated from Mohui Cave, but six of these were excavated from disturbed deposits and the relationship between the hominin teeth and the remaining two artifacts is not clear. (Location approximately 23°25′N, 107°00′E, Guangxi Zhuang Autonomous Region, China.)

## buccal
A term used to describe the outer, or lateral, aspect of the premolars and molars. The equivalent term for the canine and the incisors is labial, because the lateral aspect of these teeth is in contact with the lips, not the cheek. The measurement of the breadth of a premolar or molar tooth that is taken at right angles to its long, or mesiodistal, axis is referred to as its buccolingual breadth (i.e., the measurement runs from the outer, buccal, aspect of the tooth crown, to the inner, tongue-side, or lingual aspect of the crown). (L. *bucca* = cheek.)

## Buia
A site in the Alat Formation, Eritrea, that has yielded six hominin fossils, including a nearly complete cranium (UA 31) and an incomplete adult hip bone that have been assigned to *Homo erectus*. (Location approximately 14°49′N, 39°50′E, Eritrea.)

## Bukuran
A 1.5 Ma site within the Sangiran Dome, Indonesia, where the Hanoman 1 *Homo erectus* cranium and the Bukuran calvaria were found. (Location 07°20′S, 110°58′E, Indonesia.)

## bunodont
Teeth with low, conical crowns, as opposed to hypsodont (i.e., high-crowned) teeth. Hominin teeth are relatively bunodont, and the teeth of *Paranthropus boisei* are the most bunodont among hominins. (Gk *buno* = hill and *dont* = teeth.)

## Bura Hasuma
One of the designated subregions of Koobi Fora, Kenya. It includes localities where some of the oldest sediments in the Koobi Fora and the Kubi Algi Formations are exposed. *See also* **Koobi Fora**.

## burin
Burins are "flaked flakes," in which the relatively thick chisel-like edge is made by removing an elongated flake (called a burin spall) along one or more edges of a flake. They were originally considered to function as tools for engraving bone, antler, wood, etc., but others have suggested a wider range of functions. (Fr. *burin* = a term used by engravers and printmakers for an elongated metal chisel.)

## bushland
A biome that is dryer than woodland forest and receives 250–600 mm mean annual rainfall. Small trees are the dominant plant species, and most bushland habitats have some degree of grass cover; by definition, more than 40% is covered by bushes. Partially open habitats, such as bushland and shrubland biomes, are thought to have been the habitats of *Paranthropus* and early *Homo* between 2.5 and 2.0 Ma.

## Bushmen
*See* **San**.

## Busidima Formation
Sediments found in the Gona Paleoanthropological study area, the Dikika study area, the Ledi-Geraru study area, and Hadar, in Ethiopia, that date from 2.69–0.81 Ma. Fossil hominins recovered from the Busidima Formation have been assigned to early *Homo* and *Homo erectus*. Archeological sites in the Busidima Formation include evidence of Oldowan, Acheulean, and Middle Stone Age artifacts. [Location 10°55′–11°15′N, 40°20–35′E, Ethiopia; Busidima is the local Afar (Danakil) name for the ephemeral stream central to outcrops of the Busidima Formation.]

## Buxton-Norlim Limeworks
*See* **Taung**.

# C

## ¹³C/¹²C

A ratio measurement of stable isotopes in bone and dentine collagen and enamel apatite used for diet and paleoenvironmental reconstruction. The $^{13}C/^{12}C$ ratio works on the assumption that in herbivores the relative levels of $^{13}C$ in their collagen and apatite reflect the types of plants ($C_3$, $C_4$, or CAM) they consumed, and in carnivores the levels of $^{13}C$ in their tissues reflects the diets of their herbivore prey. For example, animals that eat plants that use the $C_3$ photosynthetic pathway (most shrubs, trees, herbs, and cool-season grasses) should have carbon isotope ratios ($\delta^{13}C$) that are lower (i.e., have less $^{13}C$) relative to animals that consume plants that use the $C_4$ and CAM photosynthetic pathways (succulents, sedges, and warm-season grasses). Carbon isotope ratios can indicate whether or not a particular environment was open (e.g., grassland or wooded savannah) or closed (e.g., forest). Carbon isotope ratios are positively correlated with temperature, and negatively correlated with relative humidity, rainfall, and elevation.

## $C_3$ and $C_4$

Two groups of plants, $C_3$ plants and $C_4$ plants, show minimal overlap in their $^{13}C/^{12}C$ ratios (commonly represented as $\delta^{13}C$ values in parts per thousand, as in the ‰ notation). The $C_3$ plant group consists of trees, herbaceous plants, and cool-season grasses, and uses the $C_3$ photosynthetic system, resulting in lower $\delta^{13}C$ values (global mean $-27$‰). This group now constitutes the majority of African plants. The $C_4$ plant group, which includes warm-season grasses such as the majority of native grasslands in the Great Plains of the USA and many tropical grass species such as maize, uses the $C_4$ photosynthetic system and has higher $\delta^{13}C$ values (global mean approximately $-13$‰). Animals that eat either $C_3$ or $C_4$ plants, or some mixture of the two, record those diet sources in their tissues. *See also* $^{13}C/^{12}C$.

## $C_3$ foods

These are either plants (e.g., trees, herbaceous plants, and cool-season grasses) that use the $C_3$ photosynthetic pathway, or the flesh of animals that browse or graze on $C_3$ plants. *See also* $C_3$ and $C_4$.

## $C_4$ foods
These are either plants (e.g., many tropical grass species such as maize) that use the $C_4$ photosynthetic pathway, or the flesh of animals that feed or graze on $C_4$ plants. *See also* **$C_3$ and $C_4$**.

## C6
An accessory cusp on the distal part of the crown of a mandibular molar. (syn. tuberculum sextum.)

## C7
An accessory cusp on the lingual side of the crown of a mandibular molar. (syn. tuberculum intermedium.)

## calcification
The process of calcium deposition that occurs during the development of hard tissues such as teeth and bones. Typically during this process an organic framework is replaced by calcium (and other associated minerals); this part of the process is referred to as mineralization. During enamel development the calcium pumped into a developing tooth is incorporated into the crystalline structure of the enamel prisms by displacing enamel matrix proteins.

## calcium carbonate
Calcium carbonate ($CaCO_3$) is particularly abundant in chalk and limestones. Water seeping through bedrock carries dissolved calcium carbonate, and when this solution reaches an air-filled cave, the release of carbon dioxide alters the water's ability to hold minerals in solution, causing its solutes to precipitate and accumulate. This is how speleothems (e.g., flowstone, stalactites, and stalagmites) are formed. Flowstones, which are formed by water flowing on the floor or walls of a cave, are generally horizontal, finely laminated layers of calcium carbonate that are found between breccia deposits in caves. Some of the fossil hominins in the southern African cave sites (e.g., Stw 573) are preserved in flowstones. Speleothems can be dated with uranium-series dating and they are also important sources of paleoenvironmental information. The terms travertine and tufa, which are used in the older literature, describe calcium carbonate formed by the precipitation of carbonate minerals. Travertine is calcium carbonate formed by the precipitation of carbonate minerals in geothermally heated hot springs and tufa is calcium carbonate formed by the precipitation of carbonate minerals in ambient-temperature water bodies (e.g., lakes, rivers and springs).

## calibrated-age methods
*See* **biostratigraphy; geochronology**.

## calibrated radiocarbon age
*See* **radiocarbon dating**.

## calibration
An adjustment made to a measurement by comparing it to a more reliable standard. The process used to correct radiocarbon dates to allow for the changes that have occurred over time to the levels of radiocarbon in the atmosphere is an example of a calibration. (Gk *kalapous* = shoemaker's last.) *See also* **radiocarbon dating**.

## calotte
The part of the cranium that can be covered by a typical hat. The unpaired bones making up the calotte are the frontal and the squamous part of the occipital; the paired bones are the parietals and the squamous parts of the temporal bones. The calotte is one of the three parts of the cranium; the others are the face and the cranial base. (Fr. *calota* = skullcap worn by a Roman Catholic priest or by an observant member of the Jewish faith.)

## calvaria
The cranium minus the face. The word calvaria is Latin feminine singular; the proper plural is calvariae. Although the term calvarium is sometimes seen in the literature, it is incorrect. In this Dictionary calvaria is used to indicate the singular except in cases where the original study used calvarium. (L. *calvus* = bald.) *See also* **cranium**.

## calvarium
*See* **calvaria**.

## canalization
A term introduced by Conrad H. Waddington to describe the tendency for development to produce a similar phenotype despite different genetic and environmental influences. Researchers distinguish between environmental and genetic canalization. Environmental canalization is the reduction in the genotype-specific environmental variance, whereas genetic canalization is the reduction in the average effect of mutations, or genetic variants. (L. *canalis* = conduit, hence "to channel.")

## cancellous bone
(L. *cancellus* = lattice; ME *bon* = bone; syn. trabecular bone.) *See* **bone**.

## candelabra model
The term used by William (Bill) Howells to describe the polycentric theory of German physical anthropologist Franz Weidenreich. Although Weidenreich had argued there was regional continuity in geographically separated populations of hominins as they evolved over time into modern humans, he also made it clear that his model incorporated interbreeding between these populations. Strictly, Howells misrepresented Weidenreich's views when he used a candelabra as a metaphor, but other polyphyletic theories of human evolution proposed in the late 19thC and early 20thC (e.g., Carleton Coon) were consistent with Howells' metaphor. *See also* **assimilation model**; **multiregional hypothesis**; **out-of-Africa hypothesis**; **replacement with hybridization**.

## Canidae
This family, which includes wild dogs and their close relatives, is one of the caniform families whose members are found at some hominin sites. *See also* **Carnivora**.

## canine dimorphism
*See* **sexual dimorphism**.

### canine eminence
Rounded area of bone that covers the anterior, or facial, aspect of the root of the upper canine. (L. *canis* = dog; syn. canine jugum.)

### canine fossa
Hollowed area posterior to the canine eminence (the bone that covers the root of the upper canine). Its superior and anterior boundary is the transverse bony buttress that runs inferiorly from the infraorbital foramen to blend with the canine eminence.

### canine jugum
Rounded area of bone covering the anterior, or facial, aspect of the root of the upper canine. (syn. canine eminence.)

### canonical analysis
A form of multivariate analysis that summarizes the variation in a large number of variables by calculating a smaller number of canonical variates or canonical axes. The axes, which describe a proportion of the overall variation and are a summary of the original variables, are oriented at right angles to each other (i.e., they are orthogonal). The amount of variation represented in each axis can be inferred from the sum of the variable loadings on each axis. (L. *canon* = rule.) *See also* **multivariate statistics**.

### Carabelli's trait
(aka Carabelli trait) An enamel feature on the mesiolingual face of a hominin maxillary molar tooth crown between the tip of the protocone and the cemento-enamel junction, or CEJ. John Robinson claimed that Carabelli's trait was more common in *Paranthropus* than in *Australopithecus*, but studies of how the trait is expressed at the enamel–dentine junction suggest that not all expressions of the trait at the outer enamel surface are the result of the same developmental pathway.

### carbon dating
*See* **radiocarbon dating**.

### carcass acquisition strategy
The various ways (e.g., hunting, confrontational scavenging, passive scavenging) that modern humans and fossil hominins acquire animal remains. Identifying carcass acquisition strategies helps characterize the ecological niche and behavioral capabilities of early hominins. Hunting refers to the active pursuit and capture of animal prey and is generally understood to be a basic component of modern human behavior. Confrontational scavenging (aka "power" or "aggressive" scavenging) is the capture of a nearly complete animal carcass from the nonhominin predators that are actively consuming it. Passive scavenging involves culling small amounts of meat or marrow from heavily ravaged animal carcasses abandoned by their initial predators. The point at which hunting emerged in the hominin behavioral record is a particularly controversial issue (e.g., the hunting vs scavenging debates that have focused on interpretations of the FLK site in Olduvai Gorge). Ethnographic studies document opportunistic passive scavenging by

some hunter–gatherer groups (e.g., the Hadza), but no modern human populations have a carcass acquisition strategy characterized by obligate passive scavenging of terrestrial mammals.

### carcass transport strategy
How and in what form animal carcasses are moved from place to place for eventual consumption. In archeology, carcass transport strategies are evaluated by examining the relationship between estimates of skeletal part frequency in a fossil assemblage and their corresponding energetic return, often calculated using an economic utility index. For example, at Olduvai Gorge, studies of skeletal part frequencies and the carcass transport strategies inferred from them have been used to evaluate the timing of hominin access to prey (primary or secondary access) and to determine whether individual fossil assemblages represent kill sites or consumption sites.

### cardioid foramen magnum
See **foramen magnum**.

### Carnivora
The order of mammals whose members consume principally meat. The proper informal term for members of this order is carnivoran, but the term usually used is carnivore. The carnivoran order, which is speciose and widely distributed, has two suborders: the Feliformia (i.e., cat-like) and Caniformia (i.e., dog-like). Common feliform families at hominin sites are the Felidae (cats), Hyenidae (hyenas), Viverridae (civets and their relatives), and Herpestidae (mongooses and relatives). Caniform families found at hominin sites include the Ursidae (bears), Mustelidae (weasels and otters), and Canidae (wild dogs). (L. *carn* = flesh and *vora*, from stem of *vorare* = to devour.)

### carnivoran
See **Carnivora**.

### carnivore
The proper informal term for members of the order Carnivora is carnivoran, but carnivore is usually used in its place. (syn. carnivoran.) See also **Carnivora**.

### carnivore guild
A carnivore guild is a group of mammalian carnivores that garner faunal resources using similar predation strategies and dental adaptations. For example, at the Miocene hominoid site of Pasalar, Turkey, four carnivore guilds have been identified on the basis of dental morphology: hypercarnivores, including bone crushers, main-carnivores (that may also have included some bone crushers), omnivores, and invertebratevores (including insectivores). It has been suggested, on the basis of archeological data and also on the rather unusual evidence of tapeworm evolution, that *Homo* may have been the basis of a hominin carnivore guild during the Plio-Pleistocene.

### carnivore modification
The spatial disturbance and modification, or destruction, of skeletal remains by carnivores. Archeological bone assemblages, particularly those from open-air sites, frequently show tooth marks or other signs of alteration by carnivores. Because carnivore ravaging can alter the

post-depositional integrity of a bone assemblage, it is important that it be identified and corrected for when developing informed hypotheses regarding hominin behavior or past environments. Actualistic studies indicate that carnivores, especially hyenas and canids, readily destroy grease-rich, low-density skeletal elements, such as long-bone epiphyses, ribs, pelves, and vertebrae. Evidence of carnivore ravaging can be detected through the identification of carnivore tooth marks on bone.

## carnivory
The consumption of flesh or other animal tissues. Although members of the order Carnivora are popularly viewed as being exclusively meat eaters (carnivores) many of them eat plant food as well and some (e.g., the grey wolf) are most accurately described as eclectic feeders. Many primates are known to engage in at least occasional meat eating, and many more are carnivorous in the wider sense of including some animal tissues (from throughout the animal kingdom) in their diets, either regularly or opportunistically. In paleoanthropology, the examination of carnivory as a hominin subsistence strategy is a primary research question. Chimpanzees, our closest living relatives, are known to hunt and consume small mammals so this capability may well have been present in early hominins. With the advent of stone tool manufacture (by at least c.2.6 Ma) hominin–faunal interactions can be demonstrated through cutmarks and the breakage patterns of fossil bones. Exactly when, and in what context, hominins began to consume more meat than contemporary chimpanzees remains a topic of lively debate. (L. *carn* = flesh plus *vorous*, from the stem of *vorare* = to devour.)

## carrying capacity
The fixed population size above which a population will decrease in size and below which it will increase in size. In logistic population growth models the carrying capacity is assumed to be a stable equilibrium.

## cartilage
One of the two connective tissues (bone is the other) that form the skeleton. Cartilage, like all connective tissues, is made up of a combination of cells and microscopic organic fibers embedded in an organic matrix. There are three types of cartilage: hyaline, fibro-, and elastic cartilage. The matrix of cartilage is mainly water plus some dissolved salts, proteins, and glycoproteins. Chondrocytes, which manufacture the matrix, are the predominant cell type in hyaline and elastic cartilage. Fibroblasts, which manufacture the organic collagen fibers, are found in all three types, but they are especially common in fibrocartilage. Collagen fibers are found in all three types of cartilage, but only elastic cartilage contains yellow elastic fibers in any significant numbers. Initially the developing skeleton is made of cartilage and most bones develop within that cartilaginous template by a process called endochondral ossification. The only hyaline cartilage in adults covers the surfaces of synovial joints and the tips of the ribs; it also forms the xiphoid process and most of the cartilaginous skeleton of the larynx. One of the consequences of aging is that cartilage becomes calcified and when it does it can fracture. (L. *cartilago* = gristle.)

## Casablanca
See **Sidi Abderrahman**; **Thomas Quarry**.

## Castel di Guido
This Marine Isotope Stage 9 open-air site in Italy has yielded hominins that show a mixture of *Homo erectus* and *Homo neanderthalensis* features, plus Acheulean artifacts. (Location 41°53′59″N, 12°16′33″E, Italy.)

## CAT
Acronym for both computer-assisted tomography and computerized axial tomography. *See* **computed tomography**.

## caudal
The term used to describe the relative position of structures in the axial skeleton that are "towards the tail": in upright hominins it refers to structures that are inferior. Thus, the axis vertebra (C2) is caudal to the atlas vertebra (C1), and the first lumbar (or L1) vertebra is caudal to the twelfth thoracic (or T12) vertebra. The opposite of caudal is cranial. (L. *cauda* = tail.)

## Caune de l'Arago
A *c.*400–350 ka cave site located at the eastern end of the Pyrenees overlooking the Tautavel plain in France; thus the hominin fossils are called "l'Homme de Tautavel" or "Tautavel Man." More than 100 hominin remains belonging to more than 26 individuals, including a face (Arago XXI), a mandible (Arago II), and a parietal bone (Arago XLVII), have been recovered along with several thousand stone flakes and retouched tools. (Location 42°50′22″N, 02°45′18″E, France.)

## Cave of Hearths
One of a number of cave complexes [others include Buffalo Cave, Gwasa Cave (aka Makapan's Cave), Historic Cave, Makapansgat, and Peppercorns] in the Makapan Valley in Limpopo (formerly Northern) Province, South Africa. Acheulean-like artifacts (*c.*700–400 ka) and Middle Stone Age artifacts (<250 ka) have been recovered, and a juvenile hominin mandible (Cave of Hearths 1) comes from the Acheulean levels. (Location 24°08′25″S, 29°12′00″E, South Africa.)

## CBA
Abbreviation for cranial base angle. *See* **basicranium**; **cranial base angle**.

## CEJ
Acronym for **cemento-enamel junction** (*which see*).

## cemento-enamel junction
(or CEJ) The junction between the inferior (lower teeth) or superior (upper teeth) limit of the enamel cap of the crown of a tooth and the cementum that covers the external surface of the root. (syn. cervix.)

## cementum
One of the three hard tissues (along with enamel and dentine) that make up a tooth. Cementum contains fibers that run from the periodontal ligament into the tooth root. At certain latitudes the microstucture of cementum may respond to seasonal environmental

changes and these cementum growth rings (annulations) have been used by zooarcheologists to infer the season at death of fauna found at hominin fossil sites. [L. *caementum* = rough (pre-dressed) stone.]

### Cenozoic
The term for the youngest (and current) era of the Phanerozoic eon. Cenozoic refers to a unit of geological time (i.e., it is a geochronologic unit) that spans the interval from 65.5 Ma to the present. Compared to the previous Mesozoic era, the Cenozoic is noted for a worldwide decline in temperatures. Although mammals were numerous and diverse throughout the Mesozoic era, the extinction of the dinosaurs allowed mammals to flourish and become the dominant large fauna of the Cenozoic. Historically the Cenozoic was subdivided into the Tertiary and Quaternary periods, but today most geologists opt to subdivide the former into two separate periods, the Paleogene and Neogene. The International Union of Geological Sciences recently, and controversially, ratified the Quaternary as the third period in the Cenozoic. (Gk *kainos* = new and *zoe* = life.) See also **Plio-Pleistocene boundary**.

### census size
See **population size**.

### center of mass
The experimentally determined or estimated location within an object of the center of its mass. This term is usually used in the sense of the center of mass of the whole body, but it can refer to the center of mass of a segment of part of the body (e.g., one of the segments of a limb).

### centimorgan
See **linkage**.

### Central Awash Complex
(or CAC) One of the major fossiliferous subregions within the Middle Awash study area, Ethiopia; the others are the Western Margin, Bouri peninsula, and Bodo-Maka. The Central Awash Complex, which is a dome-like structure that lies on the west side of the Awash River, is dissected by the Amba, Urugus, Sagantole, and Aramis drainages. Collecting areas within the Central Awash Complex that have yielded hominins include Amba East, Aramis, and Asa Issie.

### central facial hollow
A term used by Phillip Tobias to refer to central part of the face that in OH 5 is depressed relative to the forwardly positioned zygomatic bones. Similar-shaped faces are seen in other specimens of *Paranthropus boisei* and to a lesser extent in the faces of crania belonging to *Paranthropus robustus*. (syn. dished face.)

### central nervous system
See **brain**.

## central place foraging
*See* **home base hypothesis**.

## centroid
The average of a multivariate data set that is usually calculated as either the mean or median of all variables in the data set.

## centromere
*See* **chromosome**.

## Ceprano
A *c.*400 ka (Marine Isotope Stage 11) hominin calvaria (Ceprano 1) was recovered during road construction at this site that is close to Campogrande, Italy. Excavations have yielded Acheulean artifacts in overlying deposits. The Ceprano 1 calvaria has been assigned to *Homo erectus* or *Homo heidelbergensis.* (Location 41°31'42"N, 13°54'16"E, Italy.)

## Cercopithecinae
A subfamily of the family Cercopithecidae (the other is the Colobinae). The cercopithecines comprise two tribes: the Papionini or papionins (baboons and macaques) and the Cercopithecini or guenons. The paucity of late Miocene fossil remains obscures the early stages of cercopithecine evolution, but molecular evidence indicates that the two tribes split *c.*11–12 Ma. The sister clades share many features, but they can be distinguished because cercopithecins lack a hypoconulid on their lower third molars. (L. *cercopithecus* from the Gk *kerkos* = tail and *pithekos* = ape; i.e., an ape with a tail.)

## cercopithecine
An informal term for the subfamily Cercopithecinae. *See also* **Cercopithecinae**.

## cercopithecoid
An informal term for members of the primate superfamily Cercopithecoidea, also known as the Old World monkeys. *See also* **Cercopithecoidea**.

## Cercopithecoidea
Members of this primate superfamily are referred to informally as cercopithecoids or Old World monkeys (OWMs). Cercopithecoids diverged from the Hominoidea *c.*25–23 Ma and extant members are widely distributed across Africa and Asia. They can be distinguished from the New World monkeys (NWMs) of South and Central America because they have two premolars per quadrant, whereas NWMs have three premolars per quadrant and the molar crown morphology of the latter is dominated by two crests, or lophs, that run across the crown (hence the term bilophodont). Cercopithecoids have narrower noses than NWMs; this is why the latter are called platyrrhines and the former are called catarrhines. The superfamily comprises two families: one extinct, the Victoriapithecidae, and one extant, the Cercopithecidae. Cercopithecoids are relevant to paleoanthropology because they are prevalent at Plio-Pleistocene paleoanthropological and archeological sites and are used to help reconstruct the

paleohabitat at those sites. They have also been used as comparative models to help explain the events and patterns observed within human evolutionary history. (L. *cercopithecus* from the Gk *kerkos* = tail and *pithekos* = ape; i.e., an ape with a tail.)

## cerebellum

The part of the central nervous system that occupies the most posterior part of the endocranial cavity (i.e., the posterior cranial fossa). Its main function is the coordination of motor activity. The lateral hemispheres have also been shown to play a significant role in language and cognition, and their volume in hominoids is greater than that predicted by allometric extrapolation using the observed relative size relationships in monkeys. (L. diminutive of *cerebrum* = brain; so "little brain.")

## cerebral cortex

The outer layer of the cerebral hemispheres. Unless the cerebral hemispheres are sectioned, either actually or virtually (e.g., by imaging), the cerebral cortex is the only part of the cerebral hemispheres that can be inspected externally. Thus it is the only part whose surface morphology can be seen on natural and researcher-prepared endocasts. It is mainly made up of the cell bodies of neurons and glial cells and because this region of the brain does not contain a significant amount of myelin, the unstained cerebral cortex is gray-colored (hence the term "gray matter"). In modern humans the adult cerebral cortex is folded; the crest of a fold is called a gyrus and the fissure between two is called a sulcus (pl. gyri and sulci). Sulci and gyri are named according to their location (e.g., the superior temporal gyrus) or their shape (e.g., the lunate sulcus). The cerebral cortex is also subdivided into functional areas. Anatomical mapping of the cerebral cortex using cytoarchitecture (the microstructure of the cerebral cortex shows distinct layers of neuronal cell bodies that vary in size and shape) and chemoarchitecture also reveals distinct areas that are associated with different functions. The main subdivisions of the cerebral cortex of mammals include the neocortex (also called the isocortex) and the allocortex (archicortex plus the paleocortex). The neocortex is distinguished by a six-layered horizontal arrangement of cells; allocortical areas have fewer than six layers. The allocortex includes parts of the cerebral cortex concerned with the processing of olfaction (piriform cortex or paleocortex) as well as memory and spatial navigation (hippocampus, also called the archicortex). A widely used system of numbered cerebral cortical areas proposed by Korbinian Brodmann (e.g., the motor speech area comprises Brodmann's areas 44 and 45) is based on the patterns of cell layers. (L. *cerebrum* = brain and *cortex* = bark; thus the "outer covering.") *See also* **Brodmann's areas**.

## cerebral hemispheres

The cerebral hemispheres are a major component of the forebrain. Each hemisphere consists of an outer surface layer called the cerebral cortex, which is made up of the cell bodies and dendrites of neurons, and the underlying white matter, which consists of the myelin-covered processes axons of the cortical neurons. There are clusters of neuronal cell bodies buried in the white matter, the biggest of which are the basal nuclei. The cerebral hemispheres also contain a cerebrospinal fluid-filled cavity called the lateral ventricle. Named fissures or sulci mark the boundaries of the frontal, parietal, occipital, and temporal lobes of the cerebral hemispheres.

The paired hemispheres, which occupy the anterior and middle cranial fossae, are connected mainly by a substantial bundle of fibers called the corpus callosum. (L. *cerebrum* = brain, Gk *hemi* = half, and *sphaira* = surround.)

## cerebral petalia
See **petalia**.

## cerebrospinal fluid
(or CSF) A clear, straw-colored fluid, whose composition resembles that of extracellular fluid, secreted by the choroid plexuses within the ventricles. The volume of the extracerebral compartment of the CSF comprises part of the difference between the volume of the brain and the volume of the cranial cavity (i.e., the endocranial volume or ECV).

## cervix
The part of a tooth where the crown and the root meet. (L. *cervix* = neck, or a narrow region between two wider regions.)

## cf.
A term used in taxonomy to suggest that incompleteness or damage make the allocation of a specimen to a taxon problematic. Thus, a small fragment of the crown of a tooth from a 1.7 Ma site in eastern Africa with evidently thick enamel might be assigned to "cf. *Paranthropus*," "*Paranthropus* cf. *P. boisei*," or "*Paranthropus* cf. *boisei*."

## CFT
Acronym for **crown formation time** (*which see*).

## Chad Basin
The Chad Basin, in Chad, Central Africa, presently occupied by the Djurab Desert, is where the Mission Paléoanthropologique Franco-Tchadienne (French-Chadian Paleoanthropological Mission, or MPFT) has located several fossiliferous areas (e.g., Kollé, Kossom Bougoudi, Koro Toro, and Toros-Menalla). Each area has a different prefix – KL, KB, KT, and TM respectively – followed by a number for each locality within that area; further numbers are used to identify each specimen (e.g., TM 266 is the locality where TM 266-01-060-1, the type specimen of *Sahelanthropus tchadensis*, was discovered).

## *chaîne opératoire*
Literally translated from French as "operational sequence," it refers to the analysis of the technological action sequences involved in tool production. It has been most enthusiastically adopted in lithic analysis where it refers to the sequence of behaviors involved in stone tool production, use, and discard. It is close to the American concept of a lithic reduction sequence, but the intellectual origins of the *chaîne opératoire* approach are reflected in a much greater emphasis on the cognitive and cultural interpretation of tool-making behavior. *Chaîne opératoire* analysis aims to decipher the intentions of tool-makers through piece-by-piece refitting and/or mental reconstruction of action sequences based on the technological expertise of the analyst.

### character
A unit of the genotype or the phenotype used to describe and compare taxa. The condition of a character in a taxon is referred to as its character state. For example, the character "root system of the first mandibular premolar tooth" has four states among hominins: (a) a single root (1R), (b) one root divided for part of its length (1/2R), (c) one main distal root and a second mesiobuccal accessory root (2R: MB+D), and (d) two main roots, one mesial and one distal (2R: M+D). Characters can be qualitative (e.g., a sharp or a rounded boundary between the nasal cavity and the face) or they can be derived from continuous measurements. (ME *caractere* = to describe; syn. trait.)

### character conflict
When the branching pattern of a cladogram generated using the states of one character differs from the branching pattern of a cladogram generated using the states of a second character, or the states of multiple characters.

### character evolution
The pattern by which characters evolve (i.e., change state between ancestors and descendants) in a phylogeny. It is not always appreciated that the pattern of character evolution implied by a cladogram is logically equivalent to the shape of that cladogram. It is illogical to accept the phylogenetic hypothesis of a cladogram but reject hypotheses about the patterns of character evolution it contains.

### character state
See **character**.

### character state data matrix
A table in which the rows represent taxa or operational taxonomic units, the columns represent characters and the cells record the character states seen in each of the taxa (e.g., the state of the character "chin" in *Homo sapiens* is "present" while the state of the same character in *Homo neanderthalensis*, *Homo habilis*, and *Paranthropus boisei* is "absent"). The construction of a character state data matrix is the first step in a phylogenetic analysis.

### Charentian
A group of lithic assemblages within the Mousterian tradition rich (>50% of the retouched pieces) in scrapers, particularly side scrapers, but that contain few denticulates. It is also referred to as the "Quina-Ferrassie Mousterian." (Named after the French department of Charente, where several sites including La Quina are located.)

### Charnov's "dimensionless numbers"
(or DLNs) These measures, introduced by and named for the evolutionary ecologist Eric Charnov, are used in the analysis of life history. They simplify the description of relationships by describing the value of the ratio of variables rather than the values of individual variables. It has been argued that the use of DLNs allows researchers to identify cases where the DLNs are invariant and then explore the possible reasons for their invariance. Examples include

(a) exponents derived from the analysis of the relationship of life history traits to body size, (b) measures of reproductive effort, and (c) sex ratios.

## Châtelperronian

A western European technocomplex from the early Upper Paleolithic that includes many aspects of what is usually considered modern human behavior (e.g., bone tools, composite tools, and beads), but which is associated with sites that also contain *Homo neanderthalensis* fossils. Some authors have suggested that the Châtelperronian and other so-called transitional technocomplexes (e.g., Uluzzian) are evidence that Neanderthals adopted the behavior of the first modern humans that had migrated into Europe, but other researchers suggest that *H. neanderthalensis* individuals developed these techniques independently. (Fr. after the site of Châtelperron.)

## Cheboit

A *c*.6–5.7 Ma site within the Lukeino Formation in the eastern foothills of the Tugen Hills, Baringo District, Kenya, where the first evidence of *Orrorin tugenensis* (KNM-LU 335) was recovered. (Location 00°46′N, 35°52′E, Kenya.)

## Chellean

A term introduced by the French prehistorian Gabriel de Mortillet. He subdivided John Lubbock's Paleolithic period into a series of distinct "industries" (Chellean, Mousterian, Solutrean, and Magdalenian) named after sites in France. He had earlier used the term Acheulean for the same industry, but for complicated reasons he changed the type site from St. Acheul (in the Somme Valley) to Chelles (near Paris). The term Acheulean is now preferred.

## Chellean man

The term used by Louis Leakey to describe the population that included the OH 9 calvaria found at Olduvai Gorge. At the time, each stone tool industry was thought to represent a distinct race of early humans. The fossils at Olduvai were associated with the "Chellean" industry (now considered to be Acheulean or Oldowan).

## Chemeron Formation

The 5.6–1.6 Ma Chemeron Formation (previously referred to as the Chemeron Beds) is part of the Tugen Hills succession located in the Baringo region of Kenya. It lies immediately above the Kaparaina Basalts and it is separated by an unconformity from the overlying Kapthurin Formation. Several fossil hominins, including a >4.15 Ma fragment of the right side of a mandibular corpus recovered from Tabarin (KNM-TH 13150), a >4 Ma proximal left humeral fragment from Mabaget (KNM-BC 1745), and a *c*.2.4 Ma well-preserved, but isolated, right temporal bone (KNM-BC 1), have been recovered from the Chemeron Formation. (Location 00°47′N, 36°05′E, Kenya.)

## Chenjiawo

*See* **Lantian Chenjiawo**.

### chert
A sedimentary rock consisting primarily of microcrystalline to cryptocrystalline quartz. Chert is fine-grained, relatively hard, and homogeneous. This gives it excellent fracture properties (i.e., it fractures conchoidally when percussed), and not surprisingly it is the most commonly used raw material for stone tool manufacture. Many Middle Paleolithic assemblages from Europe, the Levant (Tabun), and Africa (Porc-Épic Cave) are made almost exclusively from chert. Flint is a type of dark-colored chert that was formed in marine environments.

### Chesowanja
A $c. > 1.42 \pm 0.07$ to 0.78 Ma locality/site east of Lake Baringo in Kenya. The remains of at least two hominin individuals [i.e., a partial cranium (KNM-CH 1), a partial right $M^1$ or $M^2$ crown (KNM-CH 302), and five fragments of a hominin calotte (KNM-CH 304)] as well as evidence of the Developed Oldowan have been recovered from the Chemoigut Formation at Chesowanja. (Location 36°12′E, 00°39′N, Kenya.)

### chewing
Chewing is the cyclic application of bite forces to a food item. The particular kind of chewing employed by mammals is called mastication. The side where the bite force is applied is called the "working side"; the other is the "balancing side." Starting from their maximally depressed position (i.e., maximum gape) the jaws are rapidly elevated until the teeth contact the food item. This tooth–food–tooth contact, which marks the onset of the slow part of the closing phase of the cycle, is also called the "power stroke" because it is when bite force is applied as the teeth move through the food. Coordinated movements of the tongue and jaws move food items from the mouth to the pharynx for swallowing; chewing evolved when biting was added to these cyclic jaw and tongue movements. Many tetrapods chew, but only mammals use the cyclic, rhythmic, medially directed power strokes that are characteristic of mastication. (ME *cheuan* = chew.)

### chewing cycle
See **chewing**.

### childhood
The period in modern humans that lasts from $c.3$ to $c.7$ years, post-weaning, when infants are still dependent on others. It is a unique life history stage in modern humans that is inserted between infancy and the juvenile stage. Although modern human children are weaned by the beginning of childhood, they are not able to compete effectively for food because they are small (because of their low growth rate) and their dentition is still immature. The first permanent tooth, usually a first molar or central incisor, erupts toward the end of childhood and brain growth, while still rapid, decreases in velocity toward the end of childhood. Because children are weaned relatively early and remain dependent on other adults in the group, the mothers are freed for further reproduction therefore making interbirth intervals relatively short. Some claim that evidence of childhood in the hominin fossil record extends back to $c.2$ Ma, but the evidential support for this is weak. (OE *cild* = child or infant plus *had* = condition or position.) See also **grandmother hypothesis**.

## chimpanzee
See **Pan**.

## chimpanzee archeology
The name used to describe the study of chimpanzee tool manufacture and use. Tool manufacture and use in nonhuman primates is not confined to chimpanzees, so "primate archeology" has been proposed as a more inclusive term.

## chimpanzee fossils
See **KNM-TH 45519–45522 and 46437–46440**.

## chin
See **true chin**.

## Chiwondo Beds
The Chiwondo Beds on the shore of Lake Chiwondo in northern Malawi were the focus of the fieldwork by the Hominid Corridor Research Project (HCRP). More than 145 fossil localities were discovered, with two, Uraha and Malema, yielding hominin specimens (HCRP UR 501 and 911, respectively).

## chondroblasts
See **cartilage**.

## chondroclasts
See **ossification**.

## chondrocranium
The part of the cranium that begins as cartilage and in humans and other higher vertebrates develops into bone through a process called endochondral ossification. It includes most of the occipital bone, the petrous part of the temporal bone, most of the sphenoid, and the ethmoid bones. The bony equivalent of the chondrocranium is the basicranium. (Gk *khondros* = cartilage and *kranion* = brain case.)

## chondrocytes
See **cartilage**.

## chopper
A term used to describe flaked Mode 1 stone tools particularly common in Bed I at Olduvai Gorge and other early archeological sites. A chopper is a rounded cobblestone, or angular block of stone, with a unifacially or bifacially flaked edge along part of its circumference.

## Choukoutien
See **Zhoukoudian**.

## chromatin
*See* **chromosome**.

## chromosome
A single long strand of nuclear DNA plus its associated protein scaffolding. In eukaryotes, chromosomes are found in the nucleus of the cell where they typically appear in linear forms, and are packaged into a condensed structure referred to as chromatin. Modern humans have 23 pairs of chromosomes (22 pairs of autosomes and one pair of sex chromosomes: XX in females and XY in males). Chimpanzees and gorillas have 24 pairs of chromosomes (23 pairs of autosomes and one pair of sex chromosomes). A Robertsonian translocation may have been responsible for two ape chromosomes becoming modern human chromosome 2. (Gk *chroma* = color and *soma* = body.)

## chron
Chron, which is the abbreviation of "polarity *chron*ozone," is a substantial period in the history of the Earth's geomagnetic polarity (subdivisions of chrons are called subchrons). Numbered chrons (the most recent chron is 1, with the numbers increasing as you go back in time) have replaced the eponymously named epochs (e.g., Gauss, Gilbert, etc.) that were used in older geomagnetic polarity timescales. *See also* **geomagnetic polarity time scale**.

## chronospecies
One of several terms used for a group of fossils all of which belong to the same lineage, or lineage plexus, spread across time. For example, the fossils attributed to *Paranthropus boisei* from its first appearance in the fossil record to its last appearance would make up a chronospecies. Some, but not all, researchers claim *Paranthropus aethiopicus* should be included in the same chronospecies, whereas others regard it as a separate chronospecies within the same lineage. (Gk *kronos* = time and *eidos* = form or idea; syn. paleospecies, paleontological species.) *See also* **species**.

## chronostratigraphic units
Intervals of rock accumulated through Earth's history. It is confusing that intervals of time share terms with the sequence of rock (strata) deposited in that interval. However, whereas chronostratigraphic unit subdivisions are based on a physical sequence (i.e., Lower, Middle, and Upper) geochronologic units are time-referenced (i.e., Early, Middle, and Late).

## cingulum
A bulging of the enamel just above the base of the enamel cervix (cemento-enamel junction, or CEJ). In some teeth this bulging is so well marked that it is given the name "style" (upper teeth) or "stylid" (lower teeth) with an appropriate prefix (e.g., protostylid is the part of the cingulum related to the protoconid in mandibular lower molars). (L. *cingulum* = girdle.)

## circadian
An intrinsic biological rhythm with an approximately 24 hour cycle. The development of enamel and dentine is under the influence of a circadian cell rhythm, which manifests as short-period incremental lines that can be seen in both tissues. In enamel these take the form of fine

dark microscopic cross-striations that run at right angles to the long axis of enamel prisms. (L. *circum* = around and *dies* = day.) *See also* **incremental features**.

## circaseptan

The time interval in days between long-period incremental markings (striae of Retzius) in enamel. The temporal repeat interval between the striae of Retzius (aka periodicity) is obtained by counting the number of short-period incremental markings (i.e., daily cross-striations) between consecutive striae. The range observed in modern humans is 6–12 days; the mean and median are 8 days. The range in *Pan* is 5–9 days; the mean and median are 7 days. The long-period incremental markings in dentine are called Andresen lines and within the same individual their temporal repeat interval has been found to match that in enamel. (L. *circum* = around and *septem* = seven; thus "around seven.")

## clade

A grouping of two or more taxa that contains all (no more and no less) of the descendants of their most recent common ancestor. A clade is analogous to a *make* of car (all Rolls Royce cars share a recent common ancestor not shared with any other make of car) whereas a grade is analogous to a *type* of car (luxury cars made by Cadillac, Jaguar, Lexus, and Mercedes are functionally similar yet they have different evolutionary histories and therefore have no uniquely shared recent common ancestor). The smallest clade consists of just two taxa known as sister taxa; the largest includes all living organisms. It is conventional to indicate the structure of a clade by a series of parentheses, each of which represents a sister taxon relationship. So the shorthand way of expressing the relationships of the taxa in the clade that contains *Homo sapiens* and *Homo neanderthalensis* is (*H. sapiens, H. neanderthalensis*), and for the taxa in the clade that contains *Homo*, *Pan*, and *Gorilla* it would be ((*Homo, Pan*) *Gorilla*). (Gk *clados* = branch; syn. monophyletic group.) *See also* **crown group**; **stem group**; **total group**.

## cladistic

A pattern of relationships based on taxa sharing recently evolved (i.e., derived) features, or characters, of the phenotype or genotype. Phenotypic characters are either discrete morphological features (e.g., a sharp nasal sill) or a particular range of values of a measurement. (Gk *clados* = branch; i.e., a branching pattern.) *See also* **cladistic analysis**.

## cladistic analysis

A method that uses shared derived features to generate hypotheses about the relationships among taxa. These hypotheses conform as closely as possible to the phylogenetic history of the group. The taxa used in a cladistic analysis (aka operational taxonomic units or OTUs) must be decided beforehand on the basis of alpha taxonomy. Cladistic analysis of the phenotype works by breaking morphology down into discrete features called characters. Characters, which can be nonmetrical (e.g., the number of tooth roots) or metrical (e.g., the cross-sectional area of the mandibular corpus) are assumed to be inherited independently. Several methods can be used (e.g., outgroup or ontogenetic criterion) to generate hypotheses about how to order their manifestations, or states, from the most primitive (shared by the most taxa) to the most derived (shared by the least number of taxa or found in just one taxon). The taxa are then arranged in a branching diagram that is consistent with the distribution of character states. The branching pattern that is supported by the greatest number of characters (i.e., the consensus cladogram) becomes the null hypothesis for the relationships among those taxa. That hypothesis can be tested by finding new characters and then comparing how they distribute across the original consensus cladogram. (Gk *clados* = branch, so-called because it is an analysis that results in taxa being related in a tree-like branching pattern; syn. phylogenetic analysis.)

## cladistics
*See* **cladistic analysis**.

## cladogenesis
The formation of two new species when one species splits or bifurcates into two. The splitting process is called vicariance if a geographical barrier is responsible for initiating the split. (Gk *clados* = branch and *genesis* = birth or origin.) *See also* **anagenesis**; **speciation**; **vicariance**; **vicariance biogeography**.

## cladogram
A tree-like diagram indicating the relationships among taxa. The branching is usually dichotomous (i.e., one old taxon splits into two new taxa) and the detail of the branching pattern is determined by the distribution of shared derived character states (or synapomorphies). The preferred cladogram is conventionally the most parsimonious (i.e., the pattern of branching that requires the fewest evolutionary changes). (Gk *clados* = branch and *gramma* = a letter or something written or drawn.) *See also* **cladistic analysis**.

## "classic" Neanderthals
*See* **pre-Neanderthal hypothesis**.

## classification
A subset of the activities that come under the heading of systematics. The latter is an inclusive term used for (a) assembling individual organisms into groups, (b) formalizing those groups as Linnaean taxa, (c) giving formal names to the taxa, (d) allocating the taxa to taxonomic categories, and (e) assembling the taxa into a hierarchical scheme or classification. Classification subsumes the activities involved in (c) and (d). Strictly speaking you cannot classify an individual fossil. It has to be assigned to an existing taxonomic group or if its morphology is novel to a new taxon. Then, and only then, can you proceed to classify the taxonomic group to which the fossil has been assigned. (L. *classis* = the name of the units used to subdivide the Roman army fleet, etc.) *See also* **nomenclature**; **systematics**; **taxonomy**.

## clast
A piece, or fragment, of a rock that results from a larger rock fragment being broken or reduced in some way into smaller pieces. Clasts can be as large as a boulder or as small as a grain of sand. Clastic rocks are formed from rock fragments. (Gk *klastos* = broken.)

## clavicle, evolution in hominins

The clavicle is an S-shaped bone lying superior to the first rib that articulates medially with the sternum and laterally with the acromion process of the scapula. The modern human clavicle has an anterior and inferior curve at the medial end and a superior and posterior curve laterally. Modern humans share this pronounced S shape with *Pan* and *Pongo*. *Gorilla* has a pronounced posterior curvature laterally and a relatively straight medial portion. In posterior view, modern humans are unique in having just one inverted U curvature. In contrast, apes have two curvatures that reflect the relative height of the shoulder girdle relative to the thorax. In early *Homo* the presumed primitive condition (i.e., a short clavicle and a superiorly facing scapula positioned dorsally on the thorax) was modified to the extent that, although the clavicle remained short, the scapula was more laterally situated on the thorax. In modern humans a longer clavicle is combined with a laterally oriented glenohumeral joint and a scapula that is in a more dorsal position. It has been suggested that this rearrangement would have provided the leverage needed for overhand throws.

## clay

A sediment or sedimentary rock with particles whose mean diameter is less than 4 μm. Clays are produced by the weathering of minerals and are usually transported by water or wind and deposited in low-energy environments, like standing water. They are a component of most soils. The lacustrine deposits exposed in Olduvai Gorge include a wide range of clays that reflect complex fluctuations in the water chemistry of paleo-Lake Olduvai.

## cleaver

A stone tool common among Acheulean assemblages. It is similar to handaxes in size and in some manufacturing details, but it is distinguished from them by having an unretouched distal end. Experimental evidence suggests that cleavers were used for butchery. (OE *cleofan* = to split.)

## climate

The time-averaged (>30 years) description of the Earth's weather or the average weather in a particular region (e.g., East Africa, Antarctica) of the globe. All information about past climates comes from biotic or geochemical proxies that take the place of direct observations of the four climate variables (i.e., temperature, precipitation, wind speed, and wind direction). The climate experienced in any region is the result of interactions among component climate systems (e.g., the El Niño Southern Oscillation, thermohaline circulation) and the global climate system is the product of interactions among these regional climate systems. The component climate systems, which have their own internal dynamics as does the Earth's climate system, can be influenced, or "forced," by external factors. Major climate-forcing factors include changes in the way the Earth is inclined relative to the sun, changes in the shape of the Earth's orbit around the sun, and changes in the Earth's atmosphere that affect the intensity of the sun's rays that reach the Earth (i.e., insolation). Regional climates are also influenced by the shape, or topography, of the Earth's surface (e.g., regional precipitation is influenced by rain shadow phenomena, and regional temperature by altitude). (Gk *klima* = which refers to the slope of the Earth's surface, is the origin of L. *clima* = referring to climate or latitude.) *See also* **climate forcing; insolation; paleoclimate.**

## climate forcing
Physical processes (e.g., fluctuations in the shape of the Earth's orbit and in the degree of tilt and amount of wobble of the Earth's axis, changes to ocean circulation, variations in atmospheric gases, etc.) can affect the dynamics of, and thus change or force, regional and global climate systems. In paleoanthropology there is considerable interest in climate forcing as a mechanism for environmental change and consequent evolutionary change. (Gk *klima* = which refers to the slope of the Earth's surface, is the origin of L. *clima* = climate or latitude, and L. *fortis* = strong.)

## climate models
*See* **general circulation model; models of intermediate complexity; paleoclimate**.

## climate systems
*See* **climate; El Niño Southern Oscillation**.

## climbing hypothesis
A hypothesis that summarizes several separate but related models introduced to explain the origin and evolutionary history of upright posture and bipedal locomotion within the hominin clade. Current climbing hypotheses are based on the concept of pre-adaptation and implicit in all of them is the proposition that the ancestors of the earliest hominins were predominantly arboreal and lacked adaptations for traveling on the ground in any way other than bipedally (i.e., they were unable to knuckle-walk). The hypothesis that the pre-bipedal ancestor had an upper limb too specialized to evolve any terrestrial style other than bipedalism is no longer tenable under our current understanding of great ape phylogeny because the most recent common ancestor of chimpanzees/bonobos and modern humans gave rise to both bipedal hominins and climbing/knuckle-walking panins. *See also* **bipedal**.

## cline
A gradual change in a morphological, genetic, or behavioral feature as one moves across a given geographic region (geocline) or over time (chronocline). Clines occur either because dispersed populations may have slightly different adaptations in response to local conditions, or because two separate populations have been "merged" through gene flow. Stepped clines, which are abrupt changes in a feature, are often observed at hybrid zones. In modern humans there are multiple examples of genetic clinal variation, including traits like degree of skin pigmentation or the prevalence of the B blood allele. Some have argued that the majority of modern human variation is clinal. (Gk *klinein* = to lead or bend.)

## closed basin lakes
Lakes with no inflow or outflow. They are useful for paleoenvironmental reconstruction because they represent the equivalent of a "rain barrel" in that any change in lake level can be attributed solely to variations in the amounts of meteoric precipitation and evaporation, without the complication of variations in stream flow.

## CNV
Acronym for **copy number variation** (*which see*).

## coalescent
The term used to describe an ancestral gene that gives rise to all the copies found in subsequent generations. (L. *co* = together and *alescere* = to grow.) *See also* **coalescent theory**.

## coalescent theory
A retrospective model used in molecular evolution to trace all the alleles of a specific DNA sequence back to the most recent common ancestral sequence. Thus, the evolutionary histories of all the lineages of a sequence are said to coalesce at this ancestral sequence. The coalescent process, which is affected by population size and structure, can be used to infer the demographic or selective processes that produced the resulting genealogy. For example, compared with the coalescent scheme produced under a neutral model, growing populations show a gene genealogy with long terminal branches. Populations undergoing positive selection of one lineage (a selective sweep), or those populations that have undergone a bottleneck, show a shorter coalescent time since older lineages are lost.

## coalescent time
The age of the ancestral sequence (belonging to the most recent common ancestor) that gave rise to all of the copies in subsequent generations. Self-evidently each DNA sequence potentially has its own coalescent time, so the coalescent time of one sequence should not be interpreted as providing information about the timing of the evolutionary history of a species. For example, the coalescent time for mitochondrial DNA differs from that of autosomal DNA sequences (i.e., it is much shorter) because the effective population size of mtDNA is one-fourth that of autosomal DNA.

## co-dominance
*See* **dominant allele**.

## codon
*See* **genetic code**.

## coeval
The term used for fossils that are, for all intents and purposes, the same geological age. For example, hominin fossils recovered from different localities in the same stratigraphic interval in the Turkana Basin (e.g., KNM-ER 406 and 732) are considered to be coeval. (L. *coaevus* = of the same age.)

## cognition
The mental process of "knowing." Awareness, perception, reasoning, and judgment are traditionally subsumed within cognition. Cognition also involves the ability to learn and subsequently use the knowledge that has been acquired in ways that are appropriate and adaptive. Researchers refer to different types of cognition as different types of intelligence (e.g., technical, ecological, and social). (L. *cognocere* = to learn.)

## Cohuna
A fossil skull from Australia that is presumed to be Late Pleistocene but lacks any associated radiometric dates. The Cohuna 1 skull is robust and morphologically similar (e.g., long, flattened frontal and well-developed supraorbital torus) to the Kow Swamp series found just to the east. (Location 35°55′S, 144°15′E, Australia.)

## collagen
The dominant protein in animal bone, cartilage, and connective tissue (e.g., tendons, ligaments). Collagen molecules form aggregates called collagen fibrils and the fibrils are bundled together to form collagen fibers. Collagen in bone can be used for radiocarbon dating and for stable isotope biogeochemistry. (Gk *kola* = glue and *genes* = born, for when collagen is boiled it forms gelatin, which is a glue.)

## collecting areas
*See* **Koobi Fora**.

## Colobinae
One of the two subfamilies (the other is the Cercopithecinae) of Old World monkeys within the Cercopithecidae, one of the two families within the superfamily Cercopithecoidea. In the Plio-Pleistocene, African colobines were much more diverse than they are today. Unlike modern forms, which are relatively small-bodied and arboreal, several extinct African colobines were larger and terrestrial. (L. *colobus* from the Gk *kolobos* = maimed, referring to the short thumbs that characterize the genus *Colobus*.)

## colobines
The informal name for the Colobinae, one of the two subfamilies within the Cercopithecidae. *See also* **Colobinae**.

## Combe-Grenal
A cave and talus deposit in France extensively excavated by François Bordes between 1953 and 1965. It was here that Bordes formulated the original concept of the distinctive Mousterian tool typologies and the Mousterian facies (e.g., Denticulate Mousterian, Typical Mousterian, etc.) that he believed represented individual tribes or ethnicities. Others have since questioned these interpretations, but the site remains an important reference for the Mousterian. (Location 44°48′N, 01°13′E, France.)

## comminution
A term used to describe the reduction in the particle size of food that takes place during the process of mastication. (L. *comminuere* = to reduce.)

## communication
The process by which information in the form of a message is passed from one individual to another. The various means of communication – visual, tactile, olfactory, gustatory, and auditory – correspond to the major sensory modalities. Communication has three components: (a) encoding,

(b) transmission, and (c) decoding or translation. Anthropocentric conceptualizations of communication argue that intentionality is critical and that the codification of messages and their perception is premised on mutual understanding between the sender and the receiver. This is certainly the case for the unique ways modern humans communicate via complex spoken and written language. (L. *communicare* = to share.)

## community ecology
A community is a group of organisms that live together in the same area, and ecology refers to all the interactions between an organism and its environment. Community ecology is therefore concerned with the interactions between the organisms in a specified community and their environment. Community ecology studies based on extant taxa help to reconstruct the environments inhabited by extinct hominins, and they can also be useful for increasing our understanding of speciation and extinction events. (L. *communis* = common and Gk *oikos* = house.) See also **ecology**; **synecology**.

## compact bone
See **bone**.

## competitive exclusion
Also known as Gause's Rule, the principle of competitive exclusion states that two species cannot occupy similar niches within the same ecosystem. Competitive exclusion leads either to local extinction of a species or to adaptation as one species moves into a different niche (i.e., character divergence). The competitive exclusion principle is significant to paleoanthropology because it underpinned the single species hypothesis. But competitive exclusion does not necessarily mean that two closely related species cannot coexist. For example, two or more species of guenon monkeys live closely together, but they avoid direct and sustained competition by foraging on a different mix of resources or by exploiting different areas of the canopy. Some have argued that competitive exclusion may have caused the extinction of *Homo neanderthalensis* in Europe. See also **single species hypothesis**.

## complex trait
See **epistasis**; **heritability**; **quantitative trait locus**.

## composite section
A diagram that summarizes the stratigraphy of several localities within a fossil site or several subregions within a site complex (e.g., Koobi Fora). Marker beds that are defined lithologically or chemically are used to correlate shorter sections from individual localities or subregions into a longer composite section. (L. *compositus* = to put together.)

## composite tool
Any tool made of more than one component (e.g., a spear consisting of a stone point mounted on a wooden shaft, bound together by leather, sinew, or resin). Composite tools make their first appearance in the Middle Stone Age and Middle Paleolithic (e.g., Twin Rivers and Königsaue). (L. *compositus* = to put together.)

### compound temporal-nuchal crest
A morphological feature formed when the temporal crest fuses with the superior nuchal line or nuchal crest. Such crests are common in large male gorillas and in *Australopithecus afarensis*.

### computed axial tomography
See **computed tomography**.

### computed tomography
(or CT) An X-ray-based radiological imaging method used to visualize the internal structure of bones and teeth nondestructively. As X-rays pass through a specimen their absorption and scatter (attenuation) are detected and converted into an electronic signal. If this is repeated from many different angles the spatial distribution of the attenuation values in a cross-sectional slice can be computed and reconstructed as a two-dimensional gray scale image consisting of a matrix of picture elements or pixels. Since every slice has a given thickness, each pixel corresponds to a small cube known as a volume element or voxel. Industrial micro-computed tomography (micro-CT or µCT) uses the same basic principles, but the specimen is rotated on a turntable around a fixed source/detector system. The spatial resolution of micro-CT typically ranges between 1 µm and approximately 200 µm depending on the size of the specimen, but scanning of small objects at submicrometer resolution is possible as well. With its better spatial resolution micro-CT can be used to examine structures with finer detail, such as trabecular bone and the enamel–dentine junction of teeth. Synchrotron radiation micro-computed tomography (or SR-µCT), which uses X-rays produced by a synchrotron, provides a better image quality than can be obtained with regular CT scanners and it can also capture information about microstructure (e.g., enamel microstructure). See also **Synchrotron radiation micro-computed tomography**.

### computer-assisted tomography
See **computed tomography**.

### conceptual modeling
In conceptual modeling the model is not directly based on "real" observations, but on general principles developed from observations of a wide range of animals, not just those closely related or analogous to the referent. For example, baboon behavior could be used as a literal referential model, or observations on baboons could be just one component of the information used to generate a conceptual model. (L. *conceptus* = to conceive and *modus* = standard.) See also **modeling**.

### conchoidal fracture
A rounded fracture surface of the type produced when a brittle, homogenous (glass-like) material breaks. Materials that have conchoidal fracture properties include siliceous rocks (e.g., flint, chert, chalcedony), bone, and some igneous rocks (obsidian, fine-grained basalt). The term refers to the analogy between the broken surface of a stone and the smooth, curved markings on the inner surface of a shell.

## confocal microscopy
The optics of a confocal microscope are designed so that out-of-focus information is eliminated and only information from a specified depth ("plane of interest") is collected to form the image. Researchers have developed a confocal scanning light microscope that is portable enough to take to museums yet it can still see enamel prisms in a dense solid structure such as dental enamel up to a depth of 50 μm.

## confrontational scavenging
When a nearly complete animal carcass is taken from nonhominin predators still engaged in its consumption. Modern hunter–gatherers (e.g., the Hadza) occasionally practice confrontational scavenging. *See also* **scavenging**.

## conglomerate
A rock made of large, rounded fragments (pebbles, gravel) that may be cemented together by a fine-grained matrix or mineral cement. (L. *glomus* = ball.)

## connective tissue
*See* **bone**; **cartilage**.

## consensus cladogram
*See* **cladistic analysis**; **cladogram**.

## consistency index
(or CI) A measure of the goodness of fit between a cladogram and a character state data matrix. The values for CI range between 1 and 0. A CI of 1 indicates that the character state data matrix is perfectly congruent with the cladogram (i.e., the cladogram requires no changes due to homoplasy). Homoplasy levels increase as the CI decreases toward 0. The CI value is affected by the number of taxa or the number of characters included in a character state data matrix. An alternative measure, the retention index, is not influenced by either the number of taxa or the number of characters included in a character state data matrix. *See also* **cladistic analysis**; **retention index**.

## contingency
An evolutionary perspective that stresses the effects of earlier factors or events (usually random or near random) in influencing subsequent evolutionary outcomes. For example, in small populations genetic drift will ultimately lead to the elimination or fixation of alleles, thereby reducing the population's genetic variation, but which of a pair of alleles is fixed and which is lost is due to chance. Some researchers see a role for contingency at the macroevolutionary level. Possible examples of contingency in hominin evolution include (a) small differences among the locomotor systems of eastern and southern African *Australopithecus*, (b) the appearance of several early *Homo* species (*Homo habilis*, *Homo rudolfensis*, and *Homo erectus/ergaster*) with differing morphologies and then the extinction of all save the last, (c) the evolution of *Homo neanderthalensis* morphology from later *Homo heidelbergensis* populations in higher-latitude Eurasia around the Middle/Upper Pleistocene boundary, possibly associated

with population bottleneck(s) during Marine Isotope Stage 6, and (d) the evolution of *Homo floresiensis* in Southeast Asia. It is also possible that the fixation of some or all of the characteristic morphology of *Homo sapiens* (e.g., globular neurocranium, retracted face, gracile mandible with a chin, pronounced basicranial flexion, slender limb bones, etc.) results from random factors influencing diversity in later Middle Pleistocene African populations. (L. *contigens* = to touch, but contingency now refers to an unintended event.)

### conule
A small accessory cusp of a maxillary (upper) postcanine tooth that is not one of the four main cusps (i.e., not the proto-, para-, meta-, or the hypocone). For example, the metaconule is an accessory cusp between the metacone and protocone. (L. *conule* = diminutive of cone.) *See also* **accessory cusp**.

### conulid
A small accessory cusp of a mandibular (lower) postcanine tooth (hence the suffix "-id") that is not one of the five main cusps (i.e., not the proto-, ento-, meta-, or hypoconid, or the hypoconulid). For example, the tuberculum intermedium is a conulid between the metaconid and the entoconid. *See also* **accessory cusp**; **C6**; **C7**.

### convergence
When one or more characters shared by taxa has not been inherited from their most recent common ancestor (i.e., the taxa have evolved the same character independently). Convergence is a common cause of homoplasy. *See also* **convergent evolution**; **homoplasy**.

### convergent evolution
Convergent evolution is the independent origin of similar traits in distantly related taxa (e.g., the acquisition of thick-enameled teeth in orangutans and some New World monkeys). Parallel evolution is the independent origin of similar traits in closely related taxa (e.g., coronally rotated petrous bones seen in *Homo* and *Paranthropus*, or the potentially independent evolution of bipedalism in *Australopithecus afarensis* and *Australopithecus africanus*). Convergent and parallel evolution both result in homoplasy: the appearance of similar morphology in two taxa that is not seen in the most recent common ancestor of those taxa. "Deep" convergent evolution (e.g., similar body form in whales and fishes) is relatively easy to recognize because only a minority of features is affected. Parallel evolution in closely related taxa is more difficult to distinguish from characters inherited from the most recent common ancestor because the taxa involved are more similar overall and are likely to have responded to common selection pressures in similar ways. *See also* **evolution**; **homoplasy**; **parallel evolution**.

### Coobool Creek
A site near the Coobool Crossing of the Wakool River in southern New South Wales, Australia, where researchers collected a series of hominin skeletal remains in 1950. The Coobool Creek crania are robust and morphologically similar to those from Kow Swamp, Cohuna, and Naccurie in Australia, but precise dates are not available. The anterior cranial vault shows

evidence of artificial cranial deformation, and the faces are large, robust, and subnasally prognathic. The Coobool Creek mandibles are particularly large. (Location 35°12′S, 143°45′E, Australia.)

### Cooper's Cave
A cluster of breccia-filled caves in Precambrian dolomite in the Blauuwbank Valley, Gauteng, South Africa, situated between the caves at Sterkfontein to the west and Kromdraai to the east. A hominin tooth was discovered in 1939 and since then hominins have been reported in 1995, 2000, and 2009. Uranium-series dating suggests an age of between $1.53 \pm 0.09$ and $<1.4$ Ma. With the exception of COA-1, which has been assigned to *Homo* sp., all of the hominins recovered have been assigned to *Paranthropus robustus*. (Location 26°00′46″S, 27°44′45″E, South Africa.)

### coordinate
One of a set of numbers that specifies where a point is located in space. Coordinates are used to define the locations of landmarks in methods such as geometric morphometrics and the precise location of sites, and of fossils and artifacts recovered during the course of excavations. *See also* **geometric morphometrics**.

### coordinate data
The Cartesian coordinates of landmarks (i.e., loci that are homologous across individuals in a sample). Coordinate data can be one-, two-, or three-dimensional. Two-dimensional (or 2D) coordinates are usually captured using a digitizing tablet or by measuring an image on the computer. Three-dimensional (or 3D) coordinates can be captured directly using a coordinate digitizer such as a Microscribe or Polhemus, or may be measured on surface or volumetric scans. Volumetric data can be obtained from computed tomography (CT or micro-CT) or magnetic resonance imaging (MRI or μMRI). Surface scanners provide high-resolution 3D representations of an object's surface using either laser or more traditional optical technology. Most software packages allow landmark coordinates to be measured directly on these virtual surfaces or volumetric objects.

### Cope's Rule
More a trend than a rule, it suggests that lineages tend to increase in body size over the course of evolution. The *post hoc* explanation is that the trend can be explained by directional selection (e.g., larger body size buffers against changes in the paleoenvironment, lowers the risk of predation, etc.), but several factors that impose limits on the size (e.g., birds must be light enough to fly, changes in size may be accompanied by changes in ecological niche, and that new niche may already be occupied) may oppose these pressures. A further complication is that the trend may be an artifact of the tendency for adaptive radiations to begin with small-bodied forms. Thus, if a clade begins with small-bodied ancestors, then as it diversifies to fill available niche spaces the evolution of a range of body sizes may give the appearance of a trend towards increasing body mass simply because there was a small-bodied starting point. Some workers cite recent hominins as an example of Cope's Rule insofar as the later members of the genus *Homo* are larger bodied than archaic hominins and earlier members of the genus.

## coprolite
Fossil feces or dung. Coprolites are an example of a trace fossil. (Gk *copros*=dung.) *See also* **fossil**.

## copy number variation
(or CNV) A copy number variant is a segment of DNA at least 1000 base pairs (or 1 kb) in length that is devoid of high-copy repetitive sequences (e.g., long interspersed nucleotide elements, or LINEs) and which is present in variable copy number among individuals. Copy number variants may be inter- or intrachromosomal, but the vast majority of CNVs are intrachromosomal. Studies have shown that copy number variants are common among modern humans and between modern humans and chimpanzees. It was once thought that the small amount of genetic variation that exists among modern humans (approximately 0.1% of the modern human genome) and between the modern human and chimpanzee genomes (determined to differ by as little as 1%) was primarily due to single nucleotide differences. However, it appears that genomic variation (measured in base pairs) is greater for copy number variation than for single nucleotide-level variation. Copy number variation can involve both gains or losses of genomic material and an immediate consequence of a genomic copy number gain or loss may be a corresponding increase or decrease in gene expression. Thus, intraspecific CNVs may be involved in intraspecific phenotypic variation (e.g., individual differences in disease susceptibility). Inter-specific CNVs may likewise account for between-species differences in gene expression.

## core
Any piece of stone from which flakes have been deliberately removed. Cores are present at the earliest archeological sites such as Gona in Ethiopia and they appear at most archeological sites where stone tool production has occurred. They are present in all major divisions of the Paleolithic (Early, Middle, and Later Stone Ages, and Lower, Middle and Upper Paleolithic). (ME *core*=the hard, seed-containing central part of some fruits.)

## coronal
The same term is used for the vertical plane that lies perpendicular to the sagittal plane, and for the cranial suture that is aligned at approximately 90° to the sagittal suture. Coronal is not used to refer to the dentition. (L. *coronalis*=a crown.)

## coronoid process
*See* **mandible**.

## corpus
*See* **mandible**.

## corpus callosum
The largest white matter tract (i.e., bundle of axons) in the brains of placental mammals, that runs across the midline of the brain to physically and functionally link the left and right cerebral hemispheres. In general, as brain size increases among primates the axons that traverse the corpus callosum increase in diameter and develop a greater degree of myelination so as to minimize delay in the conduction of nerve impulses between the cerebral hemispheres. However,

there are no significant differences in axon diameters within the corpus callosum between modern humans and chimpanzees. This suggests that nerve impulses take longer to travel between the cerebral hemispheres in modern humans, a factor that may have driven the increase in functional lateralization seen in modern humans. Several studies in modern humans have reported sexual dimorphism in the cross-sectional area of the corpus callosum relative to total brain size, but these results have been disputed. There is no evidence of a similar pattern of sexual dimorphism in common chimpanzees. (L. *corpus* = body and *callosum* = callous or firm.)

## correlated-age methods
See **geochronology**.

## correlation
Earth sciences An attempt to relate strata exposed in two or more different locations within the same, or different, sites. Strata can be compared using the appearance of the fossils they contain (i.e., biostratigraphy) or their chemical make-up (i.e., tephrostratigraphy). Statistics A measure of the degree of association between two variables, such that when one variable changes the other also changes. If the amount of change is the same then the correlation is perfect, if they are similar the correlation is strong, but if the changes are only a little better than random the relationship is weak. The correlation coefficient ($r$) measures the strength of the relationship. (L. *cor* = together and *relatus* = relate.)

## correlation coefficient
See **correlation**.

## correlation matrix
An $n \times n$ square matrix corresponding to a data set with $n$ variables. The value in each cell of the matrix corresponds to the correlation between the row and the column variable for that cell. The values on the diagonal of the matrix are all correlations of variables with themselves and thus they are all equal to one.

## cortex
Biology The outer covering, or layer, of a structure or organ (e.g., the outer covering of gray matter of the cerebral hemispheres), or the outer layer of dense, or cortical, bone that lies immediately beneath the periosteum of a limb bone. Archaeology The weathered exterior surface of a stone or cobble, which is often preserved on the initial flakes struck from this stone during tool production. (L. *cortex* = bark, or outer covering.)

## cortical
See **cortex**.

## cosmogenic nuclide dating
A dating system that uses the isotopes $^{26}$Al and $^{10}$Be that are produced in quartz when neutrons, protons, and muons penetrate the upper few meters of the Earth. However, if the sediments are buried at depths greater than 5–10 m the production of these isotopes

ceases (this is why it is also called cosmogenic nuclide *burial* dating). The principle of the $^{10}Be/^9Be$ system is analogous to that of radiocarbon dating. Once burial has halted the production of new $^{26}Al$ and $^{10}Be$, the different decay characteristics of the inherited $^{26}Al$ and $^{10}Be$ (the former has a shorter half-life than the latter) cause the ratio of these elements to change over time, thus providing an estimate of how long the quartz has been buried. Suitable materials include quartz and diatomite. The time range of the $^{26}Al/^{10}Be$ system is $c.1$ Ma to 4 or 5 Ma. Measurement uncertainty and uncertainty about the length of the half-lives of $^{26}Al$ and $^{10}Be$ mean that there is an effective limit of $c.100$ ka on the precision of ages derived using $^{26}Al$ and $^{10}Be$ cosmogenic nuclide dating. The time range for the $^{10}Be/^9Be$ system is 0.2–14 Ma.

## Cossack
The Cossack hominin was an approximately 40 year-old male *Homo sapiens* whose robusticity and cranial morphology (especially its low frontal and long overall skull length) is similar to the skulls and crania from other terminal Pleistocene sites (e.g., Kow Swamp, Cohuna, and Coobool Crossing) in eastern Australia. (Location 20°40'S, 117°11'E, Australia.)

## cost of transport
The cost of transport of an animal is the amount of oxygen it uses to transport a unit of its mass over a fixed distance. It is usually expressed as ml $O_2$ $kg^{-1}$ $m^{-1}$.

## Cova del Gegant
A cave directly on the ocean in Spain, in which a mandible and several isolated teeth were recovered. Due to its morphology, age, and Mousterian archeological context, the mandible has been assigned to *Homo neanderthalensis*. (Location 41°22'26"N, 01°77'22"W, Catalonia, Spain.)

## Cova Negra
One of the most complete late Pleistocene stratigraphic sequences in Spain, which has yielded numerous remains of *Homo neanderthalensis*, including abundant juvenile specimens. Archeological evidence from Cova Negra indicates sporadic, short-term hominin occupations of the site that have been interpreted as indicating a high degree of mobility among Neanderthals. (Location 38°59'N, 00°32'W, Spain.)

## covariance
A measure of association between two variables. It is similar to correlation in that positive values indicate that an increase in one variable is associated with an increase in the other, and negative values indicate that an increase in one variable is associated with a decrease in the other. The magnitude of the covariance indicates the strength of the association. Unlike correlation, covariance can take on any value (i.e., it is not limited to the range of −1 to 1).

## covariance matrix
An $n \times n$ square matrix corresponding to a data set with $n$ variables. The value in each cell of the matrix corresponds to the covariance between the row and column variable for that cell.

## covariation
How two variables change in response to each other. Covariation is typically measured as covariance or correlation.

## coxa
An old-fashioned osteological term that is best avoided. The proper name for the "hip bone" is the pelvic bone. (L. *coxa* = hip, hip bone, or hip joint.) *See also* **anatomical terminology**.

## Cradle of Humankind World Heritage Site
Located in South Africa in the province of Gauteng this World Heritage Site includes important karstic early hominin sites (e.g., Cooper's Cave, Drimolen, Gladysvale, Gondolin, Kromdraai, Malapa, Sterkfontein, and Swartkrans) that are clustered southwest of Pretoria and northwest of Johannesburg. The location was declared a World Heritage Site in 1999 thanks in large part to the advocacy and persistence of Phillip Tobias and his South African colleagues. Subsequently, the sites of Makapansgat and Taung were added as geographically separate components of the World Heritage Site. There is an interpretative center at Maropeng near Sterkfontein.

## crania
Pl. of **cranium** (*which see*).

## cranial
Used informally for fossils of the bony skull and dentition. When used formally it refers to structures that are in the direction of the head (i.e., it is the opposite of caudal). (Gk *kranion* = brain case.)

## cranial base
An informal term synonymous with the more formal term **basicranium** (*which see*). (Gk *kranion* = brain case and L. *basis* = base.)

## cranial base angle
(or CBA) The angular relationship in the sagittal plane between the anterior and posterior components of the basicranium. All of the current interpretations of the CBA are variants of the measure used by T. H. Huxley (i.e., the nasion–prosphenion–basion angle). Some use a different anterior landmark (e.g., the anterior cribriform point instead of nasion) and others use a different middle landmark (e.g., the "pituitary point" in place of prosphenion), but the choice of basion as the posterior landmark is standard. The place where the anterior and posterior components of the basicranium meet is the mid-sphenoidal synchondrosis, which lies in the floor of the hypophyseal fossa. The average CBA is more acute in adult modern humans (mean ± SD = 135 ± 3.1°) than in adult chimpanzees (157 ± 4.9°) and adult bonobos (mean = ≈148°), but the angle changes during ontogeny. For example, in modern humans the CBA is 143° in neonates, and then it becomes more acute during the first 2 years of postnatal life. This differs from the ontogenetic changes in nonhuman primates, in which the CBA becomes more obtuse throughout ontogeny. Few fossil hominin crania are complete enough to

be able to make a reliable estimate of the CBA, but what evidence there is suggests that variation in CBA within and among hominin taxa is, among other factors, affected by brain size and the relative size of the face.

## cranial base flexion
See **cranial base angle**.

## cranial fossa(e)
See **basicranium**.

## cranial vault
The roof and side walls of the cranium. The cranial vault is formed from a pair of flat bones, the parietals, the flat or squamous parts of the two temporal bones, and the squamous parts of the frontal and occipital bones. Vault bones have an inner and an outer layer of compact bone called the inner and outer tables and an intervening layer of spongy bone filled with red bone marrow, called the diploë. Diploic veins drain blood from the bones of the cranial vault to dural venous sinuses internally and also to veins on the external surface of the cranium. Veins that pass through the cranial vault and connect the veins on the inside of the cranial vault (i.e., endocranial) to the veins on the outside of the cranial vault (i.e., pericranial) are called emissary veins and they run through emissary foramina. [Gk *kranion* = brain case and L. *volvitus* = arched; syn. calvaria (*which see*).]

## cranial venous drainage
The external component of the intracranial venous system consists of superficial cerebral veins that drain the surface of the brain and these, in turn, drain into a system of venous channels called the dural venous sinuses that run between the fibrous and endosteal layers of the dura mater (the outer of the three meningeal layers). Importantly for paleoanthropology these dural venous sinuses leave impressions on the inner, or endocranial, surface of the bones of the cranium. The venous sinuses that are of most concern to paleoanthropologists are the larger dural venous sinuses within the posterior cranial fossa; namely the transverse, sigmoid, occipital, and marginal sinuses. In most modern humans and in the vast majority of chimpanzees and gorillas the venous blood draining the deep structures of the brain normally drains into the left transverse sinus and the superior sagittal sinus drains into the right transverse sinus. Each transverse sinus drains into a sigmoid sinus that runs inferiorly in a sigmoid-shaped groove from the lateral end of the transverse sinus to the superior jugular bulb, which marks the beginning of the internal jugular vein. This pattern of dural venous sinuses is known as the transverse-sigmoid system and individuals (e.g., KGA 10-525) with transverse and sigmoid venous sinuses on both sides are said to show transverse-sigmoid system dominance. In just a few percent of modern humans the venous blood from the brain drains into an enlarged midline occipital sinus, which then drains into uni- or bilateral marginal sinuses that run around the margin of the posterior quadrants of the foramen magnum. This is known as the occipito-marginal system of venous sinuses and individuals with a substantial occipital sinus and with marginal venous sinuses on both sides (e.g., OH 5) are said to show occipito-marginal system dominance. Some

individuals (e.g., A.L. 333-245 and KNM-ER 23000) have a transverse-sigmoid system of venous sinuses on one side and an occipito-marginal system of venous sinuses on the other, and a few individuals (e.g., Taung 1) have even been reported to show coexistence of the two systems on the same side. A system of emissary veins (plus some venous sinuses such as the petrosquamous sinus) connects the intra- and extracranial venous systems by passing through small foramina in the walls of the cranium. Although the incidence of occipito-marginal system dominance is distinctively higher in *Paranthropus boisei*, *Paranthropus robustus*, and *Australopithecus afarensis*, cranial venous drainage pattern is a polymorphism and it is probably not wise to depend on the presence or absence of one system or the other as a definitive taxonomic marker.

### cranium
A skull minus its mandible. The cranium consists of the calvaria (or cranial vault), the face, and the basicranium. It can also be broken down according to the way the different parts develop. The neurocranium (the parts of the cranium that surround the brain and the special sense organs) develops from the paraxial mesenchyme (i.e., the first five somites and the somite-like material rostral to the first somite) plus a small contribution from the neural crest. The viscerocranium (the part of the cranium that covers the anterior aspect of the brain, which is equivalent to the face) develops from a combination of the frontonasal process and the first pharyngeal arch. All of the viscerocranium develops from the neural crest. (Gk *kranion* = brain case.) *See also* **endocranial cavity**.

### crista paramastoidea
*See* **occipitomastoid crest**.

### critical function
This term was introduced for the morphological counterpart of the ecological concept of fallback foods. It refers to morphology that enables an animal to process its fallback food(s) as opposed to its preferred foods. For example, it has been suggested that the distinctive anterior pillars of the face of *Australopithecus africanus* may be an adaptation that allowed the premolars to be used for processing hard objects too large to be accommodated between the molars.

### *Crocuta*
The genus name of the modern spotted hyena, *Crocuta crocuta*, a member of the family Hyaenidae. These bone-crunching carnivorans are common and distinctive members of the modern African fauna and extinct forms of hyena have been implicated in the bone comminution seen at some archeological and hominin paleontological sites. Hyenas may have been important competition for the earliest tool-making and tool-using hominins. (L. *crocuta* = literally the "saffron-colored one"; the term was said to have been used by Pliny the Elder to refer to a wild animal in northern Africa.)

### Cro-Magnon
The site of Cro-Magnon was discovered in 1868 when a road crew was moving large blocks of limestone from the talus slope at the base of the steep cliffs lining the Vézère river just outside of the village of Les Eyzies, France. The modern human fossils found at the site by Édouard

Lartet (e.g., Cro-Magnon 1) were among the first in Europe and gave rise to the term "Cro-Magnon Man." (Location 44°56′N, 01°00′E, France.)

### Cro-Magnon Man
See **Cro-Magnon**.

### Cromerian
See **glacial cycles**.

### crossing over
See **recombination**.

### cross-section
Section taken at right angles to the long axis of a bone or a tooth. Cross-sections of teeth can also be in any plane that can be consistently located among observers (e.g., through the tips of cusps). When the dimensions of a cross-section of a bony structure (e.g., the femoral neck) are given, unless stated otherwise the assumption is that they are the minimum dimensions (i.e., the cross-section is at right angles to the long axis of the structure).

### cross-sectional data
Data collected from many subjects at a single point in time. This type of "broad-swath" sampling aims to gather information from large numbers of individuals of similar or different ages and character states. It is used to acquire information about, and is considered representative of, the larger population from which the sample is drawn. Longitudinal data (derived from measurements taken from a series of observations on a single individual or group of individuals) are preferable to cross-sectional data for studies that wish to examine how individuals change over time, but for obvious reasons this type of data is more difficult to collect in extant taxa and impossible to collect for fossil taxa.

### cross-sectional geometry
(or CSG) A method for investigating the mechanical adaptations of long bones that is based on mechanical engineering formulae used for modeling the bending stress experienced by straight beams under static (nonmoving) loads. A cross-section of a bone is taken at a specified location (e.g., midshaft, 66% of the length from the proximal end, etc.) and in a plane perpendicular to the long axis of the bone. Sections can be obtained either by physically sectioning the bone or by medical imagery. Second moments of area, which capture the rigidity of the section, are often used in the literature as proxies for the strength of the bone.

### cross-striations
Short-period incremental lines (i.e., circadian) in enamel at right angles to the long axis of enamel prisms. Because they reflect an approximately 24 hour cycle of ameloblast secretory activity they can be used to determine the daily secretion rate of enamel and they can be summed to estimate how many days it took to develop a tooth crown. Cross-striations are equivalent to von Ebner's lines in dentine. (L. *stria* = furrow.) *See also* **enamel development**.

## crown
The part of a tooth covered by enamel that, after eruption, projects above the surrounding alveolar bone. It is made up of an outer enamel cap and an inner core of dentine. The rest of the tooth, called the root, is made of dentine and is embedded in a bony alveolus. The boundary between the crown and the root on the surface of the tooth is called the cervix or the cemento-enamel junction (or CEJ). The boundary between the crown and the dentine within the tooth is called the enamel–dentine junction (or EDJ). (L. *corona* = crown.)

## crown area
See **crown base area**.

## crown base area
A measure of the surface area of a tooth crown that is mainly used for postcanine teeth. It has most often been expressed as the two-dimensional area of the base of the crown, rather than as a three-dimensional estimate of the functional occlusal area. The former is either computed from the product of the maximum buccolingual breadth and the mesiodistal length of the crown or measured using a device that calculates surface area from a trace of its perimeter.

## crown completion
The stage of tooth development when enamel matrix secretion is complete on all aspects of a tooth. It can be identified radiographically (when a spicule of root has formed beyond the cervix) or histologically (when enamel development has ceased).

## crown formation time
(or CFT) The time it takes to develop the whole of a tooth crown. It is the time elapsed from the onset of amelogenesis at the tip of the first (or only) cusp to be formed, until the cessation of amelogenesis at the cervix (the future cemento-enamel junction, or CEJ) of the last (or only) cusp to form. Crown formation time is typically estimated through counts of the cross-striations and striae of Retzius in the cuspal enamel (appositional enamel) and lateral enamel (imbricational enamel). If the relevant variables (e.g., daily secretion and extension rates) are held constant, teeth with larger dentine and enamel volumes will take longer to form than smaller ones. After normalizing for overall size, the modern human dentition takes longer to form than the dentitions of the extant apes and crown formation times are shorter in *Homo neanderthalensis* than they are in both living and fossil populations of *Homo sapiens*. However, in both *H. sapiens* and *H. neanderthalensis* crown formation times are significantly longer than they are in the few early hominins for which there are good data. See also **enamel development**.

## crown group
A term introduced for the smallest subset of sister taxa within a clade, or the monophyletic group that includes the living taxon, or taxa, within that clade. For example, the crown group of the hominin clade is *Homo sapiens* plus *Homo neanderthalensis*, which can be written as (*Homo sapiens*, *Homo neanderthalensis*). (L. *corona* = crown, refers to the round wreath on the very top of the head, hence the meaning of the "highest" taxon.) See also **clade**; **stem group**; **total group**.

## crural index

The ratio of the lengths of the tibia and femur (tibia length/femur length × 100). It measures the relative length of the leg below the knee. Long-limbed terrestrial quadrupeds have disproportionately elongated distal elements (high crural indices). The distal elements are the narrowest and therefore the lowest-mass portions of the lower limbs; thus by elongating the distal part of the lower limb the energetic costs of accelerating and decelerating the lower limbs during gait are minimized. Among primates the crural index varies by locomotor strategy and climate. Elongating the distal portion of the lower limb also increases the surface-area-to-volume ratio, which improves the ability to lose heat in hot, dry climates. Thus, short legs are more effective at retaining heat and low crural indices are characteristic of taxa, including *Homo neanderthalensis*, which live (or lived) in cold climates. The upper-limb equivalent is the brachial index. (L. *cruralis* = leg.)

## crypt

See **dental crypt**.

## CSF

Abbreviation of **cerebrospinal fluid** (*which see*).

## CT

Abbreviation of **computed tomography** (*which see*).

## cultural transmission

A form of social learning that gives rise to culture. In cultural transmission an individual learns, from another individual, information capable of affecting its behavior and/or other aspects of its phenotype. In the past researchers debated whether the term should be restricted to cases in which individuals learn from other individuals via imitation, but today the consensus is that limiting the term in this way is not helpful. The terms cultural transmission, social learning, and cultural learning are now used interchangeably. Anthropologists often talk about culture as if it is unique to hominins, but over the last few decades it has become clear this is not the case. Work in the laboratory and field has pointed to the existence of cultural learning in a number of nonhominins, including chimpanzees, orangutans, killer whales, humpback whales, a variety of passerine bird species, and guppies. Some of these cases are still being debated, but the notion that the songs of many passerine bird species are socially learned is no longer controversial. The adaptive significance of cultural transmission appears obvious. The ability to learn new solutions to problems from other individuals would seem to be something that natural selection can be expected to always favor. In vertical transmission a child learns from a parent; in oblique transmission an individual in one generation learns from an individual in the previous generation who is not their parent; and in horizontal transmission an individual learns from another individual in the same generation.

## culture

Numerous definitions of culture have been put forward, but most are problematic for researchers interested in human evolution. The reason is that they automatically preclude the possibility that nonhominins have culture, which affects how the issue of the origins of culture is approached. There is substantial agreement among all definitions in the sense that they hold that culture involves social learning or cultural transmission. The issues on which these definitions differ are whether all forms of social learning give rise to culture, and whether social learning is the only process involved. Some researchers argue that culture should be restricted to behaviors that result from two particular forms of social learning: teaching and imitation. Still other researchers suggest that for a behavior to be considered cultural it not only has to be socially learned but also has to show signs of elaboration through time and across generations. (L. *cultura* = cultivation.) *See also* **cultural transmission; social learning; theory of mind.**

## cursorial
Adjective used to describe animals adapted for fast terrestrial travel (e.g., running). Cheetahs and gazelles are cursorially adapted. Some researchers have suggested that the body size and shape of early *Homo* may have been adapted for endurance running. (L. *cursorius* = running.) *See also* **endurance running hypothesis.**

## cusp
A portion of a tooth crown demarcated by primary fissures and with an independent apex. Cusps are divided into primary (or main), accessory, conules, and conulids (see below). The primary cusps on the mandibular molars are the entoconid, hypoconid, hypoconulid, metaconid, and protoconid. The latter two cusps are also the primary cusps on mandibular premolars. Examples of accessory cusps on mandibular molars are the C6 and the C7. The primary cusps on the maxillary molars are the hypocone, metacone, paracone, and protocone. The latter two cusps are also the primary cusps on maxillary premolars. The C5 is an example of an accessory cusp on a maxillary molar. Enamel features that are not large and discrete enough to qualify as accessory cusps are called conules on maxillary teeth and conulids on mandibular teeth. In any event, if "cusp" is taken literally it is a misnomer with respect to hominin teeth because in all hominins the unworn cusps are blunt rather than pointed. (L. *cuspis* = point.) *See also* **cusp nomenclature.**

## cuspal enamel
Term used for the first-formed enamel over the dentine horn that makes up the occlusal aspect of a tooth and which does not show long-period growth increments at the surface. In a completed unworn tooth no portion of the cuspal enamel is visible on the surface of a tooth crown. The increments visible on the tooth surface belong to the imbricational (or lateral) enamel. Cuspal enamel forms a larger percentage of the crown volume in posterior (or postcanine) than in the anterior teeth.

## cusp morphology
The relative size and shape of the main and accessory cusps on the occlusal surface of a postcanine tooth crown. It refers to both metrical and nonmetrical assessments; most of the latter address the incidence and prevalence of accessory cusps. Nearly all of the existing research in this area has focused on the taxonomic value of interspecific differences in cusp morphology; relatively few studies have looked at the functional significance of such

differences. In modern human populations variability of expression of main and accessory cusps is usually described using the Arizona State University Dental Anthropology System (or ASUDAS). This system and its modifications have also been employed to give semiquantitative accounts of morphological details in fossil hominin dentition. Use of this system on fossils is sometimes problematic because not all of the variation in cusp morphology observed in contemporary humans can be observed in fossil hominins, and there are some features seen in fossil hominins that are not seen in contemporary modern human populations. Traditionally researchers have focused on cusp morphology at the outer enamel surface and they expressed cusp areas in two dimensions, but recently micro-computed tomography (or micro-CT) imaging has been used to investigate the morphology of the dentine horns at the enamel–dentine junction.

### cusp nomenclature

The names of the cusps of the postcanine teeth were devised by Henry Fairfield Osborn. They are consistent with what we now know is almost certainly an incorrect hypothesis (called the tritubercular theory) about the evolution of tooth crowns, but Osborn's naming scheme has outlasted his theory. Osborn's scheme begins with a reference cusp, which is on the lingual side of the crown in an upper postcanine tooth (the protocone) and on the buccal side of the crown in a lower postcanine tooth (the protoconid). He suggested that these cusps in mammals are homologous with the mesiodistally elongated main cusp in a reptile tooth, hence the prefix "proto." In the upper jaw he called the cusp on the mesiobuccal aspect of the reference cusp the paracone and the cusp situated distobuccally to it the metacone. The equivalent cusps in the lower postcanine teeth are the mesiolingually situated paraconid (which has been lost in higher primates) and the distolingually situated metaconid. The units made up of three cusps are called the trigon in the upper teeth and the trigonid in the lower teeth. In higher primates each postcanine tooth may have a distal addition to the trigon or trigonid. It is called the talon in the upper teeth and the talonid in the lower teeth. In the upper postcanine teeth the talon is represented by the distolingual hypocone. In the lower teeth it is represented by the entoconid lingually, the hypoconid buccally, and the hypoconulid (if there is one) distally.

### cutmarks

Hominin-induced modifications inflicted on bone surfaces by stone artifacts or knives during carcass processing (e.g., the defleshing, disarticulation, or skinning of animal remains). Cutmarks have a deep, V-shaped cross-section, whereas the tooth marks of large carnivores tend to be shallower and U-shaped. When archeologists identify a cutmark on bone, they are demonstrating a temporal, spatial, and behavioral association between the hominins that formed an archeological site and the associated fauna. The percentages of bones showing cutmarks as opposed to those showing carnivore toothmarks have been used as a proxy for the intensity and scope of hominin activity involved in generating a given archeological assemblage, and they have been used to make inferences about the relative timing of hominin access to animal remains.

**cytoarchitecture**
The cellular composition of any structure in the body, but in the cerebral cortex the term refers to the characteristic arrangement of neurons into layers according to their size, morphology and spatial packing density. These variables allow researchers to distinguish the boundaries of cortical areas in the brain. For example, one of the most widely used maps of the neocortex by Brodmann is based on cytoarchitecture. A Nissl staining technique, in which neural tissue is dyed with chemicals that specifically stain the tissue for cell bodies, is commonly used in the study of cytoarchitecture. (L. *cyto* = cell and *architectus* = builder.)

**cytogenetic**
See **karyotype**.

# D

## δ

Delta, or δ, is a notation used in stable-isotope biogeochemistry (e.g., $δ^{13}C$). It refers to the proportion of heavy to light isotope (e.g., $^{13}C/^{12}C$, $^{15}N/^{14}N$, $^{18}O/^{16}O$) within a given substance relative to the relevant international standard. Values of this proportion are reported as parts per thousand (‰).

## D211

The first early hominin recovered from Dmanisi. The D211 mandible, like the other hominins from Dmanisi, is dated to 1.85–1.78 Ma. It combines relatively small postcanine tooth crowns and a modern human-like molar size order with a corpus that is more robust than usually seen in modern humans or Neanderthals. This mandible and the D2282 cranium probably belong to the same individual. The consensus is that the hominins from Dmanisi sample a single hominin species that likely belongs within the genus *Homo*. Only time will tell whether the Dmanisi hominins can be accommodated within early *Homo* as *Homo habilis*, or pre-modern *Homo* as *Homo erectus*, or whether their particular combination of features justifies a new taxon. *See also* **Dmanisi**.

## daily secretion rate

The amount of enamel matrix measured in micrometers (µm) secreted by an ameloblast during each 24 hour period (i.e., the distance between adjacent cross-striations along an enamel prism). The average daily secretion rate in modern humans is close to 4 µm/day, but there is a spatial gradient with the lowest rates (approximately 2.8 µm/day) at the enamel–dentine junction (or EDJ) and the highest rates (about 5.5 µm/day) close to the enamel surface. In chimpanzees rates rise more quickly across the thickness of enamel, thus more enamel is formed in the same time. Rates are faster in the thick-enameled hyper-megadont archaic hominins, such as *Paranthropus boisei*.

## Daka calvaria

*See* **BOU-VP-2/66**.

## Dali
(大荔) A 230–180 ka Chinese site that is the source of the Dali cranium, a relatively large-brained (1120 cm³) cranium sharing some features seen in modern humans with features seen in Chinese *Homo erectus*. Flakes and scrapers found at the site have been retouched, but otherwise they are relatively primitive. (Location 34°51'57.49"N, 109°43'58.58"E, northeastern China.)

## Danakil cranium
See **UA 31**.

## Dansgaard–Oeschger cycles
Roughly 1.5 ka-long cycles of cooling and warming during the last glacial period that have been identified in the Greenland ice core and North Atlantic records.

## data matrix
Any data set organized into rows and columns. Data matrices are typically used in a multivariate analysis. The number of rows and columns may be equal or different; when they are equal the matrix is referred to as a square matrix. Some common examples of square matrices include correlation, covariance, and distance matrices.

## date
Used as a verb it refers to the act of dating, whereas as a noun it refers to a specific point in time (e.g., 1347 years BCE). Nowadays it is conventional to refer to a specimen's age rather than its date. The abbreviation ka refers to an age in thousands of years before the present (kiloannum) and Ma refers to an age in millions of years (mega-annum).

## dating
See **geochronology**.

## "Dawn man"
See **Piltdown**.

## day range
The area exploited by an individual animal or group of animals during a day (i.e., a 24 hour period). Because it is a daily subset of movements within a home range the day range is usually smaller than the home range. *See also* **home range**.

## dc
Abbreviation for a deciduous canine. Upper, or maxillary, deciduous canines are identified by a line below the abbreviation (d$\underline{c}$); lower, or mandibular, deciduous canines are identified by a line above the abbreviation (d$\bar{c}$).

## "Dear Boy"
The nickname used by Louis and Mary Leakey for the OH 5 *Zinjanthropus boisei* (now *Paranthropus boisei*) cranium. According to Phillip Tobias, the Leakeys enjoyed treating their fossils as people and spoke of them light-heartedly by their names. Apparently when

OH 5 (obviously a male and subadult) came to light, Mary Leakey knelt down and exclaimed "Oh you dear boy."

### débitage
A French term used by English-speaking lithic technologists to refer to the flakes, flake fragments, and other debris that results from the process of stone tool manufacture. (Fr. *débiter* = to cut up.)

### deciduous dentition
Refers to the teeth of the primary dentition (or milk teeth) that are shed during dental development. Deciduous tooth types are conventionally written in lower case with a "d" prefix (e.g., di and dm). In all hominins and hominids the deciduous dentition consists of two incisors (di1 and di2), a canine (dc), and two molars (dm1 and dm2, or dp3 and dp4) in each of the four quadrants of the jaw. As is the case with the adult dentition, when referring specifically to an upper, or maxillary, tooth its number is written using superscript (e.g., $di^1$ and $dm^2$); the equivalent lower, or mandibular, tooth is written using subscript (e.g., $di_1$ and $dm_2$). Deciduous molars are occasionally found as fossils, but deciduous incisors and canines are rare; one of the exceptions is the southern African site of Drimolen. First lower deciduous molars ($dm_1$) have been shown to be particularly useful for alpha taxonomy. For example, John Robinson showed that the $dm_1$s of *Paranthropus robustus* and *Paranthropus boisei* have especially complex cusp morphologies. (L. *decidere* = to fall off and *dentes* = teeth.) *See also* **permanent dentition; teeth**.

### decussation
Enamel prisms typically do not run in a straight line from the enamel–dentine junction to the surface of the tooth. Instead they twist in a wave-like manner two or three times before running straight to the surface in the outer one-third of the enamel. When cut in cross-section, bands or tracts of prisms appear to intersect at an angle, but they are just crossing over and past each other much as the component strands of a rope cross over one another. Among early hominins there are differences in the amount of decussation; it is marked in early *Homo* but much less obvious in *Paranthropus boisei*. (L. *decussis* = the crossing of two structures after the roman numeral for 10, X, in which two lines intersect.)

### Dederiyeh
A cave site in northern Syria where the remains of at least 17 *Homo neanderthalensis* individuals have been found. One, Dederiyeh 1, is a remarkably complete skeleton of an approximately 2 year-old Neanderthal child. (Location 36°24′N, 36°52′E, northern Syria.)

### deep sea temperatures
Deep sea temperatures are remarkably homogeneous and relatively stable (within 2–4 °C) through time, but subtle changes did occur between glacial and interglacial conditions. Oxygen isotopes from the shells of marine protists (calcitic foraminifera) are used to reconstruct deep sea temperatures and glacial ice volumes. The cyclicity captured in these records and their uniformity across the world's oceans provides a means of temporal correlation among ocean drilling sites. *See also* **astronomical time scale**.

## definition
The features shared by all, or at least a large majority, of the members of a taxon. The features given in a definition of a taxon include all of its attributes, not just the ones that make it distinctive. It is the latter that are emphasized in the diagnosis of a taxon. (syn. description.)

## degrees of freedom
Anatomy The number of independent axes about which movement can take place at a synovial joint. A joint like the modern human elbow joint can only flex and extend, so it is uni-axial and thus it has *one* degree of freedom. The carpometacarpal joint of the thumb (between the first metacarpal and the trapezium) has two independent axes that are determined by the shape of the joint surfaces and thus it *two* degrees of freedom. Statistics The number of values that are free to vary when calculating a particular statistic. In general, statistics tend to have as many degrees of freedom as the number of observations made ($n$) minus the number of parameters estimated in the calculation of that statistic (e.g., two parameters, the slope and the intercept, are estimated in the case of linear regression). With regard to the effect of degrees of freedom on statistical tests, as the number of values that can vary independently increases, so does the confidence in the accuracy of estimated parameters (e.g., mean, slope, intercept, etc.). Thus, if all else is held constant, statistical significance increases as the number of degrees of freedom increases.

## DEJ
Acronym for dentine–enamel junction, and one of two conventional abbreviations (EDJ is the other) used for the interface between the underlying dentine and the overlying enamel. *See* **enamel–dentine junction**.

## deleterious mutation
A mutation that causes a reduction in the fitness of an individual (e.g., it may cause disease, reduce survival in other ways, or adversely affect reproduction). Classifying a mutation as deleterious is not necessarily straightforward; some mutations may convey either benefit or harm (or both) depending upon individual circumstances. For example, a mutation in the sickle cell gene leads to sickle cell anemia if the mutant gene is inherited from both parents. However, a single copy of the mutant gene inherited from one parent confers resistance to malaria, a particular benefit in parts of the world where malaria is prevalent. Recent research has shown that some mutations in the human leukocyte antigen (HLA) system were acquired when modern humans interbred with Neanderthals and Denisovans. Modern humans benefited from the archaic HLA types, which would have been better adapted to local antigens than their own. However, some scientists believe that while broadening modern humans' capacity for immune response, some of these acquired mutations may also be linked to autoimmune disease. (L. *deleterius* = to harm and *mutare* = to change.)

## deletion
A mutation in the genome where one or more base pairs (i.e., pairs of nucleotides) have been removed or deleted from a sequence of nucleotides. These deletions can be as small as one base pair or may involve many thousands of base pairs. In a coding region any deletion that does

not involve a multiple of three base pairs causes a frameshift mutation, which is typically deleterious. Large deletions that include whole genes or a substantial portion of a chromosome are usually detrimental whereas deletions in noncoding regions are usually neutral. Insertions and deletions are commonly referred to as "indels." (L. *delere* = to abolish.)

### deltaic sediments
Sediments accumulated at the mouth of a river. They are deposited when the flow in the river reduces prior to the river entering a lake or sea. Because the water energy in a delta is very low the sediments tend to be dominated by fine-grained material (e.g., silts and clays). (Gk *delta* = fourth letter of the Greek alphabet, Δ, which is similar in shape to a river delta.)

### deme
An informal category used for a geographically localized population of closely related individuals. When applied to the fossil record the concept is sometimes referred to as a "p-deme," paleodeme, or paleocommunity-deme. The use of p-demes does not solve the problem of how many species to recognize in the fossil record, but this informal category may be a useful way to discuss the different parts of the fossil record prior to making decisions about its taxonomy. (Gk *demos* = people.)

### demography
The study of age and sex structure and/or size of a population.

### dendrochronology
See **radiocarbon dating**.

### Denisova Cave
A cave site in the Altai Republic of Siberia. The mitochondrial DNA (mtDNA) from a hominin phalanx found in 2008 is distinct from that of both Neanderthals and modern humans, and was more different from modern human mtDNA than it was from Neanderthal mtDNA. The mtDNA extracted from a molar tooth found at the site (see below) matches that recovered from the phalanx. The results of a more recent analysis of the nuclear DNA suggests that the unidentified Denisova Cave hominin and Neanderthals were sister taxa, that the former diverged from the modern human/Neanderthal lineage more than 600 ka ago. There is also evidence that it contributed approximately 4–6% of its genetic material to the genome of modern humans from Melanesia, but not to the genomes of modern humans from Africa and Eurasia. (Location 51°23′51″N, 84°40′34″E, Russia.)

### Denisovan
The informal term used for the unidentified hominin found along with Neanderthals in Denisova Cave. No formal Linnean binominal has been proposed for this taxon, which is presently only identifiable from its DNA. *See also* **Denisova Cave**

### dense bone
See **bone**.

## dental crypt
The space within the alveolar process of the maxilla or mandible within which a tooth germ develops. (L. *dens* = tooth and *crypta* = hidden.)

## dental development
A term that encompasses the initiation, morphogenesis, mineralization, and eruption of the teeth and the establishment of functional occlusion. It is a process that starts in the embryo and extends into young adulthood. In modern humans dental development takes approximately 20 years; the same process takes approximately 12 years in chimpanzees. To date, dental histologists have established that *Australopithecus* and *Paranthropus* matured on a time scale closer to that of apes than to that of modern humans.

## dental formula
The numbers of teeth in each of the four quadrants of the jaws. The formula is written beginning with the number of incisors (I or di), the number of canines (C or dc), the number of premolars (P), and ending with the number of molars (M or dm/dp). In all catarrhines, including fossil hominins and modern humans, the dental formula for the normal deciduous dentition is "2. 1. 2" (2 incisors, 1 canine, 2 molars) and for the normal permanent dentition it is "2. 1. 2. 3" (2 incisors, 1 canine, 2 premolars, 3 molars). (L. *dens* = tooth.)

## dental macrowear
Tooth wear that can be seen with the naked eye. It starts as wear facets on enamel and later the wear extends into the enamel. (L. *dens* = tooth and Gk *macros* = large.) *See also* **tooth wear**.

## dental microwear
Occlusal tooth wear that is visible only through a microscope. Dental microwear focuses on the size, number, and orientation of microscopic scratches, pits, etc. on the enamel, or on the overall complexity and the degree of directionality of the wear enamel surface. Whereas dental macrowear is a measure of the abrasiveness of the diet in the long term, dental microwear signals on fossil teeth indicate whether the food ingested in the days or weeks before death contained hard or abrasive material. Such abrasive material can be either intrinsic to the ingested foods (e.g., phytoliths) or extrinsic to the foods (e.g., adherent sand grains). Examining dental microwear can also be used to detect evidence for the eating of softer or tougher foods (e.g., polished or finely pitted surfaces).

## dental reduction
Decrease in the size of teeth between successive taxa or populations. Two examples are the reduction in crown size from the large postcanine teeth of *Australopithecus* to the smaller postcanine teeth of early *Homo*, and the further reduction in crown size from pre-modern *Homo* to *Homo sapiens*. The term is also used to describe the trend of decreasing size along the tooth row in a tooth type (e.g., the pattern of decreasing crown dimensions from M1 to M3 seen in later fossil hominins and in modern humans).

## dental wear
*See* **tooth wear**.

### dentine

The hard tissue that forms the bulk of a tooth crown and nearly all of a tooth root. Dentine is one of the three dental hard tissues (the others are enamel and cementum). Unlike bone, dentine (or ivory, which is the same thing) contains no blood vessels or included cell bodies. It consists of a mineral phase (70%) (mostly hydroxyapatite) and an organic phase made up of collagen, lipids, and noncollagenous proteins. Unlike enamel, dentine has no "grain" or planes of cleavage and it is a relatively elastic material (hence ivory was originally the material of choice for billiard or pool balls). Several classes of incremental feature are found in dentine, including long-period Andresen lines and short-period von Ebner's lines; they correspond, respectively, to long- and short-period incremental lines in enamel. (L. *dens* = tooth and *inus* = "of" or "relating to"; hence, "pertaining to a tooth.")

### dentine horn

The tallest point of the dentine underlying one of the cusps or features that make up the enamel cap of a tooth.

### dentinogenesis

The cellular activity involved in the development of dentine. (L. *dens* = tooth and *inus* = "of" or "relating to," and Gk *gena* = to give birth to.) *See also* **dentine**.

### dentition

A collective term for all of the teeth in an individual. In the context of the hominin fossil record it is used to refer to the teeth preserved in a particular specimen ("the dentition of KNM-ER 3734 comprises the roots and crowns of the left C, $P_3$, $P_4$, $M_1$, $M_2$, and the root and part of the $M_3$ crown"). It also refers to the type of dentition (i.e., deciduous, or primary; permanent, or secondary). (L. *dentes* = teeth.)

### dento-gnathic

Shorthand for "relating to the teeth and jaws." (L *dens* = tooth and Gk *gnathos* = jaw.)

### deoxyribonucleic acid

(or DNA) A nucleic acid made up of bases (the four standard bases are adenine, guanine, cytosine, and thymine) and a backbone made of a phosphate and a sugar. In DNA the 2-deoxyribose sugar alternates with a phosphate to form the backbone. These three components (base, phosphate, sugar) make up a nucleotide, and nucleotides linked by phosphodiester bonds (between adjacent sugars and phosphates) make up a single strand of DNA. The two ends of a single strand of DNA are distinctive; one is called the 5' (or "five prime") end and the other is called the 3' (or "three prime") end; replication and transcription only occur in the 5' to 3' direction. Typically, DNA is present as a double strand arranged in a double helix with the two strands being held together by hydrogen bonds that link a base in one of the DNA strands with a complementary base in the other strand (adenine always pairs with thymine and guanine always pairs with cytosine). In double-stranded DNA the two DNA molecules are antiparallel (i.e., one of the strands is in the 3' to 5' direction, while the other is in the 5' to 3' direction). Some DNA (the minority) contains genetic information in the form of genes and is called coding sequence, but the vast majority of the DNA (e.g., the DNA in the centromeres of chromosomes) is called structural or noncoding DNA. Coding DNA is transcribed into messenger RNA (or mRNA). The RNA is then either translated into a protein or it forms one of many kinds (transfer, ribosomal, or small interfering RNA) of functional RNA molecules. *See also* **transcription**.

## derived
A version or state of a character that is not its primitive character state. For example, if the character is the root system of the first mandibular premolar tooth, the primitive condition for the hominin clade is most likely two roots: a plate-like distal root and a mesiobuccal accessory root, or in shorthand 2R: MB+D. In this example there are two derived trends within the hominin clade. One is toward root reduction and simplification (i.e., a single root); the other is toward a more complex molar-like system of two plate-like roots (i.e., two roots). *See also* **autapomorphy**; **synapomorphy**.

## description
A list of features shared by all or at least a large majority of the members of a taxon. It is sometimes referred to as the definition of a taxon.

## Developed Oldowan
A stone tool tradition defined by Mary Leakey and based on artifact assemblages excavated from Beds II–IV, Olduvai Gorge, Tanzania, that were judged to be advanced relative to the Oldowan and which substantially overlapped the Acheulean industry in time. Leakey argued that the Developed Oldowan and Acheulean likely represented two distinct cultural traditions, perhaps made by *Homo habilis* and *Homo erectus*, respectively, but this interpretation has been called into question on several grounds. The Developed Oldowan is best regarded as a biface-poor variant of the Acheulean.

## development
An increase in functional ability, sometimes used interchangeably with maturity. The process involves both biological and behavioral maturity. In modern humans it is complete by the late teens. Many consider that the end point of development is the age at which successful procreation is possible. The most commonly used markers of development include secondary sexual development, skeletal maturity, and dental maturity. The dissociation between skeletal and dental maturity (e.g., as seen in the early *Homo* specimen KNM-WT 15000) creates problems for researchers wishing to estimate the age at death of individuals from extinct species based on developmental criteria drawn from extant populations.

## dexterity
Precision and/or skill when performing a function using the fingers/hands (e.g., the manufacture of an Aurignacian fine bone needle requires more dexterity than the manufacture of an Oldowan chopper). As the etymology implies, it refers to most people's preference for their right hand when performing such functions. (L. *dexter* = on the right side, or skillful.)

### di

Abbreviation for a deciduous incisor. Upper, or maxillary, deciduous incisors are identified by a superscript number ($di^1$ and $di^2$); lower, or mandibular, deciduous incisors are identified by a subscript number ($di_1$ and $di_2$).

### diachronic

Any change or process (e.g., evolution, radioactive decay) that occurs through time. (Gk *dia* = through and *kronos* = time.)

### diagenesis

The complex physical, chemical, and biological changes that occur when sediments are converted into rocks and when mineralized tissues (bones and teeth) turn into fossils. During the process organic and inorganic components of the biogenic material can be altered and replaced by chemicals from the surrounding burial environment. An understanding of the processes involved, their effects, and the degree of alteration of biogenic signals is critical for the meaningful interpretation of data obtained from archeological/fossil bone and teeth for the reconstruction of diet, habitat, migration, or climate (Gk *dia* = through, across and *genesis* = birth or origin.)

### diagenetic

*See* **diagenesis**.

### diagnosis

A list of the features or characteristics shared by all (or a large majority) of the members of a taxon that enable its members to be distinguished from those of other taxa. A diagnosis is how you tell taxa apart. A definition, on the other hand, concentrates on the morphology that members of a taxon have in common. Thus, a definition is a list of the features that binds members of a taxon together, whereas a diagnosis lists the ways a taxon differs from other taxa. (Gk *dia* = through and *gno* = to come to know.)

### diaphysis

The part of a long bone, usually equivalent to the shaft, that is formed from the primary center of ossification. (Gk *dia* = through and *physis* = growth.) *See also* **ossification**.

### diatomite

A sedimentary rock composed primarily of the siliceous skeletons (frustules) of diatoms (i.e., algae with cell walls made of two interlocking silica "valves" that belong to the class Bacillariophyceae). Diatoms occur in both marine and nonmarine settings and because some types of diatoms have limited ecological ranges they can be used for paleoenvironmental reconstruction. (Gk *diatomos* = to cut in half.)

### Die Kelders

A network of caves on the shore of the Indian Ocean, approximately 120 km/74 miles southeast of Cape Town, South Africa. Optically stimulated luminescence dates from the Middle Stone Age (MSA) sand layers point to an age of *c*.60–70 ka and electron spin resonance spectroscopic dates on teeth spanning the MSA sequence are very similar (i.e., 70 ± 4 ka assuming early uptake of uranium) and suggest the whole MSA sequence accumulated in less than 10 ka. The MSA layers contained 27 hominin specimens including 24 isolated teeth, an edentulous mandibular fragment, and two manual middle phalanges. All are morphologically similar to, but tend to be larger than, those of living Africans. The great majority of the artifacts resemble those in the later MSA levels at Klasies River. The vertebrate faunal remains largely mirror patterns observed at Klasies River and other MSA sites in southern Africa and have helped fuel debates about the hunting abilities of MSA people as well as the impact of hominin population density on the accumulation of the faunal assemblage. (Location 34°34′S, 19°21′E, Western Cape Province, South Africa).

## diencephalon
*See* **forebrain**.

## diet
The food and drink consumed by an animal. The diets of the extant higher primates have proved to be surprisingly complex in that they vary according to the abundance and availability of food items. It is useful to think of possible foods in two categories: preferred foods and fallback foods. The former are the food items animals prefer to eat in times of plenty; the latter are the food items animals eat when their preferred foods are scarce or unavailable. Diets can be broadly characterized as either ecologically specialized (i.e., stenotopic) or ecologically generalized (i.e., eurytopic). Reconstructing the diet of extinct hominins has proved to be a challenging task for in some cases the various lines of evidence result in conflicting inferences about the nature of the food items. (L. *diaeta* = way of life from the Gk *diaita* = to live one's life.) *See also* **diet reconstruction**; **fallback foods**.

## diet reconstruction
The types of foods eaten by extinct organisms can be reconstructed in several ways, but many of these lines of evidence are sensitive to the scale of the reconstruction being attempted. For example, just because tooth morphology is effective at coarse-grained levels of taxonomic and functional dietary discrimination (e.g., the carnassials of a carnivore vs the hypsodont molars of a herbivore) it does not mean tooth morphology will be as effective at discriminating at much finer-grained levels (e.g., among omnivorous extinct hominins). It is also important to consider that diet embraces possible fallback foods as well as preferred foods. The former may only be eaten on the relatively rare occasions when preferred foods are in short supply, but the ability to access them may determine the fate of individuals and thus potentially the fate of species. At the species level, lines of evidence for reconstructing diet include information from archeology, dexterity, locomotion, masticatory morphology, paleoenvironment, and paleohabitat. In other groups of large mammals observations about the food preferences of morphologically analogous extant taxa have proved to be useful sources of evidence for reconstructing the diets of extinct taxa. But some extinct hominin taxa (e.g., *Paranthropus*) have no obvious extant analogue within primates. Observations at the level of the individual (which can be combined to generate the diet of a species) are both physical and chemical. The physical methods include the detection of distinctive phytoliths and starch grains from ingested plant foods embedded in the hardened dental plaque (called calculus or tartar) that accumulates around the base of tooth crowns. Tooth wear can be

investigated at the gross and microscopic levels. Observations about gross dental wear, also called dental macrowear or dental mesowear, focus on the development of wear facets. Dental microwear focuses on the size, number, and orientation of microscopic scratches, pits, etc. on the enamel. Dental macrowear is a measure of the abrasiveness of the diet in the long term, whereas dental microwear indicates whether the food ingested in the days or weeks before death contained hard or abrasive material. Chemical methods of dietary reconstruction applied to the individual are largely based on the isotopic ($^{13}C/^{12}C$, $^{15}N/^{14}N$, $^{18}O/^{16}O$) or trace-element composition (Sr/Ca, Ba/Ca) of hard tissues, which can indicate the preferred plant types ($C_3$ vs $C_4$), the rank on the food chain, level of marine or aquatic food intake, and preferred water source. Reconstructing the diet of extinct hominins has proved to be a challenging task for in some cases the various lines of evidence result in conflicting interpretations of the nature of the food items.

## differential preservation

A term that refers to the fact that some skeletal elements are more commonly preserved in the fossil record than others. The bones and teeth that make up the hominin fossil record are made up of a mixture of organic matrix and inorganic (mineral) material. The skeletal material with the highest proportion of inorganic material usually is better preserved. Teeth are comprised of enamel (approximately 98% mineral by volume) and dentine (approximately 70% mineral by volume), and are therefore the most durable skeletal parts. Skeletal elements that are made of cortical bone (e.g., the cranium and mandible, and the shafts of long bones of the limbs) are also more dense and durable than parts with a high proportion of cancellous bone. Finally, certain parts of the skeleton are preferred by carnivores, including the hands and feet, and the epiphyses (joint areas) of long bones, so these are often missing from the fossil record. *See also* **taphonomy**.

## digging stick

A tool used for excavating soil, acquiring belowground food resources and/or planting seeds. Digging sticks may be made of wood, bark, other plant materials, or bone, and may also have stone components. Digging sticks are widely used by modern humans in traditional hunter–gatherer and agricultural societies. Digging sticks are lacking in Paleolithic archeological assemblages, with the notable exception of probable digging sticks in Members 1–3 at Swartkrans, Sterkfontein Member 5, and at Drimolen in southern Africa. These putative digging tools are made from medium–large mammal long bones, and recent analyses indicate that hominins (likely *Paranthropus robustus* and/or *Homo ergaster*) ground these tools to sharpen the ends. These tools may have been used to excavate underground storage organs, dig into termite mounds, or process fruits. The 1.7–1.0 Ma putative digging sticks, taken in combination with the use of digging tools by extant chimpanzees, raise intriguing questions about the nature of tool use by hominins prior to the advent of the flake- and core-based Oldowan stone tool culture. This has implications for our understanding of early human technological evolution, innovation, cognition, social organization, learning strategies (imitation vs emulation), and cultural transmission.

## digit

The term applies to either fingers (i.e., manual digits) or toes (i.e., pedal digits). In clinical anatomy manual digits are named and pedal digits are numbered. (L. *digitus* = finger.) *See also* **ray**.

## DIK-1-1
A child's skeleton including much of the cranium, a natural endocast, a well-preserved mandible, all the deciduous teeth except the crowns of the left lower incisors, the hyoid, both scapulae and clavicles, much of the vertebral column, sternum, many ribs, a fragment of the right humerus, a manual ray, both knee joints, both patellae, and the distal end of the left lower limb including the left foot. It was found in 2000 and thereafter at Dikika in the Afar Rift, Ethiopia, and its age is just over $c.3.3$ Ma. This remarkably complete skeleton provides exquisite detail about the morphology and life history of *Australopithecus afarensis*. Though the foot and lower limb show clear adaptations to bipedalism, the scapula provides compelling evidence that the locomotor repertoire of *Au. afarensis* included a substantial amount of climbing. *See also* **Australopithecus afarensis**.

## Dikika study area
It straddles the Awash River to the south and east of the Hadar and Gona Paleoanthropological study areas in Ethiopia's Afar Rift System section of the East Africa Rift System. Both the Hadar Formation ($c. > 3.8–2.9$ Ma) and the Busidima Formation ($c.2.7– < 0.16$ Ma) are represented. Hominins found at the site include DIK-1-1, a nearly complete juvenile skeleton. All of the hominin specimens recovered from the area are attributed to *Australopithecus afarensis*. Oldowan, Acheulean, and Middle Stone Age technologies are found in the area, but in 2009 researchers found bones dated to $c.3.39$ Ma that they claim bear stone tool cutmarks and percussion marks made prior to fossilization. Though controversial, these marks are the earliest evidence for stone-tool assisted consumption of animal tissues. (Location 11°05′N, 40°35′E, Ethiopia.)

## dimension
A property that describes how many of the three dimensions of space the variable occupies. Size variables can have a dimension equal to one (linear), two (area), or three (volume). Measures of mass are typically treated as three dimensional because mass is a property of objects in three-dimensional space and an object's mass is generally proportional to its volume. Dimension is important in scaling analyses because the isometric slope is determined by the ratio of the dimensions of the two variables contributing to the analysis. *See also* **scaling**.

## dip
The orientation in space of a geological feature such as a stratum or bed. Dip has two main components, the angle between the feature and the horizontal and the direction recorded as a compass bearing from north in the direction of the maximum angle. Strike is a bearing at 90° to the dip direction. A consideration of both the angle of dip and its direction is important during paleontological fieldwork. Knowing these allows surveying teams to identify the directions in which the strata will be relatively older or younger.

## diploë
The cancellous or spongy bone between the dense inner and outer tables of the bones of the cranial vault that contains venous blood and red bone marrow. (Gk *diploë* = doubling, from the feminine of *diplous* = two-fold or double.) *See also* **cranial vault**.

## diploid
The term used for organisms or cells that have two homologous copies of each chromosome. The term also refers to possessing two complete sets of chromosomes. Thus, the genetic complement of modern humans is 22 pairs of diploid chromosomes (also called autosomes) and one pair of sex chromosomes. The diploid number of chromosomes for modern humans is 46. (Gk *diplous* = double and *eidos* = shape or form.)

## 3′ direction
See **deoxyribonucleic acid**; **downstream**.

## 5′ direction
See **deoxyribonucleic acid**; **upstream**.

## directional selection
See **natural selection**.

## disconformity
See **unconformity**.

## discrete traits
See **nonmetrical traits**.

## discriminant function analysis
See **multivariate analysis**.

## dispersal
A change in the location of a population of organisms brought about by the physical movement of the entire population or a subset of it. Range expansion occurs when the area in which a population is found increases but the increased range still includes the original, ancestral area. For example, toward the end of the Pliocene increasing aridity led to an expansion of grasslands in Africa, and many grazing mammalian species experienced range expansion during this time period. Range shift occurs when the area in which a population is found changes location. Jump dispersal refers to when a population disperses long distances over inhospitable habitat (e.g., desert, ocean). The peopling of Oceania is one of a number of examples of jump dispersal during human evolution. (L. *dis* = apart and *spargere* = to scatter.)

## disposable soma hypothesis
One of several theories put forward to explain the evolution of senescence. The hypothesis suggests that the deterioration of the body with age is an inevitable consequence of an imbalance between the resources devoted to reproduction and those devoted to maintenance. Since the resources available to an organism are finite, if a greater proportion of that finite resource allotment is directed to reproduction then inevitably the proportion of the total resource lot devoted to repair reduces such that bodily functions will decline more rapidly with age than if fewer resources had been devoted to reproduction. (Gk *soma* = body.) See also **life history**.

## dissociation
*See* **heterochrony**.

## distal
In the limbs distal refers to structures that are in the direction of the extremity of that limb (i.e., towards the tips of the fingers or toes). With respect to the teeth, or their components, distal refers to teeth or structures within teeth that are in the direction of the back of the jaw. Thus, the elbow joint is distal to the shoulder joint; the right $M_2$ is distal to the right $P_4$ of the same dentition, and the hypoconulid of an $M_1$ crown is distal to the protoconid on the same tooth. (L. *distare* = to stand apart.)

## distal marginal ridge
A crest of enamel that delineates the distal border of the lingual face of a canine or incisor anterior tooth or the occlusal surface of a postcanine tooth crown.

## diurnal
Adjective that describes animals that are active in the day (i.e., between dawn and dusk). Dependence on daylight may have been a crucial factor in limiting most nonhuman primate ranges to the tropics for in higher latitudes there may be insufficient time for primates to maintain social bonds and fulfill their need to feed, travel, and rest. Robin Dunbar has suggested that language, which he claims is a form of social grooming essential for group bonding, served to reduce the time needed for the maintenance of social relationships. The control of fire, and thus access to light, may have helped diurnal hominins prosper in higher latitudes. (L. *diurnus* = daily.)

## divergence time
The time elapsed since two lineages separated. The genetic coalescent time of two taxa cannot be more recent than their divergence time. (L. *dis* = apart and *vergere* = to bend.)

## Djebel Irhoud
*See* **Jebel Irhoud**.

## Djebel Qafzeh
*See* **Qafzeh**.

## Djurab Desert
*See* **Chad Basin**.

## dm
Acronym for deciduous molar. Upper, or maxillary, deciduous molars are identified by a superscript number ($dm^1$ and $dm^2$); lower, or mandibular, deciduous molars by a subscript number ($dm_1$ and $dm_2$). John Robinson showed that lower deciduous molars, especially the $dm_1$s, are useful for discriminating among early hominin taxa (e.g., the $dm_1$s of *Paranthropus robustus* and *Paranthropus boisei* have especially complex crowns). Deciduous molars are sometimes referred to as deciduous premolars (or dps) because the permanent premolars develop beneath them.

## Dmanisi

An open-air site 55 km/34 miles southwest of Tbilisi, Georgia. Located on a basalt promontory at the confluence of two rivers, Dmanisi is the site of a medieval village beneath which are Bronze Age remains. In the 1980s excavations in Medieval Room XI resulted in the recovery of Plio-Pleistocene fossils, and it is now evident that Plio-Pleistocene sediments cover much of the promontory. The Plio-Pleistocene sediments are divided into two major units: Stratum A (with subunits A1–A4), which conformably overlies the Masavera Basalt, and the more superficial Stratum B (with subunits B1–B5), which is separated from Stratum A by a minor erosional disconformity. All the hominin remains and the associated fauna and artifacts are from secure Stratum B contexts. Argon-argon ages for the underlying Masavera Basalt average $1.848 \pm 0.005$ Ma and the best estimate for the age of the fossiliferous sediments, and therefore for the hominins and the archeology, is between 1.85 and 1.78 Ma. The hominins recovered include skulls (e.g., D2600/D4500, D3444/D3900), crania and calvaria (e.g., D2280), mandibles (e.g., D211), and adult and subadult associated skeletons (e.g., D2724, D4166). There is an ongoing debate about how primitive the Dmanisi remains are, with some arguing they are more primitive than *Homo erectus sensu stricto* and others arguing that they are best interpreted as being relatively small-bodied examples of that taxon. An initial review of the dental remains suggests a mix of presumed primitive features, seen in archaic hominins, and more derived features seen in pre-modern *Homo*. Most of the stone artifacts are classic Mode 1 type; retouched pieces are rare. (Location 41°20′10″N, 44°20′38″E, Georgia.)

## DNA

*See* **deoxyribonucleic acid**.

## DNA hybridization

A method for comparing DNA fragments or whole genomes that can be used to assess genetic distance. Double-stranded nuclear DNA from two samples is heated to separate it into single strands, and it is cooled to allow the single strands to come together again (or anneal). When whole genomes are being compared, the heat that needs to be injected into the system to separate the annealed single strands is used as a proxy for the number of bonds joining the strands. Because DNA strands from the same individual would be linked by the greatest number of bonds this is taken as the standard: the more distant the relationship, the fewer the bonds linking the annealed strands, so less heat is required to separate them. Whole-genome DNA hybridization has been superseded by DNA sequencing, but in the 1980s it confirmed the hypothesis that chimpanzees, bonobos, and modern humans were more closely related to each other than any of them are to gorillas. (L. *hybrida* = something made of at least two different components.)

## DNA microarray

A technology used to detect single nucleotide polymorphisms, sequence a specific region of DNA (*see* **DNA sequencing**), investigate patterns of gene expression, or identify deletions or insertions (*see* **copy number variation**). These arrays are the basis of the rapid high-throughput analysis that allows researchers to sequence DNA more rapidly than in the past.

## DNA sequencing
Methods for determining the sequence of bases present in a DNA molecule. The most common is the Sanger or chain-termination method, which can distinguish DNA strands that differ in size by as little as 1 base pair. (L. *sequi* = to follow.)

## DNH 7
A *c*.2.0–1.5 Ma almost complete adult, presumed female, skull of *Paranthropus robustus* recovered from the site of Drimolen, in South Africa. It confirms the claimed distinctions between *Paranthropus boisei* and *Paranthropus robustus*, and provides support for the hypothesis that specimens such as KNM-ER 407 and 732 represent females of *P. boisei*.

## dolomite
A limestone rock with abundant magnesium-rich carbonate minerals. Dolomite is formed in warm tropical ocean environments, and can also result from the diagenetic alteration of pre-existing limestone. The early hominin cave sites in southern Africa are breccia-filled solution cavities formed within the Precambrian dolomite that makes up much of the high veld of that region.

## domain
Part of a protein that has a specific structure and function. For example, *HOX* genes encode a protein domain known as the homeodomain, which is a highly conserved helix-turn-helix structure of 61 amino acids that can bind to enhancers of other genes and turn them on or off.

## dome
The folding of sedimentary rocks in such a way that it resembles an upside-down bowl. In its simplest symmetrical form an eroded dome results in outcrops that are oldest at the center and younger at the edge. *See also* **Sangiran Dome**.

## dominance
*See* **allele; dominant allele**.

## dominant allele
The dominant allele is the version that determines the phenotype no matter whether there are two copies (i.e., as in a homozygote) or just one copy (i.e., as in a heterozygote) present. Co-dominance can occur when both alleles are expressed in the phenotype, an example of which can be seen in human ABO blood types (e.g., AB blood type).

## downstream
A single strand of DNA has a direction determined by the numbered carbon molecules of the sugars in the DNA backbone. Downstream is in the direction toward the 3' (three prime) end; the 5' (five prime) is the upstream end. Replication and transcription only occur in the downstream (i.e., the 5' to 3') direction.

## dp
*See* **dm**.

## draft sequence
A genome sequence that is incomplete or has not been checked for errors. The first modern human genome sequence, published by the International Human Genome Sequencing Consortium in 2001, was a draft sequence in the sense that it covered only 94% of the genome and only 25% of that sequence was "finished" (i.e., sequenced with at least four-fold coverage and with a quality score of 99% accuracy).

## Drimolen
A complex of breccia-filled caves in Precambrian dolomite 5.5 km/3.4 miles northwest of Sterkfontein, in South Africa. The 2.0–1.5 Ma breccia contains many well-preserved hominins including a complete skull (DNH 7) and a mandible with an almost complete dentition (DNH 8). Nearly all of the hominins from the site have been referred to as *Paranthropus robustus* and the Drimolen sample (more than 80 specimens with an unusual number of juveniles) is the second biggest contributor to the *P. robustus* hypodigm after Swartkrans. A minority of Drimolen hominins, mainly teeth, have been assigned to *Homo* sp. (Location 25°58'08"S, 27°45'21"E, Gauteng Province, South Africa.)

## duplication
A region of DNA present in more than one copy. Copy number variants, microsatellites, and segmental duplications of chromosomes are all examples of duplications. *See also* **gene duplication**.

## dural venous sinuses
Venous channels between the two layers of the dura mater that transmit blood from the veins draining the cerebral hemispheres and from the veins draining the diploë of the bones that form the walls of the cranial cavity. Dural venous sinuses leave grooves on the endocranial surface of the cranium so their presence can be inferred from fossils. The dominant pattern of venous drainage in modern humans is superior sagittal sinus → right transverse sinus → right sigmoid sinus → right jugular bulb → right internal jugular vein. However, in *Australopithecus afarensis* and *Paranthropus boisei*, and in about 3% of modern human crania, the main venous drainage route is superior sagittal sinus → occipital sinus → marginal sinuses → jugular bulbs → internal jugular veins.

## durophagy
Examples of durophagy or "hard-object feeding" are seen in most major vertebrate groups. The diets of some early hominin taxa have been interpreted as being durophagous on the basis of morphological features that are considered adaptations to hard-object feeding (e.g., thick tooth enamel, chewing teeth with low, blunt cusps, and anteriorly positioned jaw musculature) and more recently it has been suggested that tooth chipping in some hominin taxa is also indicative of hard-object feeding. However, there is considerable debate about the relative importance of durophagy in hominins. Some researchers who support the hypothesis also suggest that any hard foods consumed were more likely to have been fallback foods and not dietary staples. (ME *dure*, from L. *dura* = hard, plus Gk *phagein* = to eat.)

# E

**ear**
*See* **external ear**; **inner ear**; **middle ear**.

**early African *Homo erectus***
An informal term used by some authors for a subset of *Homo erectus sensu lato* that others refer to as *Homo ergaster* Groves and Mazák, 1975. *See also* ***Homo ergaster***.

**Early Stone Age**
(or ESA) The oldest of the African archeological lithic stages formalized in 1929 by A. J. H. Goodwin and C. van Riet Lowe in *The Stone Age Cultures of South Africa*. Although it was developed for sequences in southern Africa the terminology has subsequently been applied to sites across sub-Saharan Africa. As used today, the ESA incorporates lithic evidence from Oldowan and Acheulean sites. The former consists largely of stone flakes and cores made using direct freehand percussion, whereas at Acheulean sites handaxes and cleavers are added to the mix. At some Acheulean sites there is evidence that the artifacts may have been shaped by hammers or billets made of organic (e.g., wood, bone, or antler) materials. Recognizing that first and last appearance dates vary locally and are subject to change as new sites are found, existing dating methods are improved, new methods are introduced, etc., the present evidence suggests the ESA began *c*.2.6 Ma and ended *c*.200 ka.

**East African Rift System**
(or EARS) The East African Rift System comprises two major branches, eastern and western. The eastern branch begins in the north in the Afar Depression, a triple junction where the EARS meets the Red Sea and Gulf of Aden oceanic rifts. It then becomes the Main Ethiopian Rift System and continues southwards into the Gregory Rift that bisects Kenya. The eastern branch splays out and terminates in northern Tanzania. The eastern branch, which is characterized by abundant volcanism both within and on its flanks, is partly occupied by numerous shallow saline lakes (e.g., Lakes Baringo and Turkana). The western branch curves from Lake Albert through large mainly freshwater lakes (e.g., Lakes Tanganyika and Malawi) to the coast in Mozambique. In this branch the lakes are deeper and the volcanism is more localized. Fossil

---

*Wiley Blackwell Student Dictionary of Human Evolution*, First Edition. Edited by Bernard Wood.
© 2015 John Wiley & Sons, Ltd. Published 2015 by John Wiley & Sons, Ltd.

and archeological sites occur along its length but they are in generally thinner and more localized sedimentary accumulations than those in the eastern branch. Most of the fossil hominin sites in East Africa occur in, or close to, the eastern branch.

## East Lake Turkana
*See* **Koobi Fora**.

## East Rudolf
The name previously used for the site now referred to as Koobi Fora or East Turkana. It was given this name because the site is on the east side of the lake that used to be known as Lake Rudolf. When the name of the lake was changed to Lake Turkana, the site name was also changed.

## East Turkana
*See* **Koobi Fora**.

## eccentricity
One of the three rhythms, or cycles, affecting the Earth's orbital geometry (i.e., the manner in which the Earth orbits the sun). Eccentricity is how much the shape of an orbit deviates from a perfect circle. The more oval the orbit is, the greater the difference between the seasons. The c.100 ka-long eccentricity cycle, which has been the dominant cyclical signal affecting the Earth's global climate for close to a million years, has a strong influence on insolation at high latitudes. (L. *eccentricus*=not having the same center.) *See also* **astronomical time scale; obliquity; precession**.

## eclectic feeder
One of several terms (others include ecological generalist, eurytope, and omnivore) used for animals that eat a diverse array of foodstuffs. Eclectic feeders might include an exceptionally wide range of plant foods in their diets (flowers, fruits, young leaves, seeds) or they might mix plant and animal foods (for example, supplementing a largely fruit-based diet with insects, small vertebrates, or eggs). Within primates, baboons are the classic eclectic feeders because they ingest a broad range of foods from cultivated crops to flamingos and juvenile antelopes. Eclectic feeding was among the adaptations that enabled Old World monkeys to inhabit a wide range of habitats, respond quickly to environmental changes (either climatic or anthropogenic) and exploit a wide geographic area. Most early hominins were likely eclectic feeders. *See also* **omnivory**.

## ecofact
A naturally occurring object at an archeological site that has not been handled or altered by hominins. Examples include seeds, shells, or bones that are introduced to site through natural methods, like wind or animal action. Although ecofacts are not directly relevant to understanding hominin behavior they can provide important information concerning the environmental context of an archeological site (e.g., pollen found in a deposit helps reconstruct past vegetation). (Gk *oikos*=house and L. *facere*=to do.)

## ecology
The term, which was first used by Ernst Haeckel in 1866, refers to all of the interactions between an organism and its environment. They include the relationships between an organism and its physical, or abiotic environment, as well as the relationships between an organism and the other organisms it comes into contact with (i.e., its biotic environment). Three subdisciplines of ecology are particularly relevant to hominin evolution: paleoecology, behavioral ecology, and human ecology. (Gk *oikos* = house and *logia* = study.)

## economy
See **efficiency**.

## ecosystem
The combination of a biotic community and its abiotic environment. It consists of dynamic interactions between plants, animals, micro-organisms, and their environment working together as a functional unit. An ecosystem can be as large as the Sahara Desert or as small as a puddle. Larger ecosystems typically comprise many different habitats. *See also* **habitat**.

## ecotone
The transitional zone between two adjacent plant communities (e.g., open grassland and forest). Ecotones can be relatively sharp transitions between adjacent habitats or a more gradual blending of habitats across a broader area. Plant and animal species found in adjacent communities may inhabit the ecotone, but some organisms may be unique to it. For example, in Africa the impala (*Aepyceros melampus*) favors the ecotone between open grassland and savanna woodland. Early hominins may have occupied the same ecotone. [*eco*(logy) plus *tone*, from Gk *tonos* = tension; i.e., "an ecology in tension."]

## ectocranial morphology
Morphological features (e.g., lines, crests, or grooves) on the outside of the cranium. They either mark where connective tissue (either the covering of a muscle, the epimysium, or the septa within a muscle, the perimysium) attaches, or they correspond to the course of a vessel. (Gk *ectos* = outside and *kranion* = brain case.)

## edaphic
Refers to anything related to soils. (Gk *edaphos* = ground or soil.)

## EDJ
Acronym for the **enamel–dentine junction** (*which see*). (syn. dento-enamel junction.)

## EDMA
Acronym for **Euclidean distance matrix analysis** (*which see*).

## effect hypothesis
A hypothesis that provides a mechanism to explain morphological trends in the fossil record. It suggests that eurytopic, or generalist, species are able to survive under a wide variety of environmental conditions, whereas stenotopic, or specialist, species can only survive under a

narrow set of environmental conditions. Thus, as environmental conditions change, stenotopic species are expected to experience higher rates of speciation and extinction than eurytopic species because the former are less well equipped to tolerate the new environment. As a result, if a clade contains both stenotopes and eurytopes the expectation is that there will be more stenotopic than eurytopic species. The preponderance of stenotopic species gives the appearance that natural selection has favored the evolution of specialized morphological characteristics. Instead, the observed species pattern (i.e., more stenotopes than eurytopes) is not the result of selection at the microevolutionary level but is the result of a macroevolutionary "effect."

## effective population size

The number of breeding individuals contributing genes to the next generation. Effective population size (or $N_e$) influences the amount of genetic drift, or change in allele frequencies due to random chance, experienced by a population. Genetic drift is weaker in large populations and stronger in small ones. Effective population size can be affected by age structure, overlapping generations, unequal numbers of males and females, skewed family size distributions, fluctuations in population size, and the geographical dispersion of populations, but all of these issues can be corrected for statistically. In modern humans, $N_e$ has been estimated at approximately 10,000. This is because, despite our current estimated population size of over 7 billion people, the effective population size of modern humans reflects the much smaller population sizes of the past, which had a narrower scope of selection for genetic types or alleles.

## efficiency

The ratio of energetic output (e.g., work done) to energetic input (e.g., energy expended). Thus a large car can be more efficient than a small car, as long as the amount of work done per unit of energy expended is greater. It is useful to make a distinction between efficiency and economy. Economy refers to the work done relative to the cost of resources required; economy cars can be less efficient than larger automobiles if they perform less work per unit energy, but they are more economical if they travel the same distance while spending less on fuel (e.g., using regular unleaded as opposed to premium gasoline). In general, larger animal species are less economical than smaller species in that they spend more to acquire the energy to power their larger bodies. However, they are generally more efficient because they use less energy per gram of their body mass.

## efficiency hypothesis

One of several hypotheses to explain the origin of upright posture and bipedal locomotion within the hominin clade. Chimpanzee quadrupedalism and bipedalism are equally energetically efficient, but modern human bipedalism is more energetically efficient than chimpanzee quadrupedalism. Thus, if the earliest hominins were adapting to increased fragmentation of forest patches at the end of the Miocene, bipedalism would have been favored over quadrupedalism because it would have allowed the earliest hominins to travel more efficiently between forest patches. A more recent version of the efficiency hypothesis suggests that hominin bipedality emerged because it was less costly energetically than knuckle-walking.

## Ehringsdorf

The Ehringsdorf fossils, which were discovered as a result of blasting in the Fischer and Kämpfe quarries in Germany, all come from the 15 mm-thick Lower Travertine that dates to c.230 ka. At least nine hominin individuals (attributed by most to *Homo neanderthalensis*) are represented by 35 fragments. They include cranial fragments, two mandibles, and postcranial remains. Mousterian artifacts have been found at the site and several hearths have been identified in the lower and middle parts of the Lower Travertine. (Location 50°58′00″N, 11°21′00″E, Germany.)

## Elandsfontein

A former dune field near South Africa's Atlantic coast where wind deflation has exposed Pleistocene–Holocene artifacts and fossils in numerous interdunal "bays." The mammalian fauna and Acheulean artifacts date mainly from the Middle Pleistocene, but later Pleistocene and Holocene artifacts and fauna (e.g., Middle Stone Age lithics and Khoekhoe pottery) have also been found. Biochronology suggests an age of between 1.0 and 0.6 Ma for the Acheulean artifacts. The Saldanha 1 calvaria and a mandibular fragment are the only hominins known from the site. (Location 33°05′S, 18°15′E, South Africa.)

## electromyography

(or EMG) A technique that measures electric activity within a muscle and then uses that as a proxy for the work performed by that muscle. For example, EMG data show that the role of gluteus medius during bipedality (i.e., to provide side-to-side balance of the trunk at the hip) is the same in apes as in modern humans, except that in modern humans it does this with the thigh extended instead of flexed as in chimpanzees.

## electron spin resonance spectroscopy dating

(or ESR) A radiation-based dating method that measures the trapped electrons resulting from radiation damage caused by the uptake of uranium and other radioactive elements into crystalline materials such as calcite (e.g., shells and coral) or apatite (e.g., bones and teeth), often at the time of their burial. The same provisos as for luminescence dating apply to ESR dating (i.e., the resetting of the "electron clock" must be unambiguously linked with hominin activity, radiation dose rate must be established, etc.). One suitable crystalline material which can be dated by this method and also directly linked with hominin activity is the tooth enamel of animals used by hominins for food (e.g., teeth from an animal skeleton that shows clear evidence of butchery) or as a source of raw materials (e.g., teeth that have been modified by hominins to form tools). The technique can provide dates over a time range of a few thousand years to more than 1 Ma. An important variable that affects the ESR method is the timing of the uranium uptake by the sample. The "early uptake model" assumes that uranium reaches its present levels soon after burial, whereas the "linear model" assumes that the rate of uranium accumulation is linear. However, neither model may accurately account for the complex history of uranium accumulation or loss in particular specimens. *See also* **luminescence dating**.

## Elephantidae

The family of extant proboscideans that includes *Loxodonta*, the African elephant, and *Elephas*, the Asian elephant.

## El Kherba

An Oldowan archeological site in Algeria that some consider a sublocality of the nearby Ain Hanech site. El Kherba was discovered and first excavated by Mohamed Sahnouni and Jean de Heinzelin in 1992–3. If it does date to the Olduvai subchron (1.95–1.78 Ma) as the excavators argue, it would be the earliest primary-context evidence for hominin occupation in North Africa. (Location 36°25′12″N, 05°49′12″E, near Sétif, Algeria.)

## El Niño

Anomalously warm waters in the eastern equatorial Pacific. This phenomenon is part of the **El Niño Southern Oscillation** (*which see*).

## El Niño Southern Oscillation

(or ENSO) A quasi-periodic (2–7 year) pattern of natural climate variability in the tropical Pacific that has far-reaching consequences around the globe. ENSO is preferred over the terms El Niño (ENSO positive or warm phase) and La Niña (ENSO negative or cold phase) because ENSO includes reference to both the ocean surface temperatures and the changes in atmospheric pressures that describe the full extent to which this ocean/atmosphere oscillation can alter climate around the globe. In Africa, ENSO influences the amount and seasonality of precipitation with regional characteristics. ENSO events and related changes in the Indian Ocean most strongly influence East and southern Africa with strengthened upper westerly winds that lead to decreased rainfall in southern Africa (December–March) and increased rainfall in East Africa (October–December). Sustained El Niño- or La Niña-like conditions may also explain global climate patterns during the warm, wet mid-Pliocene, and they may be a possible trigger mechanism for abrupt climate events such as the Younger Dryas.

## El Sidrón

The first hominin fossils from El Sidrón, Spain, were accidentally unearthed in 1994 and systematic excavations have been ongoing since 2000. The more than 1900 hominin fossils, which dominate the vertebrate fauna, were found in the Osario Gallery, one of several galleries that run at right angles to the long axis of the cave system. All of the hominins were recovered from the exposed surface of "stratum III" that dates to *c*.49–39 ka and most have been recovered during formal excavations (they have the prefix DR). All parts of the skeleton are preserved. The crania are mostly fragmentary with a bias towards young individuals. There is evidence of at least 11 individuals including one infant, one juvenile, two adolescents, and four young adults. There is ample morphological evidence (e.g., supraorbital torus, suprainiac fossa, small mastoid, shovel-shaped, labially convex incisor crowns with strong lingual tubercles, distinctive premolar occlusal morphology, taurodont molars, large joint surfaces, etc.) that the collection samples *Homo neanderthalensis*. Mitochondrial and nuclear DNA have been recovered from several specimens (two specimens that could belong to the same individual) and the extracted DNA provides evidence that evolutionary changes in the *FOXP2* gene seen in modern humans are also seen in Neanderthals. Stone artifacts include side scrapers, denticulates, a handaxe, and several Levallois points. There is copious evidence of hominin-induced bone modification, which has been interpreted as cannibalism. (Location 43°23′01″N, 05°19′44″W, Spain.)

## EMG
Abbreviation of **electromyography** (*which see*).

## emissary foramina
Holes in the cranium (e.g., the hypoglossal, mastoid, occipital, parietal, and posterior condylar foramina) that transmit emissary veins connecting the intracranial and extracranial venous systems (e.g., the veins of the scalp or the vertebral venous plexus). From posterior to anterior, the major emissary foramina of the cranial base are the occipital (occipital foramen), mastoid (mastoid foramen), condylar (condylar canal), hypoglossal (jugular foramen), and sphenoidal (emissary sphenoidal foramen). (L. *emissarium* = a drain and *foro* = to pierce, the root of *foramen* = a hole, or opening.)

## enamel
The hard, white, outer coating of the crown of a tooth. Enamel is the hardest of the three dental hard tissues (dentine and cementum are the others) and this accounts for the relative abundance of teeth in fossil assemblages. Most mature enamel (approximately 96% by weight) is an inorganic mineral called hydroxyapatite, sometimes abbreviated to apatite. The cells responsible for the manufacture of enamel, ameloblasts, secrete a matrix of proteins, water, and mineral ions. The organic phase of mature enamel (approximately 4% by weight) consists of proteins (e.g., amelogenins) that function as a "glue" to bind the crystallites together. They provide a framework or scaffold along which crystallite growth is guided. In mature aprismatic enamel, the crystallites are bound together by remnants of the organic matrix. Enamel crystallites become associated into bundles called enamel prisms, or enamel rods. The appearance of enamel prisms viewed end-on has been classified into pattern 1, pattern 2, and pattern 3 depending on their outline, area, and packing arrangement. Hominids mostly have the keyhole-shaped hexagonally packed pattern 3 enamel, but with areas of round hexagonally packed pattern 1 enamel. Short-period incremental lines in enamel form daily or even occasionally every 8–12 hours or so. These daily cross-striations run across the prisms parallel to the face of the secretory end of the ameloblast (or Tomes' process) that formed them. Long-period markings (aka striae of Retzius) have a modal value of 7 days in australopiths and 8 days in fossil *Homo*. The presence of incremental lines in enamel allows researchers to determine crown formation time and in some cases the age at death. (OF *esmail* = a vitreous coating, often white, baked on metal, glass, etc.)

## enamel decussation
See **decussation**; **enamel development**.

## enamel–dentine junction
(or EDJ) The boundary between the outer surface of the dentine core of a tooth crown and the inner surface of the enamel cap that covers it. Although the EDJ looks smooth at the gross level, under a microscope the dentine surface is scalloped to help key the enamel into the dentine. The shape of the EDJ, which is determined by the contours of the inner enamel epithelium, is generally replicated at the outer enamel surface, except where differential rates of enamel formation modify the EDJ template. Within a single species, EDJ shape (and enamel

thickness) differs among molar types, between the sexes, and among regional populations. (syn. dento-enamel junction.)

### enamel development
Enamel is formed by ameloblasts. At the apex of each developing dentine horn, the inner enamel epithelium differentiates into ameloblasts, which later become secretory ameloblasts. The rate at which differentiated ameloblasts become active secretory cells is known as the enamel extension rate. The secretory end of each cell faces inwards towards what will become the enamel–dentine junction (or EDJ). As each ameloblast moves outwards away from the EDJ it secretes matrix consisting of proteins, water, and mineral ions in its wake. Once the full thickness of enamel matrix is secreted, secretory ameloblasts become maturational ameloblasts. They alternate between pumping mineral (mostly $Ca^{2+}$ and $PO_4^{3-}$ ions) into the matrix and drawing water and degraded proteins out of it. The latter enables the crystallites to expand in diameter and pack more tightly together. Enamel maturation can take years, but by the time a tooth erupts the enamel is hard and mostly mature. Rhythmic slowing of enamel matrix secretion is preserved in the form of short (cross-striations) and long-period (striae of Retzius) incremental lines.

### enamel extension rate
See **enamel development**.

### enamel formation
See **enamel development**.

### enamel microstructure
The microstructural features of enamel reflect the cellular activity of the ameloblasts. Enamel is comprised of enamel prisms that are marked by various types of incremental lines (e.g., striae of Retzius, cross-striations, laminations, and intradian lines). Enamel prisms weave across each other and it is this "decussation" that causes Hunter–Schreger bands. Counts and measurements of the incremental features of enamel microstructure can be used to estimate daily secretion rates and, in well-preserved teeth, crown formation time.

### enamel prism
See **enamel**.

### enamel thickness
Enamel thickness is typically quantified by one-, two-, or three-dimensional measures (i.e., lengths, cross-sectional areas, or volumes). These can be absolute measures, or relative measures that scale enamel thickness to some proxy for overall tooth size. Micro-computed tomography (or micro-CT) can be used to calculate indices of relative enamel thickness in three dimensions. Linear measurements of enamel thickness have been used to support taxonomic distinctions among hominin taxa. Average enamel thickness in modern human molars has been shown to be greater in upper than in lower molars and it also increases along the tooth row from incisors through molars. Enamel thickness in hominins is known to vary from the very thick enamel of

*Paranthropus boisei*, to the thick enamel of *Australopithecus* and *Homo*, to the reportedly thin enamel seen in *Ardipithecus kadabba*, but the use of different measurement techniques presently precludes meaningful direct comparisons.

## encephalization

A relative measure of a species' brain size that tries to capture the degree to which it is larger or smaller than expected for a typical animal of the same body size. Formulae for calculating encephalization differ according to the choice of line-fitting technique and what taxa are included in the reference sample. In primates, encephalization has been linked to a number of factors including intelligence, diet, innovation rate, the degree to which the animals use binocular vision, and the total amount of visual input to the brain. (L. *cephalon* = brain from the Gk *enkephalos* = literally "marrow in the head.")

## encephalization quotient

(or EQ) The encephalization quotient, which was introduced by Harry Jerison in his 1973 book, *Evolution of the Brain and Intelligence*, was devised to account for the interspecific allometric effects of body size on brain size so that the degree of encephalization can be meaningfully compared across a range of different species. In practice, the EQ of a species is determined by calculating the ratio of its observed brain size over its "expected" brain size. Thus, a species with an EQ that is greater than 1.0 has a brain that is larger than expected for its body size and an EQ that is less than 1.0 indicates that the species has a brain that is smaller than expected. Although the exact EQ depends on the composition of the reference sample used in the analysis, without exception modern humans have been found to have the highest EQ in comparative studies of mammals and primates. Attempts to link variation in EQ to general cognitive capacities across diverse mammalian taxa have been met with mixed results. (L. *cephalon* = brain from the Gk *enkephalos* = literally "marrow in the head," and L. *quotiens* = how many times.)

## endemism

In an ecological context, endemism refers to the state of being unique to a particular geographic location. Typically, endemic types or species are more likely to develop when there is effective geographic isolation (e.g., islands). For example, lemurs (Lemuriformes) are endemic to the island of Madagascar and 50% of the primate species of Indonesia are endemic. By default larger tropical regions have more endemic species simply because they have more species and greater habitat diversity (e.g., Brazil has more endemic species than Madagascar, but it covers an area around 15 times as large). Among fossil hominin taxa the best candidate for an endemic species is *Homo floresiensis*. (Gk *endemia*, *en* = in, *demos* = people.)

## endocast

Abbreviation of **endocranial cast** (*which see*).

## endochondral ossification

Bone formation within a framework of cartilage. Endochondral ossification is a two-staged process. First, cells called chondroblasts form a hyaline cartilage model of the future bone. Cells in the middle of the cartilage model then swell and die, leaving only a mineralized shell of cartilage matrix. This mineralized cartilage is resorbed by osteoclasts at the same time as cells called osteoblasts are forming osteoid matrix on the walls of the spaces once occupied by chondroblasts. Long bones continue to grow in length because new cartilage is laid down in epiphyseal plates between the primary center of endochondral ossification in the shaft of the bone and the secondary centers in the epiphyses at either end. (Gk *endo* = within and *chondros* = cartilage.)

### endocranial
Refers to any structure within the cranial cavity (e.g., the dural venous sinuses are referred to as the endocranial venous sinuses). (Gk *endo* = within and *kranion* = brain case.) *See also* **endocranium**.

### endocranium
The cavity formed by the walls, roof, and floor of the cranium.

### endocranial capacity
*See* **endocranial volume**.

### endocranial cast
A natural or prepared cast that uses the walls, roof, and floor of the cranial cavity as a mold. The best-known natural hominin endocasts are those recovered from the southern African cave sites (e.g., Taung). Ralph Holloway developed the method that is widely used to make artificial endocranial casts. It is also possible to use the data collected from a computed tomography (or CT) scan to make a virtual endocranial cast that can then be used to manufacture a solid rendering of the cast. Only rarely are fossil crania well enough preserved to enable an accurate endocranial cast to be made and in most cases guesswork is involved when reconstructing the parts of an endocast that are missing. Thus most estimates of the average endocranial volume of a fossil hominin taxon contain a mixture of reliable and less reliable measures of endocranial volume depending on the degree to which the walls of the endocranial cavity have had to be reconstructed for individual specimens.

### endocranial cavity
Refers to the cavity formed by the walls, roof, and floor of the cranium.

### endocranial fossae
The endocranial surface of the cranial base is conventionally divided into three paired endocranial fossae. The two anterior ones accommodate the frontal lobes of the cerebral hemispheres, the middle ones the temporal lobes, and the two posterior fossae house the cerebellar hemispheres.

### endocranial morphology
Endocranial morphology comprises whatever features of the external surface of the dura mater have been preserved in the form of negative impressions on the endocranial surface of the bones of the cranial vault. These preserved features include negative impressions of sulci, gyri, and the meningeal vessels.

## endocranial volume

The volume of the brain plus the volumes of the other structures [e.g., meninges, extraventricular cerebrospinal fluid, intracranial (but extracerebral) blood vessels, cranial nerves] within the endocranial cavity. The non-central nervous system components (i.e., meninges, etc.) generally amount to approximately 15% of endocranial volume. Earl Count suggested that, for *Homo*, endocranial volume = brain mass × 1.14. The volume of the endocranial cavity used to be measured by stopping up the foramina and fissures of the cranium with cotton wool, inverting it, pouring bird seed or another type of granule into the foramen magnum and then decanting the seed or granules into a measure. Endocranial volume can also be measured by dipping a plaster or latex endocranial cast into water and measuring the volume of water displaced. Specialized software can be used to convert data obtained from computed tomography (or CT) scans into a measure of endocranial volume. In fossils, endocranial volume is often used as a proxy for brain size. (syn. endocranial capacity.)

## endurance running hypothesis

A hypothesis suggesting that modern humans' particular ability to run for long periods of time has shaped the evolution of modern human anatomy and physiology. Acclimatized modern humans can run down prey (e.g., antelope, kangaroo) by persistent pursuit until the animal collapses from exhaustion. David Carrier pointed out the "energetic paradox" that, although modern humans have a high energetic cost (ml $O_2$/kg × km) of transport (i.e., movement over distance) while running, they are able to sustain their running over much longer distances than mammals of the same body size. Energetic inefficiency and the endurance running hypothesis would seem incompatible, but the endurance running hypothesis argues that modern humans can sustain long-distance running because they have evolved unique physiological mechanisms to (a) dissipate metabolic heat and (b) store and use large reserves of energy. Dennis Bramble and Daniel Lieberman later argued that endurance running was the main driver of the musculoskeletal differences between *Australopithecus* and early African *Homo erectus*. David Raichlen has also suggested that "runner's high," a neurologically mediated feeling of euphoria experienced by some while running, may be another aspect of the modern human physiological adaptation for sustained running over long distances.

## Engis

The location of Engis 2, the earliest-discovered specimen of *Homo neanderthalensis*, which was found in 1827. (Location 50°34'N, 05°25'E, Belgium.)

## entoconid

The more distal (hence the prefix "ento-," meaning "within") of the two main lingual cusps of a mandibular (lower) (hence the suffix "-id") molar tooth crown. It forms part of the talonid. (Gk *ento* = within and *konos* = pine cone.)

## entoconulid

An accessory cusp situated mesial to the entoconid on the occlusal surface of a mandibular molar.

## environment

The sum of the biotic and abiotic conditions with which an organism interacts. The term environment is used to refer to processes at different geographical scales (e.g., local, regional, and global). Biotic and abiotic conditions often interact so the concept of environment is necessarily a complex one. For example, local vegetation (biotic) is affected by the amount of groundwater, precipitation, and soil conditions (abiotic). (OF *environ* = round about and L. *mentum* = verb-to-noun suffix.)

## *Eoanthropus dawsoni* Dawson and Woodward, 1913

After the recovery of purported hominin fossils from Piltdown, Charles Dawson and Sir Arthur Smith Woodward proposed in 1913 "that the Piltdown specimen be regarded as the type of a new genus of the family Hominidae to be named *Eoanthropus* and defined by its ape-like mandibular symphysis, parallel molar-premolar series, and narrow lower molars which do not decrease in size backwards; to which diagnostic characters may probably be added the steep frontal eminence and slight development of the brow ridges." When the Piltdown fossil was subsequently proved a fraud, the taxon name no longer had any relevance. (Gk *eos* = dawn and *anthropos* = human being, and *dawsoni* = to acknowledge the role played by Charles Dawson in the recovery of the remains of Piltdown I.)

## epigenesis

The generation of form through a series of causal interactions between genes and extragenetic, or epigenetic, factors. These epigenetic effects include heritable information (e.g., patterns of gene expression) not encoded in the organism's DNA. (Gk *epi* = around and *genesis* = birth or origin; literally the "influences around development.")

## epigenetic landscape

A visual metaphor used by Conrad Waddington to illustrate his concepts of canalization and epigenetics. In the metaphor, the development of an individual organism is likened to a ball rolling down a grooved landscape. At one level the landscape represents the developmental system, at another it represents the probability of different phenotypic outcomes. Genes determine the shapes of the valleys and ridges. According to this metaphor mutations can shift both the mean (the locations of valleys) and also the variance (the steepness of the sides of the valleys) of the phenotypic outcomes produced by a particular genotype or developmental configuration.

## epigenetics

Conrad Waddington's original definition of epigenetics is that it includes all of the factors, intrinsic or external/environmental, that act on and influence the genetic program for the development of an organism. Under this definition intrinsic influences include gene–gene interactions and hormones; external influences include diet and temperature. In modern genetics, epigenetics refers to information (e.g., patterns of gene expression) that is heritable, but which is not encoded in the DNA of the organism. Examples of epigenesis include the propensity of offspring to inherit levels of energy extraction from food in the gut. If a previous generation was relatively short of food, their gut biochemistry is programmed to extract more energy. If this propensity is passed on to a successive generation that has ready access to processed, calorie-rich food, and if they continue to extract the maximum amount of energy from the food they ingest, then this can result in high levels of obesity within that generation. (Gk *epi* = around and *genesis* = birth or origin; literally the "influences around development.") See also **nonmetrical traits**.

## epigenetic traits
See **epigenetics**; **nonmetrical traits**.

## epimerization
A change from one of two possible forms of a chiral (asymmetrical) molecule to the other; it usually refers to the switch from the L, left-handed or levo, form to the D, right-handed or dextra, form. For example, in ostrich egg shell dating one quantifies the ratio of L-isoleucine which has been naturally converted over time to D-alloisoleucine. Because that reaction is temperature-dependent, if average temperature can be estimated independently of sample age, the method can provide ages for material that is difficult to date using other methods.

## epiphyseal plate
The radiotranslucent cartilaginous plates between the diaphysis (formed from a primary ossification center) of a long bone and the epiphyses that form from the secondary ossification centers. (Gk *epi* = around and *physis* = growth.)

## epiphysis
The part of a bone that develops from a secondary ossification center. Epiphyses are found at the ends of long bones, but many bony processes (e.g., acromion process of the scapula, transverse process of a vertebra and the tibial tuberosity) are epiphyses. (Gk *epi* = around and *physis* = growth.)

## epistasis
Interactions among genes that influence a complex trait. The phenotypic expression of a gene can be affected (e.g., masked) by one or more modifier genes. For example, the ABO locus on chromosome 9 controls the production of the blood group antigens that determine blood type. A separate locus on chromosome 19 controls the production of the H antigen, which is an intermediary in the production of A and B antigens. In other words depending on the genetic code at the ABO locus the H antigen is converted to an A or B antigen; in individuals with O blood type it is unaltered. If an individual is homozygous recessive at the H locus, they will not produce H antigen, and thus will appear phenotypically O regardless of their ABO genotype. In this case, the H locus masks the ABO locus. Epistasis also describes the phenomenon whereby the effect of a mutation on the phenotype depends on its context (i.e., where in the genome it occurs). In classical Mendelian genetics, epistasis referred to nonallelic dominance relationships. Epistasis increases the complexity of adaptive, or fitness, landscapes and thus may have been involved in the evolution of complex morphological and behavioral traits during the course of human evolution. (Gk *ephistanai* = to place upon, from *epi* = around and *histanai* = to place.)

## epistemology

The study of the nature, validity, and scope of knowledge. Discussions of what types of knowledge can, and cannot, be deduced from the hominin fossil record are epistemological discussions. Colloquially, epistemology has been defined as "how we know what we know or what we think we know." (Gk *episte* = to understand and *logos* = to learn.)

## epoch

Epoch is used in at least two senses that are relevant to paleoanthropology. In the first of the two senses, it can be used geochronologically or chronostratigraphically. In geochronology, an epoch is a subdivision of a period and it is itself subdivided into ages. For example, the Pliocene, Pleistocene, and Holocene are all epochs within the Tertiary period. The chronostratigraphic equivalent of an epoch is a series. In the second of the two senses, an epoch is a prolonged period in the past when either a normal or a reversed magnetic field predominated. The latter use has been superceded by the term "polarity chronozone," which is usually shortened to "chron." (Gk *epokhe* = a point in time.)

## equid

The informal name for the perissodactyl family that comprises the horses and their allies.

## Equidae

The family of perissodactyls that comprises the horses and their allies.

## equifinality

The principle that different processes can produce similar, if not identical, patterns or results. The term was coined in 1949 by Ludwig von Bertalanffy in the context of general systems theory. The concept became widely applied in the zooarcheological and taphonomic literature in the 1980s, often in cases where it is unclear which of several taphonomic scenarios created the patterns observed in a fossil assemblage (e.g., the dominance of limb elements at FLK-*Zinjanthropus* and FLKN levels 1–2 at Olduvai Gorge could be the result of hominin activities or carnivore ravaging). The principle of equifinality also applies to adaptive explanations for distinctive morphology (e.g., is the large chest cavity of *Homo neanderthalensis* an adaptation to the cold, or to high activity levels, to both, or to something else?) (L. *aequi* = equal and *finis* = end.)

## Equus Cave

A cave in the western face of the Oxland tufa fan on the Gaap escarpment at Buxton, Northern Cape Province, South Africa, very close to the site of Taung 1. The site contains abundant vertebrate fossils in four stratigraphic units designated 1A, 1B, 2A, and 2B. 1A contains Later Stone Age artifacts; the others contain Middle Stone Age (or MSA) tools. Eight isolated hominin teeth and a left mandibular corpus with $M_2$–$M_3$ may derive from the MSA levels. (Location approximately 27°37′S, 24°38′E, Northern Cape Province, South Africa.)

## era

A unit of geological time (i.e., a geochronologic unit). An era is next in order of magnitude below an eon (e.g., the Phanerozoic eon is divided into three eras – the Paleozoic, Mesozoic, and Cenozoic – each marked by major faunal turnovers). Eras, which usually span hundreds of millions of years, are subdivided into periods. The chronostratigraphic unit equivalent of an era is an erathem. (L. *aera* = counters used for calculations.)

### erathem
The chronostratigraphic unit equivalent of an era; erathems are subdivided into systems.

### error
*See* **Type I error**; **Type II error**.

### eruption
The process of tooth movement through the alveolar bone, past the gumline (gingival emergence) and into its functional position. The assessment of the stage of eruption is typically based on either hard or soft tissue evidence. The hard tissue evidence measures the position of a tooth from a relatively stable landmark (e.g., the bony alveolar margin) usually from radiographs. The extrinsic stain on the teeth of many wild collected primates that forms when they erupt into the oral cavity is a way to judge how much of a tooth was subgingival and how much was beyond gingival emergence at the time of death. Studies of modern humans show there is only about 10–12 months between alveolar and gingival eruption of a lower first molar and that the gingival eruption of molar teeth is very close in time to their coming into functional occlusion. It is self-evident that the progress of tooth eruption in the hominin fossil record has to be based on evidence from the hard tissues. The sequence of tooth eruption has also been reported to differ among fossil hominins, great apes, and modern humans. (L. *erumpere* = to break out.)

### ESA
Acronym for the **Early Stone Age** (*which see*).

### ESC
Acronym for the **evolutionary species concept** (*which see*).

### ESR
Abbreviation for **electron spin resonance spectroscopy dating** (*which see*).

### Ethiopian Rift System
The Ethiopian section of the East African Rift System comprises three main components, the Afar Rift System (also known as the Afar Depression) in the northeast, the Main Ethiopian Rift System in the middle, and what some call the Omo rift zone in the south. The Ethiopian Rift System contains many productive hominin sites, including the Dikika, Gona, Hadar, and Middle Awash study areas in the Afar Rift; Chorora, Gadeb, Gademotta, Kesem-Kebena, Konso-Gardula, and Melka Kunturé in the Main Ethiopian rift, and Fejej, Omo-Shungura, and the Usno Formation in the Omo rift zone.

### ethology
The scientific study of animal behavior. Primate ethology generates information about the behavior of living primates that provides the context for developing hypotheses about the behavior of early hominins and the evolution of modern human behavior. (Gk ethos = *character* and *logia* = study.)

## Euclidean distance

A multivariate measure of distance that can be applied to the distance between objects in space or the morphological distance between objects or groups. Euclidean distance is equal to the square root of the sum of squared differences in all variables. However, morphological distance is more typically measured using distances that take into account the standard deviation of variables and their correlation with each other (e.g., Mahalanobis distance).

## Euclidean distance matrix analysis

(or EDMA) A technique for comparing the two- or three-dimensional shape and form of complex objects such as crania, mandibles, and postcranial elements. A series of landmarks is identified for an object and its form is represented by a square matrix of pairwise Euclidean distances between landmarks. The rows and columns of the matrix are the landmarks and the value for each cell in the matrix corresponds to the physical distance in two- or three-dimensional space between that pair of landmarks. Hypothesis tests in EDMA typically rely on statistical methods called bootstrap procedures to generate confidence intervals for scaling ratios, difference values, etc. There is a debate in the literature between proponents of EDMA and Procrustes analysis regarding the validity and/or appropriateness of these procedures. Debates focus on (a) the power and Type I error rates of the methods, (b) the consistency within a method for estimating mean form and/or shape, and (c) any bias in shape estimation.

## European hypothesis

See **Homo heidelbergensis**.

## European mammal neogene

A system of biozones for dating Miocene and Pliocene sites within the age range of 24 Ma to $c.2$ Ma. The zones of the European mammal neogene (or MN) system run from MN1, the oldest, at the beginning of the Miocene, to MN17 in the Plio-Pleistocene. Some of the MN zones span several millions of years (e.g., MN5 between $c.17$ and $c.14$ Ma) whereas others span less than a million years (e.g., MN15 between 4 and 3.5 Ma). The MN system has replaced the classical system of European mammalian biozonations (e.g., Vallesian, Turolian, Ruscinian, Villafranchian, Cromerian, etc.).

## eurytope

A species adapted to a broad range of environmental conditions or to a broad ecological niche (e.g., bush pig), or having morphology that is interpreted as being part of a eurytopic adaptation (e.g., the molars of a bush pig). Most early hominin species were ecological generalists rather than specialists, particularly with respect to diet. (Gk *eurus* = broad or wide and *typos* = place.)

## eurytopic

See **eurytope**; **eurytopy**.

## eurytopy

The condition of being ecologically generalized. A eurytopic (aka generalist) species is one that can use or consume a wide range of ecological resources, and thus can live in a range of habitats. With respect to diet, with exception of *Paranthropus boisei*, most early hominin species were generalists. (Gk *eury* = broad or wide and *topos* = place.) *See also* **effect hypothesis**; **stenotopy**; **habitat theory hypothesis**; **turnover-pulse hypothesis**.

## event
A short period of the opposite field direction during a prolonged period when either a normal or a reversed magnetic field predominated. The term "event" has been superseded by "subchron," as "epoch" has been superseded by "chron." For example the Olduvai subchron is a normal event within the Matuyama reversed chron.

## evo-devo
Abbreviation of **evolutionary developmental biology** (*which see*).

## evolutionary developmental biology
Often abbreviated to "evo-devo," the term evolutionary developmental biology was first used by Brian Hall in 1992, but the relationship between development and evolution has a long history. Within evolutionary biology, interest in development as the missing element of the modern synthesis reached a turning point in the 1970s and 1980s with works by Gould, Raff and Kaufman, Hall, and others. To some, the modern discipline of evolutionary developmental biology is mainly concerned with the ways in which development relates to evolutionary explanation, while others regard it more inclusively as being about the evolution of developmental mechanisms. Either way, the field of evo-devo, which represents a novel synthesis of evolutionary and developmental biology, has influenced both fields and it has the potential to illuminate the molecular basis of the important morphological changes that have made the hominin clade distinct.

## evolutionary scenario
A complex hypothesis that attempts to explain why a phylogeny looks the way it does. It provides explanations for speciation and extinction events and for patterns of morphological evolution. (L. *scaena* = scene.)

## evolutionary species concept
(or ESC) An attempt by George Gaylord Simpson to add a temporal dimension to the biological species concept. Simpson suggested that under the evolutionary species concept a species is "an ancestral-descendant sequence of populations evolving separately from others and with its own evolutionary role and tendencies."

## evolutionary taxonomy
During the first half of the 19thC most naturalists held that the evidently hierarchical organization of taxa revealed the plan of a divine creator. In his book *On the Origin of Species* Charles Darwin suggested that the so-called "natural order," which previous generations of naturalists had uncovered was exactly what would be expected if species had evolved by descent with modification. Evolutionary taxonomy is one of the approaches that taxonomists have developed in response to Darwin's insight; the other is phylogenetic systematics. Both approaches hold that classifications should reflect descent. However, they differ over the

validity of paraphyletic taxa and the importance of incorporating information about adaptation into classifications. Phylogenetic systematics insists that taxonomic groups include all the descendants of the most recent common ancestor of that group and therefore rejects paraphyletic taxa. In contrast, evolutionary taxonomy recognizes that one or more of the descendants of a given ancestor may be so adaptively different (i.e., they are in a different grade) from the ancestor or the other descendants that they warrant allocation to a distinct group. Thus, unlike phylogenetic systematics, evolutionary taxonomy accepts both monophyletic and paraphyletic taxa.

# E

### evolvability
A term coined by Gunther Wagner that refers to the degree to which a species or structure responds to natural selection with evolutionary change. Key determinants of evolvability are the extent to which genetic variation is expressed at the phenotypic level, the rate of mutation, and the extent to which variation in the trait under selection is correlated with variation in other traits. (Ge. *Evolutionsfähigkeit* = the "ability to evolve.") *See also* **canalization**; **modularity**; **morphological integration**.

### exaptation
A feature or trait that did not originate as a direct consequence of natural selection acting on its current function. Instead, the trait was co-opted to perform its current functional role after having been initially nonfunctional or adapted (through natural selection) to perform a different role. The classic example is that feathers most likely evolved in the ancestors of birds as an adaptation related to thermoregulation, but, once present, feathers were co-opted for flight and display and thus can be thought of as exaptations for those two functions. An example of an exaptation in modern humans and later hominins is the opposable thumb. A mobile, grasping thumb evolved in our earliest primate ancestors as an adaptation for grasping branches, and that basic design, plus subtle but important modifications, was co-opted for tool use by later members of the hominin clade. Exaptation explains the related term preadaptation, which describes the phenomenon in which a trait appears to be well designed for a functional role it does not currently perform. The term exaptation was part of a broader critique of what Stephen Jay Gould and Richard Lewontin called the "adaptationist programme," namely the tendency for evolutionary biologists to ascribe to natural selection a nearly limitless ability to explain the evolution of form.

### exon
The portion of a gene that undergoes transcription and eventually forms part of a functional messenger RNA (mRNA) molecule (as opposed to the intron part of a gene, which is removed from the RNA). It gets its name because it is a portion of the part of the gene that is "expressed."

### expensive tissue hypothesis
(or ETH) An evolutionary hypothesis proposed by Leslie Aiello and Peter Wheeler in 1995 to suggest that the metabolic requirements of relatively large brains are offset by a corresponding reduction of the gut. Because gut size is highly correlated with diet, relatively small guts imply high-quality, easy-to-digest food. This led to the conclusion that the brain size increase that

occurred during hominin evolution could not have been achieved without incorporating increasing amounts of high-quality animal-based foods in the diet, irrespective of the selective factors resulting in large hominin brains. Dietary change was proposed to have occurred with the appearance of *Homo ergaster* c.1.8 Ma. This is based on the claimed marked increase in brain size, the smaller teeth and jaws, the emergence of more modern human-like body proportions implying a smaller gut, and the increased evidence for scavenging and/or hunting in the archeological record, which have all been associated with the emergence of *H. ergaster*. The basic premise of an energetic trade-off with brain size has stood the test of time. However, new research suggests that it might involve features other than, or in addition to, diet and gut size, such as muscle mass or locomotor efficiency. The expensive brain hypothesis allowed Karin Isler and Carl van Schaik to provide a unifying explanatory framework, arguing that the costs of a relatively large brain in mammals in general must be met by any combination of increased total energy turnover or reduced energy allocation to another expensive function such as digestion, locomotion, or growth and reproduction. Application to the fossil record has also been complicated by evidence for animal-based foods in the diet predating significant brain expansion, the suggestion that cooked food played an important role in hominin brain evolution, and new interpretations of the body proportions of early *Homo*. However, these factors do not diminish the importance of the ETH in focusing research on energetic trade-offs in hominin evolution.

## exploratory analysis
The process of using multiple statistical techniques to identify patterns in a data set, as distinct from hypothesis testing in which a specific relationship identified *a priori* is tested.

## exponential population growth
A model of population growth in which a population increases (or decreases) in size by a constant factor per unit time.

## exposure
A place where cross-sections of strata have been exposed by erosion or by faulting and thus these are places where the fossils contained in those strata may be partly or fully exposed to view. (syn. outcrop.)

## expressivity
When different levels of expression of a gene are reflected in differences in the phenotype, and specifically when individuals with the same genotype vary in their phenotype. For example, two individuals who have the same genotype for a gene influencing weight may differ in their weight (i.e., their phenotypes differ) because the expression of the gene could be affected by factors such age, diet, illness (i.e., epigenetic factors), sex, or by interactions with other genes.

## extant
Means literally "still standing" and refers to taxa that are alive today. (L. *ex* = out and *stare* = stand; syn. living, surviving.)

### extant species
One of the many terms used to describe a contemporary species that is defined on the basis of criteria that can be observed, such as evidence of interbreeding, or the lack of any evidence of interbreeding between individuals belonging to that species and individuals belonging to any other species. (syn. neontological species, biological species.) *See also* **species**.

### extension rate
The rate at which fully differentiated ameloblasts or odontoblasts become active secretory cells. Extension rates are usually measured along the enamel–dentine junction (or EDJ) in longitudinal sections of teeth. Initial extension rates close to the cusp tip are fast (about 20–30 µm/day) in all tooth types, but the rate falls to around 3–6 µm/day in the cervical two-thirds of the crown. Extension rate can be calculated directly by inspection, or it can be inferred from the angle between the long-period incremental lines or accentuated lines in the enamel and the dentine and the EDJ. Low angles imply a rapid extension rate and high angles a relatively slow extension rate. *See also* **dentine; enamel development**.

### external auditory meatus
The passage between the ear lobe (auricle or pinna) laterally and the tympanic membrane medially. The lateral one-third is made of a U-shaped fibrocartilaginous tube that is continuous laterally with the elastic cartilage of the pinna. The medial two-thirds is the bony part of the external auditory meatus. Some researchers have suggested that the size and shape of lateral end of the osseous part of the external auditory meatus are useful for discriminating among early hominin taxa (e.g., *Australopithecus anamensis* or *Kenyanthropus platyops*). (L. *ex* = outer, *audire* = to hear, and *meatus* = passage.)

### external ear
The external ear is made up of the earlobe (auricle or pinna) and the external auditory meatus. The skeleton of the lateral one-third of the external auditory meatus is fibrocartilage and the pinna is made up of elastic cartilage; neither of these structures fossilizes. The medial two-thirds of the external auditory canal is bony and is made up of contributions from the squamous and tympanic plate components of the temporal bone. (L. *ex* = outer.)

### extinct
The term means literally "to extinguish" and it refers to all taxa (e.g., *Australopithecus afarensis*) that are no longer alive. (L. *ex* = out and *stinguere* = to put out, or quench.)

### extractive foraging
Any sequence of behaviors that enables organisms to extract embedded foodstuffs (e.g., fruits enclosed in hard shells, underground storage organs buried in the soil, marrow within the trabecular cavities of bones, or the soft tissues within the exoskeleton of insects). In hominin evolution, tool use is key to this behavior, but in nonhominins extractive foraging can take place without the use of tools (e.g., the aye-aye uses a highly modified finger to dig insects out of bark). The importance of extractive foraging has been discussed widely in the primate cognition literature as it has been argued that it takes greater mental ability to remove or

process embedded foods than foods that are not embedded. Thus, extractive foraging might have acted as a selective pressure for the evolution of the brain and cognition in primates and hominins. Of the nonhuman primates, chimpanzees, capuchins, and the aye-aye all use extractive foraging and they are also relatively highly encephalized. However, there are a number of limitations to the extractive foraging model of primate, and perhaps hominin, intelligence. One of the most important of these is the disjunction between observations of extractive foraging and measured intelligence in a range of animals and birds. Nonetheless, the hypothesis that extractive foraging was central to hominin food procurement and that it allowed hominins access to a wider range of foods, enabling competition with other members of their ecological communities, is an attractive one.

## Eyasi

A site consisting of at least two fossiliferous localities on the shore of Lake Eyasi, Tanzania. Ludwig Kohl-Larsen collected at the locality in the 1930s, recovering fossil evidence of three partial hominin crania. Subsequently, it has been revisited several times by researchers trying to clarify the stratigraphy. Some of these studies have uncovered additional hominin cranial fragments to make a total of seven hominin individuals. All of the hominins have been assigned to archaic *Homo sapiens* and they are considered to be Late Middle or early Upper Pleistocene in age. Electron spin resonance spectroscopy and uranium-series ages for a wildebeest tooth found 5 m from a hominin frontal range between 200–100 ka. Archeological evidence found at the site includes an early Sangoan or Njarasan late Acheulean industry as well as an overlying Middle Stone Age industry. (Location 03°32′26″S, 35°16′05″E, Tanzania.)

# F

### face
One of the three main components of the cranium (the others are the vault, or calotte, and the basicranium.) There is no formal definition of the face, but a good working definition is that it is the part of the cranium that extends from the alveolar plane of the upper jaw inferiorly, superiorly to the point where the frontal bone is narrowest (i.e., minimum frontal breadth), and then posteriorly across the supraorbital region. The face is made up of paired and unpaired bones. The paired bones are the nasal, maxillary, palatine, and zygomatic; the unpaired bones are the anterior part of the frontal (the flat, or squamous, part of the frontal belongs to the vault), the ethmoid, and the vomer. (L. *facies* = face.)

### facetted platform
A platform on a flake marked by multiple facets where small flakes have been removed to shape or modify the striking platform. Many (but not all) Levallois flakes have facetted platforms.

### facial mask
The term used for the parts of the facial skeleton that are visible when the cranium is viewed from the front (i.e., anteriorly, or in norma frontalis).

### facial visor
The term used for the antero-superiorly facing, plate-like infraorbital surface of the face in *Paranthropus boisei*. Its mostly relief-free morphology contrasts with the morphology of the same part of the face of *Paranthropus robustus*, which displays, from anterior to posterior, an anterior pillar, a maxillary trigon, and a zygomaticomaxillary step.

### facies
In earth science it refers to a lateral, different-looking, subdivision of the same stratigraphic unit that formed under different conditions. For example, at lake margins, sedimentary strata often grade from sandy clays near the lake margin to coarser sands further inland. The lateral variability implied by the definition used above has also been applied to archeological assemblages (e.g., Bordes used the term to describe variation within the Mousterian that he attributed to contemporaneous but distinct cultural groups). (L. *facies* = face.)

## facultative biped
See **bipedal**.

## FAD
Acronym for **first appearance datum** (*which see*).

## fallback foods
The foods that an animal turns to when its preferred foods are not available. Optimal foraging theory suggests that when animals forage they favor foods that are (a) abundant, (b) energy-rich, (c) easily accessed, and (d) easily processed, so that, all things being equal, an individual gets the greatest yield per unit of energy expended on feeding (i.e., animals maximize their feeding efficiency). But when preferred foods are unavailable for either predictable (e.g., seasonality) or unpredictable (e.g., drought, volcanic eruption) reasons, animals have to turn to foods that are either less energy-rich or that cost more energy to procure and process. Some have argued that the critical ability to process fallback foods may have shaped the evolution of the masticatory system (especially enamel thickness) in at least some primates. Researchers have suggested that the need to process plant underground storage organs, a possible early hominin fallback food, may have been an important factor in the emergence of the derived masticatory morphology of megadont and hyper-megadont early hominins. Fallback foods may explain mismatches between two of the lines of evidence (i.e., dental microwear and gross morphology) about diet.

## fault
A planar fracture through the Earth's crust such that the rocks above and below the fracture move relative to each other. There are three main groups of faults that are named for the direction of movement in their fault plane: dip-slip (vertical), strike-slip (horizontal), and oblique-slip (vertical and horizontal). A dip-slip fault can be classified as either normal or reverse. If the hanging wall (i.e., the rocks above the fault plane) is displaced downwards (i.e., down the dip) relative to the rocks of the footwall (i.e., the rocks below the fault plane) it is called a normal fault. If the hanging wall is displaced upwards (i.e., up the dip) relative to the rocks of the footwall the fault is referred to as a reverse fault. A strike-slip fault is characterized by a vertical fault surface and footwalls that move laterally left (sinistral faults) or right (dextral faults). Plate boundaries are formed by transform faults, a type of strike-slip fault. Oblique-slip faults contain significant aspects of both dip-slip and strike-slip faults.

## fauna
Biology A term that refers to all of the animals in an ecosystem or, when combined with a taxonomic modifier (e.g., mammal/ian, vertebrate, avian), to a particular component of the animal life in that system. Paleontology Refers to (a) all of the animal remains from a particular site (e.g., the Swartkrans fauna) or (b) all of the animals known from a defined time period (e.g., Pliocene fauna) or region (e.g., British fauna). The fauna at a paleoanthropological site is described in the form of a faunal (or taxon) list that catalogues, in rank taxonomic order and to the minimum determinable (i.e., the most specific) level, all the taxa to which faunal remains

recovered from the site have been assigned. In the absence of other more precise and reliable dating methods the faunal list can be used as an indication of the age, as well as the paleoenvironment, of the fossil site. (L. *Fauna* = in Roman mythology the sister of the Faunus, the god of nature and *faunus* = a mythological creature that has the head of a man and the body of a goat.) See also **biochronology**.

### faunal assemblage

A group of associated animal bones or fossils, generally found in a particular spatial and temporal context, which can be treated as a unit for analysis (e.g., the accumulated fauna from all of the localities in Bed I of Olduvai Gorge would be regarded as a "faunal assemblage"). *See also* **assemblage**.

### faunal break

A discontinuity that results in an apparently abrupt faunal change. Within Bed II at Olduvai Gorge a change in the fauna just above the Lemuta Member is referred to as the "faunal break." *See also* **Olduvai Gorge**.

### faunal list

*See* **fauna**.

### Fauresmith

A southern African stone artifact industry characterized by the variable presence of diminutive handaxes, elongated flakes, or blades, points, and Levallois technology. The Fauresmith has traditionally been considered a regional variant of the later or terminal Acheulean or as intermediate between the Acheulean and the industries of the Middle Stone Age. Evidence from Wonderwerk Cave and elsewhere suggests a Middle Pleistocene age (e.g., older than *c.*285 ka) for the Fauresmith industry.

### favored place hypothesis

A model that suggests Early Stone Age sites are a byproduct of hominin transport and discard behavior. Over time, the occasional discard of artifacts at "favored places" (i.e., frequently visited foraging areas where hominins would find and consume foods, rest, carry out social activities, sleep, etc.) would lead to the passive accumulation of a localized store of raw material. This would reduce the need for artifact transport while foraging in the immediate area and over time stone and debris from multiple butchery events would form dense archeological concentrations.

### FEA

Acronym for **finite element analysis** (*which see*).

### feature

When used by archeologists it refers to any nonportable object (e.g., hearths, dwellings) made, modified, or used by hominins. Dental morphologists sometimes use feature as an inclusive term for the cusps, crests, ridges, and pits on the outer enamel surface.

## fecundity
A measurement of the potential reproductive capacity of an organism. Fecundity can be gauged generally by an individual's age in relationship to the average reproductive life span for that taxon or more precisely by the number of gametes (eggs or sperm) it possesses at a given time during its life span. In modern human females fecundity reaches a peak in the early 20s and declines thereafter (it reaches zero at menopause), whereas modern human males do not show such clear age-related patterns of diminishing reproductive capacity. (L. *fecundus* = fruitful.)

## feeding
An inclusive term for the ingestion and intra-oral processing (including chewing) of food items. Feeding occurs in sequences of gape cycles in which the lower jaws are cyclically depressed and elevated relative to the upper jaws and the tongue is moved to position the food item(s) for chewing and swallowing. In primates a feeding sequence, which starts with ingestion, is followed by cycles in which the food is manipulated, investigated, and transported to the molar tooth row for chewing. A feeding sequence ends when the bolus of food is swallowed. *See also* **chewing**.

## Fejej
Site in the northeast of the Turkana Basin, in southern Ethiopia, that contains fossiliferous sedimentary rocks ranging in age from the Oligocene through the Miocene and into the early Pliocene. There are several localities in the site, one of which, FJ-4, has a minimum age of 4.18–4.0 Ma, based on argon-argon dating, potassium-argon dating, and biostratigraphy. Six heavily worn hominin mandibular teeth and tooth fragments and an unworn right $P_4$ from this locality have been attributed to *Australopithecus afarensis*, but the dentition is in the size range of *Australopithecus anamensis* and some researchers have assigned the Fejej hominins to that taxon. Artifact assemblages attributed to the Oldowan Industry were found at FJ-1, a separate, later, locality. (Location 04°37′24″ N, 36°27′36″ E, Ethiopia.)

## Feldhofer Grotte
*See* **Kleine Feldhofer Grotte**.

## feldspar
The most abundant group of minerals in the Earth's crust. Feldspars are generally of igneous or metamorphic origin but they may also form authigenically (i.e., formed in place) within sediments. Alkaline feldspars (i.e., ones rich in sodium and potassium) are good targets for isotopic dating. *See also* **argon-argon dating; potassium-argon dating**.

## felid
The informal name for the Felidae family (the cat family), one of the feliform families whose members are found at some hominin sites. *See also* **Carnivora**.

## Felidae
One of the feliform families whose members are found at some hominin sites. The Felidae include two subfamilies: the Felinae (cheetahs and other small–medium-size cats) and the Pantherinae (lions, leopards, and tigers). Felids first appear in the fossil record during the Oligocene. *See also* **Carnivora**.

### feliform
Informal term for taxa within the suborder Feliformia, one of the two suborders in the order Carnivora.

### Feliformia
One of the two suborders in the order Carnivora. Common feliform families at hominin sites are the Felidae (cats), Hyenidae (hyenas), Viverridae (civets and their relatives), and Herpestidae (mongooses and relatives).

### FEM
Acronym for finite element models. *See* **finite element analysis**.

### fertility
This term is used in a number of different ways in the paleoanthropological literature, but in demography it is a population-level measurement of the number of births divided by the number of childbearing lives. On a per-individual basis, fertility is the number of live births a woman can be expected to have over her entire reproductive career.

### fibrocartilage
*See* **cartilage**; **joint**.

### field theory
A model of dental morphogenesis that suggests there are developmental fields within which regional morphogenetic cues influence the development of specific tooth types (i.e., incisors, canines, premolars, and molars).

### finger
*See* **digit**; **ray**.

### finished sequence
A genome that has been sequenced (usually multiple times) and checked for errors. For example, when first completed, only 25% of the modern human genome had a "finished sequence" (i.e., at least four-fold coverage with an accuracy of 99%).

### finite element analysis
(or FEA) An advanced computational technique used by engineers to predict the response when physical systems are subjected to known loading conditions. The most common anthropological application of FEA is to biomechanical problems concerning the stress and strain experienced by bones and teeth when they are exposed to various types of forces. In FEA, an object of complex geometry (e.g., a bone) is typically modeled as a mesh of tetrahedra or hexahedra (these are the "elements" referred to in the name of the method) joined at nodes (e.g., the vertexes of the tetrahedra). The elements are assigned the material properties (e.g., stiffness) of a given substance (e.g., cortical bone), forces are applied to nodes on the model, certain nodes are constrained from moving, and displacement is calculated at each

node. Strains and stresses throughout each element are obtained by differentiating the element displacement field using the known or estimated material property for that element. The result is a characterization of stress and strain patterns across the object as a whole. It is critical that all finite element models be subjected to a validation study in which results from the FEA are compared to experimentally derived strain data. If there is good correspondence between the experimental and the FEA data then the model can be used to test biological hypotheses (e.g., early hominin facial biomechanics).

## finite element method
See **finite element analysis**.

## finite element modeling
See **finite element analysis**.

## fire
See **pyrotechnology**.

## first appearance datum
(or FAD) The date of a taxon's earliest appearance in the fossil record. For various reasons, the FAD of a taxon almost certainly postdates the taxon's origination in, or migration to, that region. Just how much earlier the origination or migration occurred is determined by two factors. The first is any error in the methods used to date a site, the second is the nature of the relevant fossil record prior to the FAD. This second problem is often described by the old adage, "absence of evidence is not evidence of absence." Hominins are such a rare component of the mammalian faunal record that researchers need to find a substantial number of nonhominin mammalian fossils (at least several hundred) without finding *any* evidence of a particular hominin taxon before it can be reasonably assumed that hominin taxon was not part of the faunal assemblage being sampled. In such cases (e.g., *Homo* at Omo-Shungura), the FAD has a small confidence interval ($c.100$ ka), but if, as is the case at Koobi Fora, there is a major unconformity spanning several hundred thousand years (i.e., there is no fossil record during that time) prior to the FAD of *Homo* then the FAD of *Homo* at that location has a large ($c.900$ ka) confidence interval.

## FISH
Acronym for **fluorescence *in situ* hybridization** (*which see*.)

## fish
Fish bones are common in early hominin sites. Most, but not all, were deposited post-mortem through natural water transport that promotes rapid burial and preservation. Fish bones can provide information on hominin predation and other aspects of hominin behavior as well as biogeographic and environmental information relevant for the reconstruction of hominin paleoecology. For example, the ecology of the fish taxa helped to reconstruct the paleoenvironment of the late Miocene Lothagam site as being close to a large and slow-moving river with nonbrackish, well-vegetated back bays and swamps. This reconstruction indicates the

availability of potable water and sources of river margin foods (e.g., sedges and catfish) to hominins. Evidence for fish as prey is rare in early hominin sites. This could be because it *is* rare to prey on fish, but it could also be because of taphonomic processes (fish bones are less dense than bones of terrestrial animals) and/or if the whole fish was consumed it would leave few traces. Nevertheless, fish bones with tooth and/or cutmarks have been reported from Koobi Fora and Olduvai Gorge. The dominant fish taxon in these and other Pleistocene archeological sites is the large catfish *Clarias*, which is readily caught in shallow or receding waters in Africa today. While most Pleistocene hominins could only catch the most accessible fish, improved technology, especially boats, in the latest Pleistocene and Holocene resulted in increased numbers and greater taxonomic diversity of fish in later archeological sites.

## Fish Hoek

At this cave site in South Africa, which was first investigated in 1925, Middle Stone Age (or MSA) Stillbay deposits lie below a Later Stone Age midden and beneath the former are the Howieson's Poort layers, the source of the Fish Hoek 1 hominin skeleton. Late Acheulean (Stellenbosch culture) artifacts were found 6 m below the surface of a talus deposit. The MSA layers are apparently beyond the range of radiocarbon dating, but optically stimulated luminescence dates at other sites suggest the Stillbay dates to *c.*70 ka and the Howieson's Poort to *c.*65–60 ka. (Location 34°07'S, 18°24'E, Western Cape Province, South Africa.)

## fission track dating

A radiogenic dating method that exploits the tendency of $^{238}U$ to undergo spontaneous disintegration or fission. The fragments produced by the fission leave a trail or "track" of damage in the crystal that can be seen when thin sections of the rock are etched with chemicals. First the "natural" tracks (i.e., the ones that have accumulated since the formation of the crystal) are counted, then the specimen is subjected to a neutron dose large enough to drive all the remaining $^{238}U$ to undergo fission at which time the "induced" tracks are counted. The ratio of the two track counts and the neutron dose are sufficient to calculate the age of the crystal. Rocks have to be rich in uranium (e.g., natural glasses, zircon, sphene, and apatite) in order to be suitable for fission track dating. The method works best in rocks older than 300 ka. (L. *fissus* = to split.)

## fissures

<u>Neurology</u> In adult brains, the outer cortical layer of the cerebral and cerebellar hemispheres is deeply folded. The blunt crests of the folds are called gyri; the fissures between them are called sulci. <u>Dentition</u> Features on the outer enamel surface of teeth are also called fissures. Primary fissures separate the main cusps and secondary fissures delineate the smaller accessory cusps. (L. *fissura* = cleft.)

## fitness

The capability of an individual to reproduce so that its genes are represented in the next generation. Fitness is measured as the proportion of genes contributed to the next generation by an individual relative to the contributions of other individuals. It is equal to the relative probability of survival and reproduction for a genotype. Since fitness is a relative term the genotype with the best fitness is usually assigned a value of 1 and other genotypes have values less than 1. The concept may also be referred to as genetic, or Darwinian, fitness.

## fitness-generating function

A numerical function that describes the genetic fitness of a particular combination of traits by taking into account all the fitness trade-offs related to that phenotype. Michael Rosenzweig suggested that grades or genera share a similar fitness-generating function (or G-function) and the reason that a new adaptive type emerges is that its new G-function has better fitness trade-offs than the G-function of the taxon it supercedes. For example, the ability to consistently and efficiently manufacture small sharp stone flakes useful for cutting the hide of a large mammal could allow the user more efficient access to important and valuable food resources than would be achieved by hominins that had to wait until nonhuman predators had penetrated the carcass. This would be a behavioral example of a new G-function.

## fitness landscapes

A metaphor that allows researchers to visualize the complexity of factors influencing fitness and the ways in which an individual or species might improve their fitness. The peaks of the three-dimensional landscape represent regions of higher genotypic fitness; the valleys are regions of lower fitness. Sewall Wright and others make the point that these landscapes are complex. Some of that complexity is due to interactions among genes. Other parts of the complexity are due to phenomena such as epistasis (i.e., when the effect of a mutation on the phenotype depends on where in the genome it occurs). Many of the theoretically possible pathways on the metaphorical fitness landscape are inaccessible for one reason or another; these restrictions (e.g., additional cusps on mandibular postcanine teeth tend to occur in the talonid, the distal part of the crown, and not in the more mesial trigonid) are referred to as evolutionary constraints. George Gaylord Simpson applied a similar principle, which he called "adaptive landscapes," to population-level dynamics.

## fixation

In a population where more than one form of a gene (i.e., allele) exists, fixation occurs when, for one or more of a number of reasons (e.g., random genetic drift, natural selection, or as a consequence of small breeding populations) variation in allelic distribution declines to the point where only one allele remains. Once this happens the only ways a population can be released from fixation for this allele are (a) mutation and (b) the immigration of new alleles.

## flake

An inclusive term for any fragment removed from a core made of stone (or much less commonly bone). The term can be applied to complete flakes as well as to the various flake-like fragments (e.g., debris, débitage) that are produced during the process of fracturing stone to make flakes (i.e., knapping). Flakes may also be produced in the absence of hominin activity (e.g., when rocks are knocked together in the bed of a river), but this post-depositional flaking does not normally show the distinctive morphology associated with conchoidal fracture. *See also* **conchoidal fracture**.

## flake-blade

A term used in descriptions of southern African Middle Stone Age sites such as Klasies River. Whereas blades are artifacts that are more than twice as long as they are wide, a flake-blade is an elongated flake with dorsal scars that suggest several parallel (or subparallel or convergent) flakes have been previously removed.

## flake types

A typology used to describe stone flakes produced during knapping based on the occurrence of cortex (original exterior surface of the raw material) on the surface of the flake. The distribution of flake types in an assemblage provides information about the knapping methods used and what stages within a reduction sequence are represented. The manufacture of experimental stone tools enables researchers to generate expectations for the distribution of flake types. This allows assemblages that include all stages of reduction to be distinguished from those in which only part of the reduction sequence is represented, or from assemblages in which certain types of flake have been preferentially removed (for example, because they are themselves used as tools). With respect to knapping methods, cortical platforms are associated with "unifacial" reduction in which percussion is consistently against the original raw material surface. This contrasts with the noncortical platforms that are associated with "bifacial" or "polyfacial" methods. In the latter the "scars" left by previous flake removals are used as the platforms for subsequent flake removals. *See also* **actualistic studies**.

## flint

*See* **chert**.

## flora

A collective term for the plant life occurring in a specific area or at a particular time. The fossil records of plants at paleoanthropological sites tend to be sparser and less well known than the faunal records. The flora at hominin sites can be studied by searching for fossilized pollen, fossilized leaves, vines, woods, and fruits, by looking at phytoliths (i.e., the silica components of plants) or by looking at plant leaf wax biomarkers. (L. *Flora* = Roman goddess of flowers.) *See also* **palynology; trace fossil**.

## Florisbad

An old spring eye 40 km/24 miles north-northwest of Bloemfontein in what was the Orange Free State of South Africa (now Free State Province). Robert Broom described the early nonhominin fossils in 1912, but not the Florisbad 1 hominin cranium, which was found in subsequent excavations. Biostratigraphy of the fauna suggests a Late Middle to Upper Pleistocene age for the hominin. Electron spin resonance spectroscopy dating applied to teeth (both faunal and the hominin $M^3$) produced an age of $259 \pm 35$ ka and optically stimulated luminescence dating suggests an age of 300–100 ka.

## Florisbad 1

A fragmentary hominin cranium comprising the frontal, both parietals, both nasals, the right zygomatic and maxilla, part of the left maxilla, and the right $M^3$. Its original taxonomic allocation was to *Homo helmei*, then to *Homo florisbadensis*, and most recently to *Homo sapiens*. The supraorbital torus is large (but smaller than in the hominin crania from Kabwe and Saldanha) and, according to some, shows incipient separation of its medial and lateral segments. The forehead rises more steeply than in Kabwe or Saldanha and the broad but short face bears a canine fossa. The cranial vault is wide. The thick vault bone bears two circular tooth marks, suggesting the individual was killed or scavenged by a large carnivore. The cranium is transitional in morphology between *Homo heidelbergensis* and *Homo sapiens* (e.g., it is intermediate with respect to vertical facial dimensions, brow ridge size, and inclination of the frontal bone), whereas inferred primitive features (e.g., a large brow ridge, receding frontal, and a broad base to the vault) distinguish it from anatomically modern humans. If *H. helmei* can be satisfactorily diagnosed, which is debatable, Florisbad 1 would be the holotype of this species.

## flowstone
See **calcium carbonate**; **speleothem**.

## fluorescence *in situ* hybridization
(or FISH) A type of *in situ* hybridization method that uses fluorescently labeled probes to detect and locate specific DNA sequences. *See also* ***in situ* hybridization**.

## fluorine dating
Fluorine from groundwater leaches into buried materials over time, so that older bones have higher fluorine levels, thus providing a relative date. Fluorine dating is best known for the role it played in helping to expose the Piltdown fraud. The animal fossils at the Piltdown site were found to have fluorine levels that varied between 1.6 and 3%, whereas the levels in the alleged hominin fossils claimed to be from Piltdown were similar to those found in contemporary bones (i.e., about 0.2%). Joseph Weiner and Wilfrid Le Gros Clark collaborated with Kenneth Oakley to apply a more sensitive form of fluorine analysis to the allegedly ancient hominin fossils. These analyses showed that the fluorine levels in the cranial bones and the mandible were 0.1 and 0.03%, respectively. These low and different fluorine levels implied that the cranial bones and the mandible of Piltdown I were not the same age and suggested that the mandible was almost certainly modern. These results, along with other lines of evidence showing the cranial bones and the mandible had been stained and that the cusps of the teeth in the mandible had been filed down, were sufficient for Weiner and his colleagues to declare that the Piltdown hominin fossils were modern and not ancient and that their association with the Piltdown site was the result of an elaborate hoax. (L. *fluor* = the name of a group of minerals used to make fluxes.) *See also* **Piltdown**.

## fluvial
Relating to a river or stream. Thus, a fluvial paleoenvironment is one involving a river or stream, and fluviatile sediments are ones laid down by rivers or streams. (L. *fluvius* = river.)

## fluviatile
*See* **fluvial**.

## fMRI
Abbreviation for **functional magnetic resonance imaging** (*which see*).

## folivory

Literally, leaf eating. Animals that eat leaves are called folivores. Leaves provide less net energy than fruits or insects because the complex cellulose of leaves requires specialized digestive processing and bacterial colonies in the fore- or hindgut to break it down. Folivorous primates tend to have shearing crests on their teeth that mechanically process leaves much as scissors slice through paper. Soft-tissue digestive adaptations to folivory in primates can include a foregut with a large, complex, multichambered stomach (e.g., colobine monkeys) or a hindgut with either an enlarged caecum (e.g., in some strepsirhines) or a long colon (e.g., howler monkeys and gorillas). Among extant great apes, folivory features prominently only in the diets of the gorilla. Thus far folivory has not been seriously proposed as a dietary specialization of extinct hominins. (L. *folium* = leaf and *vorous*, from the stem of *vorare* = to devour.) See also **Jarman/Bell principle**.

## Fontana Ranuccio

An open-air site southeast of Rome, Italy, in the Sacco-Liri valley that was discovered during mining. It has produced three *Homo erectus*-like hominin teeth plus elephant bone implements and small flint handaxes and scrapers that are likely Acheulean. The archeological levels have been dated to *c.*460 ka. (Location 41°45′35″N, 13°16′03″E, Italy.)

## fontanelle

Fontanelles are places on the newborn cranium where the bone of the cranial vault is not yet completely ossified. They were called fontanelles because they transmit the pulsations of the intracranial arteries via the cerebrospinal fluid and the venous blood in the dural venous sinuses. The largest are the rhomboid-shaped anterior fontanelle at bregma, where the frontal (metopic), coronal, and sagittal sutures meet, and the triangular posterior fontanelle at lambda between the sagittal and the lambdoid sutures. (L. *fontanelle* = small fountain or spring.)

## Fontéchevade

Cave site in the Charente region of France excavated by several amateurs since the 1870s and then by Germaine Henri-Martin from 1937 to 1954. It is one of the reference sites for the Tayacian. Two hominin fragments recovered at the site were thought to be quite old, with one (Fontéchevade I) resembling *Homo neanderthalensis* and the other, a partial calotte (Fontéchevade II), appearing more modern; this conjunction formed part of the evidence used to support the pre-sapiens hypothesis. However, a program of excavation and a redating effort have shown that these fossils are more recent; the modern human-like fossil could even be an early Upper Paleolithic *Homo sapiens* individual. The lowest level was originally thought to date to the last interglacial (Marine Isotope Stage 5e) based on the warm-climate fauna and the crude stone tools, but recent radiocarbon dating and electron spin resonance spectroscopy dating suggest an age of *c.*33–60 ka (i.e., Marine Isotope Stage 3) for the top of this level. (Location 45°40′43″N, 00°28′47″E, France.) See also **pre-sapiens hypothesis**.

## foot

The foot, which consists of seven tarsal bones, five metatarsals, and 14 phalanges (three for each toe, except for the big toe, or hallux, which has two), plus any sesamoid bones, is divided into the hindfoot, midfoot, and forefoot. The hindfoot consists of the calcaneus and the talus, which articulate together to form the subtalar joint. The midfoot consists of the rest of the tarsal bones (navicular, three cuneiforms, and cuboid) and all of the metatarsals. The forefoot consists of the phalanges or toe bones. The hindfoot and midfoot articulate at the transverse tarsal joint (i.e., between the talus and the navicular and the calcaneus and cuboid). The articulation between the midfoot and the forefoot comprises the five metatarsophalangeal joints (i.e., between the head of the metatarsal of each pedal ray and the base of the corresponding proximal phalanx). Articulations within the midfoot are more complex. The navicular has facets for each of the three cuneiforms and often also the cuboid. The lateral cuneiform articulates with the cuboid laterally. The first metatarsal articulates with the medial cuneiform and the second metatarsal articulates with the intermediate cuneiform. The second metatarsal is recessed within the tarsal row, so it also has facets with the medial cuneiform and the lateral cuneiform as well. The third metatarsal articulates with the lateral cuneiform, while the fourth and fifth metatarsals articulate with the cuboid. In modern humans the fourth metatarsal is also recessed within the tarsal row and contacts the lateral cuneiform. The modern human foot differs considerably from the ape foot, reflecting a shift from a grasping appendage to a propulsive one. The most apparent differences are that in apes such as the chimpanzee only the lateral side of the foot touches the ground while standing bipedally and the big toe points away from the other toes. In contrast the modern human foot is positioned with the bottom (i.e., plantar) aspect of the foot towards the ground and the hallux is adducted (i.e., it is in line with the longitudinal axes of the other digits). Adaptations for habitual bipedality include a robust calcaneus with a lateral plantar process. The modern human calcaneocuboid joint is unique among living taxa in possessing an interlocking facet that stabilizes the joint during the push-off phase of walking. The elongated tarsals of a modern human foot increase the lever arm for efficient bipedal push-off. Metatarsal robusticity also differs, with chimpanzees having a $1>2>3>4>5$ formula and modern humans a $1>5>(3, 4)>2$ formula. The metatarsals of the modern human foot have a distinct torsion (i.e., the long axis of the metatarsal head is not in line with the long axis through the articular surface at the base of the metatarsal) so that during the stance phase of the walking cycle the bases of the metatarsals can form a transverse arch while all of the metatarsal heads lie squarely on the ground. Compared to all apes, modern human feet have relatively short and straight phalanges, a longer Achilles tendon, and a thicker plantar aponeurosis (fascia). *See also* **bipedal**; **walking cycle**.

## foot arches
Structural configurations that elevate the midfoot. Modern humans, chimpanzees, gorillas, and orangutans possess a transverse arch, a mediolaterally oriented structure at the tarsometatarsal junction. Modern humans differ, however, from all other extant primates in also possessing a longitudinal arch that runs the length of the foot. It serves both to stiffen the foot to make it an effective lever and to allow ligaments and the plantar aponeurosis to store elastic energy during the stance phase of walking. *See also* **foot**; **longitudinal arch**; **walking cycle**.

## foot function
The modern human foot is stiff and straight whereas the extant ape foot is more flexible and adapted for a wider range of motion. Modern humans and nonhuman apes are plantigrade (i.e., they strike the ground with their heel) whereas monkeys are either digitigrade or

semiplantigrade. During modern human walking forces are greatest at heel strike and again at push-off, producing a characteristic "double-humped" force/time curve in the walking cycle. During the stance phase of walking the modern human midfoot is normally elevated due to the longitudinal arch that is maintained by ligaments and the plantar aponeurosis; in contrast, the ape midfoot contacts the ground during the stance phase. As the stance phase proceeds the modern human foot undergoes eversion (i.e., the plantar surface faces outwards) due in part to muscle activity that helps transfer the weight-bearing function towards the robust hallux for push-off. Eversion also keeps the transverse axes of the midtarsal joints roughly parallel and allows for some midfoot mobility, which is important for keeping the foot in contact with uneven substrates. Activity of the calf muscle (i.e., the gastrocnemius–soleus complex) at push-off plantarflexes the foot at the ankle and this causes inversion of the hindfoot. This motion twists the calcaneus against the cuboid into a close-packed position and it is this that helps elevate the midfoot, whereas in apes such as the chimpanzee the midfoot remains in contact with the ground after heel lift. This mobility at the calcaneocuboid joint, combined with dorsiflexion at the fourth and fifth tarsometatarsal joints and rotation within the tarsals, is called the "midtarsal" or "midfoot" break. As the heel lifts in modern humans the fulcrum shifts directly to the metatarsophalangeal joints and the resulting dorsiflexion of the phalanges tightens the plantar aponeurosis (known as the "windlass effect") and helps to convert the foot into a stiff and effective lever. *See also* **foot movements; midfoot break; walking cycle**.

## foot movements
The primary movements of the foot are dorsiflexion (extension) and plantarflexion (flexion) at the ankle (talocrural) joint (this occurs around an approximately mediolateral axis), inversion and eversion that takes place primarily at the subtalar joint (this occurs around an approximately anteroposterior axis), and internal and external rotation of the foot that occurs around an approximately superoinferior axis. Pronation is a complex movement that combines external rotation, dorsiflexion, and eversion; supination is a combination of internal rotation, plantarflexion, and inversion. At the beginning of the stance phase of walking the hindfoot pronates slightly but during midstance the foot supinates, thus increasing its stability. After midstance slight pronation helps the foot absorb the forces of body weight in the arch and hindfoot supination at the end of stance phase, together with the "windlass effect," promotes efficient propulsion by the plantarflexors. The toes can also move sideways. Adduction is movement of the toes (other than the second digit) toward the second digit; abduction is movement of the toes away from the second digit. The second digit can itself be abducted away from its normal longitudinal axis.

## footprints
*See* **Koobi Fora hominin footprints; Laetoli hominin footprints**.

## forager
One who engages in a subsistence strategy based primarily on the collection of wild plant foods, fishing, and the hunting and scavenging of animals. A predominantly foraging (aka "hunting and gathering" or "hunter–gatherer") lifestyle is practiced by the Hadza of Tanzania, the Aché of Paraguay, the San of southern Africa, the Inuit of Canada and Alaska, and the Mbuti of the

Democratic Republic of the Congo, among others. The movements of hunter–gatherers, which tend to forage in small, relatively egalitarian groups, depend on the spatial distribution and seasonal availability of resources. In regions with a relatively low and even resource distribution, foragers live in mobile groups that move seasonally to exploit resources as they become available. Where resources are spatially and/or temporally more clumped (e.g., in regions with seasonal fish or large mammal migrations) hunter–gatherers reside in larger groups that do not need to move as frequently. Food storage is more likely in the latter context, where seasonal abundance may provide surpluses that can be stored for later consumption. Hunting and gathering was the predominant mode of subsistence before plant and animal domestication began $c.10$–$12$ ka BP. Paleoanthropologists and archeologists analyze the ways in which modern foragers produce and use tools in an attempt to understand how tool use might have developed in hominin evolution. The use by contemporary hunter–gatherers of wooden digging sticks and carrying devices used to transport infants, both of which are typically constructed out of material that does not survive into the archeological record, should remind researchers of the danger of interpreting the behavior of early hominins solely on the basis of durable (i.e., lithic) evidence. The behavioral ecology of extant hunter–gatherer groups provides an indication of what the hunting and gathering adaptation might have been like for more ancient forms of *Homo*, but the extent to which this adaptive complex (i.e., sexual division of labor, extensive food sharing, and central place foraging) extends back into the Pleistocene is debated. (OF *fourrage* = to search for food.)

## foraging

An inclusive term for the search for, and acquisition of, food items. With respect to human evolution researchers use information about the foraging behaviors of extant organisms to develop referential models from which inferences can be made about the functional morphology of extinct hominin taxa. Optimal foraging theory involves balancing the estimated energy expended in the search for food with the net energy yielded by ingesting the food. *See also* **extractive foraging**; **forager**; **modeling**. (OF *fourrage* = to search for food.)

## foraging efficiency

The amount of energy obtained relative to the amount of time (and therefore energy) spent searching for and handling food. Optimal foraging theory assumes that the objective of all foragers is to maximize foraging efficiency. Indications of a decline in foraging efficiency include an expansion of diet breadth and increased utilization of low-ranking prey. Foraging efficiency can be affected by numerous factors including environmental change, technological innovation, and resource depression. The detection of changes in foraging efficiency through time in archeological contexts is central to understanding the factors mediating prehistoric human subsistence strategies.

## foramen magnum

The largest opening in the basicranium. This midline opening in the occipital bone transmits the spinal cord together with the meninges that cover it, the spinal contributions to the accessory (XI) cranial nerves, the vertebral and spinal arteries, and the thin-walled veins that connect the internal vertebral venous plexuses with the dural venous sinuses within the endocranial cavity. Two important midline cranial landmarks, basion anteriorly and opisthion posteriorly, help locate the foramen magnum. Important foramen magnum variables involve its *location* relative to the long (i.e., the sagittal) axis of the neurocranium or to transverse axes (e.g., bi-carotid, bi-tympanic, etc.) defined by bilateral landmarks, its *orientation* relative to horizontal planes such as the Frankfurt Horizontal, its *shape* described either nonmetrically or metrically, and its *size* captured either by linear dimensions or by estimating or measuring its surface area. The head-balancing index and the condylar position index are among the best known of many attempts to locate the foramen magnum, or a proxy (e.g., porion or the occipital condyles) along the midline axis of the cranium. Both of these indices require a well-preserved cranium, so researchers looked for methods to locate the foramen magnum with respect to transverse axes within the basicranium. The nonhuman great ape with the most anteriorly situated foramen magnum is *Pan paniscus* and there is overlap, albeit very limited, between the ranges of variation in *P. paniscus* and some modern human populations. Among adult fossil hominins the position of the basion relative to the bi-tympanic line is most anterior in *Paranthropus boisei*. When the fossil record of a taxon is limited to the cranium at least one researcher suggested that the foramen magnum may be an appropriate and convenient proxy for body weight, but subsequent research has shown that foramen magnum area instead has a relationship with endocranial volume that is independent of its relationship via body weight. (L. *foramen* = an opening and *magnus* = great.) *See also* **cranial venous drainage**.

### foraminifera

An order of protozoans with many holes or foramina in their hard calcareous ($Ca_2CO_3$) shells. The calcium carbonate in the shells is the source for the oxygen isotopes used for paleoclimate reconstruction and for the temporal correlation known as orbital tuning. The $^{18}O/^{16}O$ ratios of the shells of deep sea, or benthic, foraminifera contain a temporal record of deep sea temperature and glacial ice volume. (L. *foramen* = an opening.)

### Forbes' Quarry

A limestone quarry on the north face of the Rock of Gibraltar. In 1848, a hominin cranium, Gibraltar 1, was found at the site and it shows most of the distinctive suite of Neanderthal characteristics. With hindsight it was the first adult cranium of *Homo neanderthalensis* to be recovered. (Location 36°08′N, 05°20′W, Gibraltar, UK.)

### force

The source of change to an object's momentum. A force, which has both a magnitude and a direction (i.e., it is a vector), has an equal and opposite reaction force. Thus, if the foot applies a force to the ground the ground applies a reaction force of equivalent magnitude (and opposite direction) to the foot. Anatomical systems (e.g., the upper limb in tool making and the lower limb in running) must be constructed in a way that enables them to both apply and resist forces. (L. *fortis* = strong.)

### force plate

A device that measures force using electronic devices that convert, or transduce, a force into an electric signal (i.e., they are load cells). The simplest force plates measure only the force normal (i.e., perpendicular) to and in the center of the plate, much like a bathroom scale does. Most force plates used in biomechanical research calculate the force at the center of pressure and provide all three main directional components of force (normal, side-to-side, and fore–aft) over time. Most applications in biomechanics concern the force exerted by (and on) the hindlimbs (and, if relevant, the forelimbs) during gait. *See also* **kinetics**.

## forcing
*See* **climate forcing**.

## forebrain
The largest of the three developmental subdivisions of the brain. Also known as the prosencephalon, it consists of two subdivisions, the diencephalon and the telencephalon. The diencephalon consists of subcortical structures including the thalamus and hypothalamus. The telencephalon refers to the cerebrum and includes the cerebral cortex, white matter, and basal ganglia. In primates in general, but especially in the great apes and particularly in modern humans, most of the forebrain consists of the cerebral hemispheres. It is the part of the brain that undergoes the most expansion during human evolution. However, this is an extension of a positively allometric trend seen across a range of mammals whereby increases in brain mass size are generally accompanied by preferential growth of the forebrain. (ME *fore* = the front and OE *braegen* = brain, most likely from the Gk *bregma* = the front of the head.)

## forest
An ecosystem dominated by trees and other woody vegetation. Within this broad category there are several distinct forest biomes (e.g., tropical dry, tropical wet, temperate deciduous, montane, and gallery or riparian forests) relevant to hominin evolution. A forest is defined as a stand of trees, which can vary from 10 to more than 50 m in height, with interlocking crowns. Forests can have several layers, including shrub and ground layers; in some forest biomes a grass layer may be present. The majority of African forests are evergreen or semi-evergreen although localized patches of deciduous forests do exist. (L. *foris* = outside.)

## form
Form refers to those aspects of an object that remain when all of its attributes other than size and shape are removed (i.e., when location in space, rotation, etc., are ignored). Form is often studied through the use of geometric morphometrics and/or Euclidean distance matrix analysis. Size and shape are of great interest in human evolution. The analysis of similarity and variation in form is used in alpha taxonomy, phylogenetic reconstruction, sexual dimorphism, developmental scaling, investigating the hormonal control of growth, and paleoecological reconstructions.

## formation
An inclusive term used in stratigraphy to refer to a set of strata whose upper and lower boundaries are evident. Formations are divided into members (e.g., the Koobi Fora Formation is divided into eight members) and members into beds.

## Fossellone

One of the large cave sites in the Monte Circeo complex in Italy. A fragmentary immature mandible and three teeth have been recovered from the Mousterian levels. Previously known as Circeo IV, this individual (now called Fossellone 3) belongs to *Homo neanderthalensis*. At least two other individuals (Fossellone 1 and 2) were found in the Aurignacian layers. (Location, 41°14′N, 13°05′E, Italy.)

## fossil

The word fossil now means a relic, or trace, of a formerly living organism, but in the past the term fossil was also applied to inorganic materials such as rocks and crystals. The modern interpretation includes "true" fossils (i.e., remnants of the organism itself, in the form of either hard or soft tissues) and "trace" fossils (aka ichnofossils) that provide direct evidence an organism has been at a particular place at a particular time long after the individual, or its hard tissues, have disappeared. Footprints (e.g., the Laetoli and Ileret hominin footprints) are an example of a type of trace fossil that preserves the impression of a soft tissue (i.e., the skin of the sole of the foot). The natural endocranial casts that faithfully reproduce the inner (i.e., endocranial) surface of the missing parts of the brain case of the southern African fossil hominins (e.g., Taung 1) are another example of a trace fossil that preserves the impression of hard tissue. (L. *fossilis* = to dig.)

## fossilization

The process whereby organic materials (usually bones or teeth but also other materials like wood) are transformed into a fossil. During life the bones and teeth – the structures that make up the vast majority of the hominin fossil record – are made up of a mixture of organic matrix and inorganic material. The dense cortical bone that forms much of the cranium and mandible, and the outer cortical bone of the long bones of the limbs, comprises about 70% by volume mineral, or inorganic, material and around 20% by volume organic material. The inorganic component is mostly a complex of chemicals called hydroxyapatite; the organic material consists predominantly of collagen embedded within a matrix of other organic components (as well as the cells that resorb bone and secrete bone matrix). The balance (approximately 10% by volume) is made up of water. Dentine, which forms the core of mammalian teeth, has a chemical composition similar to that of bone. Enamel, which is the hard, white, outer covering of teeth, consists of about 95% by weight inorganic material and less than 1% by volume organic material. It is the hardest biological tissue known and is almost pure apatite. During fossilization the organic component of hard tissues degrades and is replaced either totally or partially by minerals from the surrounding rock; fossils are effectively "bone-shaped" or "tooth-shaped" rocks. Enamel, dentine, and bone may take up elements (e.g., iron and manganese) from the surrounding deposits that stain them brown or black. The chemicals in the inorganic component are exchanged with those in the rock to different degrees and in different ways. Some cells actually mineralize and preserve during life within bone such that fossil bones can contain evidence of cells and even cell structure. The chances that the bones and teeth of a dead animal will end up in the fossil record are vanishingly small. Most of the animals that die in the wild are eaten by scavengers, their soft tissues are consumed or degraded by insects and bacteria, and once the soft tissues that link the bones at joints are lost the bones are disarticulated and scattered. They are then reduced to small taxonomically unrecognizable fragments by a combination of the destructive effects of diurnal heating and cooling and by being trampled underfoot by large mammals. In all but a very few cases these processes effectively eliminate the soft and hard tissues thus removing any evidence that individual had existed and denying it a place in the fossil record. If an animal *does* survive as a fossil it is usually because the remains were covered with sediment of some kind soon after death. The sediment protects the skeletal remains from damage and the moisture it contains provides a medium for the exchange of chemicals between the skeleton and the sediments surrounding it, which permits fossilization. Intentional burial increases the chance that hard tissues will fossilize and this partly explains the increase in the number of hominin fossils in later time periods. Larger population sizes likely also contribute to this increase in numbers of fossils. Some of the organic material in bones is retained long enough to make it useful for dating (*see* **amino acid racemization dating**; **radiocarbon dating**) and for helping to identify what type of animal the bone comes from (e.g., protein and DNA analysis).

## Foster's island rule
*See* **island rule**.

## founder effect
The term refers to the reduction in genetic diversity that occurs when a subset of the individuals belonging to a larger population forms the basis of (or "founds") a smaller subpopulation. The new subpopulation is likely to have allele frequencies that differ from those of the parent population and, since this sampling event is likely to change the gene frequencies compared to the parental populations, this phenomenon is often referred to as intragenerational genetic drift. In modern humans, founder effects can be seen in subpopulations such as the Amish, the Finns, and the French Canadians, each of which has an unusually high incidence of particular inherited diseases. Founder populations usually do not survive, but occasionally they do and evolve into a new species via the process of peripatric speciation. The founder effect very likely contributed to the genetic distinctiveness of the subpopulations of early hominin taxa that formed the basis of new species. *See also* **peripatric speciation**.

## fovea
A depression on an otherwise smooth surface. In mandibular postcanine teeth the fovea anterior is a depressed area of enamel between the mesial marginal ridge and the trigonid crest, and the fovea posterior is a depressed area of enamel between the ridge connecting the entoconid with the hypoconulid and the distal marginal ridge. There are comparable depressions on the upper teeth. There is a fovea on the head of the femur (the fovea capitis) where the ligament of the head of the femur is attached. The central portion of the macula region of the retina of all haplorhine primates is called the fovea centralis retinae. This part of the retina has an extraordinarily high packing density of cone photoreceptors and makes high visual acuity possible. (L. *fovea* = a small pit; pl. foveae.)

## FOXP2
An abbreviation of "*f*orkhead bo*x* *P2*," FOXP2 is a transcription factor that regulates genes involved in the development of the brain, lungs, and gut. Evidence from songbirds and mice implicates FOXP2 expression in the brain with the production of vocalizations. In modern

humans, expression of FOXP2 is important for the normal development of the neural circuits underlying language. There are two amino acid substitution differences between the FOXP2$^{human}$ protein and the equivalent protein in chimpanzees, FOXP2$^{chimp}$. Analysis of ancient DNA has indicated that Neandertals and possibly even Denisovans share these substitutions with modern humans. Researchers have speculated that these differences may have been involved in the evolution of complex language in modern humans.

## *FOXP2*

*FOXP2* is a gene in modern humans (NB: modern human genes are always italicized) that encodes FOXP2, a transcription factor that is important for the development of the brain, lung, and gut. Recent evolutionary changes in the *FOXP2* gene sequence are thought to be involved in the production of modern human language. *See also* **FOXP2**.

## frameshift mutation

A mutation within the coding sequence of a gene that involves changes other than the insertion or deletion of whole codons (i.e., other than multiples of three nucleotides). Such mutations disturb the reading frame of the gene and because of this they typically disrupt the messenger RNA (mRNA) produced by the gene, which can lead to a truncated version of the protein being translated, a protein with disrupted function, or no protein at all.

## frontal lobe

The anterior part of the cerebral hemisphere; the parietal lobe is immediately posterior to it. The frontal lobe is important for controlling movement and in planning and coordinating behavior. It consists of two main parts, the primary motor cortex (M1) located posteriorly in the precentral gyrus, and the prefrontal cortex (PFC) located anterior to the motor areas. The frontal lobe also includes supplementary and premotor areas. M1 and other motor areas are important for controlling the voluntary movements of the body, while the PFC is associated with executive functions (e.g., decision making, goal-oriented behavior). Broca's area, which is associated with speech, is also located in the frontal lobe in the inferior frontal gyrus. Although neuroscientists traditionally believed that enlargement of the frontal lobe in modern humans was responsible for the enhanced cognitive abilities in our species, recent studies have shown that the frontal lobe of modern humans comprises approximately the same proportion of the cerebral hemisphere as it does in the great apes.

## frontal trigon

The region between the supraorbital tori and the temporal crests or raised temporal lines. It can be either flat or hollowed and it is especially evident in some hyper-megadont archaic hominin crania (e.g., OH 5).

## frugivory

Literally, fruit-eating. Most primates are frugivores to a greater or lesser extent. Primate frugivores may eat ripe or unripe fruits and seeds. Hominoids favor ripe fruits, whereas some monkeys can tolerate unripe fruits. Some have argued that the ability to eat unripe fruit is one important reason for the adaptive radiation of the Old World monkeys and it may have helped them take over many of the niches previously occupied by hominoids in the late Miocene and Pliocene. In general frugivorous primates have larger brain sizes and larger home ranges relative to body mass than folivorous primates. (L. *frug-*, *frux* = fruit, and *vorous*, from the stem of *vorare* = to devour.)

## functional genomics
The study of basic functional aspects of the genome. Functional genomics subsumes research topics such as the mechanisms of gene expression during development, gene network interactions, gene–protein interactions, and the epigenetic processes responsible for proteome production.

## functional magnetic resonance imaging
(or fMRI) A method commonly used to investigate brain activity *in vivo* in modern human subjects and occasionally in macaque monkeys. It maps regional metabolic changes by tracking the different magnetic properties of oxygenated and deoxygenated hemoglobin present in the blood during regional activity. *See also* **magnetic resonance imaging**.

# G

### Gadeb
A site near the western edge of the South-East (or Somali) Plateau that borders the Afar Rift System in Ethiopia. An age range of c.1.5–0.78 Ma is supported by tephra correlated with samples from the Turkana Basin. No hominins have been found, but artifacts that occur at a number of localities and stratigraphic levels in the Gadeb area have been provisionally attributed to the Developed Oldowan (but containing handaxes) and to the Acheulean. (Location 07°05′N, 39°21′E, Ethiopia.)

### Gademotta
An artifact-bearing area on the flanks of a collapsed caldera approximately 5 km/3 miles west of Lake Ziway in the Main Ethiopian Rift. No hominins have been found, but the Middle Stone Age (or MSA) strata has been dated to $c.280 \pm 8$ ka, which makes Gademotta one of the oldest known MSA sites. (Location 08°03′N, 38°15′E, Ethiopia.)

### gait
The way an organism (e.g., hominin, horse, etc.) uses its limbs for locomotion. For example, the difference between a bipedal running gait and a bipedal walking gait is that in the latter one foot is in contact with the ground at all times. (ON *gata* = path.)

### Galería
One of the fossiliferous cave fillings in the Sierra de Atapuerca, a series of eroded, karstic, limestone hills 14 km/9 miles east of Burgos in northern Spain, which is permeated by sediment-filled cave systems. The largest is the Cueva Mayor-Cueva del Silo system, where the Sima de los Huesos is located, but just 1 km/0.6 miles away, in the Trinchera del Ferrocarril, is the Galería (TG) site with 17 m of sediment. Excavations at Galería, which is very close to Gran Dolina (TD), began in 1979. The sediments range from $503 \pm 95$ ka at the base to $185 \pm 26$ ka at the top. A hominin mandible fragment (AT76-T1H) was discovered on the surface in 1976 and a cranial vault fragment was found subsequently. (Location 42°21′09″N, 03°31′08″W, Spain.) *See also* **Atapuerca**.

### Galería del Osario
*See* **El Sidrón**.

## Galili

A site 30 km/18 miles to the west of the Mulu basin, Ethiopia. Unpublished argon-argon ages suggest that the sediments at Galili date from $c.5$ Ma to $c.3.4$ in the lower (fossiliferous) part. Hominins found at the site include six hominid teeth and a proximal femur fragment. None of the archeological evidence from the site was found *in situ*. (Location 09°46′N, 40°33′E, Somali Region, Ethiopia.)

## gallery forest
*See* **forest**.

## Gamedah
One of the named drainages/subdivisions within the Bodo-Maka fossiliferous subregion of the Middle Awash study area, Ethiopia.

## gamma taxonomy
It is common for reference to be made to alpha taxonomy, but you very seldom read, or hear, any reference to "beta" or "gamma" taxonomy. Whereas alpha taxonomy results in species being "characterized and named" and beta taxonomy refers to phylogenetic reconstruction, gamma taxonomy refers to the "analysis of intraspecific variation and to evolutionary studies." (Gk *gamma* = the third letter in the Greek alphabet and *taxis* = to arrange or "put in order.") *See also* **systematics**; **taxonomy**.

## gape
*See* **chewing**.

## Garba
*See* **Melka Kunturé**.

## Garusi
The name used by Ludwig Kohl-Larsen for the site now known as Laetoli. *See also* **Laetoli**.

## Garusi 1
The Garusi 1 maxilla was an important part of the evidence that led Don Johanson and colleagues to link the hominins from Hadar and Laetoli and to combine them into a single hypodigm as *Australopithecus afarensis*.

## Garusi River Series
*See* **Laetoli**.

## Gause's Rule
Also known as the principle of competitive exclusion, this states that two species cannot stably occupy an identical niche within the same ecosystem. *See also* **competitive exclusion**; **niche**.

## Gauss chron
*See* **geomagnetic polarity time scale**.

## gazelle
*See* **Antilopini**.

## gene
A region of DNA that contains information that specifies (a) a protein (or the RNA product it encodes) or (b) regulates the expression of messenger RNA (mRNA). Thus, a gene includes both coding and noncoding DNA. Genes typically include a promoter region that is usually upstream (at the 5′ end) of the gene sequence and it is where transcription is initiated and regulated. Most genes also have multiple exons and introns. Over evolutionary history genes (or parts of genes) can be highly conserved (i.e., they are resistant to change because they are subject to strong purifying selection) if change is highly detrimental, or they may be less conserved if some change is compatible with preservation of protein function. Genes that have been subject to directional selection (e.g., *FOXP2*, which encodes a transcription factor linked to phenotypes of speech and language) have been used to reconstruct adaptive changes in the hominin clade. Gene regions, such as introns or third positions of codons, that are less conserved may effectively evolve neutrally and thus can be used to reconstruct population history. (Gk *genes* = to be born.) *See also* **exon**; **intron**; **neutral theory of molecular evolution**.

## gene duplication
If a gene is duplicated and if the additional copy of the gene is not selected against (e.g., because of changes in the amount of protein produced) it may be free to mutate and that mutation may result in a change in function. This is one way novel genes are created over deep evolutionary time. For example, the alpha- and beta-globin genes evolved via gene duplication from a myoglobin-like gene. Genes may also be duplicated because they are located in regions of the genome that are subject to copy number variation. For example, the salivary amylase gene, *AMY1*, which produces the enzyme responsible for starch hydrolysis in the mouth, has been duplicated such that five to seven copies are present on average in modern humans while only two copies are present in chimpanzees. It has been suggested that this is because modern humans, particularly agricultural populations, typically have diets that are high in starch, whereas this is not usually the case with chimpanzees.

## gene expression
The process whereby the information encoded by a gene is used to make a gene product (e.g., a protein or one of the many kinds of RNA). Gene expression can be regulated in many ways (e.g., enhancers and repressors) and at different stages of transcription and translation. Gene expression may change during development, it may differ among the tissues of the body, and it may change according to environmental influences.

## gene family
A group of related (i.e., paralogous) genes that arose over time via gene duplication and thus are descended from a common ancestral gene. Members of the same gene family, also known as a multigene family, often have similar functions. For example, all the genes in the globin and *HOX* gene families are involved, respectively, in heme transport and development.

## gene flow
The movement of alleles of genes from one population to another. Along with genetic drift, mutation, and natural selection, gene flow (also known as gene migration) is one of the four forces of evolution. High levels of gene flow cause two populations to become more similar (i.e., it decreases the genetic distance between them). In modern humans, culture, language, and geography are factors that may restrict gene flow. Gene flow can result when substantial numbers of people simultaneously migrate from one area to another in a single event, or it can result when there are many episodes of smaller numbers of people moving from one place to another.

## gene genealogy
*See* **coalescent**.

## gene networks
Gene networks (aka gene regulatory networks) are sets of genes that interact (usually via their products) to perform some physiological or developmental function. This is one of the core concepts of systems biology. The impetus behind the concept is that gene function can only be understood within the context of the larger system within which it operates.

## gene pool
All of the alleles present in a population.

## general circulation model
(or GCM) A type of climate model that simulates dynamic interactions between the atmosphere, cryosphere, ocean, and land surfaces. The model is run until it reaches equilibrium and then it can be used to simulate climate variations which result from external forcings (e.g., insolation) that cause departures from the equilibrium state.

## generalist
*See* **eurytope**.

## generation time
The average interval between the birth of an individual and the birth of their offspring, or the average age of mothers giving birth in a given population.

## genetic assimilation
*See* **canalization**.

## genetic bottleneck
*See* **bottleneck**.

## genetic code
The genetic code is a three-nucleotide system (in DNA and RNA) that encodes genetic information. Each sequence of three nucleotides (also referred to as a triplet codon) defines either an amino acid or a punctuation message (i.e., the beginning or end of a messenger RNA or

protein sequence) for the processes of translation and transcription. Since there are four nucleotides there are 64 possible three-nucleotide sequences, but there are only 20 commonly occurring amino acids. This means that the genetic code is redundant with many amino acids such as serine and leucine (as well as the "stop" punctuation) encoded by several different codons. The code used in the modern human genome is commonly referred to as the standard or universal genetic code, and is shared among most organisms, but the genetic code can vary (e.g., the mitochondrial genome differs very slightly from the universal code and the mitochondrial code in mammals differs slightly from that in *Drosophila*).

### genetic correlation
A value that indicates how much of the genetic influence on two traits is common to both. Different traits in the same organism (e.g., length and mass) are often correlated. Typically, such correlations (also called phenotypic correlations) have genetic and environmental components. Genetic correlation is reflected in the proportion of variance that two traits share due to genetic causes (i.e., pleiotropy and gametic phase disequilibrium between genes affecting different traits). If the genetic correlation is greater than zero this suggests the two traits are influenced by common genes.

### genetic distance
A statistical measure of the relationships among DNA sequences, protein sequences, or other genetic materials. Genetic distances can be used to construct a phylogeny or to visually depict by other means the relationships between alleles, species, or populations. The relationships among the DNA sequences, protein sequences, or populations determine genetic distance, but there are circumstances in which two different sets of relationships can result in the same genetic distance.

### genetic drift
A random change in gene frequencies between generations. It occurs because the gametes produced by each parent contain a haploid genome that is a random assortment of chromosomes (i.e., for each type of chromosome an individual receives only one of the two chromosomes of that type within each parent). Genetic drift can also occur because of the founder effect. Genetic drift is more likely to be a significant factor in small populations because they have a smaller effective population size and it may have played an important role in shaping the diversity of past hominin populations. Along with gene flow, mutation, and natural selection, genetic drift is one of the four forces of evolution. (syn. random drift, drift.) *See also* **founder effect**.

### genetic hitchhiking
*See* **selective sweep**.

### genetic map
A map of the genome generated using a set of markers. The distance between markers is calculated not by the physical distance (that would generate a physical map) but by observing the number of recombination events that occur between them. For example, if two markers on the

same chromosome are typically separated after each meiosis they are likely located on opposite ends of the chromosome, whereas two markers that are tightly linked (i.e., typically inherited together) are likely to be close to each other on the same chromosome. The distance between markers in a genetic map is measured in centimorgans. One centimorgan equals the distance between markers in which there is a 1% recombination frequency (i.e., one recombination event per 100 meioses). *See also* **linkage**.

## genetic variation
The genetic diversity present in a species. Genetic variation is created through mutation but it is affected by genetic drift, natural selection, and gene flow. Patterns of genetic variation can thus inform researchers about the relative importance of these forces of evolution. They can also be used to infer demographic features of a species or population such as its effective population size, its potential hybridization patterns, or the likelihood that the population is expanding, stable, or contracting.

## gene tree
A diagram that depicts the phylogenetic relationships among alleles or between homologues of the same gene. Note that the tree for one gene can differ from that of another gene, or from the structure of a multigene tree, or from the species tree. Several gene trees, as well as phylogenies based on many genes, are consistent with the hypothesis that chimpanzees and modern humans are more closely related to each other than to any other taxa.

## gen. et sp. indet.
An abbreviation of "indeterminate genus and species." When dealing with fragmentary specimens it is sometimes possible to identify the higher taxon to which a specimen belongs, but it may not be possible to identify its genus and, therefore, its species, with certainty. In that case the taxonomic allocation researchers should use is gen. et sp. indet. For example, although some researchers are confident assigning the associated skeleton KNM-ER 1500 to *Paranthropus boisei*, others are more cautious and refer it to Hominidae gen. et sp. indet.

## genome
The complete DNA sequence of an organism. Traditionally, the term genome is used to refer to the complete set of chromosomes in a gamete (i.e., one set of chromosomes) so somatic cells in primates, which have two sets of chromosomes, have a diploid genome. Organelles such as mitochondria have their own DNA complement that can also be called a genome (e.g., the mitochondrial genome).

## genomic imprinting
In diploid organisms whether a gene is expressed or not depends on the parent from whom the individual inherited that gene's allele. For some genes, gene expression only occurs from the allele inherited from one parent. Genetic imprinting occurs in the germline cells (typically through methylation) and it determines whether it is the allele inherited from the mother or the one inherited from the father that is expressed in the offspring. Genomic imprinting is present in all marsupial and placental mammals.

## genomics
An umbrella term that subsumes all of the activities (e.g., mapping, sequencing, etc.) involved in the study of the genome.

## genotype
A term introduced for the genetic composition of an organism (as opposed to its physical appearance or phenotype). The term is also used for the combination of allelic states of a locus (or of a set of loci) on a pair of homologous chromosomes (e.g., a person with type A blood could have the genotype AO or AA for the ABO locus on chromosome 9). (Gk *genes* = to be born and L. *typus* = image.)

## genotype–environment interaction
A technical term used in quantitative genetics to refer to the phenotypic effect of the interaction between variation in genotypes and variation in the environment (i.e., $G \times E$). This term encapsulates the consensus view that variation in traits is caused by interactions between genotypes and their environment. Interactions between genotypes and their environment occur both prenatally (through the *in utero* environment) and in postnatal life.

## genus
The next to lowest category in the original classificatory system introduced by Linnaeus. According to Ernst Mayr a genus is as a group of species of common ancestry that shares both homogeneous and distinctive adaptations. Some researchers have proposed that a genus should be defined as a species or monophylum whose members occupy a single adaptive zone. This definition differs from Mayr's concept in that it excludes paraphyletic taxa and it does not require the adaptive zone to be unique or distinct. The same researchers suggested two criteria for assessing whether or not a group of species has been correctly assigned to a genus. First, the species should belong to the same monophyletic group as the type species of that genus. Second, the adaptive strategy of the species should be closer to the adaptive strategy of the type species of the genus in which it is included than it is to the strategy of the type species of any other genus. (Gk *genos* = birth; pl. genera.)

## genu valgum
See **valgus**.

## geochronologic units
The geologic time scale has a dualistic character in which intervals of time share terms with the sequence of rock (strata) deposited in that interval. Geochronologic units are time-referenced and thus subdivisions are temporal (e.g., Early, Middle, and Late), whereas their rock equivalents are chronostratigraphic units and their subdivisions are physically sequential (e.g., Lower, Middle, and Upper). In the hierarchy of terminology for the geologic time scale, geochronologic units are (from larger to smaller) eons, eras, periods, epochs, and ages (e.g., *Australopithecus afarensis* fossils from Laetoli date to the Pliocene epoch). (Gk *geo* = earth and *kronos* = time.)

## geochronology

The umbrella term used for methods that provide ages for fossils or rock layers. Dating methods traditionally have been divided into two categories: absolute and relative. Absolute dating methods are mostly applied to the rocks in which the hominin fossil was found or to nonhominin fossils recovered from the same level. This is why researchers must take great care to preserve any evidence that links a fossil to a particular rock layer. Absolute dating methods either rely on knowing the time it takes for natural processes, such as atomic decay, to run their course or they relate the fossil horizon to precisely calibrated global events such as reversals in the direction of the Earth's magnetic field. This is why absolute dates can be given in calendar years. Relative dating methods mostly rely on matching nonhominin fossils found at a site with equivalent evidence from another site that has been reliably dated using absolute methods. If the animal fossils found at site A are similar to those at site B, site A can be assumed to date to approximately the same age as site B. Compared to absolute dating methods, relative dating methods only provide approximate ages for fossils. But researchers involved in dating have largely abandoned the absolute/relative dichotomy. Instead, geochronological methods are now more usually divided up on the basis of the nature of the method and how the results of those methods are expressed. For example, sidereal dating methods (e.g., dendrochronology) are based on counting annual events; isotopic dating methods (e.g., argon-argon dating, radiocarbon, etc.) measure changes in the absolute amounts or ratios of isotopes; radiogenic dating methods (e.g., electron spin resonance spectroscopy dating, fission track dating, etc.) rely on measuring the effects of radioactive decay; chemical and biological dating methods (e.g., amino acid racemization) rely on measuring the progress of time-dependent processes; geomorphic dating methods (e.g., sedimentation rates) rely on processes that affect the landscape; and correlation dating methods (e.g., biostratigraphy, magnetostratigraphy, stratigraphy, tephrochronology) rely on relationships established using differences that are themselves time-independent. Numerical-age dating methods (e.g., sidereal, isotopic, and radiogenic) provide an age in years; calibrated-age and relative dating methods (e.g., chemical, biological, and geomorphic) provide an approximation of the real age and correlated-age dating methods (e.g., biostratigraphy, magnetostratigraphy, stratigraphy, tephrochronology) provide an age by relating the fossil or horizon to another independently dated sequence. (Gk *geo* = earth and *kronos* = time.)

## geomagnetic anomaly

A boundary between two prolonged periods during which the direction of the Earth's magnetic field has been consistent. These boundaries (aka magnetozones) are between what were previously referred to as epochs and events, but contemporary geomagnetic researchers now use numbered chrons and subchrons. Because the direction of the Earth's magnetic field reverses at these boundaries or anomalies, they are also known as polarity reversals or field reversals. (Gk *geo* = earth, *magnes* = magnet, and *anomalos* = uneven; syn. anomaly.) *See also* **geomagnetic polarity time scale**.

## geomagnetic dating

The use of the magnetic polarity of a stratum, or a sequence of magnetic polarities of a longer section, to date sediments or igneous rocks. *See also* **geomagnetic polarity time scale**.

## geomagnetic polarity time scale

(or GPTS) Time scale based on past changes in the direction of the Earth's magnetic field. The liquid core of the Earth behaves like a dipole magnet that generates a geomagnetic field. In the present day a magnetized needle will point to the North Pole (this is called the normal direction). However, at the beginning of the 20thC Bernard Brunhes, a French geophysicist known for his pioneering work in paleomagnetism, showed that when a magnetized needle was placed in the magnetic field preserved by iron oxide in some older lavas the needle pointed to the South Pole (the reversed direction). Motonori Matuyama demonstrated that the Earth's magnetic field had been reversed as recently as the early Quaternary. Scientists were initially unsure whether the lavas had been laid down during a global magnetic reversal ("field reversal"), or whether they had subsequently reversed their primary normal magnetism ("self reversal"). However, when potassium-argon dating began to be applied to the reversed rocks it became clear that "field reversal" was a reality. The first published land-based paleomagnetic timescale was relatively crude but within a short time the basic structure of the GPTS (i.e., long periods of either normal or reversed magnetism, then called epochs, punctuated by shorter periods of the opposite magnetic direction, then called events) was established. Researchers then began the process of integrating the land-based record with magnetic anomalies detected in the ocean floor, and by the late 1960s a geomagnetic reversal time scale was established. It recognized four epochs, each named after pioneer physicists (i.e., listed in order starting with the most recent: Brunhes, Matuyama, Gauss, and Gilbert); events within epochs were named for the locations where they were first identified (e.g., Jaramillo, Cobb Mountain, Olduvai). Confusingly, any change of magnetic direction (normal to reversed, or reversed to normal) is referred to as a "reversal" of the geomagnetic field. The latest version of the GPTS has dropped the eponymous terminology in favor of a numerical sequence. Epochs have been replaced by numbered chrons (the most recent chron is 1) and events by numbered subchrons. A letter postfix is used to identify chrons that have been added since the scheme was adopted (e.g., 3A) and if a chron has both normal and reversed components these are identified by lower case n or r (e.g., 3An and 3Ar). Subchrons within a chron are also numbered and their magnetic directions are identified using the same abbreviation system (e.g., the first reversed subchron in 3An is 3An.1r). The precision of the GPTS has been substantially increased by matching changes in geomagnetism with the increasingly detailed record of oscillations in the $^{18}O/^{16}O$ record. This, in turn, takes advantage of the correlations possible between such continuous geological records and the predicted ages derived from calculations using the frequencies of the Earth's orbital components (i.e., orbital eccentricity and the tilt and precession of the Earth's axis). This enables the ages of geological transitions to be tuned to the astronomical signal, a process called "orbital tuning." Geomagnetic data are not by themselves sufficient evidence to provide ages for strata containing fossils. One of the other dating methods is needed to provide an absolute date so that the sequence of changes in magnetic direction in the sedimentary sequence can be matched with the GPTS. When this is done the detailed sequence of magnetic reversals preserved in a section of sediments can then be used to date fossiliferous strata within it. One of the first events to be recognized was a brief episode of normal magnetic direction at c.1.9 Ma within the Matuyama reversed epoch. What was originally called the Olduvai event (now the Olduvai subchron or 1r.1n) was identified in the basalt underlying the fossiliferous sediments at Olduvai Gorge, Tanzania; it was the first time geomagnetic data were used to help date a hominin fossil site. The Cenozoic epoch boundaries are now tied to the GPTS chron system. The Miocene-Pliocene boundary is in chron C3 at 5.3 Ma, and the Pliocene/Pleistocene is either in chron C2 at 2.6 Ma or at the chron C2/chron C1 boundary at 1.8 Ma depending on whether the old or new definition is used. (syn. magnetic anomaly time scale, magnetic polarity timescale, magnetostratigraphy.)

## geomagnetic reversal time scale
*See* **geomagnetic polarity time scale**.

## geometric mean
A method for calculating the average of a set of numbers. It differs from the arithmetic mean because instead of summing $n$ values and dividing by $n$, the geometric mean calculates the product of $n$ values then takes the $n$th root of the result. The geometric mean may also be calculated as the antilog of the arithmetic mean of logged values. The geometric mean of a set of values is always less than or equal to the arithmetic mean of those values. The geometric mean is used to combine multiple measurements of an organism or structure into a single measurement that is a proxy for its overall size. *See also* **size**.

## geometric morphometrics
Used for the analysis of form, this is a family of methods that preserve the original geometry of the measured objects during all stages of the analysis. Geometric morphometric methods also visualize shape differences and shape changes using deformation grids (i.e., morphing of two-dimensional images or three-dimensional surfaces). Geometric morphometrics is based on the coordinates of landmarks. Two-dimensional coordinates are usually captured using a digitizing tablet or by measuring an image on the computer; three-dimensional data can be captured directly using a coordinate digitizer such as a Microscribe or Polhemus, or they can be measured on surface or volumetric scans that provide three-dimensional representations of an object's surface by using either laser or more traditional optical technology. Most software packages allow landmark coordinates to be measured directly on these virual surfaces or on volumetric objects. Statistical analysis in geometric morphometrics is performed using Procrustes shape coordinates (or their equivalent). Results of multivariate methods (e.g., principal component analysis, multivariate regression, and partial least squares) that preserve this well-defined metric (i.e., the Procrustes metric) in shape space can be visualized as actual shapes or shape deformations in the geometry of the original specimens. The weightings for linear combinations of the original variables can be visualized as shape deformations called relative warps. Many geometric morphometric analyses employ randomization tests (e.g., permutations or bootstrapping) rather than parametric methods to assess the statistical significance level of a given hypothesis.

## geophyte
*See* **underground storage organ**.

## "George"
Nickname given by Louis and Mary Leakey to OH 16 from Olduvai Gorge.

## Gesher Benot Ya'akov

The Gesher Benot Ya'akov (or GBY) deposits stretch for approximately 3.5 km/2 miles along both sides of the Jordan River in Israel. No hominins have been found at the site, but GBY has at least 13 archeological horizons that preserve fauna, plant remains, and Acheulean handaxes and cleavers that date to Marine Isotope Stages 18–20 (i.e., more than 790 ka). It also provides what is currently the earliest reliable evidence for (a) the use of fire, (b) nut-cracking, and (c) a system of spatial organization. (Location 33°00′28″N, 35°37′44″E, Israel.)

## gestalt

A pattern that has properties or meaning(s) that are greater than the sum of its components (e.g., the distinctive overall appearance of the face of *Paranthropus boisei*). (Ge. *stellen* = place.)

## gestation length

The period of time between ovulation and birth. Gestation length varies widely among primate taxa. Within primates, mouse lemurs have a gestation period of about 60 days, lemurs around 133 days, macaques from 145 to 185 days (it varies across species), and great apes between about 240 and 250 days. Modern humans average 267 days. Gestation length is an important component of reproductive investment. An active area of discussion has involved the question of extended gestation length in *Homo neanderthalensis*. Some researchers proposed that, based on aspects of pelvic morphology, Neanderthal gestation lasted 12 months. This was hotly debated, however, and evidence from subsequent finds of pelvic elements led to it eventually being discounted. The most recent reconstructions of Neanderthal pelves and neonatal head size suggest that brain size at birth was comparable in Neanderthals and *Homo sapiens*.

## G-function

*See* **fitness-generating function**.

## Gibraltar

A large limestone outcrop ("the Rock") on the southern coast of Spain that juts into the Mediterranean. The cave sites in this outcrop preserve evidence of some of the latest-surviving *Homo neanderthalensis* in Europe, and its southern location and relative isolation on the far end of the Iberian peninsula are consistent with the hypothesis that it is a refugium. Two Neanderthal fossils were found in the late 19thC and early 20thC. The first, a cranium from Forbes' Quarry (Gibraltar 1) on the north side of Gibraltar was discovered during blasting for limestone in 1848. It was the first Neanderthal fossil to be discovered, but the importance of the find was not recognized until 1862, some 2 years before the *H. neanderthalensis* fossils from Kleine Feldhofer Grotte were recognized as a distinct species. This discovery led several researchers to begin more concentrated examination of other Paleolithic sites on the Rock, and in 1926 Dorothy Garrod found the second Gibraltar fossil at Devil's Tower, along with animal bones, charcoal, and other evidence of human occupation. Several other caves with Mousterian levels have since been discovered and documented, including Ibex Cave, Vanguard Cave, and Gorham's Cave. *See also* **Gorham's Cave**

## Gibraltar 1
See **Forbes' Quarry**; **Gibraltar**.

## Gilbert chron
See **geomagnetic polarity time scale**.

## glabella
The most anteriorly projecting point in the midline of the cranium between the superior orbital margins. It is especially projecting in *Paranthropus*, and in particular in *Paranthropus boisei*. (L. *glabellus* = smooth and hairless.)

## glacial cycles
The term "ice age" was apparently introduced by Goethe, but it soon became clear that it was not one monolithic event but a series of colder glacial periods punctuated by warmer interglacial periods. These cold/warm cycles were originally identified and defined on the basis of lithological evidence, and before the use of Marine Isotope Stages the antiquity of a horizon or a site was reflected by it being linked with one of these named glacial cycles. There are several regional terminologies for glacial cycles, but the one with historical priority is the scheme developed for use in the Alps that names the glacial periods, from recent to the oldest: Würm, Riss, Mindel, Günz, and Donau. The interglacial periods are named after the adjacent glacial periods, the older one first (i.e., Riss-Würm, Mindel-Riss, Günz-Mindel, and the Donau-Günz). Geologists in northern Europe used a different terminology, with separate terms for the glacial and interglacial cyles (Würm = Weichselian; Riss-Würm = Eemian; Riss = Saalian; Mindel-Riss = Holsteinian; Mindel = Elsterian; Günz-Mindel = Cromerian and the Bavelian; Günz = Menapian; Donau-Günz = Waalian, etc.). The British Isles had yet another scheme, with the Ipswichian interglacial equivalent to the Riss-Würm/Eemian, the Hoxnian interglacial equivalent to the Mindel-Riss/Holsteinian, and the Anglian glaciation equivalent to the Mindel/Elsterian. Around the time this latter scheme was introduced "cold stage" began to replace "glacial" and "temperate stage" replaced "interglacial." Subsequently, all of the named stages or cycles have been superseded by numbered Marine Isotope Stages. See also **Marine Isotope Stages**.

## glacial ice volume
Glacial ice volume refers to the volume of ice locked up in continental glaciers and it may be usefully measured in sea-level-rise-equivalent units. As the volume of continental ice has waxed and waned during glacial (cold) and interglacial (temperate/warm) cycles, the changing volume of continental ice has dramatically altered sea levels. The volume of continental ice locked up in the East Antarctic Ice Sheet is equivalent to approximately 65 m of sea-level rise and that in the West Antarctic Ice Sheet and the Greenland ice cap is equivalent to approximately 8 and 6 m, respectively. All the other ice caps and valley glaciers are estimated to be equivalent to about 0.5 m of sea-level rise. At the Last Glacial Maximum sea levels were approximately 120 m lower than they are today and this may have provided opportunities for hominins to migrate out of Africa across the narrow, shallow, channel at the southern end of the Red Sea, and out of Southeast Asia into Australasia. Lowered sea levels also provided access to new coastal habitats (e.g., around southern Africa) that are now inaccessible because they

are below sea level. Glacial ice volume can be roughly estimated from geomorphological evidence on land, ice rafting in the oceans, and most importantly the oxygen isotope ratio of benthic foraminifera.

### glacial period
See **glacial cycles**.

### Gladysvale
A cave system within Precambrian dolomite 21 km/13 miles northeast of Sterkfontein and approximately midway between Drimolen and Gondolin, in South Africa. The deposits span a period between $c.2.4$ Ma and 14 ka. The GVED deposits, which represent an unroofed paleocave, date to 780–560 ka, whereas deposits within the cave are older, and those in the Peabody Chamber are younger than the GVED. A single handaxe was recovered from a deposit below a flowstone recording the magnetic reversal at the Brunhes-Matuyama boundary at $c.780$ ka. The first hominin finds, isolated teeth from breccia dumps, were assigned to *Australopithecus africanus*; a single phalanx assigned to *Homo* sp. was recovered from later *in situ* excavations. A probable hominin hair found in a hyena coprolite was reported from GVID sediments dated to $c.200$ ka. (Location 25°53′42″S, 27°46′21″E, Gauteng Province, South Africa.)

### glass
An amorphous solid usually formed by the rapid cooling of magma. Volcanic glass forms when magma is quenched (i.e., it is rapidly cooled). It occurs in large volumes as obsidian and in small, microscopic volumes it forms the matrix of most lavas. Fragmented glass explosively erupted from a volcano is referred to as vitric tephra. Because the chemistry of a glass closely approximates the composition of the parent magma at the time of cooling, vitric tephras record a geochemical fingerprint of an eruption. Widely dispersed products of such an eruption can thus be correlated geochemically to provide an isochronous marker and in some cases these markers can be dated by absolute methods (e.g., by argon-argon dating of associated potassium-bearing feldspars). *See also* **obsidian**; **tephrostratigraphy**.

### gnathic
Anything that relates to the upper (maxilla) or the lower (mandible) jaw. (Gk *gnathos* = jaw.)

### Gomboré
See **Melka Kunturé**.

### gomphoses
See **joint**.

### gomphothere
The informal name for the taxa within the family Gomphotheriidae, which contains all the proboscideans with shovel-shaped tusks. (Gk *gomphos* = joint, *ther* = combining form, wild beast.) *See* **Proboscidea**.

## Gona Paleoanthropological study area
The Gona Paleoanthropological study area is located in the west-central section of the Afar Rift System, Ethiopia; it is west of the Hadar study area and north of the Middle Awash study area. Initial fieldwork at Gona was carried out by Hélène Roche and later by Jack Harris in the 1970s and collaborative work by Jack Harris and Sileshi Semaw began in 1987. Systematic excavations at Kada Gona began in 1992 and these resulted in the recovery of the oldest known stone artifacts dated to 2.6 Ma. The early Pliocene deposits exposed in the Gona section of the Western Margin have yielded hominins attributed to *Ardipithecus ramidus*. The eastern and southeastern portions of the study area in the Kada Gona, Ounda Gona, and Dana Aoule drainages contain sediments more than 80 m thick containing Plio-Pleistocene fossil fauna and Oldowan artifacts. In addition, several Acheulean occurrences with estimated ages between 1.7 and 0.5 Ma have been documented in these drainages and some of these sites have also yielded hominin fossils including a female pelvis and a cranium discovered in 2006. The Gawis drainage and the area the east of it contain late Acheulean, Middle Stone Age, and Later Stone Age sites. (Location 11°07–08′N, 40°18–19′E, Ethiopia.)

## Gona pelvis
See **BSN49/P27**.

## Gondolin
An abandoned limeworks situated on a cave system within Precambrian dolomite near Broederstroom, 24 km/15 miles northwest of Sterkfontein, South Africa. Two *in situ* sequences, GD 2 and GD 1/3, are recognized at the site of the old cave along with extensive *ex situ* dumpsite deposits. Both of the hominin teeth were recovered from breccia blocks from one of these dumpsite deposits, Gondolin Dump A (GDA). GD 1 is the likely source of the breccia blocks in Gondolin Dump A that yielded the two hominin teeth. Fauna from GD 2 indicates an age equivalent to that of Swartkrans Member 1 (>1.8 Ma) and paleomagnetic analysis indicated that the GD 2 fossil bearing siltstone preserves a normal magnetic polarity consistent with the Olduvai normal polarity subchron between 1.95 and 1.78 Ma; GD 2 is slightly older than GD 1, although both likely accumulated close to 1.8 Ma. Two hominin teeth, GDA-1, a possible *Homo* left lower molar, and GDA-2, a large *Paranthropus robustus* left $M_2$, have been recovered from the site. (Location 25°49′50″S, 27°51′51″E, North West Province, South Africa.)

## Gongwangling
See **Lantian Gongwangling**.

## Gorham's Cave
Part of a complex of four sea caves (the others are Bennett's, Hyena, and Vanguard) at the base of the eastern face of the Rock of Gibraltar. The four caves, which are within the youngest of five blocks of Jurassic limestone that form the Rock, are filled with wind-blown sands mixed with organic material that were deposited during periods of lowered sea levels when the coast was up to 4.5 km/2.8 miles away from the caves. The deposits span the period from 55 to 15 ka

(i.e., the sequence includes the Last Glacial Maximum). Gorham's Cave, which has the deepest deposits (18 m) preserves the most complete sequence. Systematic excavations have exposed about 29 m² area of the floor of the inner cave. There are four main occupation levels, the lowest of which contains 103 flint, chert and quartzite stone tools attributed to the Mousterian technocomplex, plus discrete lumps of charcoal and a presumed hearth. (Location 36°07'15"N, 05°20'31.35"W, Gibraltar, UK).

## Gorilla

What we know as the gorilla was first described by the Reverend Savage in 1847. It was initially interpreted as a new species of chimpanzee called *Troglodytes gorilla* (at that time the generic name *Troglodytes* was used for chimpanzees); the genus *Gorilla* was not introduced until the 1860s. Paul Matschie and others in the early 20thC described a number of supposed new species of gorillas, but in 1929 Coolidge sank them all into a single species, *Gorilla gorilla*, with two subspecies (the western gorilla, *G. g. gorilla*, and the eastern gorilla, *G. g. beringei*). In the 1960s it became clear that eastern gorillas are divisible into two subspecies, the true mountain gorilla (*G. g. beringei*) and an eastern lowland or Grauer gorilla (*G. g. graueri*). Nowadays it is more usual to recognize two species of gorilla, *Gorilla gorilla* (western) and *Gorilla beringei* (eastern), based on their morphological and genetic differences. Two subspecies are recognized in each region. The Cross River gorilla (*G. g. diehli*) and the more widespread *G. g. gorilla* in the west; and the Grauer gorilla (*G. b. graueri*) and the mountain gorilla (*G. b. beringei*) in the east.

## graben

In structural geology a graben is a geological structure consisting of two (or more) inwardly directed normal faults, which form a trough-like feature. Because of their trough-like nature, grabens are locations where potentially fossil-bearing sediments are preferentially deposited and preserved and thus grabens make obvious targets for geological and paleontological exploration. The East African rift valleys are examples of grabens. *See also* **Afar Rift System**; **East African Rift System**.

## gracile australopith

*See* **robust australopith**.

## grade

Julian Huxley introduced this term for what he called an "anagenetic unit." Whereas a clade is analogous to a make of car (e.g., all Rolls Royce cars share a recent common ancestor not shared with any other make of car), a grade is analogous to a type of car (e.g., luxury cars made by Mercedes, Jaguar, and Lexus are functionally similar yet they have different evolutionary histories and therefore have no uniquely shared recent common ancestor). Thus, it is a category based on what an animal looks like and does, rather than on what its phylogenetic relationships are. Thus "leaf-eating monkey" is a grade that contains both Old and New World monkeys, but it is not a clade because the two groups of monkeys from the Old and New Worlds are components of different larger clades. Grades may also be clades, but they do not have to be. (L. *gradus* = stage.) *See also* **genus**.

## graminivory
Literally, the consumption of grass. This is a highly specialized dietary strategy and the only primates that follow it are found within the genus *Theropithecus*. Today *Theropithecus gelada* live in the Ethiopian Highlands where their diet consists almost entirely of the leaves, rhizomes, stems, and seeds of grasses. (L. *gramen* = grass and *vorous*, from stem of *vorare* = to devour.) *See also* **seed-eating hypothesis**.

## grandmother hypothesis
A model of human evolution introduced by Kristen Hawkes and colleagues that emphasizes the important role played by postmenopausal women in helping to provide food for their daughter's children. It is an attempt to explain why, in modern humans, female fertility and mortality are decoupled so that females have a long post-reproductive lifespan. The hypothesis suggests that post-reproductive adults and particularly grandmothers can continue to indirectly influence their own reproductive success (i.e., inclusive fitness) by contributing to the reproductive success of their daughters. Grandmothers achieve this by foraging for foods such as tubers that their grandchildren would find difficult to access on their own. This assistance increases the chances of survival for existing offspring and it also allows daughters to wean infants early, and move on to the next pregnancy, thus shortening the interbirth interval.

## Gran Dolina
One of the fossiliferous cave fillings in the Sierra de Atapuerca, a series of eroded, karstic, limestone hills 14 km/8.5 miles east of Burgos in northern Spain. It is permeated by sediment-filled cave systems. The largest is the Cueva Mayor-Cueva del Silo system where the Sima de los Huesos is located. Just 1 km/0.6 miles away in the Trinchera del Ferrocarril is the Gran Dolina (TD) with 18 m of sediment. Seven levels (TD3–4, 5–7, and 10–11) have produced stone tools and in 1994 more stone tools and hominin fossils were recovered from the Aurora stratum (a conspicuous layer within the TD6 lithostratigraphic level). The hominin fossil elements from the TD6 sample represent at least 10 individuals. Researchers now recognize six lithostratigraphic layers within the Aurora stratum, which they refer to as the Aurora archaeostratigraphic set (or AAS). The age of TD6 is $731 \pm 63$ ka. The hominin fossils recovered from the Gran Dolina cave have been assigned to a new species, *Homo antecessor*, on the basis of a hitherto unknown combination of a modern human-like midfacial morphology (e.g., a canine fossa and an acute zygomaticoalveolar angle) and a primitive *Homo ergaster*-like dental morphology. However, critics suggest that the "modern" facial morphology is due to the immaturity of the ATD6-69 facial fragment. The archeological evidence is mainly small Mode 1 artifacts, many of which have been used for butchery and woodworking. The cutmarks on some of the hominin specimens shows that the hominins were deliberately defleshed. Bone tools have been recovered from the *c.*400 ka TD10-1. (Location 42°21′09″N, 03°31′08″W, Spain.)

## granivory
Literally, the consumption of seeds. This term applies principally to rodents and insects that are exclusive seed-eaters. Some living primates (e.g., *Theropithecus*) consume seeds, particularly of flowering plants and sometimes grasses, as part of their diet but no living primate has an exclusively granivorous diet. (L. *granum* = grain and *vorous*, from stem of *vorare* = to devour.) *See also* **graminivory**; **seed-eating hypothesis**.

## grassland

A biome with mean annual rainfall between 250 and 800 mm, a high rate of evaporation, periodic severe droughts, and a rolling to flat terrain. A grassland is a biome where most animals are either grazing or burrowing species, and where there is less than 10% cover of woody plants. The majority of modern African grasslands are dominated by $C_4$ grasses, but prior to 1.7 Ma most of the East African fossil localities characterized as grasslands were a mix of $C_4$ and $C_3$ plants; modern $C_4$ grasslands are a more recent phenomenon. Most grasslands require periodic fires for maintenance, renewal, and elimination of encroaching woody growth. Production in grasslands is positively correlated with precipitation. Whereas grasslands are thought to have at one time covered about 40% of the land surface on Earth, today they cover less than 12%. Much of the secondary grasslands in Africa would revert to woodland if not for overgrazing and fire. Grasslands tended to dominate at many hominin sites in Africa between 2.0 and 1.0 Ma, as evidenced by carbon isotopes in soils and deposits of fossils from mammalian grazers. *Homo erectus* first appeared during this period and is thought to have existed in grasslands and lightly wooded habitats. After 1.0 Ma many of these grasslands became bushlands and woodlands.

## Gravettian

An early Upper Paleolithic technocomplex found throughout Europe, dated to a particularly cold phase (28–19 ka) during the Last Glacial Maximum. It is characterized by the appearance of diagnostic stone tools including burins and straight-backed points, the exploitation of large herd mammals, particularly mammoths, the expansion of long distance trade as evidenced by shells, and the production of ivory and clay stylized figures of women (known as Venus figurines) and animals. Sites in central and eastern Europe (e.g., Pavlov, Dolní Věstonice) show further technological specialization, including meat-storage pits, bone burning for fuel, and construction of shelters out of animal bones. A similar technocomplex from the same time period, but which is confined to central and southwestern France, is known as the Perigordian.

## gray matter

A term that describes regions of the central nervous system that mainly comprise the cell bodies of neurons. There is gray matter in the cerebral cortex, the cerebellar cortex, the nuclei of the brain and the brain stem, and in the central portion of the spinal cord.

## gray scale image

See **computed tomography**.

## gray scale values

See **computed tomography**.

## grazer

An animal that feeds on ground-level vegetation (e.g., terrestrial or aquatic grasses). Grazers tend to have a relatively broad snout (e.g., extant hippopotami) so that the anterior teeth can trap as much grass as possible, whereas browsers have a narrower snout (e.g., modern giraffes) that allows them to select items in bushes and trees. In tropical environments, the vegetation that grazers feed on (i.e., grasses) uses the $C_4$ photosynthetic system. Thus grazing animals have bones and enamel that have more $^{13}$C relative to the bones of browsing animals that consume plants using the $C_3$ photosynthetic system. See also **$C_3$ and $C_4$**; **grazing**.

## grazing

The consumption of the leaves and shoots of monocotyledonous plants (i.e., grasses and sedges having a single embryonic seed leaf or cotyledon) that grow at, or slightly above, ground level. Many antelopes, particularly the alcelaphines (i.e., wildebeest and allies that belong to the tribe Alcelaphini) and the antilopines (i.e., gazelles and allies that belong to the tribe Antilopini) graze on grasses. A grazing diet in extinct animals is usually inferred through studies of tooth morphology or stable isotope, usually carbon, geochemistry. Grazing on tropical grasses that have a $C_4$ photosynthetic pathway leaves a stable carbon isotopic signal that can be detected in tissues, including the enamel of fossil teeth. Grazing is important for understanding the environmental context of hominin evolution. Adaptive radiations of grazing ungulates occurred along with the reduction of forest and more wooded landscapes in Africa. The spread of grazing as a dietary strategy in the faunal community thus helps document changes in the environments inhabited by early hominins, and some hominin taxa (e.g., *Paranthropus boisei*) have a carbon isotope $C_4$ signature that is consistent with a grazing-like niche. See also **$^{13}$C/$^{12}$C**; **stable isotopes**.

## great ape

An informal taxonomic category that includes extant and fossil taxa. The extant "great ape" taxa are chimpanzees/bonobos, gorillas, and orangutans; the fossil taxa are all the extinct forms that are more closely related to chimpanzees/bonobos, gorillas, and orangutans than to any other living taxon. In the traditional, pre-molecular, taxonomy the informal term "great apes" was equivalent to the family Pongidae. (OE *great* = thick or large and *apa* = ill-bred and clumsy; literally the "large and clumsy" ones, as opposed to the "small and clumsy" ones, known as the "lesser apes." Before the apes had been investigated scientifically and appreciated on their own terms, "apes" were regarded as being "clumsy" because they lacked dexterity.) See also **ape**.

## Green Gully

A soil pit 14.5 km/9 miles northwest of Melbourne and 1.6 km/1 mile south of Keilor, Australia. In 1965 hominin bones were accidentally uncovered during commercial sand-extraction operations and the direct radiocarbon dating of collagen from the skeletal suggests they are c.6500 years BP. Most researchers suggest that the modern human remains from Green Gully represent two individuals, Green Gully 1, the partial skeleton of an adult female that includes portions of the cranium, thorax, and right upper limb, and Green Gully 2, the partial skeleton of an adult male that includes most of the left upper limb, a fragmentary pelvis, and both lower limbs. Archeological evidence includes large and small quartzite scrapers and quartzite fabricators. (Location 37°44′S, 144°50′E, Victoria, Australia.)

## Grenzbank zone

A 1 m-thick boundary layer between the underlying Sangiran Formation and the overlying Bapang Formation in the Sangiran Dome of central Java, Indonesia. This stratum is a calcareous conglomerate rich in fossils, including those of hominins. Sites where this formation is exposed include Blimbingkulon, Mojokerto, and Sangiran. Published argon-argon dating suggests the lowermost Bapang Formation hominins are 1.51–1.47 Ma and the youngest hominins from the Sangiran Formation are $c.$ 1.6 Ma; thus the Grenzbank is $c.$ 1.6–1.51 Ma. Hominins recovered from the Grenzbank Zone include BK 7905, BK 8606, Sangiran 6 ("Meganthropus A"), and Sangiran 9 ("Mandible C"). (Ge. *grenzbank* = "border bed.")

## Grimaldi

A complex of limestone caves and rockshelters at the base of a cliff about 1 km/0.6 miles from the Italy/France border on the Mediterranean coast; it is also known as the "Balzi Rossi" or "Baousse Rousse" caves after the reddish color of the cliff. The caves and rockshelters that have yielded hominin fossils are, from west to east, Grotte des Enfants, Riparo Mochi, Barma del Caviglione, Riparo Bombrini, Barma Grande, Baousso da Torre, and Grotta del Principe. As many of the caves were explored before rigorous excavation techniques were developed, there are many unanswered questions about their stratigraphy and chronology. However, direct radiocarbon dating of a number of the skeletons gives an age of $c.$ 24 ka for the Gravettian layers and $c.$ 11 ka for the Epigravettian. All of the complete or near-complete skeletons known from the Paleolithic deposits in the caves belong to *Homo sapiens* and they appear to have been intentionally buried. Archeological evidence found at the site includes Mousterian assemblages from the lower Middle Paleolithic levels, while Aurignacian and Gravettian assemblages are known from the Upper Paleolithic. In addition to Gravettian lithic technology, a number of "Venus" figurines have been found in the Upper Paleolithic levels of the caves, but there is some confusion as to which figurines come from which cave. (Location 43°47′N, 07°37′E, Italy.)

## Grotta Breuil

One of the caves in the Monte Circeo complex in Italy, it is a large karstic chamber located on the southwestern face of the promontory, just above the present sea level. The few hominin remains, and an abundance of associated stone tools and a rich faunal assemblage, are interpreted as representing a late occurrence of *Homo neanderthalensis* in the Italian peninsula. The hominin remains are dated to late Marine Isotope Stage 3 (i.e., $c.$ 45–35 ka). A local Mousterian industry was found associated with the remains. (Location 41°14′N, 13°05′E, Italy.)

## Grotta di Lamalunga

A cave system in karstic limestone near Altamura, Bari, Apulia in Italy, where a hominin skeleton (Altamura 1) covered in a variably thick coating of calcareous material was found in 1993 by speleologists.

## Grotta Guattari

One of the many caves in the Monte Circeo complex in Italy, it contained the most iconic of the hominin remains recovered from the Monte Circeo promontory. Hominins found at the site include Guattari 1 (formerly Circeo I), a fairly complete cranium, and two fragmentary mandibles, Guattari 2 and 3 (formerly Circeo II/A and III/B), all attributed to *Homo neanderthalensis*. Archeological evidence found at the site has been interpreted as a local Mousterian industry. It was thought that Guattari 1 was the subject of a cannibalistic ritual, because of the damaged foramen magnum and the "circle of stones" surrounding the skull and other evidence. This is most likely not the case, since more thorough study in the 1980s suggests the cave was primarily a hyena den and was only secondarily occupied by hominins. (Location 41°14′N, 13°05′E, Italy.)

## Grotte des Enfants
The westernmost of the caves in the Grimaldi complex; it is also known as the Grotta dei Fanciulli. See also **Grimaldi**.

## Grotte des Fées
See **Arcy-sur-Cure**.

## Grotte des Pigeons
This large (>400 m$^2$) cave near the village of Taforalt in the Beni Snassen mountains of Morocco was formed in Permo-Triassic dolomitic limestone. The approximately 10 m-thick archeological sequence consists of Aterian and Iberomaurusian artifacts as well as fossils within multiple layers of silt, silty loams, interstratified speleothems, and anthropogenic ashes and charcoal. The Aterian levels contain early shell beads with an age estimate of 82 ka, whereas the Iberomaurusian strata include more than 180 Late Pleistocene (c.17–13 ka) hominin burials. Archeological evidence found at site includes Middle Paleolithic tools such as scrapers, small Levallois cores, and thin bifacially worked (Aterian) foliate points, well-defined hearths, and 13 small (<2 cm) *Nassarius gibbosulus* shells transported more than 40 km/25 miles to the site. Some of the shells may have been deliberately pierced, 10 show wear patterns suggestive of "string damage," and hematite is present on 10 of them. Older excavations recovered a number of typical Aterian tanged or pedunculate points and other retouched implements. The amount of cedar (*Cedrus*) charcoal recovered fluctuates throughout the Pleistocene. Dates from some of the Iberomaurusian strata suggest the presumed modern human occupations coincided with Heinrich events. (Location 34°48′38″N, 02°24′30″E, Morocco.)

## Grotte du Hyène
See **Arcy-sur-Cure**.

## Grotte du Loup
See **Arcy-sur-Cure**.

## Grotte du Renne
See **Arcy-sur-Cure**.

## ground reaction force
See **force**.

## group
Formal geological term for a stratigraphic package comprised of several formations (e.g., in the Turkana Basin, the Omo Group includes the Koobi Fora, Nachukui, and Shungura Formations).

## growth

Growth is broadly defined as an increase in size. During ontogeny, growth occurs in a number of dimensions, including mass (which represents the sum of lean and fat mass, water, organ weight, and skeletal weight), proportion, and linear dimensions. In modern humans pronounced growth in stature occurs shortly after puberty; this is unique in comparison to nonhuman primates. The end point of growth is the size attained by adulthood, but this point may be reached at different times for different dimensions.

## growth plate

A cartilaginous plate (aka the epiphyseal plate) located at the ends of long bones in individuals who are still undergoing growth (e.g., juveniles, individuals with estrogen insensitivity). The chondrocytes of the growth plate proliferate and turn into cells that eventually ossify and contribute to proximal–distal bone growth. Under the influence of steroid hormones, the cartilage cells of the growth plate cease to proliferate and the whole area becomes ossified. The rate at which ossification occurs, and the relationship of this ossification to chronological age, can vary greatly among individuals but methods have been devised that assess bone age based on the degree of ossification of the bones of the wrist and hand (e.g., the Gruelich–Pyle and the Tanner–Whitehouse methods). The states of fusion of epiphyses from fossil hominins are used to estimate chronological age, but the efficacy of such methods will vary depending on the reference population used and the skeletal elements preserved.

## growth spurt

See **adolescence**.

## GSI-Narmada

This calvaria, which preserves most of the right side and part of the left parietal, is the only significant fossil hominin evidence from the Indian subcontinent. See also **Hathnora**.

## gubernacular foramen

The opening of a gubernacular canal at the surface of the alveolar process. The opening communicates with the dental crypt that contains the developing tooth germ. See also **gubernaculum**.

## gubernaculum

With respect to the development of the hard tissues, gubernaculum refers to a cord of fibrous tissue that connects the outer layer of the dental follicle (the fibrous sac surrounding the developing tooth germ in the bony crypt) with the mucous membrane that lines the mouth or oral cavity. Its function is to guide or direct the erupting tooth to its proper location in the alveolar process. The bony gubernacular canals containing the gubernacular cords of the permanent anterior teeth open on the lingual side of the erupted deciduous teeth at a gubernacular foramen, but those of the permanent molars (and sometimes the premolars) rise directly upwards through the opening in the roof of the bony crypt. During eruption the gubernacular cords decrease in length, increase in thickness, and become replaced at the surface by an epithelial plug, through which the tooth cusp(s) eventually emerge into the mouth. (L. *gubernaculum* = a rudder or helm.)

## guild
In ecology the term guild is used to describe a group of species that exploit resources in a similar way. Overall body size and functional morphology-based interpretations of the types of resources consumed are considered when defining a guild. For example, dental morphology was used to identify four carnivore guilds at the Miocene hominoid site of Pasalar, Turkey. (ON *gildi* = payment. Craftsmen or tradesmen had to pay dues to belong to their appropriate guild.)

## Günz
*See* **glacial cycles**.

## gyri
Pl. of **gyrus** (*which see*).

## gyrus
The visible part of a fold in the cerebral (or cerebellar) cortex that can be seen on the surface of the brain (e.g., the pre-central gyrus is where the motor cortex is situated). (Gk *gyros* = circle, or ring.)

# H

### ²H/¹H
The relative ratio of these two stable isotopes of hydrogen (or δD, where D stands for deuterium, the ²H isotope) is used in stable-isotope biogeochemistry. Animal δD, which is most commonly measured in hair or fur, feathers, or bone collagen, reflects the water they drink and the food they eat. Plants get all of their hydrogen from water. Water sources (e.g., groundwater, surface water, and rain water) have different δD values, thus plants using different water sources (e.g., due to differences in growing season or rooting depth) will have different δD values. Evapotranspiration also increases δD values in leaves because the lighter isotope (¹H) has a higher vapor pressure and evaporates first so that the remaining leaf water is enriched in ²H. This effect is exacerbated in windy or arid regions and is reduced in humid regions.

### *h* or *h²*
Abbreviation of narrow-sense heritability. *See* **heritability**.

### *H* or *H²*
Abbreviation of broad-sense heritability. *See* **heritability**.

### habitat
The place where an organism, population, or community lives. Almost all habitats are shared by a range of taxa that together contribute to the community ecology of that habitat. Habitat differs from niche in that the former stresses location while the latter stresses environmental variables. (L. *habitare* = to inhabit.)

### habitat mosaic
*See* **mosaic habitat**.

### habitat theory hypothesis
An overarching theory proposed by Elisabeth Vrba that seeks to explain how environmental change influences evolutionary patterns in the fossil record. Habitat theory embraces no fewer than seven related hypotheses: (a) most species survive periods of environmental change by passively "tracking" the changing geographical distribution of their preferred habitat (e.g., as savanna conditions spread, so do the ranges of savanna-adapted mammals); (b) areas with

---

*Wiley Blackwell Student Dictionary of Human Evolution*, First Edition. Edited by Bernard Wood.
© 2015 John Wiley & Sons, Ltd. Published 2015 by John Wiley & Sons, Ltd.

irregular topography (i.e., peaks and valleys) will show higher rates of species' ranges being split (this phenomenon is called vicariance) than will areas of gentle topography (i.e., smooth gradations in elevation) because irregular topography encourages habitat fragmentation; (c) during periods in which global temperatures vary strongly by latitude, species living in equatorial latitudes should be more prone to vicariance, speciation, and extinction as a result of cyclical environmental change than species living at higher latitudes; (d) during extreme cooling trends, species' extinctions should be concentrated at low latitudes because warm equatorial-type habitats are fragmenting or disappearing; conversely, extinctions should be concentrated at high latitudes during extreme warming trends; (e) speciation does not occur unless it is "forced" by environmental change; thus, most of the turnover in lineages (i.e., the extinction of species and the appearance of new ones) occurs simultaneously in multiple taxa in what is known as a "turnover-pulse"; (f) major global climatic changes are responsible for the majority of mammalian speciation events; (g) species whose ecological resources persisted even as the Earth oscillated between environmental extremes experienced low vicariance, speciation, and extinction rates. Persistent species will include both ecological generalists (i.e., eurytopes), which can exploit a variety of resources, as well as ecological specialists (i.e., stenotopes), the preferred resources of which do not diminish in the face of environmental change. Conversely, species for which ecological resources disappeared during periods of environmental change will have experienced more vicariance and thus higher speciation and extinction rates. *See also* **effect hypothesis; eurytope; macroevolution; stenotope; turnover-pulse hypothesis.**

## Hadar

Shorthand for the "Hadar Research Project area" that is situated to the west of the Awash River in northeastern Ethiopia. The first fossil hominins were found at Hadar in 1973 and research is still ongoing. The Hadar Formation is 3.4–3.0 Ma and the overlying Busidima Formation spans from *c*.2.4 to less than 2.0 Ma. The vast majority of the many hominins found at Hadar come from the Hadar Formation and have been attributed to *Australopithecus afarensis* on the basis of their dental and gnathic similarities to specimens from Laetoli (including the type specimen, LH 4). Since 1973 the site has yielded approximately 400 *Au. afarensis* specimens and the Hadar sample constitutes approximately 90% of all known *Au. afarensis* fossils. Teeth, jaws, and crania are the most abundant specimen types, but postcranial fossils make up about 40% of the site sample. The Hadar hominin sample includes A.L. 288-1 ("Lucy"), a remarkably complete *Au. afarensis* associated skeleton, and A.L. 444-2, a 75–80% complete *Au. afarensis* adult skull. The material from A.L. 333 provides a rich assortment of adult and juvenile skull, skeletal, and dental remains, but it probably does not sample a single biological population of *Au. afarensis*. A few hominin fossils have been attributed to *Homo*, including A.L. 666-1, which has been referred to as *Homo* aff. *Homo habilis*. Archeological evidence found at the site includes Oldowan artifacts. (Location 11°10′N, 40°35′E, Ethiopia.)

## Hadar Formation

An incised channel of the Awash River has exposed a substantial thickness of sediments that can be mapped throughout the Hadar region. Maurice Taieb and his colleagues initially designated four formations – Hadar, Meschelle, Leadu, and Haouna-Lédi – but at least two of these formations (Hadar and Meschelle) are coeval, and all four are now rationalized into a single unit called the Hadar Formation that is subdivided into four members (i.e., the Basal, Sidi Hakoma, Denen Dora, and Kada Hadar Members). Each of the members above the Basal Member is defined as the sediments between the base of a designated tuff and the base of an overlying designated tuff. The three designated tuffs forming the stratigraphic divisions are the Sidi Hakoma Tuff (SHT), Triple Tuff-4 (TT-4), and the Kada Hadar Tuff (KHT). The oldest member, the Basal Member, includes sediments that date from an unconstrained lower age of less than 4 to 3.42 Ma, and the youngest, the Kada Hadar Member, dates from 3.20 to *c*.2.9 Ma. The fossil hominins recovered from the Hadar Formation have all been assigned to *Australopithecus afarensis*. The Hadar Formation has provided evidence of bone modification (cutmarks and percussion marks) but no stone tools have been recovered. See also **Busidima Formation**; **Gona Paleoanthropological study area**.

### Hadley circulation

(or Hadley cell) A description of the general circulation of the atmosphere in the tropics. The equator-ward air, supplied by the trade winds, that rises in the Intertropical Convergence Zone (or ITCZ) accounts for the high rates of precipitation in the ITCZ, and the sinking air in the subtropics accounts for the global distribution of arid regions.

### Hadza

One of the world's few remaining populations of hunter–gatherers. They live in a 4000 km$^2$/1544 sq. mile region of northern Tanzania, south of the Serengeti, in a savanna woodland habitat around the shores of Lake Eyasi. Out of a total population of around a thousand individuals approximately 300–400 depend on hunting and gathering wild foods for over 90% of their diet; the remaining members of the tribe combine foraging with trading and begging or gaining income from the tourist trade. The Hadza reside in camps with an average size of about 30 individuals. The frequent movement between camps has been linked with the Hadza's lack of traditional land rights or sense of individual ownership of natural resources. Camps move every 2 months or so in response to seasonal availability of water and foods. During the dry season (i.e., June to November) camp size is relatively large, due to the limited availability of drinking water and thus the greater concentration of animals near watering holes. During the wet season (i.e., December to May) drinking water is more freely available in many areas and camp sizes are smaller. The Hadza diet consists of a wide variety of plant and animal foods. There is a distinct sexual division of labor. Men typically go hunting alone with poisoned-tip arrows and take a wide variety of birds and mammals; they also collect baobab fruit and honey. Women forage in groups and primarily focus on gathering plant foods such as baobab, berries, and tubers (underground storage organs); even quite young children forage successfully. Observations about the Hadza, who collect and consume foods in an ecosystem that may have been quite similar to that of our early hominin ancestors, allow researchers to test hypotheses about evolutionary ecology. (etym. "Hadza" is short for "Hadzabe," the word that identifies the Hadza people in their own Hadzane language.) See also **forager**; **hunter–gatherer**.

### haemoglobin

See **hemoglobin**.

## Hahnöfersand

A hominin skull fragment was recovered from this island along the bank of the River Elbe just downstream of Hamburg, Germany, in 1973. Supposedly associated with Pleistocene fauna, its initial radiocarbon age of 36.3 ka ± 300 years suggested this specimen was the earliest anatomically modern human in Europe, but it has recently been redated by a more precise AMS radiocarbon dating method to the Holocene. (Location 53°32′50″N, 09°43′38″E, Germany; etym. name of the island.)

## half-life

*See* **radiometric dating**.

## hammer

A hammer refers to an object used to fracture something. Hammers may be made of a variety of materials, including stone, bone, antler, or wood. *See also* **hammerstone**.

## hammerstone

A rock, often oval in shape, held in the hand and used for a variety of pounding tasks. In hard-hammer percussion, a hammerstone is used to strike flakes off a core. In combination with an anvil, a hammerstone may be used to flake rocks (bipolar technique) or to break open foods such as nuts, marrow bones, or shellfish. Hammerstones often develop characteristic pitting on one or both ends.

## hand

The primate hand can be divided into the wrist, palm, and the digits. The latter include the thumb (the first digit, also known as the pollex) and the medial four fingers (digits two through five). There are 27 bones in the modern human and African ape hand: eight carpal bones in the wrist, five metacarpals in the palm, three phalanges (proximal, intermediate, and distal) in each finger, and two phalanges (proximal and distal) in the thumb. All other primates, apart from some lemurs, have an additional carpal bone, the os centrale. The primate hand differs from that of most other mammals in its prehensile or grasping ability. An opposable thumb is also generally considered to be a specialized feature of primates, but only catarrhines have a truly opposable thumb. Opposability is accentuated in the modern human hand compared to other primates, especially apes, because the modern human thumb is relatively long and the fingers are relatively short. In contrast, grasping by the flexed fingers against the palm or by the thumb against the side of the index finger is accentuated in apes. Their fingers are remarkably long and their thumb relatively short. Because the ape hand is specialized for suspension and climbing, modern human hand proportions are most similar to those of terrestrial Old World monkeys, like the baboons and mandrills. *See also* **digit**; **opposable thumb**; **os centrale**.

## handaxe

Large (>10 cm in length) bifacially flaked, oval- or teardrop-shaped stone tools that are characteristic of the Acheulean industrial complex. Handaxes can be made on cobbles, nodules, or large flakes and they were probably used for a wide variety of tasks, including butchery and perhaps woodworking. Handaxes occur in Africa and parts of Eurasia from *c.*1.6 Ma to perhaps as recently as 0.16 Ma.

## hand, evolution in hominins

Much of what is known about early hominin hand evolution is inferred from only a few fossils from different individuals and thus our understanding of overall hand proportions and function in early hominins is limited. The hand of the last common ancestor of modern humans and *Pan* is most parsimoniously assumed to be similar to that of an African ape, although fossil evidence from the time of divergence is scarce. The earliest evidence of hominin hand morphology is two phalanges associated with *Orrorin tugenensis* (c.6 Ma). In these specimens, the quite broad adult thumb distal phalanx foreshadows the morphology of later hominins. However, the researchers who wrote the initial descriptions of two nearly complete hands of *Ardipithecus ramidus* (4.4 Ma) suggest that those specimens represent the morphotype of the last common ancestor of panins and hominins. The *Ardipithecus* hand is interpreted as lacking evidence of the specializations for suspension, climbing, or knuckle-walking that are found in extant African apes, and instead is said to be similar to the hands of more generalized, arboreal, and quadrupedal Miocene apes (e.g., *Proconsul*). The *Ardipithecus* thumb is large and robust and the metacarpals and phalanges are relatively short, such that the hand proportions are more similar to those of Old World monkeys than to extant African apes. The researchers who discovered and analyzed these fossils suggest that the morphology of the chimpanzee is too derived to be a good model for the hand of the stem hominin. Several hand fossils of *Australopithecus afarensis* from Hadar, Ethiopia (c.3.2 Ma), indicate these hominins had relatively short fingers compared to the length of the thumb and a more derived capitate–trapezium–second metacarpal articulation; both lines of evidence suggest the ability to use some modern human-like precision grips. However, compared to later hominins and modern humans the phalanges are more curved and the thumb is gracile with a more curved articulation with the trapezium. Hand fossils associated with *Au. africanus* have a similarly derived morphology. Several isolated hand bones from Swartkrans and associated hand fossils from Olduvai Gorge (OH 7) are consistent with a mosaic pattern of hominin hand evolution, but the taxonomic attribution of these fossils (i.e., *Paranthropus* or early *Homo*) is uncertain as both taxa are found at each site. Hominin hand fossils from Swartkrans show a combination of features seen in modern humans and in the extant apes. Thirteen bones are attributed to the immature OH 7 hand, including carpals, metacarpals, and phalanges. Again, some features of the hand are similar to African apes (e.g., the morphology of the scaphoid and the large flexor sheath ridges on the phalanges), while other aspects are more modern human-like (e.g., the remarkably flat and broad metacarpal facet on the trapezium and the expanded apical tuft of the distal phalanx of the thumb). The hands of later *Homo* species such as *Homo antecessor*, *Homo neanderthalensis*, and early *Homo sapiens* are well represented in the fossil record and their morphology can be generally described as fully modern human-like. However, there are remarkably few fossils attributed to *Homo erectus* and this gap in the hominin fossil record makes it unclear exactly when and how a modern human-like morphology evolved. Nearly complete hands from Shanidar, in Iraq, and Kebara, in Israel, show that the biggest differences between Neanderthal and modern human hands are the robusticity of the bones and slight variations in the orientation of the carpometacarpal joints. The Neanderthal hand morphology is interpreted as being well adapted for a strong grip. The morphology of early modern human hands from Qafzeh and Skhul is interpreted as being more similar to that of modern humans than to Neanderthals, suggesting that these hominins were capable of the full repertoire of modern manipulative abilities and tool making. The wrist bones of *Homo floresiensis* are remarkably primitive and have more morphological similarities with African apes and early hominins (e.g., OH 7) than with later *Homo*. Thus, this primitive morphology brings into question not only the taxonomy of *H. floresiensis* but also current ideas about what type of morphology is necessary for the manufacture and use of tools.

## hand, function

The modern human hand is capable of both prehensile (e.g., pinching or gripping an object using precision grips and/or power grips) and nonprehensile (e.g., pushing, lifting, or tapping) movements. All of these movements are accomplished through complex, integrated actions of the numerous joints, muscles, and ligaments within the forearm, wrist, and hand. Muscles that move the joints of the hand can be divided into extrinsic and intrinsic muscles. Extrinsic muscles have proximal attachments outside of the wrist or hand, and most of these are attached proximally to the medial and lateral condyles of the humerus, and distally to the digits. Intrinsic muscles have proximal attachments within the hand, most attaching to the carpals and inserting on the digits or running between two rays (e.g., the interosseous muscles). All of the articulations among the carpal bones (i.e., intercarpal joints), between the carpals and metacarpals (i.e., carpometacarpal joints), between the metacarpals and proximal phalanges (i.e, metacarpophalangeal joints), and between phalanges (i.e., interphalangeal joints) are synovial joints. The radiocarpal, midcarpal, metacarpophalangeal, and interphalangeal joints are relatively mobile, whereas the intercarpal joints within each carpal row and some carpometacarpal joints are more stable. The functional axis of the hand is the third ray (i.e., the middle finger); side-to-side movements of the digits (i.e., abduction and adduction) are referenced to this axis. *See also* **abductor**; **adductor**.

## hand, tool use and tool making

The key factor that enables hominins to use and make tools is the length of the thumb compared to the fingers and the resulting ability to form a precise grip between the end of the thumb and distal pad of the index finger. However, experimental research with modern humans making or using Oldowan tools suggests that three types of precision grip are used: (a) the pad-to-side grip between the thumb and the side of the index finger, (b) the "baseball," or three-jaw chuck, grip, and (c) a pinch grip between the thumb and the four finger pads. The three precision grips allow for effective tool use and tool making while minimizing muscle fatigue and stress on the hand joints. Researchers have looked in the hominin fossil record for skeletal morphologies that in modern humans are linked with these grips. Kinematic studies of modern humans making stone tools have shown that the ability of the modern human wrist to achieve greater degrees of extension relative to the wrist of chimpanzees may have played a key role in the evolution of hominin stone tool production. Electromyographic studies of modern humans making Oldowan tools have shown that muscles on both the radial (thumb) side and ulnar (fifth digit) side are used most consistently. Apes can be trained to use crude stone tools thus showing that modern human hand morphology is not necessary for this behavior, but perhaps a key adaptation in hominins was the ability to forcefully manipulate stone with one hand (instead of two). Modern human hand function during tool use and tool making differs from that of nonhuman primates in two ways: (a) modern humans are able to apply more force to pinch, rather than hold, an object securely with precision grips, and (b) because of the increased mobility of the digits, modern humans are able to accommodate the thumb and fingers to the shape of objects. In contrast to modern humans, the relatively short thumb and long fingers of chimpanzees are not able to create a stable, firm pinch of an object or to control it by the pads of cupped thumb and fingers.

## Hanging Remnant
*See* **Swartkrans**.

## haplogroup
A group of similar haplotypes that share a common ancestor. Haplogroups are commonly used in studies of mitochondrial and Y-chromosome DNA. (Gk *haplo* = single and It. *gruppo* = group.)

## haploid
Organisms or cells (e.g., sperm and egg cells, called gametes or germline cells) that have only one copy of each chromosome. The Y chromosome is also referred to as haploid since there is only one present in a normal male cell. (Gk *haplo* = single and *oid* = like or resembling.)

## haplotype
An abbreviation of "haploid genotype," haplotype refers to a specified sequence of DNA that is confined to one chromosome. The term haplotype is also used to refer to sets of single nucleotide polymorphisms (or SNPs) or short tandem repeats (or STRs) found together (or linked) on the same chromosome (or the same part of a chromosome). (Gk *haplo* = single and *typos* = figure or model.)

## Happisburgh
Site in the Hill House Formation (or HHF) just south of Mundesley on the Norfolk coast of eastern England that was re-exposed by a reduction in the level of the North Sea. This site, which is between 0.99 and 0.78 Ma, contains *in situ* cores, flakes, and flake tools made of flint. No handaxes were found. (Location 52°49′36″N, 01°31′58″E, England.)

## Hardy–Weinberg equilibrium
A principle in population genetics that suggests that if mating is random (the technical term is panmictic), the allele frequencies in the population will not change from generation to generation (i.e., they are in equilibrium) unless the population is affected by one or more of the four forces of evolution (mutation, genetic drift, natural selection, and gene flow). If mating is not random (e.g., if there is assortative mating or endogamy/exogamy) the genotype frequencies will change, but not in the way predicted by the Hardy–Weinberg equilibrium. (etym. named after the English mathematician Godfrey Hardy and the German physician Wilhelm Weinberg.)

## Hargufia
A series of localities in the Bodo-Maka fossiliferous subregion of the Middle Awash study area, Ethiopia, that have produced Mode 1 and Mode 2 Acheulean artifacts. The site is equivalent in age to Olduvai Gorge Bed IV and Olorgesailie.

## Harris lines
Dense lines seen in plain radiographs of the long bones of juveniles, subadults, and young adults. Harris lines, which are orientated at right angles to the long axis, are the result of a slowing of cartilage growth caused by an insult such as a serious systemic infection.

## HARs
An acronym for **human accelerated regions** (*which see*).

## Hathnora
A group of paleontological and archeological localities at Hathnora in the Narmada Basin, India. Four (Surajkund 1, 2, and 3, and Hathnora 1) are in the Surajkund Formation and one (Hathnora 2) is in the Baneta Formation. (Location 22°49′14″N, 77°51′14″E, India.) *See also* **GSI-Narmada; Hathnora 1**.

## Hathnora 1
One of several localities near the village of Hathnora on the northern bank of the Narmada River in Madhya Pradesh, India. Biostratigraphy and electron spin resonance dating suggest a minimum age of *c.*50 ka and a maximum age of more than *c.*200 ka. It is the source of the Hathnora 1 (aka GSI-Narmada) calvaria, the only significant evidence of fossil hominins from the Indian subcontinent.

## Haua Fteah
This site (also known as Hawa Ftaih) is located 60 m above sea level on the northern slopes of Gebel el Akhdar on the Mediterranean coast of Libya, east of the Gulf of Sirte. The site preserves one of the longest archeological chronologies in North Africa with deposits spanning the Historic period through the Middle Paleolithic. Two hominin mandible fragments (Haua Fteah I and II) recovered from early Middle Paleolithic levels were assigned to *Homo sapiens rhodesiensis*; the modern taxonomic equivalent would be *Homo heidelbergensis* or *Homo rhodesiensis*. (Location 32°55′12″N, 22°07′48″E, Cyrenaica, Libya.)

## Hayonim
Cave site located in western Galilee, Israel, on the left bank of a tributary of the Nahal Yassof river. Fragmentary hominin remains have been recovered from Acheulean and Mousterian layers; the latter includes early Middle Paleolithic blade tools. (Location 32°56′N, 35°13′E, Israel.)

## Hazorea
This undated open-air site in the middle of a field in northern Israel contains flint tools "typical of the Lower Paleolithic" and two nearly complete hominin occipital bones that share morphological affinities with both *Homo erectus* and the Steinheim cranium. (Location 32°39′N, 35°06′E, Israel.)

## HCRP UR 501
This adult mandibular corpus, which is preserved in two nearly equally complete parts and includes the crowns of the premolars and the first two molars, was found in the Chiwondo Beds at Uraha in Malawi. Though small, enough morphology is preserved on this mandible to

be sure that it does not belong to an archaic hyper-megadont hominin. Instead, it resembles mandibles such as KNM-ER 1802 that have been assigned to early *Homo*. If it does belong to an early *Homo* taxon, it extends the range of that taxon more than 1000 km/620 miles to the south. Its allocation to *Homo rudolfensis* has been called into question.

## hearth

When used by paleolithic archeologists, the term hearth refers to visible evidence for the controlled use of fire. Fire use may be recognized through sediment colorization caused by high heat and oxidation, or by the presence of large amounts of charcoal or wood ash. Often, but not always, evidence of attempts to control a fire takes the form of shallow pits or stone barriers (e.g., a ring of stones) recoverable through careful excavation. In the absence of such features it is difficult to be sure whether a fire was deliberately set by hominins or was a natural fire (e.g., lightning setting fire to a tree resulting in a burned out tree stump). (OE *hoerth* = hearth.)

## heel strike

*See* **walking cycle**.

## Heinrich events

Ice-rafting events identified by Hartmut Heinrich on the evidence of "dropstones" (i.e., pebbles) found among sediments on the floor of the North Atlantic Ocean. Sea ice often forms in shallow waters and incorporates sediments from the continental shelf. Subsequent movement of sea ice can transport these sediments and the dropstones fall to the sea floor when the sea ice melts. Six Heinrich events have been identified during the last glacial (i.e., 60–17 ka). They are numbered H1–H6, with H6 being the oldest, but some consider the Younger Dryas to be the most recent Heinrich event in which case it would be H0. Climate shifts coinciding with Heinrich events have been reported in paleoclimate records well beyond the Atlantic. For example, cold sea surface temperatures in the North Atlantic associated with increased ice rafting may have altered rainfall patterns in Africa.

## *Helicobacter pylori*

Over half of all modern humans are infected by this Gram-negative bacterium, which is found in the stomach. It is harmless in the short term, but its spiral structure allows it to burrow into the mucosa of the upper gastrointestinal tract, where it causes peptic ulcers in both the stomach and duodenum and it is a risk factor for gastric cancer. *Helicobacter pylori* is highly polymorphic, and it is possible to match the proliferation of the more than 300 *H. pylori* strains to human migrations. The geographical patterns of genetic isolation by distance (or IBD) assessed using *H. pylori* and modern human genetic diversity are a good match, both showing a continuous loss of genetic diversity with increasing geographic distance from East Africa. Simulations suggest that *H. pylori* migrated from East Africa *c.*58 ka, a time that is consistent with the estimated departure of modern humans from that region. Researchers have also used the distribution of *H. pylori* haplotypes to investigate the history of the peopling of the Pacific. They found that "a distinct biogeographic group" called hpSahul split from the main Asian populations of *H. pylori c.*31–7 ka and moved into Sahul (i.e., New Guinea and Australia) and they suggest those populations remained isolated there for *c.*23–32 ka. Much later, *c.*5 ka, a second biogeographic group of *H. pylori*, hpMaori, split off as a subpopulation of hpEastAsia and that subpopulation subsequently spread into Melanesia and Polynesia. (Gk *helix* = spiral, plus L. *bacter* = rod and *pylorus* = gatekeeper, after the pyloric valve in the stomach, where this bacterium is found.)

## helicoidal wear plane
The pitch (i.e., the transverse orientation) of the occlusal wear plane is not the same along the tooth row. Wear is normally greater on the lingual, or tongue, side of the maxillary teeth in the anterior part of the postcanine tooth row, with the posterior part of the tooth row either showing less emphasis on the lingual side or more wear on the buccal, or cheek, side. Phillip Tobias proposed that this twist in the pitch of the occlusal wear plane was not seen in hominins until the emergence of *Homo* but more systematic analyses suggest that the twisting nature of the occlusal wear plane has no taxonomic significance. (Gk from *helix* = spiral.)

## hematite
Hematite ($\alpha$-$Fe_2O_3$), the principal ore of iron, is the primary component of (red) ochre, a common colorant found as an exotic material at many Late Pleistocene sites from southern Africa to France. (Gk *haimatites* = blood-like stone.) *See also* **ochre**.

## hemispheric dominance
Hemispheric dominance refers to an asymmetry of the structure or function of the cerebral hemispheres of the brain. For example, there is strong population-wide leftward functional hemispheric dominance among modern humans in the neural representation of language. Similarly, the modern human brain also displays anatomical hemispheric dominance as evident by petalias and by the leftward asymmetry of the planum temporale in the temporal lobe. *See also* **petalia**.

## hemizygous
The possession of only one copy of an allele or chromosome in a diploid organism. All mammals and thus all primates are diploid organisms and males are hemizygous for both the X and Y chromosomes. If a gene is deleted on just one of the two autosomal chromosomes, then the individual is hemizygous for that gene. Because males are hemizygous for the X chromosome, recessive X-linked phenotypes are more common in males, as are X-linked recessive diseases and disorders such as hemophilia and color blindness.

## hemoglobin
The major protein in red blood cells that carries oxygen from the lungs to the tissues. It is made up of two alpha-globin and two beta-globin protein subunits, each of which contains a heme group. Hemoglobins were some of the earliest and most extensively studied proteins in regional populations of modern humans. [Gk *haîma* = blood plus L. *globulus* = globe and *in* = pertaining to; abbreviation of hematinoglobulin (i.e., hematin plus globulin).]

## herbivore
An animal that consumes plants. (L. *herba* = grass, plus *vorous*, from the stem of *vorare* = to devour.) *See also* **herbivory**.

## herbivory

The consumption of plants. The term herbivory subsumes diets that focus on a type (i.e., grass, or graminivory) or specific part (i.e., fruit, or frugivory) of plants. Herbivory can be considered as a form of predation in which an organism consumes a plant. (L. *herba* = grass, plus *vorous*, from the stem of *vorare* = to devour.)

## heritability

The proportion of phenotypic variation in a population that is attributable to genetic variation among individuals. The heritability of a particular trait may differ among populations because of environmental factors. For example, the heritability of a complex trait such as stature might be high in a population that is well nourished and has access to health care and lower in a population where malnourishment and disease are common (i.e., the conditions that most likely prevailed in early hominins). This is because in the latter population the environmental variance accounts for a higher proportion of the phenotypic variance. Broad-sense heritability ($H$ or $H^2$) reflects *all* possible genetic contributions to the phenotypic variance (e.g., additive genetic variance, dominance, epistasis, and parental effects). Narrow-sense heritability ($h$ or $h^2$) refers to the portion of the phenotypic variation that is additive (aka allelic) and does not take into account variation due to epistatic, dominance, or parental effects. (L. *hereditare* = to inherit, from *heres* = heir.)

## heritable

See **heritability**.

## herpestid

The informal name for the Herpestidae, a feliform family. *See also* **Carnivora**; **Herpestidae**.

## Herpestidae

The animal family that includes mongooses and their close relatives. *See also* **Carnivora**.

## Herto

A locality on the Bouri peninsula in the Middle Awash region of Ethiopia. Prospecting in 1997 and 1998 led to the discovery of eight fossil hominins from the Upper Herto Member, including three partial crania (BOU-VP 16/1, 2, and 5). All have been assigned to a subspecies of *Homo sapiens*: *Homo sapiens idaltu*. The stone tools found in the Upper Herto Member include handaxes, cleavers, and other bifaces as well as Levallois flake tools. Along with the slightly earlier Omo I and II from the Omo-Shungura, the Herto remains are the earliest evidence for modern humans in Africa. (Location 10°15′55″N, 40°33′38″E, Ethiopia.)

## heterochrony

Changes in the timing of the appearance, or the rate of development, of a feature in the descendant relative to the developmental timetable in the immediate ancestor. Even when such shifts in rate and/or timing are apparently minor, they can produce significant alterations in morphology. Analyses of heterochronic transformations focus on the ontogeny of size and size-related changes and explore the idea that selection may act on size through growth rates or timing. Detailed studies of ontogeny permitting a comparative analysis of the rate and duration of growth are relatively rare; most heterochronic studies focus on adult morphologies and their presumed evolutionary transformations. It is important to discriminate between the results of heterochrony and the processes that generate the results. (Gk *hetero* = other and *kronos* = timing.) *See also* **acceleration**; **neoteny**; **pedomorphosis**; **peramorphosis**.

## heteroplasmy
This term refers to the mixture of several different sequences of mitochondrial DNA in the same cell or individual. In the debate about "mitochondrial Eve," the assumption that mitochondrial DNA was inherited strictly from the mother was questioned since occasional examples of heteroplasmy due to a paternal contribution were found in other animals. If this were true in modern humans, it would affect the coalescence time for mitochondrial DNA (i.e., the age of "mitochondrial Eve"). However, subsequent research has failed to provide any evidence of a significant paternal contribution to human mitochondrial DNA. (Gk *hetero* = other and L. *plasma* = form or shape.) *See* **mitochondrial DNA**.

## heterosis
The technical term for hybrid vigor, heterosis refers to the increased fitness of hybrid offspring compared to their parents. Heterosis is usually ascribed to two phenomena: dominance, in which harmful recessive alleles from one parent are masked by dominant alleles from the other, and overdominance or heterozygous advantage, in which the combination of recessive and dominant alleles produces even greater advantage than either of the homozygous forms. In the skeleton heterosis is typically manifested as increased size. The role of hybridization in human evolution is currently poorly understood, though hotly debated (e.g., modern humans and Neanderthals). (Gk *heteroios* = different in kind; syn. hybrid vigor.) *See also* **hybridization**.

## heterozygosity
*See* **heterozygous advantage**.

## heterozygous advantage
The circumstance in which it is advantageous to be heterozygous (i.e., have different versions of the allele on the two chromosomes) because the heterozygote has a higher fitness. An example that may have been relevant to human evolution is the sickle cell (S) allele of the beta-globin gene. In the homozygous form this allele is deleterious because it causes sickle cell anemia, but in regions where malaria is endemic the heterozygous form protects against malaria. In such an environment, homozygotes for the wild-type allele are at a disadvantage because they are more susceptible to malaria. Thus, both alleles (S and s) are maintained in the population. (Gk *hetero* = other and *zygotos* = yoked; syn. heterozygote advantage.) *See also* **hemoglobin**.

## Hexian
(和县) The site known as Hexian (aka Longtandong or "Dragon Pool Cave") is situated on the northern slope of Wangjiashan ("hill of Wang's family") just south of the town of Taodian in Hexian County, Anhui Province, China. The most recent dates suggest that Hexian is *c.*400 ka, which is equivalent to the base of layer 3 of Zhoukoudian Locality 1. The hominins found at

the site include a well-preserved young adult hominin calvaria of a presumed male (Hexian 1, or PA 830), a parietal fragment, a mandible fragment, and isolated teeth. (Location 31°53′03″N, 118°12′18″E, eastern China.)

## High Cave, The
*See* **Mugharet el 'Aliya**.

## higher taxon
Any taxon above (i.e., more inclusive than) the level of the genus. Thus, the subtribe Australopithecina and the tribe Hominini are examples of higher taxa, whereas the species *Paranthropus boisei* is not. (Gk *taxis* = to arrange or "put in order.")

## high-ranking prey
High-ranking prey are those taxa or food resources that within an optimal foraging theory framework provide higher-than-average energetic returns. The prey-choice model predicts that high-ranking prey should always be pursued when encountered and it assumes that abundances in archeological contexts approximate their abundances on past landscapes. Body size influences prey rankings because larger taxa (e.g., large ungulates) provide greater energetic returns than small taxa, but ethnographic evidence suggests that other factors (e.g., prey mobility) may play an important role in explaining modern human, and presumably hominin, foraging decisions and prey rankings. *See also* **low-ranking prey**; **optimal foraging theory**.

## high-survival elements
Parts of the skeleton with a high proportion of dense cortical bone (e.g., long bone shafts, parts of the cranium, and mandible) that are resistant to destructive taphonomic processes. Because high-survival elements provide the most accurate representation of those bones processed by hominins (e.g., dense cortical bone is most likely to preserve cutmarks), their study can be particularly useful when examining carcass transport strategies in bone assemblages that have been subjected to destructive processes. *See also* **skeletal element survivorship**.

## hindbrain
The most posterior, or caudal, of the three swellings, or vesicles, that give rise to the brain. The hindbrain (or rhombencephalon) is comprised of the medulla oblongata, the pons, and the cerebellum. The cerebellum is a dorsal outgrowth from the same embryonic hindbrain subdivision that contains the pons. (OE *hinder* = behind and OE *braegen* = brain, most likely from the Gk *bregma* = the front of the head.) *See also* **brain**.

## hippocampus
A seahorse-shaped structure located in the medial part of the temporal lobe of the brain, adjacent to the amygdala. The hippocampus is involved in functions such as spatial orientation, navigation, emotions, and the storage of declarative memory (i.e., the parts of the memory concerned with persons, places, and objects). (Gk *hippos* = horse and *kampos* = sea monster.)

## hippopotamid
Informal name for a member of the artiodactyl family comprising the hippopotami. *See also* **Artiodactyla**.

## Hippopotamidae
The artiodactyl family that includes hippopotami and their close relatives. *See also* **Artiodactyla**.

## histology
The study of microscopic anatomy. Histology often involves generating a section of a tissue to visualize structures using light or transmission electron microscopy. For hard-tissue histology, a replica can be made for study with scanning electron microscopy, or fractured or polished surfaces can be examined with confocal microscopy. For example, traditional approaches to understanding enamel development and dentine development involve preparing thin sections (approximately 100 μm) to view them by transmitted or polarized light microscopy.

## HLA
Acronym for human leukocyte antigens.

## Hoedjiespunt
A brown hyena den in Middle and Late Pleistocene dunes that date to Marine Isotope Stage 5. It has provided Middle Stone Age artifacts, evidence of shellfish collection, and four permanent teeth of a juvenile hominin and a fragmentary right hominin tibia that may belong to the same individual. (Location 33°00′S, 17°57′E, Western Cape Province, South Africa.)

## Hofmeyr
The Hofmeyr 1 skull was recovered in 1954 from a dry channel bed of the Vlekport River. The specimen was severely damaged in 1970 and comparison with photographs taken in 1965 reveal the loss of the anterior part of the face, the anterior teeth, and part of the ramus of the mandible. The results of a multivariate analysis show that Hofmeyr 1 falls closest to the centroid for European crania associated with the Upper Paleolithic, whereas it is at the edge of the range of variation of recent sub-Saharan Africans. These results fit predictions of the out-of-Africa hypothesis for the origins of modern humans. (Location 31°34′S, 25°58′E, South Africa.)

## Hohle Fels Venus
An exquisitely carved ivory figurine recovered in six fragments from the southern German Aurignacian site of Hohle Fels. The figurine is a lifelike depiction of a female modern human except that it lacks a head, its limbs are de-emphasized and it has exaggerated secondary sexual characteristics. It is in the tradition of the Venus figures from the Gravettian period (e.g., the Willendorf Venus). The c.35 ka calibrated radiocarbon years age of the Hohle Fels Venus is at least 5 ka older than any other figurines and is the earliest known figurative image of a modern human.

## Holocene
The youngest (and current) epoch of the Neogene Period within the Cenozoic Era. The onset of the Holocene (11.7 ka) follows the termination of the Younger Dryas cooling event (aka stadial). The Holocene was thought to be a period of relatively monotonous interglacial conditions but proxy-based paleoclimate reconstructions suggest there was substantial climatic variability within the Holocene. For example, the early Holocene was much wetter than at present, with an abrupt transition to drier conditions c.5 ka. Historical records document climate fluctuations during the Medieval Warm Period (c.AD 1100–1300) and the Little Ice Age (c.AD 1400–1900). These temperature excursions in mid- and high latitudes are coincident with changes in aridity in some low-latitude regions. (Gk *holos* = whole and *kainos* = recent.)

## holotype
According to the International Code of Zoological Nomenclature a holotype is the specimen listed as the type specimen in the original description of a taxon (e.g., OH 7 is the holotype of *Homo habilis*). (Gk *holos* = whole and *typus* = image.) See also **lectotype**; **paratype**; **type specimen**.

## Holstein
See **glacial cycles**.

## Homa peninsula
See **Kanam**; **Kanjera**.

## home base hypothesis
A model for Oldowan site formation developed by Glynn Isaac, who claimed that food sharing provided the selective milieu for the further evolution of cognition and for the emergence of modern human language. The home base hypothesis incorporated elements that were claimed to distinguish extant hunter–gatherer and African ape adaptations, including (a) the use of a central place from which individuals dispersed and returned on a daily basis, (b) a sexual division of labor whereby males hunted or scavenged for animal tissue and females gathered plant food resources, (c) delayed consumption, and (d) transport of food to the base where food sharing and social activities took place.

## *homeobox* genes
A family of genes involved in the regulation of anatomical development (morphogenesis). They are called *homeobox* genes because of a short 180 base pair DNA sequence that encodes part of a protein molecule called the homeodomain. Homeodomain molecules can bind DNA, so proteins that include a homeodomain can act as transcription factors and thus help regulate the gene expression of other genes. *Homeobox* (also abbreviated to *Hox*) genes control spatial positioning within the anterior–posterior and proximal–distal body axes. In modern humans, the 39 *HOX* genes are found in four clusters on chromosomes 2, 7, 12, and 17 [NB: if a gene name is written with all capital letters it means it is a modern human gene (e.g., *HOXA1*); the homologous gene in mice would be *HoxA1*]. (Gk *homoio* from *homos* = the same and "box" apparently because these genes have a small enough sequence that it can be written in a "box" in a manuscript.)

## homeodomain
See **homeobox** genes.

## home range
The total area exploited by an individual animal or a group of animals. Home range size varies between species and even between populations of the same species. It is influenced by a variety of factors including the productivity of the habitat and the dietary strategy of the animal. For example, insectivorous and frugivorous primates tend to have relatively larger home ranges than primates that rely on more ubiquitous resources, such as grass or leaves. Body mass of a population (or group) is an important determinant of home range size. Larger animals tend to need absolutely more food, so will have larger home ranges. Estimates of the home ranges of extinct hominins use a model derived from modern human foraging populations.

## hominid
Informal or vernacular name for a species (or an individual specimen belonging to a taxon) within the family Hominidae. (L. *homin* = man or human being.) See **Hominidae**.

## Hominidae
In 1825 John Gray, the Keeper of Zoology at the original British Museum in South Kensington, London, England (i.e., before the natural history collections were housed in a separate Natural History Museum), suggested that Linnaeus' mammalian orders should be broken down into family, tribe, and subtribe. As part of this rationalization he introduced the family Hominidae for modern humans. Originally the family Hominidae excluded all the nonhuman great apes, but given the copious morphological, molecular, and genetic evidence for a close relationship between modern humans and chimpanzees/bonobos, some researchers now take the view that the modern human clade should be recognized at a lower level than that of the family. In this more inclusive interpretation the living representatives of the family Hominidae are the orangutans, gorillas, chimpanzees/bonobos, and modern humans. Therefore, the fossil representatives of the family Hominidae would be all the extinct taxa that are more closely related to the extant taxa listed above than to any other living taxon. Some researchers still use the family name Hominidae for the modern human twig of the tree of life, but most are moving towards a usage that is consistent with the molecular and morphological evidence (i.e., recognizing the modern human clade at the level of the tribe as the Hominini or even at the level of a subtribe as the Hominina). (L. *homin* = man or human being.)

## hominin
The informal or vernacular term for a specimen that belongs to a taxon within the tribe Hominini, and thus is more closely related to modern humans than to any other living taxa. For example, *Paranthropus boisei* is a hominin taxon and KNM-ER 406 is a hominin fossil. (L. *homin* = man or human being.)

## Hominina
If the tribe Hominini is interpreted to include both the clade that contains modern humans and the clade that contains extant chimpanzees/bonobos, then some researchers discriminate between the two clades at the level of the subtribe. In which case the clade that contains modern

humans would be called the Hominina and the clade that contains extant chimpanzees/bonobos would be called the Panina. (L. *homin* = man or human being.) *See also* **Hominini**.

## homininan

The informal term for individuals or taxa within the subtribe Hominina. *See also* **Hominina**.

## hominin footprints

Footprints are trace fossils (aka ichnofossils) that provide a record of fossilized behavior on a relatively short time scale (in some cases on scales as short as minutes to days). They provide information about the soft tissues of the plantar surface of the foot and the manner in which the foot makes contact with the ground during the gait cycle. Modern human footprints are distinctive because they begin with a deep, rounded heel print, followed by an impression made by the lateral side of the foot, then a deep impression under the metatarsal heads as the center of pressure moves anteromedially. The sequence ends with a relatively deep impression for the hallux in line with the other toes. Within the forefoot, the deepest part of the print often occurs below the medial metatarsal heads and hallux where the highest toe-off forces occur; the shallowest impression typically occurs under the medial side of the midfoot where the longitudinal arch, if well developed, does not penetrate the substrate to the same extent as other parts of the foot. However, human footprint morphology shows substantial variation, which can be attributed to certain aspects of foot anatomy, gait mechanics, and the properties of the substrate such as its firmness, particle size, and moisture content. Footprints are rare in the hominin fossil record. The only hominin footprints from the Pliocene are those made by three individuals walking in a wet ash deposited $c.$ 3.66 Ma in Laetoli, Tanzania. The Laetoli hominin footprints show compelling evidence that bipedal gait was well established in the hominins that made them, but ambiguities remain regarding interpretations of evidence for a longitudinal arch and of the nature of gait. From the early Pleistocene, three 1.53–1.4 Ma hominin footprint trails from Koobi Fora, Kenya, show many of the characteristics of modern human footprint morphology noted above. Fossilized footprints are more common in the late Pleistocene and Holocene. A set of footprints in the Roccamonfina volcanic ash in southern Italy that dates to 385–325 ka appears fully modern human-like. Footprints from $c.$ 117 ka sediments in South Africa and from 19–23 ka layers at Willandra Lakes in Australia resemble those made by modern humans living in those regions today. *See also* **Koobi Fora hominin footprints**; **Laetoli hominin footprints**.

## Hominini

The tribe that many researchers use for modern humans and all the fossil taxa more closely related to modern humans than to any other living taxon. Some researchers use it to refer to the clade that contains both modern humans and chimpanzees/bonobos. (L. *homin* = man or human being.) *See also* **Hominidae**.

## hominoid

The vernacular term for a member of the superfamily Hominoidea. The living hominoids (i.e., modern humans, chimpanzees/bonobos, gorillas, orangutans, and gibbons) all belong to the same clade. Fossil hominoids are all the taxa that are more closely related to living hominoids than they are to any other living taxon. *See also* **Hominoidea**.

## Hominoidea

The superfamily comprising the living taxa of modern humans, chimpanzees/bonobos, gorillas, orangutans, and gibbons, and all the fossil taxa that are more closely related to these taxa than to any other living taxon. (L. *homin* = man or human being and *oides* = resembling.)

## Homo

See *Homo* Linnaeus, 1758.

## *Homo* Linnaeus, 1758

The history of the interpretation of the genus *Homo* since its introduction in 1758 in the tenth edition of Carl Linnaeus' *Systema Naturae* has been one of episodic relaxation of the criteria used for deciding what taxa should be included within the genus, with each episode resulting in one or more extinct taxa being added to *Homo*. As originally conceived by Linnaeus, the genus *Homo* incorporated two species, *Homo sapiens* (i.e., modern humans) and *Homo troglodytes* (a mythical creature). The first extinct hominin species, *Homo neanderthalensis*, was added to *Homo* by William King in 1864. The inclusion of *H. neanderthalensis* within *Homo* meant expanding its definition to include archaic and derived morphology (e.g., discrete and rounded supraorbital margins, midline facial projection, a distinctive parietal and occipital morphology, and robust limb bones with relatively large joint surfaces) that are either not seen at all, or not seen in this combination, in *H. sapiens*. The next modification to the interpretation of the genus *Homo* occurred in 1908 when Otto Schoetensack added *Homo heidelbergensis* to the genus. This meant that *Homo* now embraced at least one individual with a mandible that had a robust mandibular corpus and lacked a true chin. In 1921 Sir Arthur Smith Woodward added *Homo rhodesiensis* to accommodate fossils from South Africa, and in 1932 Oppenoorth added *Homo soloensis* based on discoveries made at Ngandong in Java, Indonesia. The addition of the Ngandong fossils meant that crania substantially more robust than those of modern humans and *H. neanderthalensis* were now included in *Homo*. Nonetheless, the endocranial volumes of all the crania belonging to *H. neanderthalensis*, *H. rhodesiensis*, and *H. soloensis* are still within, or close to, the modern human range.

When Franz Weidenreich formally proposed that two existing extinct hominin species, *Pithecanthropus erectus* and *Sinanthropus pekinensis*, should be merged into a single species and transferred to *Homo* as *Homo erectus*, their addition resulted in *Homo* subsuming an even wider range of morphology. The addition of the *H. erectus* hypodigm at this time, even before the discovery of the small *H. erectus*-like crania from East Africa and Dmanisi, substantially increased the range of endocranial volume within the genus *Homo*. Compared with pre-1940 interpretations of *Homo*, the inclusion of fossils attributed to *H. erectus* meant that *Homo* now included individuals with a smaller neurocranium, a lower vault, a broader cranial base relative to the vault, and more complex premolar roots. Their crania also have a substantial, shelf-like torus above the orbits and there are often both sagittal and angular tori, although the expression of some, or all, of this morphology may be size-related. In *H. erectus* the occipital sagittal profile is sharply angulated, with a well-marked supratoral sulcus and the inner and outer tables of the vault are thickened. The cortical bone of the postcranial skeleton is generally thick. The long bones are robust, and the shafts of the femur and the tibia are flattened from front to back relative to those of other *Homo* species. However, all the postcranial elements of

*H. erectus* are consistent with a habitually upright posture and obligate long-range bipedalism. It was not long before other genera were transferred to *Homo* (e.g., *Meganthropus* in 1944 by Ernst Mayr, *Telanthropus* in 1961 by John Robinson, and *Atlanthropus* in 1964 by Wilfrid Le Gros Clark). In 1964, when Louis Leakey, Phillip Tobias and John Napier added the remains of *Homo habilis* from Olduvai Gorge (e.g., OH 7, 8, 13, 16) to *Homo*, it meant that Le Gros Clark's 1955 diagnosis of *Homo* needed to be amended. In order to accommodate *H. habilis* in the genus *Homo*, Leakey and his colleagues reduced the lower end of the range of brain size to 600 cm$^3$. They claimed that the other criteria (e.g., dexterity, erect posture, and a bipedal gait) did not need to be changed because the functional capabilities of *H. habilis* were interpreted as being consistent with these functional requirements. The next significant addition to *Homo* fossils was the recovery of KNM-ER 1470 from the Upper Burgi Member of the Koobi Fora Formation and the subsequent proposal to subsume *Pithecanthropus rudolfensis* into *Homo* as *Homo rudolfensis*. Morphologically, the latter taxon presented an apparently unique mixture of a relatively large *Homo*-like neurocranium, and a broad *Paranthropus*-like face, but it lacked the distinctive combination of small anterior and large postcanine teeth typical of *Paranthropus*, especially *Paranthropus boisei*. As a consequence, from 1972 onwards the genus *Homo* subsumed a substantially wider range of facial morphology than it did prior to the discovery of KNM-ER 1470. In due course other cranial specimens from Koobi Fora (e.g., KNM-ER 1590, 1802, 1805, 1813, 3732) and Olduvai (e.g., OH 62) were added to the early *Homo* hypodigm, as were cranial and postcranial fossils from Members G and H of the Shungura Formation, Members 4 and 5 at Sterkfontein, Member 1 at Swartkrans, the Chemeron Formation, Uraha, Hadar, the Nachukui Formation in West Turkana, and Dmanisi in Georgia.

These additions to "early *Homo*" subsumed a wide range of cranial and dental morphology. For example, the endocranial volumes of the specimens range from just less than 500 to around 850 cm$^3$. The mandibles also vary in size, but all have relatively robust bodies and premolar teeth with relatively complex crowns and roots. The discovery of OH 62 provided the first unequivocal postcranial evidence for *H. habilis*, so it is significant that OH 62 has been interpreted as having limb proportions that are at least as ape-like as those of individuals attributed to *Australopithecus afarensis* and limb cross-sectional geometries that are unusual in a committed biped. It is also likely that the KNM-ER 3735 associated skeleton belongs to *H. habilis*, even though its shoulder morphology is like that of earlier hominins. Some researchers have suggested that the inclusion of *H. habilis* and *H. rudolfensis* into *Homo* is unjustified on the basis that to do so weakens the case for recognizing *Homo* as a distinct grade.

The most recent species to be added to *Homo*, *Homo floresiensis* broadened the morphological scope of *Homo* even further. The specimens attributed to this species were recovered from deposits in the Liang Bua cave on the Indonesian island of Flores (and are given numbers beginning with LB1) and were initially dated to between approximately 74 and 18 ka. Its small brain and body size makes *H. floresiensis* a particularly significant addition to *Homo*. The endocranial volume of *H. floresiensis* is considerably smaller than that of the most primitive species assigned to *Homo* (e.g., adult endocranial volume in *H. habilis* ranges between 509 and 687 cm$^3$). This is despite the stature estimates of 106 cm for LB1 and 109 cm for LB8 that are only slightly smaller than the estimated stature of 118 cm for the *H. habilis* OH 62 partial skeleton.

Some researchers interpret *Homo* much more inclusively. For example, in 1972 John Robinson proposed that *Australopithecus* be sunk into *Homo*, and in 2003 Darren Curnoe and

Alan Thorne proposed that there were only three species in the hominin clade and they included all three in *Homo*, as *Homo ramidus*, *Homo africanus*, and *Homo sapiens*. In 1998 Morris Goodman and colleagues made the radical proposition that only one genus, *Homo*, be used for the whole of the hominin clade. In summary, there is no consistent definition of the genus *Homo*, and researchers have variously chosen to define *Homo* based on inferred behavior (e.g., *Homo habilis*), morphology (e.g., *Homo erectus*), time period (e.g., *Homo floresiensis*), or in relationship to other species (by subsuming all hominins into *Homo*). (L. homin = man or human being.) *See also* **genus**.

### *Homo antecessor* Bermúdez de Castro et al., 1997

Between 1994 and 1996 more than 80 hominin fossils were recovered from the Gran Dolina cave in the Sierra de Atapuerca. Their discoverers considered them too primitive to be allocated to either *Homo neanderthalensis* or *Homo heidelbergensis* so they assigned them to a new species, *Homo antecessor*. They claimed that *H. antecessor* combines a modern midfacial morphology (a canine fossa, an acute zygomaticoalveolar angle, and a modern human nasal morphology) with primitive *Homo ergaster*-like dental morphology. Critics of the decision to recognize a new taxon suggest that the "modern" facial morphology is due to the immaturity of the ATD6-69 facial fragment. The hypodigm, which currently represents at least 11 individuals, shows evidence of cannibalism.

### *Homo erectus* (Dubois, 1893*)

Hominin species established by Eugène Dubois to accommodate fossil hominins recovered in 1891 at Trinil, Java, Indonesia. In his initial publication of the Trinil remains Dubois referred the skullcap to *Anthropopithecus erectus*. The name reflected Dubois' initial conviction that he had discovered the remains of a fossil ape (at that time *Anthropopithecus* and not *Pan* was the preferred genus name for chimpanzees), but he changed his mind and a year later he transferred the new species to a new genus, *Pithecanthropus*. The discovery of the Trinil calotte was significant because its small cranial capacity (about 940 cm$^3$) and primitive shape (e.g., low brain case, and quite sharply angulated occipital region) differed significantly from the only two hominin taxa, *Homo sapiens* and *Homo neanderthalensis*, known at the time. Discoveries made by Ralph von Koenigswald at Sangiran, also in Indonesia, were added to the hypodigm of *Pithecanthropus erectus* and in 1940 Franz Weidenreich formally proposed that the hypodigms of *Sinanthropus pekinensis* from Zhoukoudian and *P. erectus* from Java should be merged into a single species as *Homo erectus pekinensis* and *Homo erectus javanensis*, respectively. Subsequently, the hypodigms of other genera were transferred to *H. erectus* (*Meganthropus* in 1944 by Ernst Mayr, *Telanthropus* in 1961 by John Robinson, and *Atlanthropus* in 1964 by Wilfrid Le Gros Clark). Some researchers interpret *H. erectus* as a chronospecies that evolves through time. Others argue that there are potential species-level differences between the Sangiran/Trinil hypodigms and the more recent hypodigms from Sambungmacan and Ngandong. (*NB: the year of publication is usually given as 1892, but this is incorrect. The publication was a report by Dubois of excavations carried out in 1892, but it did not become a public record until 1893, so this should be the year of its publication). (L. homin = man or human being and erectus = to set upright.)

## Homo ergaster Groves and Mazák, 1975

A hominin species established by Colin Groves and Vratislav Mazák to accommodate fossil hominins recovered from Koobi Fora that in their judgment did not belong in the hominin taxa that existed at the time. Most other researchers acknowledge these differences exist, but many are inclined to view them as regional variations within *Homo erectus*. (L. *homin* = man or human being and *ergo* = work.)

## Homo floresiensis Brown et al., 2004

A hominin species erected by Peter Brown and colleagues in 2004 to accommodate LB1, a partial adult hominin skeleton, and LB2, an isolated left $P_3$ recovered by a research team led by Mike Morwood in 2003 from the Liang Bua cave on the Indonesian island of Flores. More material belonging to LB1 and evidence allocated to LB4–9, including LB6, a second partial skeleton that lacks a cranium, was recovered in 2004. The hypodigm now includes close to 100 individually numbered specimens that are estimated to represent fewer than 10 individuals. The taxon was immediately controversial for at least two reasons. First, its initial estimated geological age of between *c.*17 and *c.*74 ka substantially overlaps in time with evidence of modern humans in Southeast Asia. Second, while its discoverers and describers acknowledged its small overall size (the stature of LB1 is estimated to be 106 cm and its body mass to be roughly between 25 and 30 kg), its especially small brain (around 420 cm$^3$) for an adult hominin and its primitive morphology were most parsimoniously interpreted to be evidence of a novel endemically dwarfed pre-modern *Homo* or a transitional hominin taxon. Other researchers have suggested that no new taxon needs to be erected because they claim *H. floresiensis* was sampled from a population of *Homo sapiens* – most likely related to the small-statured Rampasasa people who live on Flores today – that was afflicted by either an endocrine disorder or one or more of a range of syndromes (e.g., Laron syndrome) that include microcephaly. Both explanations, a new species and a rare pathology, are exotic, but those who support a pathological explanation for the individuals represented by LB1–15 need to explain what pathology results in an early *Homo*-like cranial vault, primitive mandibular, dental, carpal and pedal morphology, and a brain that while very small apparently has none of the morphological features associated with the majority of types of microcephaly. Initially it was suggested that *H. floresiensis* was a dwarfed *Homo erectus*, but the burden of subsequent analyses suggest that it may be more closely related to a transitional hominin such as *Homo habilis*. (L. *homin* = man or human being and *flores* = the name of the island, Flores, where the type site is located.)

## Homo georgicus Gabounia et al., 2002

A hominin species established in 2002 by Gabounia and colleagues for an adult mandible recovered from Dmanisi in 2000. Most researchers recognize that the vast majority of the hominin fossils recovered from Dmanisi belong to a taxon that shares features with both *Homo habilis* and *Homo erectus* (and with *Homo ergaster* if that taxon is recognized). At this stage, few researchers are willing to assign this material to a novel taxon; however, *Homo georgicus* is the species name that has priority if that need arises in the future. The workers who introduced the new taxon saw it as intermediate between *H. habilis* and *Homo rudolfensis* on the one hand and *H. ergaster* on the other. The most recent interpretation reduces *H. georgicus* to below the level of a subspecies. (L. *homin* = man or human being and Georgia is the name of the country within which the Dmanisi site is located.)

## *Homo habilis* Leakey, Tobias and Napier, 1964

A hominin species established by Louis Leakey, Phillip Tobias, and John Napier in 1964 to accommodate fossil hominins (OH 4, 6, 7, 8, and 13; OH 14 and 16 were not included as paratypes, but were "referred" to *Homo habilis*) recovered between 1959 and 1963 from Beds I and II, Olduvai Gorge, Tanzania. Leakey and colleagues claimed that brain size, cranial, and dental morphology, and inferences about dexterity and locomotion, both distinguished the new taxon from *Australopithecus africanus* and justified its inclusion in *Homo*. Subsequent discoveries at Olduvai (e.g., OH 24, 62, and 65) and from other sites (e.g., Koobi Fora: KNM-ER 1470, 1802, 1805, 1813, 3735; Sterkfontein: Stw 53; Swartkrans: SK 847; Hadar: A.L. 666-1) have also been assigned to this taxon. The hypodigm has a relatively wide range of cranial and dental morphology. Endocranial volume ranges from just less than $500\,cm^3$ to $c.800\,cm^3$. All the crania in this group are wider across the base of the cranium than across the vault but facial morphology varies, with KNM-ER 1470 having its greatest width across the midface whereas KNM-ER 1813 is broadest across the upper face. Mandibles vary in size and robusticity, with those from the larger individuals having robust mandibular bodies. Postcanine teeth vary in size and crown morphology. Some mandibular premolars and molars (e.g., OH 7) are buccolingually narrow, whereas larger mandibular premolar teeth within the hypodigm (e.g., KNM-ER 1802) have buccolingually broader crowns, complex talonids, and more complex root systems. Some researchers consider the cranial variation within a single taxon to be excessive in scale and unlike the pattern of intraspecific variation seen in the *Pan/Homo* clade. They suggest *H. habilis* subsumes two taxa: *Homo habilis sensu stricto* and *Homo rudolfensis*. (L. *homin* = man or human being and *habilis* = handy.)

## *Homo habilis sensu lato*

A taxon whose hypodigm includes all the specimens referred to under *Homo habilis*. This is the sense in which *H. habilis* is used by researchers who are content that variation within that hypodigm is compatible with a single-species interpretation. See also *Homo habilis*.

## *Homo habilis sensu stricto*

A taxon whose hypodigm includes a subset of the specimens referred to under *Homo habilis*. This is the sense in which *H. habilis* is used by researchers who suggest that the variation within that hypodigm is not compatible with a single-species interpretation; they generally use *Homo rudolfensis* for the second taxon.

## *Homo heidelbergensis* Schoetensack, 1908

A hominin species created by Otto Schoetensack to accommodate a robust hominin mandible found in 1907 in a commercial sandpit at Mauer, near Heidelberg, Germany. The Mauer mandible lacks a chin and exhibits other traits Schoetensack interpreted as primitive, including a broad ramus and an anteroposteriorly deep mandibular symphysis, but these are combined with modern human-like dental proportions, including reduced canines. In light of this evidence Schoetensack concluded that the Mauer mandible's mosaic of primitive and derived traits was sufficient to distinguish it from *Homo sapiens*, *Homo neanderthalensis*, and *Homo erectus* (then known as *Pithecanthropus erectus*). Others have suggested that *H. heidelbergensis* might be the most appropriate species name for a group of Afro-European hominin fossils (e.g., Arago, Bodo, Kabwe, Mauer, Ndutu, and Petralona) that others were referring to as "archaic" *Homo sapiens*. In 2009 Aurélien Mounier and colleagues provided the most detailed morphological and taxonomic analysis of the holotype of *H. heidelbergensis* in which they concluded that there are morphological grounds for recognizing *H. heidelbergensis* as a taxon separate from *H. neanderthalensis*, *H. sapiens*, and *H. erectus*, and they supply both a definition and a differential diagnosis for the taxon, and consider the mandibles from Arago, Montmaurin, Sima de los Huesos, and Tighenif to be conspecifics of the Mauer mandible. Opinions about the relationships of *H. heidelbergensis* differ. The "Afro-European hypothesis" suggests that *H. heidelbergensis* was a geographically widespread and diverse species that gave rise to *H. neanderthalensis* in Eurasia and *H. sapiens* in Africa. The "European hypothesis" suggests that *H. heidelbergensis* is restricted to Europe and is only ancestral to *H. neanderthalensis*. In this hypothesis modern humans are descended from an African taxon that would be *Homo rhodesiensis* if that taxon's hypodigm included Kabwe and Tighenif and *Homo mauritanicus* if it included the latter but not the former. The "accretion model" suggests that "*H. heidelbergensis*" is an early stage in the evolution of *H. neanderthalensis* and thus *H. heidelbergensis* would be a junior synonym of *H. neanderthalensis*. Until recently, *H. heidelbergensis* was considered the oldest hominin taxon in Europe; but if *Homo antecessor* is a valid taxon then this would antedate it, as would a hominin tooth from Orce that some believe represents a distinct taxon. (L. *homin* = man or human being and *heidelbergensis* = because the holotype comes from the Grafenrain commercial sandpit, which is southeast of Heidelberg, Germany.) *See also* **pre-Neanderthal hypothesis**.

## *Homo helmei* Dreyer, 1935

The taxon name assigned to a partial cranium (Florisbad 1) discovered in 1932 in Florisbad, South Africa. Some researchers interpret Florisbad 1 as intermediate in morphology between *Homo heidelbergensis* and *Homo sapiens*. Its vertical facial dimensions, brow ridge size, and inclination of the frontal bone distinguish it from the former, yet it retains primitive features (e.g., a large brow ridge, receding frontal, and broad vault) that distinguish it from anatomically modern humans. While some suggest that Florisbad 1 could serve as the holotype of the taxon *Homo helmei*, most researchers take the view that there is no satisfactory diagnosis for such a taxon. Other specimens that have been suggested as belonging to *H. helmei* include Jebel Irhoud, Ngaloba (aka LH 18), Omo II, and Singa.

## homoiology

Despite the meaning of the Greek word on which it is based, homoiology was coined by Remane in 1961 for the part of the phenotype whose morphology is influenced by the type and level of activity undertaken (i.e., by environmental factors). Homoiology contributes to variation within a taxon and in theory homoiology can contribute to phylogenetically misleading morphological similarities among taxa when different genotypes express the same or similar phenotypes because of similar functional demands and not because of shared recent evolutionary history (i.e., they were present in the most recent common ancestor of the taxa concerned). Homoiologies are the result of a more inclusive phenomenon referred to as phenotypic plasticity. (Gk *homoios* = similar.) *See also* **homoplasy; phenotypic plasticity**.

### *Homo (Javanthropus) soloensis* **Oppenoorth, 1932**

A hominin taxon introduced for the calvaria (Ngandong 1, 2, and 3) recovered from Ngandong, Java, Indonesia in 1931. Oppenoorth compares the new calvaria with "Neandertalers," the "Rhodesia-skull," and "Wadjak man." He does not specifically refer to Ngandong 1 as the holotype, but it is evident from the text that he regards it as such. Since then researchers have either relegated the taxon to a subspecies of *Homo sapiens* as *Homo sapiens soloensis*, a subspecies of *Homo neanderthalensis* as *Homo neanderthalensis soloensis*, or it has been subsumed into *Homo erectus* as either a junior synonym of *H. erectus* or as a recognized subspecies of that taxon (e.g., *Homo erectus soloensis*). Recent analyses have concluded that the Ngandong hominins are more derived than the Bapang-AG sample of *H. erectus*. (L. *homin* = man or human being, *Java* = from the name of the island where the fossils were found, Gk *anthropos* = human being, and *solo* = refers to the river Solo in whose banks the site is located.)

### homology

A term introduced by Richard Owen in 1843 that now refers to similar structures, genes, or developmental pathways shared by more than one taxon that are inherited from their most recent common ancestor. For example, canine teeth are homologous across all primates, but the eyes of vertebrates and cephalopods are not. In genetics, the term homologous refers to DNA or protein sequences that share a common ancestry. Homologies are the basis for assembling taxa into clades or monophyletic groups. For example, within the hominin clade the outcome of the debate about whether *Paranthropus* is a monophyletic group rests on resolving whether the distinctive facial and dental morphologies shared by taxa currently included in *Paranthropus* are homologies or are the result of parallel evolution. Some people recognize degrees of homology, whereas others argue that structures are either homologous or not. The term "deep homology" has been introduced for cases where even though organisms do not share the same morphology they do share the same genetic regulatory cascade (e.g., common photoreceptor precursors that result in differently structured eyes in arthropods, squids, and vertebrates). (Gk *homologos* = agreeing, from *homo* = the same and *logos* = proportion; (adjective: homologous). *See also* **homoplasy**.

### *Homo neanderthalensis* **King, 1864**

A hominin species established in 1864 by William King for the partial skeleton recovered in 1856 from the Kleine Feldhofer Grotte in the part of the Düssel valley in Germany named after Joachim Neander. King's conclusion about the taxonomic implications of the specimen's distinctive morphology differed from that of Rudolf Virchow, who considered it that of a recent individual affected by pathology, and from the conclusions of scientists such as Hermann Schaaffhausen, Charles Lyell, and Thomas Henry Huxley who, while arguing for the fossils' antiquity and evolutionary significance, refrained from assigning the Kleine Feldhofer Grotte hominin remains to a separate species. Discoveries made earlier, such as the infant's cranium from Engis, Belgium (1828), and the partial cranium from Forbes' Quarry, Gibraltar (1848), were subsequently recognized as belonging to the same species. In the following half-century further *Homo neanderthalensis* remains were discovered at other European sites including La Naulette and Spy (Belgium), Šipka (Moravia), Krapina (Croatia; with over 900 specimens representing multiple individuals), and Malarnaud, La Chapelle-aux-Saints, Le Moustier (lower shelter), La Ferrassie, and La Quina, among others, in France. By the early 20thC *H. neanderthalensis* was by far the best-known extinct hominin species. In 1924–6 the first *H. neanderthalensis* remains were found outside of western Europe at Kiik-Koba in the Crimea. Later discoveries were made at Tabun cave on Mount Carmel in the Levant, at Teshik-Tash in central Asia, and at two sites in Italy: Saccopastore and Monte Circeo. Further evidence was added after WWII, first from Iraq (Shanidar), then from additional Levantine sites in Israel (Amud and Kebara), and Syria (Dederiyeh). New fossiliferous localities continue to be discovered in Europe and Western Asia (e.g., Saint-Césaire and Moula-Guercy in France, Zaffaraya in Spain, Mezmaiskaya in Russia, Vindija in Croatia, and Lakonis in Greece). To date, Neanderthal remains have been found throughout much of Europe below 55°N, as well as in the Near East and Western Asia, but no fossil evidence has been found in North Africa.

Many mid-20thC workers considered Neanderthals to be only subspecifically distinct from modern humans, as *Homo sapiens neanderthalensis*, but in the past decade or so there has been an increasing acceptance that the Neanderthals are morphologically distinctive, so much so that many consider it unlikely that such a derived form could have given rise to the morphology seen in modern humans. There is, however, another school of researchers who point to, and stress, the morphological continuity between the fossil evidence for *H. sapiens* and the remains others would attribute to *H. neanderthalensis*. The distinctive features that distinguish *H. neanderthalensis* from both earlier (e.g., *Homo erectus*, *Homo heidelbergensis*) and contemporary taxa (i.e., *H. sapiens*) include cranial (e.g., large, rounded discrete brow ridges, projecting midface, inflated cheeks, small mastoid process, suprainiac fossa, and occipital bun), mandibular (long corpus, retro-molar space, and asymmetric mandibular notch), dental (e.g., large shovel-shaped incisors, occlusal morphology of molars and premolars, a high incidence of taurodontism), and postcranial (e.g., long clavicle, teres minor groove extending onto the dorsal surface of the scapula, large infraspinous fossa, long, thin pubic ramus, and large joints) morphology. There are also reports that the ontogeny of *H. neanderthalensis* differs in several ways from that of modern humans.

The taxon *H. neanderthalensis* is currently the only extinct hominin for which there is sound ancient DNA evidence. In 1997 Mathias Krings and his colleagues announced they had recovered short fragments of the hypervariable region I (HVRI) of the mitochondrial DNA (or mtDNA) control region from a piece of bone taken from the right humerus of the Kleine Feldhofer Neanderthal 1 type specimen. Krings and colleagues then sequenced a 340 base pair (bp) segment of a different part (HVRII) of the mtDNA control region of the type specimen. Two laboratories confirmed the extraction of a 345 bp sequence of HVRI from a rib fragment belonging to a Neanderthal from Mezmaiskaya. Krings and colleagues reported sequences obtained from a Neanderthal individual (Vi75; now Vi 33.16) from layer G3 at Vindija, Croatia. Subsequently, mtDNA was extracted from the humeral shaft (NN 1) from a second individual at the Neanderthal type site and from four more Neanderthal individuals (Vi77 and Vi80, Engis 2 and La Chapelle-aux-Saints) as well as from five fossil *Homo sapiens* individuals. The Vi80 sequence was identical to that of the sequence reported for Vi75 and none of the modern humans yielded any amplification products with the Neanderthal primers. More recently the complete mtDNA of a specimen from Vindija has been sequenced along with the full mtDNA sequences of five individuals (Neanderthal 1 and 2, Sidrón 1253, Vi33.25, and Mezmaiskaya) and these studies suggest that genetic diversity within *H. neanderthalensis* was substantially less than that seen in modern humans. In 2010, Richard Green and colleagues compared sequence data from three individuals (Vi33.16, 25, and 26) from Vindija with sequences from Neanderthal specimens from El Sidrón, Kleine Feldhofer Grotte, and Mezmaiskaya and with the sequenced nuclear genomes of five modern humans: two from sub-Saharan Africa, one each from Papua New Guinea and France, and one Han Chinese. The results show that whereas the modern humans from sub-Saharan Africa displayed no evidence of any Neanderthal DNA, the three modern humans from outside of Africa showed similar, low amounts (between 1 and 4%) of shared DNA with Neanderthals. These results are compatible with either a deep split within Africa between the population that gave rise to modern Africans and a second one that gave rise to present-day non-Africans plus Neanderthals, or with the hypothesis that there was hybridization between Neanderthals and modern humans soon after the latter left Africa, perhaps in western Asia. Fred Smith's assimilation model and Bräuer's African hybridization and replacement model of modern human origins appear most compatible with the current morphological and genetic evidence on Neanderthal/modern human diversity.

Nuclear and mtDNA have been used to generate estimates of the date of divergence of *H. sapiens* and *H. neanderthalensis*. A recent estimate using mtDNA and based on an assumed divergence time of c.7–6 Ma for the modern human and chimpanzees/bonobo lineages suggests a divergence date of 660 ± 140 ka. Estimates of the divergence time of the ancestral modern human and Neanderthal populations based on the nuclear genome range from 440 to 270 ka; 440 ka is based on an early (8.3 Ma) divergence date for chimpanzees/bonobos and modern humans and 270 ka on a later (5.6 Ma) divergence date. The earliest fossils that most researchers would accept as *H. neanderthalensis* are from Marine Isotope Stage (MIS) 5 (i.e., c.130 ka), but anything beyond that date is contentious. For example, some interpret the fossil evidence from Swanscombe (MIS 11) and the Sima de los Huesos (possibly as early as MIS 13) as showing enough *H. neanderthalensis*-like morphology to justify those remains being included in *H. neanderthalensis*, whereas others see a distinction between these specimens, which they would include in *H. heidelbergensis* and later "true" Neanderthals that they claim do not appear until MIS 6. (L. *homin* = man or human being and *neanderthalensis* = literally "of Neander's valley," the Neander in question being Joachim Neander, a 17thC non-ordained Lutheran ecclesiastical poet and composer.)

## homonymy

When two or more available names of different taxa have the same spelling. This may happen when species names that were initially used in different genera later find themselves in the same genus because the two original genera were combined. An example in paleoanthropology involves the Omo 18-1967-18 mandible, which was made the holotype of a new genus and species *Paraustralopithecus aethiopicus* by Arambourg and Coppens in 1968. However, many researchers regard Omo 18-1967-18 as belonging to either *Australopithecus* or *Paranthropus*. But in 1980 Tobias designated the Hadar hypodigm of *Australopithecus afarensis* species as *Australopithecus aethiopicus*, so according to the International Commission on Zoological Nomenclature rules relating to homonymy if Omo 18-1967-18 is included in *Australopithecus* it cannot retain its species name as

*Australopithecus aethiopicus*. However, if it is included in *Paranthropus*, it can retain its species name as *Paranthropus aethiopicus*.

## homoplasy

Resemblances between taxa due to processes other than descent from a recent common ancestor. Such resemblances can be mistaken for shared derived similarities, or synapomorphies, which are the principal evidence for phylogeny. Thus homoplasy complicates attempts to estimate phylogenetic relationships, and, if there are enough of them, homoplasies can make the generation of a reliable phylogenetic hypothesis problematic. There are several forms of homoplasy: analogy, convergence, parallelism, reversal, and homoiology. Analogy and convergence are both caused by adaptation to similar environments. In the case of analogies, natural selection operates on different developmental processes; in the case of convergences it acts on the same developmental processes. Parallelisms result from aspects of ontogeny that limit phenotypic diversity, but which have no necessary connection with the demands of the environment. Parallelisms are, in other words, byproducts of development and not adaptations. There is a continuum between parallelism and convergence, but convergence usually occurs across greater phylogenetic distances. A fourth type of homoplasy is reversal. Most cases of reversal are probably due to natural selection, but recent work on silenced gene reactivation suggests that some reversals may also be neutral with regard to adaptation. The final form of homoplasy, homoiology, is attributed to nongenetic factors (e.g., activity-induced bone remodeling). (Gk *homos* = same and *plasis* = mold.) *See also* **cladistic analysis**; **homoiology**.

### *Homo rhodesiensis* Woodward, 1921

A pre-modern *Homo* taxon introduced by Sir Arthur Smith Woodward for the cranium and limb bones recovered from the Broken Hill lead mine at Kabwe in what is now Malawi. Woodward reasoned that a new species of *Homo* was needed for the specimen because it was not as primitive as *Homo erectus* and not as derived as either *Homo sapiens* or *Homo neanderthalensis*. Many researchers regard *Homo rhodesiensis* as a junior synonym of either an inclusively defined *H. sapiens* or *Homo heidelbergensis*. However, the taxon is used by researchers who interpret *H. heidelbergensis* as an exclusively European pre-modern *Homo* taxon. (L. *homin* = man or human being and *rhodesiensis* = from what was in 1921 the British protectorate of Northern Rhodesia – now Zambia – a country named after Cecil Rhodes.)

### *Homo rudolfensis* (Alexeev, 1986) Groves, 1989 *sensu* Wood, 1992

Hominin species formally established by Alexeev as *Pithecanthropus rudolfensis* and transferred to *Homo* by Colin Groves. The differences between *H. rudolfensis* and *Homo habilis sensu stricto* include that the crania of the former taxon are wider across the midface instead of the upper face, the mandibles are larger and more robust. It was claimed that the crowns of the postcanine teeth are larger and broader, but new evidence suggests that the mandibles such as KNM-ER 1802 may have been misassigned to *H. rudolfensis*. Some researchers consider *H. rudolfensis* to be a more likely ancestor for *Homo erectus* than *H. habilis sensu stricto*, but many researchers prefer to include *H. rudolfensis* within a more inclusively interpreted *Homo habilis sensu lato*. (L. *homin* = man or human being and *rudolfensis* = after the old name, Lake Rudolf, for Lake Turkana.)

## Homo sapiens Linnaeus, 1758

A hominin species established by Linnaeus to accommodate modern humans. The initial discovery of a fossil modern human was made in 1822–3 in Goat's Hole Cave in Paviland, Wales, but the first widely accepted evidence that modern humans were ancient enough to have fossilized representatives came when skeletal remains were discovered by workmen at the Cro-Magnon rock-shelter at Les Eyzies de Tayac, France, in 1868. Further discoveries of *H. sapiens*-like fossils were made elsewhere in Europe including at Mladeč (1881–1922), Předmostí (1884–1928), and Brno (1885). In Asia and Australasia similar fossils were recovered from Wadjak, Indonesia (1889–90), the Zhoukoudian Upper Cave, China (1930), Niah Cave, Borneo (1958), Tabon, Philippines (1962), and the Willandra Lakes, Australia (1968 and thereafter). In the Near East comparable fossil hominins have been recovered from sites such as Skhul (1931–2) and Djebel Qafzeh (1933 and 1965–75). The first African fossil evidence of populations that are difficult to distinguish from modern humans came in 1924 from Singa in Sudan. Thereafter comparable evidence has come from Dire-Dawa, Ethiopia (1933), Dar-es-Soltane, Morocco (1937–8), Border Cave, South Africa (1941–2 and 1974), Omo-Kibish, Ethiopia (1967), Klasies River Mouth, South Africa (1967–8), and Herto, Middle Awash study area, Ethiopia (1997). With the exception of Omo-Kibish (*c.*190 ka) and Herto (*c.*170 ka), there is no firm evidence to suggest that any of the above sites are likely to be more than 150 ka, and most are probably less than 100 ka.

The assembly of morphological criteria for "modern humanness" has proved to be a surprisingly difficult task. With respect to the cranium, little progress has been made since William (Bill) Howells suggested that anatomically modern human crania share a universal loss of robustness with variation in shape largely located in the upper face, and particularly the upper nose and the borders of the orbits. The postcanine teeth of anatomically modern humans are notable for the absolutely and relatively small size of their crowns, and for a reduction in the number of cusps and roots. The postcranial skeleton of *H. sapiens* has limbs that are long relative to the trunk, elongated distal limb bones, a relatively narrow trunk and pelvis, and low estimated body mass relative to stature. In summary, compared to their more archaic immediate precursors, anatomically modern humans are characterized postcranially by their reduced body mass, linear physique, and distinctive pelvic shape that includes a short, stout, pubic ramus, and a relatively large pelvic inlet.

In Africa there is fossil evidence of crania that are generally more robust and archaic-looking than those of anatomically modern humans, yet they are not archaic or derived enough to justify being allocated to *Homo heidelbergensis*. Specimens in this category include Irhoud 1 from North Africa, LH 18 from East Africa, and Florisbad 1 and Cave of Hearths 1 from southern Africa. There is undoubtedly a gradation in morphology that makes it difficult to set the boundary between anatomically modern humans and *H. heidelbergensis*, but the variation in the later *Homo* (i.e., post-*Homo erectus*) fossil record is too great to be accommodated in a single taxon. Researchers who wish to make a distinction between fossils such as Florisbad 1 and LH 18 and subrecent and living modern humans either do so taxonomically by referring the former specimens to a separate species, *Homo helmei*, or they distinguish them informally as "archaic *Homo sapiens*." A few researchers have suggested that *H. sapiens* should be a much more inclusive taxon. For example, because they could see no obvious morphological discontinuity between *H. sapiens* and *H. erectus*, Milford Wolpoff and colleagues recommended the boundary of *H. sapiens* be changed to incorporate *H. erectus*, thus echoing a similar proposal made earlier by Ernst Mayr. This interpretation has received little support, but at least the authors made an explicit statement about the scope of the morphology they were prepared to subsume within *H. sapiens*. (L. *homin* = man or human being and *sapiens* = to be wise.)

### *Homo sapiens idaltu* White et al., 2003
A subspecies of *Homo sapiens* established because the Herto hominins were judged to be morphologically just beyond the range of variation seen in anatomically modern *Homo sapiens*, but not within the range of other known fossil hominins.

### *Homo sylvestris*
This is not a formal species name. It was included in the discussion when the genus *Homo* was introduced in 1758 by Carl Linnaeus in the tenth edition of his *Systema Naturae*. It apparently referred to a mythical nocturnal cave-dwelling form from Java, Indonesia, but the discovery of *Homo floresiensis* suggests that *Homo sylvestris* perhaps might not have been mythical after all. (L. *homin* = man or human being and *silva* = forest.)

### *Homo troglodytes* Linnaeus, 1758
The second of two species of *Homo* (the first being *Homo sapiens*) coined by Linnaeus in the key tenth edition of *Systema Naturae*. He defined it as "*Homo nocturnus*" (aka "nocturnal humans"), and as evidence for the existence of this species he cited "Homo sylvestris Orang Outang" from Jakob de Bondt, an early 17thC Dutch physician in Java and "Kakurlacko" from Kjoeping, a Swedish traveller, and evidence from Dalin, the president of the Swedish Royal Academy. de Bondt may have been describing a real orangutan, or maybe just an exceptionally hairy female; Kjoeping and Dalin were describing albinos on the Indonesian island of Ternate, who hid away from the light in caves. Thus, *Homo troglodytes* Linnaeus, 1758 is, in part or in whole, a synonym of *H. sapiens*. (L. *homin* = man or human being and *troglodyta* = cave dwellers, from the Gk *troglos* = cave and *dutai* = those who enter.)

### honing
When the upper canine and the anterior lower premolar rub together to form a blade-like edge on one or both teeth. This honing complex, which is seen in living and fossil great apes, is assumed to be the primitive condition for hominins. Hominins are in part defined by the absence of a honing complex and its loss has been related to a shift in behavior where males do not need to display, or sometimes fight with, their canine teeth to get access to their preferred mate. (OE *hān* = stone, from a whetstone used to sharpen knives or tools.) *See also* **sexual selection**.

### Hopefield
*See* **Elandsfontein**.

### hornfels
A rock that resembles fine-grained basalt. It typically forms from high-temperature baking and recrystallization of a pre-existing rock, a process that usually removes features such as bedding planes. Because it is even more brittle than basalt, it makes a poor raw material for knapping, and thus tends to be used at sites where better raw materials are unavailable (e.g., Florisbad).

## horst
An area of relative uplift between a pair of normal faults; the adjacent subsided area is known as a graben. In heavily faulted areas (e.g., parts of the East African rift valley such as the faulted margins of the Olorgesailie paleolake basin) substantial environmental heterogeneity can occur over relatively short lateral distances. Horst formation can expose otherwise buried strata (e.g., Bouri Formation in the Middle Awash study area).

## Hortus
A cave site located high in a limestone massif in southeast France. The main deposition occurred during the Würm I and II cold phases. The remains of between 20 and 36 Neanderthal individuals were recovered, as well as abundant Mousterian artifacts from the lower levels of the site. There was no evidence of intentional burial and a high proportion of the individuals were young adults. The majority of the bones are maxillae or mandibular fragments, and it has been suggested that these were accumulated as the result of cannibalism, but there is no unambiguous evidence of cutmarks. (Location 43°47′N, 03°50′E, France.)

## Hotelling $T^2$ test
A multivariate generalization of the *t* test used to determine whether a vector of means of two or more continuous variables differs between two groups (e.g., do the centroids of two sets of cranial measurements differ from each other?).

## housekeeping gene
A gene expressed in connection with the basic functions of a cell (e.g., transcription, translation, energy production, and cell division).

## HOX
An abbreviation of *homeobox*. See **homeobox genes**.

## HP
Acronym for the Howieson's Poort industry. See **Fish Hoek**.

## human
This term should be reserved for referring to species within the genus *Homo*, or to the species *Homo sapiens*, or to individuals within those species, but many use it to refer to any taxon in the hominin clade, or all of the members of the hominin clade.

## human accelerated regions
(or HARs) Regions of the genome that show relative conservation in vertebrates, but which show changes in the lineage leading to modern humans subsequent to the most recent common ancestor of modern humans and chimpanzees and bonobos.

## human anatomical terminology
See **anatomical terminology**.

## humerofemoral index

Ratio of the lengths of the humerus and femur calculated as 100×(humerus length/femur length). The earliest evidence of a modern human-like humerofemoral index (i.e., a relatively longer femur) comes from associated limb bones from Bouri, Middle Awash study area, Ethiopia. *See also* **intermembral index**.

## hunter–gatherer

A group of people, or a member of such a group, whose primary subsistence strategy is based on the hunting or gathering of wild plant and animal resources. Prior to the advent of domestication, virtually all hominin groups were hunter–gatherers and it is presumed that our more recent hominin ancestors (i.e., members of the genus *Homo*) had a hunter–gatherer subsistence economy. Present-day hunter–gatherers are of particular interest to paleoanthropologists because they provide analogues for understanding past hominin subsistence strategies, behavioral patterns, and social interactions. *See also* **forager**; **Hadza**; **San**.

## Hunter–Schreger bands

Microscopic three-dimensional features of tooth enamel caused by successive bundles of enamel prisms running in different directions. They are best seen under reflected or polarized light. There are differences in Hunter–Schreger band morphology among hominins (e.g., *Paranthropus* vs early *Homo*).

## hunting

The pursuit and killing of wild animals, predominantly for food but also for hides or furs, or to fulfill social or religious obligations, and sometimes for trade. Among ethnographically documented hunter–gatherers worldwide, hunting was, and remains today, an important economic, social, and political activity. In these groups most hunting is done by men. Why men devote substantial time and effort to hunting big game remains controversial. Meat from large game provides important nutrients (protein, fat, vitamins) and energy, and when cooked it is highly valued for its taste. But large animals are difficult to find and kill, and ethnographically documented hunters such as the Hadza (Tanzania), San or Bushmen (Botswana, Namibia), and Australian Aborigines fail to make a kill far more often than they succeed. The origin and evolution of big-game hunting are topics of ongoing debate. Chimpanzees, our closest living primate relatives, hunt other mammals (mostly other primates) and this suggests that the most recent common ancestor of modern humans and chimpanzees/bonobos may have hunted. For nearly half of the known hominin fossil record (i.e., from about 6–5 Ma until about 2.6 Ma) we have no archeological sites or places where demonstrably butchered animal bones (i.e., bones with cutmarks made by stone tools) are concentrated together with stone tools. Thus for the earlier part of the hominin record we can only assume that our ancestors occasionally killed and consumed smaller animals in the manner modern chimpanzees use today. After about 2.6 Ma, however, we begin to find sites with concentrations of butchered remains of many different species of ungulates, many considerably larger than the hominins themselves, indicating that hominins had begun transporting carcasses or portions of carcasses from the place where they were acquired to a central place on the landscape where they were further processed and consumed. However, there is considerable controversy about whether these large game animals were hunted or were acquired by scavenging carcasses killed and abandoned by large predators.

## hunting and gathering
See **forager**; **hunter–gatherer**.

## hyaline cartilage
See **cartilage**; **joint**.

## hybrid
The offspring resulting from the interbreeding of individuals from two different recognized species (interspecific hybridization) or from the interbreeding of individuals from two different subspecies, demes, or allotaxa (intraspecific hybridization). Hybridization can occur in areas, called hybrid zones, where the ranges of two species overlap [e.g., the ranges of the gelada (*Theropithecus gelada*) and the Hamadryas baboon (*Papio hamadryas*) in the Awash region of Ethiopia]. Hybridization is a feature of reticulate evolution and Cliff Jolly has suggested that hominin species may have arisen as the result of hybridization as well as by classical processes such as allopatry. (L. *hybrida*, var. of *ibrida* = mongrel.) *See also* **reticulate evolution**; **speciation**.

## hybridization
Evolution Interbreeding between individuals from genetically differentiated lineages or between local demes. *See also* **hybrid**. Genetics and molecular biology An outdated molecular biology technique where two complementary strands of DNA are joined together, then driven apart by heating. (L. *hybrida*, var. of *ibrida* = mongrel, Gk *izein* = to make.)

## hybrid zone
See **hybrid**.

## hydroxyapatite
(aka apatite) The dominant salt in the mineral portion of bone, cementum, dentine, and enamel.

## hyenas
Hyenas belong to the mammalian family Hyaenidae within the order Carnivora. The group includes the striped (*Hyaena hyaena*), the spotted (*Crocuta crocuta*), and the brown (*Parahyaena brunnea*) hyenas and the aardwolf (*Proteles cristata*). In the Plio-Pleistocene hyenas were a prominent member of the carnivore guild and if, as some researchers suggest, hominins became members of that guild as scavengers and/or hunters, then hyenas and hominins would have been in competition. Hyenas, like modern humans, are known to accumulate large collections of bones (i.e., faunal assemblages). In the absence of rigorous taphonomic investigations, hyena-accumulated bone concentrations can potentially be misidentified as the product of hominin activities, leading to erroneous behavioral interpretations. In addition to accumulating faunal assemblages, hyenas are also known to modify or destroy skeletal remains discarded by

modern humans and probably by early hominins; this phenomenon is called carnivore ravaging. (Gk *huaina* = swine.)

### hyoid
A bone in the midline of the neck between the mandibular symphysis and the manubrium of the sternum which, together with the mylohyoid muscle, forms the floor of the mouth and the base of the tongue. Hyoid bones occur occasionally in the hominin fossil record (e.g., Dikika, Kebara, Sima de los Huesos). Some researchers claim that the morphology of the hyoid can be used to make inferences about the capacity for spoken language, but this claim has been disputed. (Gk *hyoeides* = shaped like the Greek letter upsilon.)

### hypervariable region
See **mitochondrial DNA**.

### hypocone
The main cusp on the distolingual aspect of a maxillary (upper) mammalian molar. In some teeth its tip is lower than the tips of the meta-, para-, and protocone hence the prefix meaning "below." It is part of the talon. (Gk *hypo* = under, below and *konos* = pine cone.)

### hypoconid
The main buccal cusp distal to the protoconid on a mandibular (lower) (hence the suffix "-id") molar tooth crown. Its prefix is a misnomer in hominins because its tip is not usually lower than those of the other main cusps. It is part of the talonid. (Gk *hypo* = under, below and *konos* = pine cone.)

### hypoconule
An accessory cusp on the occlusal surface of a maxillary molar.

### hypoconulid
The main cusp at the distal end of a mandibular (lower) molar crown; it is wedged between the hypoconid on the buccal side and the entoconid on the lingual side. It is part of the talonid. Most modern humans do not have a hypoconulid but it is found in archaic hominins and in most pre-modern *Homo* taxa. (Gk *hypo* = under, below and *konos* = pine cone.)

### hypodigms
All the fossil evidence assigned to a taxon, but the term is usually used to refer to all the fossil evidence assigned to a species. (Gk *hypo* = under, below and *deik* = to show.)

### hypoglossal canal
A bony canal that passes through the basioccipital bone just in front of the occipital condyles (hence its other name, the anterior condylar canal). It transmits the motor axons of the twelfth cranial (hypoglossal) nerve, which innervate all of the intrinsic muscles of the tongue (i.e., the muscles that make up the tongue itself) plus all but one of the muscles that move the tongue around in the mouth. Because modern humans have larger hypoglossal canals than

apes, some researchers have suggested that the size of the hypoglossal canal can be used to indicate whether a fossil hominin might have had the ability to speak. (Gk *hypo* = under, below and *glossa* = tongue.)

## hypoplasia

Literally "arrested growth," hypoplasia refers to areas on teeth where the production of enamel matrix by ameloblasts has been disrupted by a nonspecific stressor (e.g., systemic infection, a deficient diet, or starvation) experienced during tooth formation, resulting in a depression on the surface of a tooth. Furrow defects (i.e., linear bands encircling a tooth crown), which are the most common type of hypoplasia, are the result of an interruption to enamel matrix production that affects many ameloblasts. In pit defects the disturbance is more localized and does not include all simultaneously formed enamel. Hypoplasias usually affect all the teeth in which enamel matrix formation is occurring, so they can be used to compare the relative ontogeny of teeth along the tooth row. (Gk *hypo* = under, below and *plasis* = growth.)

## hypostotic

*See* **nonmetrical traits**.

## hypothesis testing

The process of testing a specific relationship which has been identified *a priori*, as opposed to exploratory analysis in which multiple techniques are used to explore possible patterns within a data set.

## hypothetical common ancestor

When a cladistic analysis groups two taxa or two clades together as sister taxa or sister clades, certain assumptions can be made about the character states of the common ancestor of the two taxa or clades. These assumptions amount to a prediction about the morphology of the most recent common ancestor of the two taxa or clades. Researchers who are skeptical about the chances of ever finding an actual ancestor prefer to describe the predicted taxon as a hypothetical common ancestor.

## Iberomaurusian

A term that refers to Late Pleistocene artifacts (and to the hominins that made them) found at coastal sites in Morocco, Algeria, and Tunisia that date from *c.*18 ka to 11 ka. Iberomaurusian stone artifact assemblages are dominated by bladelets and microliths and there are evident technological and behavioral differences between the Iberomaurusian and its regional antecedents. Iberomaurusian strata often unconformably overlie Aterian ones, suggesting that the area may have been temporarily abandoned by hominins until settlement by Iberomaurusian populations. (etym. a combination of *Ibero-*, as in the Iberian Peninsula, and L. *Mauritania* = the name of a Roman province that is coincident with what is now Morocco and Algeria.)

## ice rafting

The process of sediment transport via sea ice. Icebergs that calve off from continental glaciers carry bedrock-derived sediments into the ocean and the sea ice that forms in shallow waters incorporates sediments from the continental shelf. Both are transported by wind and ocean currents. The morphology of ice-rafted sediments can be distinguished from other forms of transport (e.g., fluvial or aeolian). The extent of sea ice transport increases during times of glaciations, and it was the sediments derived from sea ice that provided evidence of abrupt climate change in the form of Heinrich events.

## ichnofossils

Direct evidence (e.g., footprints, tracks, burrows, leaf impressions, etc.) that an organism has been in a place at some time in the past. The Laetoli hominin footprints are examples of ichnofossils. (Gk *ikhnos* = footprint; syn. trace fossil.) *See also* **trace fossil**.

## ICZN

Acronym for the **International Code on Zoological Nomenclature** (*which see*).

## -id

If this postfix is used after the name of a cusp it means it is on a mandibular (lower) tooth. Thus, metaconid is the term for the cusp mesial to the protoconid for the lower teeth whereas metacone is the term for the equivalent cusp in an upper tooth. (L. *-id* = source or origin.)

## identification
The process of allocating individual fossil specimens to an existing taxon or to a new taxon. Classification only applies to taxa; you identify an individual fossil, but classify the taxon to which it belongs. *See also* **classification; systematics.**

## Ileret
One of the main subregions of the Koobi Fora site complex. *See also* **Koobi Fora.**

## imbricational enamel
Enamel formed after the cuspal (or appositional) enamel has been formed. The striae of Retzius within the imbricational enamel emerge on the crown surface as depressions between parallel ridges called imbrication lines, which look like the edges of the tiles on a tiled roof. (L. *imbricatus* = roof made of tiles; syn. cuspal enamel; perikymata.) *See also* **enamel development; incremental features.**

## imbrication lines
*See* **imbricational enamel; perikymata.**

## imitation
*See* **social learning; theory of mind.**

## immunochemistry
A branch of biochemistry that investigates the components and chemical reactions of the immune system. Morris Goodman used the immune reactions in the blood serum of one species caused by the blood of another species to investigate the relationships among higher primates. He showed that the patterns of immune reactions between specific antibodies and serum albumin were virtually identical in modern humans and chimpanzees, indicating that these species were very closely related. Two closely related techniques, immunocytochemistry and immunohistochemistry, use antibodies to label molecules (usually proteins) in intact cells and in fixed tissue, respectively. (L. *immunis* = an individual who is exempt from public service and *chemista* = alchemist, a person skilled in the art of alloying metals.)

## immunological distance
*See* **molecular clock.**

## *incertae sedis*
The phrase indicates that a particular specimen is difficult to assign to any taxon, or that a taxon is difficult to assign to a higher taxonomic group. Certain fossils resemble several known species and are thus difficult to assign. For example the Ceprano 1 calvaria somewhat resembles *Homo erectus*, specifically later Lower Pleistocene African specimens such as Daka (Bouri), as well as *Homo heidelbergensis*, so it was initially assigned to *Homo incertae sedis*. [L. *incertae sedis* = "of uncertain (taxonomic) position."] *See also* **taxonomy.**

## incremental features

Physical features in tissues such as enamel, dentine, and cementum that record their growth at regular, predictable time intervals. This is because their growth is appositional in nature (i.e., one layer of tissue is deposited upon a previous layer). Microscopic markings representing intrinsic temporal rhythms in hard-tissue secretion may be annual (e.g., cementum annulations, lines of arrested growth in bone), long-period (i.e., greater than 1 day; e.g., striae of Retzius/perikymata in enamel or Andresen lines/periradicular bands in dentine), short-period (i.e., about 24 hours or circadian; e.g., cross-striations in enamel, von Ebner's lines in dentine), or less than 24 hours (e.g., intradian lines in enamel and dentine). Mineralized tooth tissues are never replaced so these increments of growth are preserved throughout life. For a long time it was widely believed that growth increments were not preserved in bone but recent reports show the presence of annual/seasonal and, more recently, possible long-period growth increments in the bone tissues of many vertebrates, including mammals. (L. from *increscere* = to increase.) *See also* **enamel microstructure**.

## *incurvatio mandibulae*

A midline depression on the external aspect of the symphysis of the mandible between the alveolar process and the base. There needs to be an *incurvatio mandibulae* for a mental protuberance to qualify as a true chin, which is one of the hallmarks of *Homo sapiens* that distinguish it from other members of *Homo*.

## incus

One of the three auditory ossicles in the middle ear. The incus occasionally survives into the fossil record (e.g., the SK 848 incus belonging to *Paranthropus robustus*). *See also* **auditory ossicles; middle ear**.

## indel

Shorthand for insertion(s) and deletion(s) in the genome. *See* **insertion; deletion**.

## independent contrasts

*See* **phylogenetically independent contrasts**.

## Indonesia

Indonesia was a geographical entity before it became a political one. It includes the large and small islands that extend eastward from the Malay Peninsula. The large islands include Borneo, Java, the Moluccas, Sulawesi, Sumatra, and Timor; the small islands include Bali, Flores, and Sumba. Fossil hominin taxa found in Indonesia include *Homo erectus*, *Homo floresiensis*, and *Homo sapiens*. (L. *Indus* = Indian and Gk *nesos* = island, literally "Indian Islands"; equivalent to the Malay Archipelago.)

## industrial complex

A higher-rank archeological taxonomic term that refers to multiple related industries (e.g., the Acheulean industrial complex refers to sites across the world that have handaxes and other Acheulean tools).

## industry
A spatially and temporally constrained collection of archeological assemblages, considered by some to represent the range of material items produced by a prehistoric group of people. The precise boundaries of an industry are rarely defined and usage of the term varies among authors.

## inferior nuchal line
A roughened line on the occipital bone that marks the attachment of the fascia between the biggest nuchal muscle (semispinalis capitis) posterosuperiorly and a smaller nuchal muscle (obliquus capitis superior) anteroinferiorly.

## infraorbital visor
*See* **facial visor**.

## inner ear
A cavity medial to the middle ear within the petrous part of the temporal bone. It consists of a series of complex intercommunicating cavities referred to as the bony labyrinth. The bony labyrinth is filled with a fluid called perilymph and suspended in the perilymph is the membranous labyrinth, a smaller endolymph-filled facsimile of the bony labyrinth. The walls of the membranous labyrinth contain the hair cells that detect sounds, posture, and motion. The membranous labyrinth is divided into three functional components: anteriorly the snail-shaped cochlea for sound, then the saccule and utricle for posture, and posteriorly the semicircular canals for motion. The relative size and shape of the semicircular canals of the bony labyrinth have proved to be diagnostic variables for reconstructing taxonomy, and also informative of function, among the living great apes and hominins. *See also* **bony labyrinth; semicircular canals**.

## inner table
*See* **cranial vault**.

## innominate
The old name for what the *Terminologia Anatomica* now refers to as the hip or pelvic bone. (L. *innominatus* = no name.)

## insectivory
Literally, the consumption of insects. Insects are rich in energy and protein but many insects are rare, cryptic, and difficult to catch. Insectivory is most common in the smallest primates, which require high-quality food because of their need for relatively more energy per unit body mass than large primates and because they are inefficient processors of low-quality foods. Obligate insectivores have teeth with high, sharp cusps that facilitate the penetration of the insect exoskeleton. Modern humans and chimpanzees are known to consume insects and insect products (e.g., honey) opportunistically and it is likely that early hominins would have done the same.

## insertion

A mutation in the genome where one or more base pairs (i.e., nucleotides) have been added (i.e., inserted) into a sequence of nucleotides. Insertions range in size from one to many thousands of base pairs. In a coding region any insertion that is not a multiple of three base pairs causes a frameshift mutation; they are typically deleterious. The significance of insertions for human evolution is two-fold. First, insertions can add functionally important genetic material (e.g., whole genes). Second, insertions can be used to understand the relationships among groups (e.g., Alu insertions, which have been used to study population history, argue for a larger population size in Africa than in the rest of the world over most of human evolutionary history). Insertions and deletions are commonly referred to by the abbreviation "indels."

## in situ

Fossils or artifacts found in their parent horizon either during the course of an excavation or in the process of eroding out of that horizon. For example, all the Dmanisi fossils have been recovered from excavations and KNM-ER 1813 was found eroding out of the sediments of the Upper Burgi Member of the Koobi Fora Formation. (L. *in* = in and *situs* = place.)

## in situ hybridization

A family of methods used in molecular biology that attaches labels to the probes used to identify discrete sections of the DNA molecule. For example, one method uses fluorescent labeling to identify probes designed to detect specific DNA sequences; another uses radiography.

## insolation

The amount of solar radiation reaching the Earth's surface. Solar radiation is more intense at or near the equator than at or near the poles. Changes to the shape of the Earth's orbit (i.e., eccentricity), or to the tilt (i.e., obliquity) or steadiness of that axis (i.e., precession) all take the Earth (or parts of the Earth in the case of changes in the tilt angle) nearer to or further away from the sun. In doing so, they alter insolation patterns and affect the climate. (L. *insolare* = in the sun.)

## interbirth interval

The time between the birth of one offspring and birth of the following offspring of a different gestation. The length of the interbirth interval is one of the primary life history variables that factor into discussions about "fast" and "slow" life histories. *See also* **grandmother hypothesis**; **life history**.

## interglacial

*See* **glacial cycles**.

## intermembral index

Ratio of upper limb length relative to lower limb length, calculated as 100 × (humerus length + radius length)/(femur length + tibia length). This index, which is the most common way researchers measure and compare limb proportions in primates and hominins, reflects major differences in locomotor strategies. Large-bodied primates (e.g., orangutans) tend to have high intermembral indices (i.e., their long upper limbs allow them to reach across gaps between trees and hold onto multiple supports to distribute their weight). In contrast, many small-bodied primates (e.g., galagos) have low intermembral indices because long lower limbs are mechanically advantageous for leaping. Modern humans are an exception to this trend, for although their body mass is large they have long lower limbs that increase the speed and energetic efficiency of bipedal walking and running. *See also* **humerofemoral index**.

### International Code of Zoological Nomenclature
The International Code of Zoological Nomenclature (or ICZN) sets out rules and recommendations that help "promote stability and universality in the scientific names of animals and to ensure that the name of each taxon is unique and distinct." If you name a new species you need to abide by the rules of the ICZN and it is also sensible to follow its recommendations. The ICZN is the responsibility of the International Commission on Zoological Nomenclature whose mission is "to create, publish and, periodically, to revise the International Code of Zoological Nomenclature." The ICZN also considers and rules on specific cases of nomenclatural uncertainty; their rulings are published as "Opinions" in the *Bulletin of Zoological Nomenclature*.

### interproximal wear
(aka approximal or interstitial wear) Wear between tooth crowns that occurs when adjacent teeth move slightly against each other in the mouth during chewing. In the early stages of interproximal wear between two teeth, most of the wear occurs in the form of a concave wear facet on the proximal (i.e., mesial) face of the crown of the distal tooth, hence the emphasis in the term on proximal wear. But if the teeth wear against each other long enough a substantial wear facet also appears on the distal face of the crown of the mesial tooth. Some taxa (e.g., *Paranthropus boisei*) show marked interproximal wear that results in laterally and vertically extensive, more or less flat, interproximal wear facets. Marked interproximal wear can reduce the mesiodistal length and the overall occlusal area of a tooth crown; thus it needs to be taken into account when comparing the size and shape of worn and unworn tooth crowns. *See also* **tooth size**; **tooth wear**.

### interspecific allometry
*See* **allometry**; **scaling**.

### interstadial
*See* **glacial cycles**.

### interstitial growth
*See* **ossification**.

### interstitial wear
*See* **interproximal wear**.

### intertropical convergence zone

(or ITCZ) The region where the trade winds converge and the air rises. The ITCZ migrates between the Tropics of Capricorn and Cancer as it follows the overhead sun with the seasons; its location is also influenced by contrasts between land and sea temperature. The position of the ITCZ influences the pattern of seasonal precipitation maxima in Africa (e.g., East Africa receives most of its precipitation from April to May and from September to November in association with the biannual movement of the ITCZ and with the seasonally reversing winds).

### intracranial venous drainage

See **cranial venous drainage**; **intracranial venous system**.

### intracranial venous system

The intracranial venous system consists of superficial cerebral veins that drain the surface of the brain, which in turn drain into a system of venous channels called the dural venous sinuses. The latter, which run between the fibrous and endosteal layers of the dura mater (the outer of the three meningeal layers), leave impressions on the endocranial surface of the cranium. See also **cranial venous drainage**.

### intramembranous ossification

See **ossification**.

### intraspecific variation

Variation within species due to differences in size, shape, or (almost always) a mixture of the two. The factors contributing to intraspecific variation in a fossil sample include: (a) ontogeny, differences due to fossils coming from individuals at different stages of development; (b) geography, differences due to fossils coming from different regions of a species' range; (c) time, differences due to directional or random changes through time; (d) sex, primary or secondary sexual differences in the hard tissues, which is called sexual dimorphism and in some extinct hominin taxa this can be a major component of intraspecific variation; and (e) intrasexual variation, differences between same-sex individuals within a species that are not due to (a)–(c). Homoiology (i.e., morphological variation due to differences in activity levels among individuals) is another factor that can drive intrasexual variation.

### introgression

Gene flow between hybridizing lineages. Introgression can result in reticulate evolution (i.e., a "web-like" rather than a "tree-like" pattern of relationships through evolutionary time). (L. *introgredi* = to step in.) See **hybrid**; **hybridization**.

### intron

A noncoding region of a gene that lies between exons. During transcription, introns are removed by a process called splicing. Introns may contain regulatory elements such as enhancers. (Abbreviation of *intr*agenic regi*on*.)

## intrusive
Fossils or artifacts that have been secondarily introduced into sediments either naturally (e.g., by burrowing rodents) or deliberately (e.g., by ritual burial). (L. *intrudere* = to thrust in.)

## Isampur Quarry
One of the few Acheulean sites on the Indian subcontinent. At this site all the stages of biface production, from mining to the refining of handaxes, were found. The site dates to 1.27 Ma–730 ka and indicates the early presence of hominins, probably *Homo erectus*, in this region. (Location 16°30′N, 76°29′E, India.)

## Isenya
A large *c.*0.7–0.6 Ma open-air Acheulean site characterized by its dense archeological levels (particularly its abundance of cleavers) and rich fossil fauna. (Location 01°40′S, 36°50′E, Kenya.)

## Ishango
A site complex on the right bank of the Semliki River where it exits from Lake Edward approximately 6 km/3.5 miles south of Katanda, in the Democratic Republic of the Congo. The main artifact- and bone-bearing levels at Ishango, which date to *c.*25–20 ka BP, were deposited on an eroded bench of Plio-Pleistocene deposits. The majority of the hominin remains are anatomically modern, but some molar teeth are strikingly large and may be fossils of early *Homo* derived from the Plio-Pleistocene deposits. Archeological evidence includes a microlithic Later Stone Age industry, with bone artifacts that include harpoons. (Location 00°08′S, 29°36′E, Democratic Republic of the Congo.)

## Isimila
A rich *c.* > 330 ka open-air Acheulean site that is an important example of landscape-scale excavations undertaken in order to explore spatial and temporal variation in the Early Stone Age archeological record. (Location 07°53′48″S, 35°36′12″E, Tanzania.)

## island rule
Island populations of mammals tend towards either gigantism or dwarfism depending on the original size of the organism. According to the island rule, if a mammal was smaller than a rabbit it will get larger and if it is larger than a rabbit then it will tend to dwarf. The logic is that in the absence of mainland predators organisms isolated on islands no longer need to be small in order to be cryptic. Larger organisms, especially herbivores, get smaller because they have access to less forage on an island and in the absence of mainland predators there is no need for a large body mass to avoid predation. The island rule is the most compelling explanation for the small size of *Homo floresiensis*.

## isochron
Refers to events happening at the same time, or to the morphology of a complex surface (e.g., a land surface or the outer enamel surface of a developing tooth) at a point in time.

### isometry
A change in size without an associated change in shape. (Gk *iso* = equal and *metron* = measure.) *See also* **allometry**.

### isotope
One of two or more atoms that have the same atomic number but different atomic masses. Because they have the same atomic number (i.e., the same number of protons) they occupy the same place in the periodic table (hence the name "isotope") and have the same element name, but because they differ in their number of neutrons they will have a different numeric superfix (e.g., $^{13}C$ and $^{12}C$) to reflect differences in atomic mass. Many isotopes of the lighter elements such as carbon, nitrogen, and oxygen do not decay. Instead their proportions are determined by processes such as photosynthesis (e.g., $^{13}C/^{12}C$), climate or diet (e.g., $^{2}H/^{1}H$, $^{13}C/^{12}C$, $^{15}N/^{14}N$, $^{18}O/^{16}O$, and $^{87}Sr/^{86}Sr$). These so-called "stable isotopes" are used by researchers to reconstruct diet, habitat, and migration in modern humans and extinct hominins. (Gk *iso* = equal and *topos* = place.)

### isotropy
Isotropy is used in reference to the material properties of a substance (e.g., bone or enamel) or to dental microwear textures. Materials are isotropic when their material properties, particularly their various measures of material behavior (e.g., stiffness, as reflected by the elastic modulus or Young's modulus; shear stiffness as reflected by the shear modulus; the relationship between axial and lateral strains, as reflected by Poisson's ratio) are the same in all directions. In dental microwear studies the term isotropy is used when the microscopic wear features are not consistently aligned in any given direction (i.e., feature alignment varies without patterning). (Gk *iso* = equal and *tropus* = direction.) *See also* **dental microwear**.

### Iwo Eleru
A fragmentary skull and associated skeleton was excavated from the base of the sequence of a large rock-shelter in the rain forest of western Nigeria. Although evidently modern human and recent (*c.* < 11 ka), the relatively archaic morphology of the skull as seen using multivariate analysis (it does not cluster with recent African crania) suggests that the origin of modern humans may have been a complex evolutionary process. (Location 07°26′N, 05°08′E, Nigeria.)

# J

### jackknife
A resampling procedure, similar to the bootstrap, used to calculate a confidence interval and/or the degree of bias for any sample statistic (e.g., mean, variance, median, correlation coefficient, regression slope, etc.).

### Jakovec Cavern
Part of the Sterkfontein cave system containing breccia equivalent in age to Sterkfontein Member 2. Twelve hominins attributed to *Australopithecus africanus* were recovered from the Jakovec Cavern in 1995.

### Jaramillo
*See* **geomagnetic polarity time scale**.

### Jarman/Bell principle
Based on observations of living species, this general principle suggests that large-bodied animals can exist on low-quality diets because they require less energy per unit body mass than smaller-bodied ones. However, the food consumption of large-bodied animals is still absolutely greater than that of smaller-bodied animals although the relative energy intake of large-bodied animals is lower compared to small-bodied animals; if they focused on less abundant high-quality foods (e.g., ripe fruits and insects) large-bodied animals might not be able to meet their daily energy needs. The Jarman/Bell principle provides a useful basis for investigating the role of energetics in hominin evolution.

### jaw
Refers to either the maxilla (i.e., upper jaw) or the mandible (i.e., lower jaw). (ME *jowe* or *joue* = cheek.)

### Jebel Irhoud
(alternate spellings: Jebel Ighoud, Djebel Irhoud) This North African cave site has produced the remains of at least four hominin individuals. They were originally attributed to *Homo neanderthalensis* (or were at least considered to possess some Neanderthal traits), but reanalysis suggests that similarities to early European fossils may be due to shared ancestral traits. The

Irhoud material, which includes two mostly complete crania (Irhoud 1 and Irhoud 2) and a juvenile mandible with a well-preserved dentition (Irhoud 3), displays similarities with early modern humans from the Near East. (Location 31°51′36″N, 08°52′12″W, Morocco.)

## Jebel Qafzeh
See **Qafzeh**.

## Jinniushan
(金牛山) This fossil-hominin-bearing locality (Site A) is one of several breccia-filled fissures located in an isolated karst tower locally known as Jinniushan (Golden Ox Hill), approximately 20 km/12 miles southeast of the city of Yingkou in Liaoning Province, China. In 1984 a $c. > 200$ ka partial hominin skeleton of a male about 20 years of age was discovered in layer 7 (not in layer 6 as originally reported) of Site A. (Location 40°34′40″N, 122°26′38″E, northeastern China.)

## joint
A discontinuity between two bones. Joints are classified according to whether the tissues that separate the bones are solid (i.e., fibrous and cartilaginous joints) or have a cavity (i.e., synovial joints). These joint types are further distinguished into subtypes based on the kind of bones, cartilage, and membranes involved. (ME *joint* = junction.)

## jugal
A cranial bone in amphibians, reptiles, and birds. In mammals it is the equivalent of the zygomatic bone. It is an arcane term that is best avoided, but it is still used for some linear measurements (e.g., bijugal breadth). (L. *iugum* or *jugum* = yoke.) See also **malar**.

## jugular bulb
See **cranial venous drainage**.

## jump dispersal
A type of dispersal that occurs when a subpopulation of a larger population of organisms crosses an existing barrier to disperse. Jump dispersal may be an important process leading to allopatric speciation, and involuntary jump dispersal may have been how hominins reached islands (e.g., Flores) in South Asia. Once hominins were able to build seagoing craft they were able to undergo jump dispersal by choice.

## junior synonym
A name for a taxon that has been superseded by a different name that has historical priority. The junior synonym may be an objective synonym (when two taxa have been erected with the same type specimen) or a subjective synonym (when two taxa, formerly thought to be different, have been synonymized). For example, the names *Homo neanderthalensis* King, 1864, *Homo primigenius* Schwalbe, 1903, and *Palaeanthropus europaeus* Sergi, 1910 were all given to a taxon whose type specimen was the associated skeleton from the Kleine Feldhofer Cave. Consequently the latter two names are objective junior synonyms of *H. neanderthalensis*.

The name *Australopithecus transvaalensis* was given by Robert Broom to a specimen from Sterkfontein and later he transferred it to a new genus *Plesianthropus*. Most authorities now consider the Sterkfontein fossils to represent the same species as the Taung skull (i.e., *Australopithecus africanus* Dart, 1925). As a result, the specific name *transvaalensis* is considered a junior subjective synonym of *africanus*, and the generic name *Plesianthropus* is considered a junior subjective synonym of *Australopithecus*.

## juxtamastoid eminence

A bony process seen in some, but not all, modern human crania. When present, it lies between the digastric fossa laterally and the groove for the occipital artery medially and it probably owes its presence to bone being pushed up between those structures, especially when the occipital artery takes a course lateral to the occipitomastoid crest and the occipitomastoid suture. A juxtamastoid eminence is present in early hominin crania such as SK 27 and SK 847 and the combination of a large juxtamastoid eminence and a small mastoid process is one of the distinctive features of *Homo neanderthalensis*.

## juxtamastoid process

*See* **juxtamastoid eminence**.

# K

### K/Ar
See **potassium-argon dating**.

### 23 ka world
See **astronomical time scale**.

### 41 ka world
See **astronomical time scale**.

### 100 ka world
See **astronomical time scale**.

### *K*-selection
See **r/*K*-selection theory**.

### ka
The abbreviation for an age determination or age estimate expressed in thousands of years. For example, "the date of the anatomically modern human crania from the Omo is *c*.190 ka." The term kyr or k.y. should be used to describe periods of time as in "30 kyr have elapsed since anatomically modern humans first reached Europe." (L. *kilo* = thousand and *annum* = year.) See also **age estimate**.

### Kabuh Formation
See **Bapang Formation**.

### Kabwe
Fossils and artifacts were discovered at this site in Zambia, formerly known as Broken Hill, in 1921 during lead and zinc mining operations at Number 1 Kopje, a dolomitic outcrop containing an approximately 30 m-long, clay- and breccia-filled cavity. The more than 23 large mammal species from Kabwe (e.g., extinct large baboons, saber-toothed cat, warthog, giraffid, and giant buffalo) are similar to fauna from Bed IV at Olduvai Gorge, and thus the assemblage is estimated by some authors to be as old as 1.3 Ma, but the precise association of the extinct taxa

with the hominin remains is unclear and such a date is at odds with the Middle Stone Age attribution of the (equally poorly provenanced) artifacts recovered with the fossils. Kabwe 1 (British Museum of Natural History registration no. E686 and formerly Broken Hill 1), a nearly complete cranium, became the holotype for *Homo rhodesiensis*, but is attributed by many authors to *Homo heidelbergensis*. Additional hominin remains recovered from Kabwe include both cranial and postcranial fossils. Artifacts associated with the hominin fossils include flakes with facetted platforms, discoidal and perhaps Levallois cores, hammerstones and possible anvils, as well as three shaped bone tools. (Location 14°27′S, 28°26′E, central Zambia.)

## Kabwe 1

This cranium, which is the holotype of *Homo rhodesiensis*, preserves the worn maxillary dentition except for the right $I^2$ and the crown of the right $M^3$, and was discovered at the site of the same name in 1921. Kabwe 1 (endocranial volume around 1280 cm$^3$) resembles the hominin fossils Bodo 1, Arago, Petralona, and Saldanha, and many regard it as consistent with the hypothesis that these specimens all belong to the hypodigm of *Homo heidelbergensis*, or *H. rhodesiensis* if the former taxon is interpreted as having some *Homo neanderthalensis* autapomorphies. If Kabwe 1 is the same age as Olduvai Gorge Bed IV then it is possibly the earliest evidence for *H. heidelbergensis* and/or *H. rhodesiensis*.

## Kalahari Dune Expansion

The Kalahari desert in southern Africa has undergone several expansions during hyper-arid periods, creating the "Mega Kalahari." Fossil dunes from the Mega Kalahari dated by luminescence techniques indicate arid conditions between 115 and 95 ka, 46 and 41 ka, 26 and 20 ka, and post-20 ka. Comparable periods of aridity are also seen in the paleoclimate reconstruction from the Tswaing Impact Crater.

## Kalambo Falls

An important archeological site where the Kalambo Falls Formation sediments crop out immediately upstream of the falls of the Kalambo River, Zambia. Although there is no fauna, the waterlogged conditions at the site preserve abundant organic remains including wood and pollen. The composite stratigraphic sequence of the different excavations provides a crucial benchmark for the cultural historical succession of southern and East Africa. Stone artifacts found at Kalambo Falls range from Early Stone Age (i.e., Acheulean), Middle Stone Age (i.e., Lupemban), Later Stone Age, as well as "intermediate" industries (i.e., Sangoan and "Magosian"). Most are made from locally available quartzite, mudstone, or silcrete. Their ages are *c*.180 ka for the Acheulean levels and *c*.75 ka for the Sangoan levels. Preserved pollen suggests fluctuations in the mosaic of forest and swamp environments, an interpretation supported by preserved seeds, tree bark, and wood charcoal. There are wooden artifacts in the Acheulean and Sangoan levels. (Location 08°35′S, 31°14′E, Zambia.)

## Kanam

A system of gullies on the northern margins of the Homa Peninsula, in the Winam Gulf, Lake Victoria, Kenya. In 1932 a fossilized hominin mandible was found in a Kanam West erosion gully by a member of Louis Leakey's East African Archaeological Expedition. The projection that had been interpreted as a true chin on the mandible was the result of a benign bone tumor, thus discounting the original hypothesis that this was an early example of an anatomical trait usually seen only in modern *Homo sapiens*. (Location 00°21′S, 34°30′E, Kenya.)

## Kanapoi

The site, located southwest of Lake Turkana and just north of the Kakurio River, Kenya, was discovered in 1965. Three stratigraphic intervals are represented in the Kanapoi exposures. All lie beneath the *c*.3.4 Ma Kalokwanya basalt and all but one of the hominin fossils were recovered from sediments dated to between 4.17 ± 0.03 and 4.07 ± 0.02 Ma by argon-argon dating. The first hominin fossil recovered from Kanapoi was a distal left humerus found in 1965. Eight more hominin fossils were recovered in 1994 and a further 24 more were recovered between 1994 and 1997. The initial humerus was regarded as *Australopithecus* sp. but a mandible found in 1994, KNM-KP 29281, was made the holotype of *Australopithecus anamensis*, and this is the taxon to which the 1965 humerus and all the subsequent (i.e., 1994 and thereafter) discoveries have been allocated. (Location 02°19′N, 36°04′E, Kenya.)

## Kanjera

The site consists of several collection areas [i.e., Kanjera North (KN) and South (KS)] in a series of erosion gullies on the northern margin of the Homa Peninsula in the Winam Gulf, Lake Victoria, Kenya. Louis Leakey's 1932 East African Archaeological Expedition had recovered anatomically modern-looking hominins. The apparent association of anatomically modern humans with Acheulean tools, along with Hans Reck's discovery of an anatomically modern skeleton (OH 1) erroneously associated with Pliocene fauna at Olduvai Gorge, formed the basis of Louis Leakey's view that modern humans had a long, separate, lineage in East Africa extending back to Oldowan times. Subsequent excavations at Kanjera have shown the hominins to be from much more recent contexts or intrusive burials into geologically older strata. The KN sequence comprises three units: the Kanjera Formation (*c*.1.5–0.5 Ma), the Apoko Formation (<0.5 Ma), and the Black Cotton Soil (terminal Pleistocene-Holocene). All of the hominins except Kanjera 3 (probably an intrusive burial into the Kanjera Formation) are derived from the Black Cotton Soil, but sediments containing Oldowan artifacts at KS extend back to *c*.2.0 Ma. (Locations KN: 00°20′04″S, 34°32′10″E; KS: 00°20′22″S, 34°32′16″E, Kenya.)

## Kapcheberek

See *Orrorin tugenensis*.

## Kapsomin

See *Orrorin tugenensis*.

## Kapthurin Formation

The Middle Pleistocene portion of the Tugen Hills sequence, Kenya. An approximately 125 m-thick sequence of tuffaceous, predominantly lacustrine, and fluvial sediments exposed over an area of about 150 km²/58 sq. miles, west of Lake Baringo, Kenya. Hominin [e.g., two mandibles (KNM-BK 67 and KNM-BK 8518) and an associated skeleton (KNM-BK 63–66)] and chimpanzee fossils (KNM-TH 45519–45522 and KNM-TH 46437–46440) and Acheulean and

Middle Stone Age (MSA) artifacts have been recovered from the Kapthurin Formation. Argon-argon dating of lavas and tephra and tephrostratigraphic correlation by electron probe microanalysis suggest an age bracket of 780 to less than 200 ka, with the lower Kasurein Basalts within K1 dated to $610 \pm 40$ ka and upper portions of K4 (the Bedded Tuff Member) dated to $235 \pm 20$ ka. The Kapthurin Formation sequence is unusual in showing complex temporal patterning in Acheulean and MSA artifacts that is inconsistent with simple replacement scenarios during the early appearance of MSA lithic technology. (Location 00°47′N, 36°05′E, Kenya.)

## Karari Ridge
One of the main subregions of the Koobi Fora region. *See also* **Koobi Fora**.

## karst
Limestone deposits that have been affected by water rich in carbonic acid. The mildly acidic ground water produces small (e.g., flutes), medium (e.g., sinkholes), or large-scale (e.g., limestone pavements) surface features. Underground, the water action leads to fissures and solution cavities and when breccia and bones fill these solution cavities the process results in cave sites like those in the Blaauwbank valley (e.g., Sterkfontein, Swartkrans). (Ge. *karst* = kras, after the "kras" region in Slovenia not far from Trieste, where this geological feature is common.)

## karstic
A landscape typically dominated by small, medium, and large-scale karst features on the surface and by channels and cavities within the rock. *See also* **karst**.

## karyotype
The complement of chromosomes found in an individual or normally found in a taxon. The karyotypes of modern humans and the great apes have 46 and 48 chromosomes, respectively. Banding patterns revealed by Giemsa staining (also called G-banding) or other dyes or fluorescently labeled probes show that sometime during hominin evolution chromosomes 2a and 2b found in the great apes fused to form modern human chromosome 2. Prior to the availability of large amounts of DNA data, differences in chromosomal number and pattern were used to determine species relatedness. (Gk *karyon* = nut, kernel and *typos* = impression, mark, figure, or original form.)

## Katanda
Three sites exposed on the eastern bank of the Semliki River, in the Democratic Republic of the Congo. Their age is contested but none of the dating results suggest an age younger than 50 ka. The Katanda sites contain 12 worked bone tools, including seven barbed points or harpoons, five unbarbed pointed pieces, and one "knife," plus approximately 16,000 lithic artifacts composed primarily of quartz, quartzite, and chert flaking debris, including discoidal and single- and multi-platform cores. (Location 00°06′N, 29°35′E, Democratic Republic of the Congo.)

## KBS
Acronym for the "Kay Behrensmeyer site" at Koobi Fora and used for the KBS Member in the Koobi Fora Formation, Koobi Fora, Kenya.

## Kebara

This cave, located on the western escarpment of Mount Carmel in Israel, includes Middle and Upper Paleolithic layers. The Middle Paleolithic levels have been dated at 62–48 ka, the transition between the Upper and Middle Paleolithic is *c.*45 ka, and the Upper Paleolithic levels are between *c.*36 and 32 ka. Excavations have recovered the remains of at least 15 individuals from the Upper Paleolithic layers and 29 individuals from the Middle Paleolithic layers. The latter, which include a nearly complete skeleton of an infant (KMH1) as well as the nearly complete torso of an adult (KMH2), all belong to *Homo neanderthalensis*. The Upper Paleolithic levels include Bronze Age, Lower Natufian, Kebarran, Aurignacian, and Ahmarian industries. The Middle Paleolithic levels contain Levantine Mousterian tools. Charred macrobotanical remains and phytoliths from the Middle Paleolithic levels have been interpreted as evidence of broad-spectrum plant foraging by Neanderthals, as well as for the controlled use of fire. (Location 32°34′25″N, 34°58′08″E, Israel.)

## Kedung Brubus

An exposed river-bed deposit 35 km/22 miles southeast of Trinil, Indonesia. It was here, on November 24, 1890, that Eugène Dubois discovered Kedung Brubus 1, a right mandibular fragment containing part of the canine and $P_3$ roots that most likely represents a juvenile *Homo erectus*. As poorly preserved as it is, Kedung Brubus 1 was the first early hominin to be recovered from Java, Indonesia. The fauna from Kedung Brubus, including the hominin mandible, comes from the sandstones of the Trinil beds of the Bapang (formerly Kabuh) Formation. Argon-argon dating of volcanics from breccias and tuffs at and near Kedung Brubus suggest a latest Pliocene/earliest Pleistocene age for the deposits. (Location 07°30′S, 111°54′E, Indonesia.)

## Keilor

An adult, probably male, cranium and a fragmentary femur were discovered in 1940 during quarrying operations on a river terrace about 16 km/10 miles northwest of Melbourne, Australia. The specimen was recovered from a silt deposit, below a minor disconformity. Radiocarbon dating of the bone suggests an age of 12 ka while dates from the adhering carbonates cluster around 6.8 ka. The Keilor cranium is anatomically modern human, but it is unlike its more robust contemporaries from Kow Swamp in that it is relatively orthognathic, and has moderately sized teeth and a curved frontal region. (Location 37°42′S, 144°51′E, Australia.)

## *Kenyanthropus platyops* Leakey et al., 2001

A hominin genus and species established in 2001 by Meave Leakey and colleagues to accommodate cranial remains recovered from the *c.*3.5 Ma Kataboi Member at Lomekwi, West Turkana, Kenya. The initial report lists the holotype cranium (KNM-WT 40000) and the paratype maxilla. There are an additional 34 specimens (e.g., three mandible fragments, a maxilla fragment, and isolated teeth) but the researchers reserved their judgment about the taxonomy of most of these remains. The main reasons Leakey and colleagues did not assign the Lomekwi material to *Australopithecus afarensis* are its reduced subnasal prognathism, anteriorly situated zygomatic root, flat and vertically orientated malar region, relatively small but thick-enameled molars, and the unusually small $M^1$ compared to the size of the $P^4$ and $M^3$. Some of the morphology of the new genus, including the shape of the face, is *Paranthropus*-like yet it lacks the postcanine megadontia that characterizes *Paranthropus*. It has been suggested that the KNM-WT 40000 cranium, which is permeated by fine cracks and is distorted, belongs to *Au. afarensis* but the diagnostic morphology is unaffected by the changes that have occurred postmortem. (Gk *anthropos* = human being, *platus* = flat, and *opsis* = face; etym. *Kenya* = Kenyan.)

## key innovations
Evolutionary innovations that have allowed a lineage to exploit a new adaptive zone. Taxa that share these innovations are normally recognized as occupying the same grade.

## keystone resource
A food resource that has a major impact on several species within an ecological community. For example, during periods of low fruit availability primates, marsupials, birds, and many other frugivorous animals rely on figs for a significant portion of their energy. [ME *key(e)* = central and *stan* = stone; it refers to the central wedge-shaped stone in an arch that locks the other stones in place.]

## KHS
See **Omo-Kibish**.

## Kibish
See **Omo-Kibish**.

## Kibish Formation
A formation in the Lower Omo Basin, Ethiopia. Its main outcrops occur along the Omo river, where it skirts the west side of the Nkalabong mountains about 100 km/60 miles north of the northern end of the present-day Lake Turkana. Single crystal argon-argon dating provides ages for Mb. I and Mb. III tuff feldspars of $196 \pm 2$ and $104 \pm 1$ ka respectively and radiocarbon dating on molluscs from Mb. IV gives ages in the range of *c*.3250 to *c*.9500 years ago. The three hominin individuals, Omo I, an associated skeleton including a fragmentary skull of a young adult, Omo II, an adult calvaria, and Omo III, fragments of an adult cranial vault, are judged to have come from Mb. I and their ages have been estimated at *c*.195 ka. They are currently the earliest evidence for *Homo sapiens*.

## Kiik-Koba
This Middle Paleolithic cave site in the Crimea has never been absolutely dated, but the presence of cold-phase fauna and the nature of the stone tools at the site suggests an age around the end of Marine Isotope Stage (MIS) 5 or the beginning of MIS 4 for the layer containing the hominins. The site preserves the remains of an adult hominin, Kiik-Koba 1, whose grave was disturbed in antiquity and was subject to the removal of all but a few teeth, portions of both hands, the right lower leg, and both feet. The remains of a young child less than a year old (Kiik-Koba 2) have also been recovered. Both individuals are assigned to *Homo neanderthalensis*. Three Middle Paleolithic layers were recovered, and the site is the type site for the Kiik-Koba Mousterian. (Location 45°03′N, 34°18′E, Crimea; etym. Turkik phrase for mountain goat cave.)

## Kilombe

An Acheulean site complex on the western flank of the Rift Valley, Kenya. Reversed paleomagnetization in layers immediately overlying the main artifact horizon indicates a probable age of more than 0.8 Ma. The extensive exposures of the main biface horizon allow comparisons of contemporaneous local occurrences at locations more than 100 m apart. In some areas of the site Developed Oldowan-like small bifaces are found alongside large classic Acheulean specimens. Several obsidian bifaces indicate transport over long distances. As at Olorgesailie, some artifact occurrences at Kilombe do not include bifaces. (Location 00°06′S, 35°53′E, Kenya.)

## kinematics

A branch of biomechanics that describes the motion of objects (kinematics sounds like, but should not be confused with, kinetics, which describes the effects of forces on objects, particularly the effects of forces on the motion of objects). Kinematics includes information on the position, velocity (speed) and acceleration (rate of change of velocity) of an animal, anatomical element (e.g., the hand), or object. Kinematic data are collected using motion analysis equipment, accelerometers, or other movement-measuring devices and these data are often collected in concert with data on kinetics. (Gk *kinesis* = movement and *kinein* = to move.)

## kinetics

A branch of biomechanics describing the effects of forces on objects, particularly the effects of forces on the motion of objects, such as the forces between the foot and the ground during walking. Kinetic data are collected with force plates, pressure pads, or with other force-measuring devices, and such data are often collected in concert with data on kinematics. (Gk *kinesis* = movement.) *See also* **kinematics**.

## Klasies River

The Klasies River Main Site on the Tsitsikamma coast of South Africa between Klasies River and Druipkelder Point comprises several caves and shelters (Caves 1 and 2, Shelters 1a, 1b, and 1c). The excavations revealed a long Middle Stone Age (or MSA) sequence capped by a dense Later Stone Age shell midden and underlain by a culturally sterile storm beach formed during a period of high sea level. The storm beach at the base of the section likely formed in Marine Isotope Stage 5e (i.e., *c.*125–115 ka) or during a previous interval of high sea level. The MSA faunal remains in the Sands-Ash-Shell (SAS) are *c.*80–*c.*60 ka (the pre-Howieson's Poort MSA is *c.*72.1 ka, the Howieson's Poort strata is *c.*64 ka, and the post-Howieson's Poort MSA levels are *c.*58 ka). The hominins from the MSA levels, several of which look as if they have been defleshed and/or burned, are remarkable for several reasons. They show substantial differences in overall size, suggesting either that this population had a higher level of sexual dimorphism than that seen in recent modern humans, or that the collection samples more than one morphologically distinct population. Many argue that the Klasies River hominins represent the earliest anatomically modern humans from southern Africa, but the degree to which the specimens are "anatomically modern" has been debated. Shellfish as well as mammal and tortoise bone occur throughout the sequence. Hearths, shell middens, and a highly fragmented fossil fauna with abundant cut and percussion marks attest to the use of the site as a living area. Analyses of the faunal remains from the MSA of Klasies and other sites in southern Africa have fueled lively debates over the hunting abilities of MSA versus Later Stone Age people, as well as the relative roles played by taphonomy and population density in shaping the faunal record. (Location 34°06′S, 24°24′E, Eastern Cape Province, South Africa.)

## Klasies River Mouth
See **Klasies River**.

## Kleine Feldhofer Grotte
A limestone cave in Germany, long since destroyed by mining, where the original hominin associated skeleton (Neanderthal 1; an adult calotte plus 15 postcranial bones) was discovered by miners in 1856. The Neanderthal 1 associated skeleton was made the holotype of *Homo neanderthalensis* King, 1864. No faunal or archeological evidence from the Kleine Feldhofer Grotte was reported and there seemed to be no prospect that such evidence, and thus information about the context and dating, could ever be obtained. However, archival research by Ralf Schmitz and Jürgen Thissen enabled what remained of the sediments from the Kleine Feldhofer Grotte at the base of the south wall of the Neander valley to be identified. Excavations of these sediments in 1997 and 2000 resulted in the recovery of fauna, and fragments of hominin bone, some of which can be fitted onto the original Neanderthal 1 calotte. The balance of the hominin remains recovered by Schmitz and his colleagues has been attributed to two additional individuals (Neanderthal 2 and 3). Researchers have recovered short fragments of mitochondrial DNA from the humerus of the Neanderthal 1 type specimen, from what might be the right tibial shaft of the same specimen (NN 4), and from the shaft of a right humerus (NN 1) of a second individual. Micoquian artifacts typical of late Middle Paleolithic as well as Upper Paleolithic artifacts from the Gravettian were also recovered. Direct radiocarbon dating of Neanderthal 1 and 2 suggest an age of *c*.40 ka. (Location 51°14′36″N, 06°57′04″E, Germany.) See also **Homo neanderthalensis**.

## Klipgat
See **Die Kelders**.

## knapping
The act of intentionally fracturing stone by percussion for the production of flakes. (ME *knappen* = apparently it refers to the sound made when flakes are struck from a core.)

## KNM-ER 992
This *c*.1.52 Ma subadult mandible from the Ileret region of Koobi Fora, Kenya, is broken close to the midline, but it is otherwise well preserved and includes both sides of the corpus, part of the right and much of the left ramus, and the crowns and roots of the canines to the $M_3$s on both sides. Along with KNM-ER 820, KNM-ER 992 provided the first evidence of a *Homo erectus*-like mandibular morphology that was in several ways more primitive than that of *H. erectus* from Java and Zhoukoudian. It is the holotype of *Homo ergaster*.

### KNM-ER 1470

This *c*.2.03 Ma adult cranium from the Karari Ridge region of Koobi Fora, Kenya, lacks most of the base. The reassembled cranium comprises separate calvarial and facial components; there is a join at nasion but it is not possible to determine the precise facial angle with any reliability. Most of the surface bone of the specimen is missing and the specimen has suffered from some plastic deformation. The cranial vault is similar in shape to that of *Homo habilis*, but larger (its endocranial volume is approximately 750 cm$^3$). The face is more orthognathic than that of *H. habilis* and it is widest not in the upper face as in the case in *H. habilis*, but in the midface. It is either a large-bodied presumed male of an inclusively interpreted *H. habilis* (in which case *H. rudolfensis* is a junior synonym of *H. habilis*) or it is the lectotype of *H. rudolfensis*, a second species of early *Homo*. The discovery of the smaller, but similarly morphologically distinctive, maxilla, KNM-ER 62000, has strengthened the case for a second early *Homo* taxon.

### KNM-ER 3228

This *c*.1.9 Ma right pelvic bone from the Upper Burgi Member of the Koobi Fora Formation is missing the pubic and ischial rami. Attributed to early *Homo*, it is presently the earliest pelvic bone that suggests a form of bipedalism similar to that of *Homo sapiens*. It shares with modern humans functionally significant features such as the relative lengths of the ischial body and tuberosity, a sagittally expanded iliac blade, a large acetabulum, and an acetabulosacral buttress. However, KNM-ER 3228 also resembles archaic hominin pelvic bones in other ways (e.g., an iliac blade that diverges somewhat laterally relative to lower parts of the hip and a somewhat protuberant anterior superior iliac spine region).

### KNM-ER 62000

See **KNM-ER 1470**.

### KNM-KP 271

This distal humerus, the first hominin to be recovered from Kanapoi, Kenya, was puzzling because, notwithstanding its antiquity, it was closer morphologically to modern humans than is a temporally much younger distal humerus from Kromdraai in South Africa that belongs to TM 1517, the holotype of *Paranthropus robustus*. If KNM-KP 271 does belong to *Australopithecus anamensis* then this taxon had a remarkably modern human-like distal humeral morphology.

### KNM-KP 29281

This *c*.4.15 Ma mandible from Kanapoi, Kenya, is the holotype of *Australopithecus anamensis*. It preserves the corpus with all the tooth crowns present and undamaged. The distinctive symphyseal morphology of this mandible was one of the reasons for creating a new species for the Kanapoi hominins.

### KNM-LT 329

Discovered in 1967, this *c*.5–4.5 Ma mandible from Lothagam, Kenya, was among the first tantalizing pieces of evidence (other evidence included the Kanapoi humerus KNM-KP 271, the Lukeino molar KNM-LU 335, and the Tabarin mandible KNM-TH 13150) of a hominin taxon, or multiple hominin taxa, that was, or were, older than *Australopithecus afarensis*. This specimen is currently assigned to *Australopithecus* sp. or *Ardipithecus* sp.

## KNM-LU 335

These lower molars from the site of Cheboit in the Tugen Hills, Kenya, were recovered from the *c*.6 Ma Lukeino Formation and were, with hindsight, the first evidence of *Orrorin tugenensis* to be discovered.

## KNM-TH 13150

Discovered in 1984, this 4.4 Ma mandible from Tabarin, in Kenya, may have been the first recovered evidence of *Ardipithecus ramidus*, though its assignment to this taxon is still questioned. If it does belong to this taxon, it would extend its known geographical range.

## KNM-TH 45519–45522 and 46437–46440

These teeth, consisting of incisors and molars probably from the same individual, were found at the site of Kapthurin in the Tugen Hills, Kenya. They were recovered from the Kapthurin Formation and are *c*.545 ka. They are the first, and presently the only, fossil evidence of the panin clade, the extant sister clade of hominins.

## KNM-WT 15000

It is difficult to exaggerate the importance of the KNM-WT 15000 skeleton since before its discovery relatively little was known about the postcranial skeleton, ontogeny, and life history of *Homo erectus*. This remarkably complete *c*.1.5 Ma associated skeleton of an adolescent male (the modern human age equivalent is about 12 years old) was recovered by excavation from Nariokotome III, a locality in West Turkana, Kenya. It provided conclusive proof of the essentially modern human limb proportions and limb morphology of a *H. erectus*-grade hominin and in doing so helped researchers make more sense of the isolated limb bones recovered from sites such as Koobi Fora, Kenya.

## KNM-WT 17000

This *c*.2.5 Ma cranium was found at the Lomekwi 1 locality in West Turkana, Kenya, in 1985. There are few well-preserved crania in this time period, and the discovery of KNM-WT 17000 prompted a re-evaluation of the relationships among the more primitive hominin taxa such as *Australopithecus africanus* and *Australopithecus afarensis*, and the later megadont and hypermegadont archaic hominin taxa such *Paranthropus robustus* and *Paranthropus boisei*. Its discoverers interpreted it as belonging to "an early *A.* [*Australopithecus*] *boisei* population," but other researchers were more impressed with the differences between KNM-WT 17000 and the *P. boisei* hypodigms. Many allocate it to what was originally called *Paraustralopithecus aethiopicus* and which is now called either *Australopithecus aethiopicus* or *Paranthropus aethiopicus*. Some suggest that the taxon to which KNM-WT 17000 belongs might be a link between *Au. afarensis* and the common ancestor of *P. robustus* and *P. boisei*.

## KNM-WT 40000

This *c*.3.5 Ma relatively complete but plastically deformed cranium was recovered from the Kataboi Member of the Nachukui Formation at locality LO-6N at Lomekwi in West Turkana, Kenya. It is in two parts, one preserving the neurocranium, the supraorbital region, and part of the cranial base, the other preserving the rest of the face. The flat and wide face is *Paranthropus*-like yet it lacks the postcanine hyper-megadontia that characterizes *Paranthropus boisei* and *Paranthropus aethiopicus*. The KNM-WT 40000 cranium is the holotype of *Kenyanthropus platyops*.

## knockin
*See* **knockout**.

## knockout
A genotype, usually of a model organism (e.g., mouse) in which a particular gene is nonfunctional. The ability to knock out specific genes revolutionized developmental biology and genetics and in 2007 Mario Capecchi, Martin Evans, and Oliver Smithies were awarded the Nobel Prize for Physiology or Medicine for their pioneering work on this technology. A disadvantage of gene knockouts, however, is that they only tell researchers what happens to the organism when the targeted gene does not perform its function. Interpreting such results can be like trying to figure out how a car works by taking components out one by one and seeing what each deletion does to the car's performance. To overcome these limitations, other related strategies have been developed. One is the use of multiple knockouts of related genes. Another is to replace the targeted gene with an allele that performs an altered function rather than one that is inactive; this is known as the knockin approach. A third is to make a knockout (or a knockin) that is conditional on the expression of some other gene. Such conditional knockouts only disrupt the function of the targeted gene in specific tissues or at specific times in development and thus can provide much more precise information about gene function. Mice have been used as model organisms to test the predicted relationship between brain size and cranial shape in hominins.

## knuckle-walking
*See* **hand, evolution in hominins**.

## Konso
Previously known as Konso-Gardula, this site on the western flank of the Ethiopian Rift System was discovered in 1991 during the Paleoanthropological Inventory of Ethiopia. Field work conducted beginning in 1991 resulted in the discovery of 1.45–1.4 Ma hominin fossils that were assigned to *Homo erectus* (e.g., KGA 10-1) and excellently preserved specimens of *Paranthropus boisei* (e.g., KGA 10-525), as well as Acheulean artifacts. (Location 05°20'N, 37°25'E, Ethiopia.)

## Konso-Gardula
*See* **Konso**.

## Koobi Fora
Koobi Fora (formerly known as East Rudolf and East Turkana) is a substantial area of outcrops and exposures on the northeastern shore of Lake Turkana, in Kenya. It is one of several fossil sites (the others include Lothagam, Omo, and West Turkana) within the Lower Omo and the Turkana Basin and its size and complexity make it comparable to the larger study areas in

Ethiopia (e.g., Hadar, Middle Awash). The three original formations (Kubi Algi, Koobi Fora, and Guomde) have since been rationalized into a single unit called the Koobi Fora Formation within which eight members are recognized. It is overlain by the Galana Boi Beds. The fossiliferous exposures were initially subdivided into three named subregions: Ileret in the north, Koobi Fora midway, and Allia Bay to the south, but this has been superseded by one that recognizes seven subregions. Their names, together with the numbered collecting areas within each subregion, are given below in order from north to south: Il Dura (14, 15, 41, and 42), Ileret (1–12), Il Naibar Lowlands (116–17, 136–7), Karari Ridge (105, 118, 129, 131), Koobi Fora Ridge (102–15), Bura Hasuma (107, 110, 119–21, 123–4), and Sibilot (202–4, 207, 212, 250–2). All of the vertebrate fossils recovered were recorded and numbered consecutively in a field book. Until 1970 hominin fossils from Koobi Fora had the prefix *FS* (acronym for field: surface) followed by a number (e.g., what are now KNM-ER 406 and 407 were *FS*-158 and *FS*-210, respectively). Since 1970 each hominin from Koobi Fora has had the prefix KNM for the Kenya National Museum (now National Museums of Kenya) where they are curated, ER for East Rudolf, and then the field number (e.g., KNM-ER 1813). Fossil hominins recovered from Koobi Fora have been assigned to *Paranthropus boisei* (e.g., KNM-ER 406, 407, and 732), *Homo erectus* or *Homo ergaster* (e.g., KNM-ER 992, 3733, and 3883), *Homo habilis* (e.g., KNM-ER 1478, 1501, and 1813), *Homo rudolfensis* (e.g., KNM-ER 1470 and 62000), and probably also *Australopithecus afarensis* (KNM-ER 2602). Several hominin trackways (aka footprint trails) have also been uncovered.

The Koobi Fora region has abundant carefully excavated early Pleistocene Oldowan and Developed Oldowan archeological sites. Extensive lateral sediment exposures combined with tephrostratigraphic correlations have permitted landscape-scale studies of early archeological and environmental variation. There are a few sites with Acheulean and Middle Stone Age artifacts and their geological contexts are in many cases poorly understood. There is a mid-to-late regional Holocene archeological record, including sites characterized by fishing economies as well as early domesticates (primarily cattle) in the region. (Location 03°35′–04°25′N and 36°10–30′E, Kenya.) *See also* **Koobi Fora Formation**.

## Koobi Fora Formation

A substantial thickness of sediments on the northeastern shore of Lake Turkana, Kenya. The strata from which fossil hominins have been recovered at Koobi Fora were originally assigned to the Kubi Algi, Koobi Fora, and the Guomde Formations, all overlain by the Galana Boi Beds. These three formations have since been rationalized into a single unit called the Koobi Fora Formation within which eight members are recognized. Each member is defined as the sediments between the base of one designated tuff and the base of the designated tuff immediately overlying it. The members, starting at the base, are the Lonyumun, Moiti, Lokochot, Tulu Bor, Burgi, KBS, Okote, and the Chari Members. The Lonyumun Member, the oldest, dates from 4.35 to 4.0 Ma and the Chari Member, the youngest, dates from 1.39 to 0.7 Ma. The sediments have been dated using magnetostratigraphy, biostratigraphy, and various radiogenic methods. Fossil hominins recovered from the Koobi Fora Formation have been assigned to, or represent, *Paranthropus boisei*, *Homo erectus* or *Homo ergaster*, *Homo habilis*, *Homo rudolfensis*, and probably also *Australopithecus afarensis*. There are also several footprint trails. The Koobi Fora Formation includes evidence of Oldowan, Developed Oldowan,

Acheulean, and Middle Stone Age sites and artifacts. There is also a mid-to-late Holocene archeological record, including sites characterized by fishing economies as well as early domesticates (primarily cattle) in the region.

### Koobi Fora hominin footprints

Hominin footprints dating to *c*.1.53–1.4 Ma have been found at Koobi Fora, Kenya, on at least three surfaces at two localities: FwJj14E located at Ileret and GaJi10 on the Koobi Fora Ridge. Compared to modern humans, the FwJj14E hominin footprints suggest a narrower heel and forefoot, a wider instep (midfoot), and less pronounced arch elevation. But when compared to the Laetoli hominin footprints the FwJj14E footprints suggest a more contracted (i.e., narrower) and deeper instep, and the narrowest point of the instep is closer to the heel. Although other taxa (*Homo habilis*, *Paranthropus boisei*) cannot be ruled out as the makers of the prints the morphology and size of the Ileret hominin footprints are most consistent with the large size and tall stature seen in *Homo erectus* or *Homo ergaster*. See also **Homo erectus**; **Koobi Fora**.

### Koro Toro

The region in Chad, Central Africa, where researchers have identified Pliocene fossiliferous sediments that date to 3.58±0.27 Ma. Four hominins have been collected from Koro Toro: three from locality KT 12 and one (a maxillary fragment) from KT 13. The three hominins from KT 12 include a mandibular fragment (KT 12/H1) that is the holotype of *Australopithecus bahrelghazali*. (Location 16°00′21″N, 18°52′34″E, Chad.)

### Kossom Bougoudi

One of the "fossiliferous areas" within the Chad Basin, in Chad, Central Africa; others include Kollé, Koro Toro, and Toros-Menalla. See also **Chad Basin**.

### Kow Swamp

A site 3.2 km/2 miles south of Leitchville, Victoria, Australia, where researchers recovered 17 individuals that date from the Last Glacial Maximum (i.e., *c*.22–19 ka) and that make up what is currently the largest known single population of Late Pleistocene *Homo sapiens*. Their remarkable robusticity initially led researchers to link these modern humans directly to earlier *Homo erectus* specimens from Java, Indonesia, but another explanation is that their robusticity was an adaptation to the deteriorating climatic conditions. (Location 35°55′S, 144°19′E, Australia.)

### Krapina

This sandstone rockshelter in Hušnjakovo Brdo (Hill), Croatia, which dates to Marine Isotope Stage (MIS) 6 or the beginning of MIS 5e (i.e., it is *c*.130 ka), has yielded around 900 hominin fossils of *Homo neanderthalensis* and about 1200 lithic Mousterian artifacts. The hominin fossils, many of which are fragmentary, represent a minimum of approximately 70 individuals. The Krapina hominin sample is an important source of information about intraspecific variation and the ontogeny of Neanderthals. (Location 46°10′N, 15°52′E, Croatia; etym. the name of a nearby town.)

## Kromdraai

A complex of breccia-filled caves in Precambrian dolomite a mile or so east of Sterkfontein in the Blaauwbank valley in Gauteng Province, South Africa. In 1938 a local schoolboy, Gert Terblanche, found a hominin cranium that was to become the type specimen (TM 1517) of *Paranthropus robustus*. There are two fossiliferous localities: Kromdraai A (the "faunal site" or KA) and Kromdraai B (the "hominid site" or KB). Three members (1–3) are recognized in Kromdraai B and the majority of the hominin fossils have come from KB Member 3 (i.e., between 1.78 and 1.65 Ma) and with only one exception they have been assigned to *P. robustus*. Although all of the Kromdraai hominins are given different specimen numbers some are very likely from the same cranium, skull, or skeleton (e.g., MLD 6 and 23 almost certainly belong to the same individual and likely to the same cranium as MLD 1; MLD 4, 18, and 28 likely come from the same skull; the MLD 2 mandible and the MLD 7 and 8 ilium fragments probably belong to the same individual). Developed Oldowan/Early Acheulean artifacts have been recovered from KA and residue analysis suggests the hominins were using them to process bone. (Location 26°00′00″S, 27°45′00″E, South Africa.)

## KT 12/H1

This *c*.3.6 Ma mandible fragment from Koro Toro, Chad, Central Africa, is the holotype of *Australopithecus bahrelghazali*.

## Kulkuletti

A site in the Gademotta Formation that is on the flanks of a collapsed caldera about 5 km/3 miles west of Lake Ziway in the Main Ethiopian Rift System. The younger (*c*.280 ka) levels at Kulkuletti along with those at Gademotta between them provide one of the oldest and richest samples of Middle Stone Age (or MSA) artifacts. Among more than 8000 recovered artifacts are Levallois flakes, points, and cores (many of them large), unifacially and bifacially retouched points, as well as blades and blade cores. No faunal evidence has been recovered. (Location 08°03′N, 38°15′E, Ethiopia.)

## Kuseralee Dora

See **Ardipithecus ramidus**.

## k.y.

*See* **kyr**; *see also* **age estimate**.

## kyphosis

The term kyphosis (and its antonym lordosis) refer to the angular relationship within or between sections of a compound structure (e.g., the vertebral column) or between two planes (e.g., the anterior and posterior components of the cranial base). Thus, the parts of the adult modern human vertebral column that are concave anteriorly (i.e., the thoracic and sacral) are referred to as being kyphotic, whereas the parts of the adult modern human vertebral column that are concave posteriorly (i.e., the cervical and lumbar regions) are referred to as being lordotic. With respect to the cranial base, organisms or individuals with a smaller, more acute (i.e., more modern human-like) cranial base angle are said to have a more kyphotic basicranium, whereas organisms or individuals with a larger, more obtuse (i.e., more chimpanzee-like) cranial base angle have a more lordotic basicranium. (Gk *kyphos* = bent forwards.)

### kyr

The abbreviation for a period of time expressed in thousands of years, as in "30 kyr have elapsed since anatomically modern humans first reached Europe." The result of an age determination, or an age estimate, should be expressed differently (e.g., "the age of the anatomically modern human crania from the Omo is *c.*190 ka."). Despite this, many contemporary reports use the abbreviation "kyr." (L. *kilo* = thousand.) *See also* **age estimate; ka.**

# L

## L. 338y-6
This *c*.2.4 Ma incomplete juvenile calvaria, which was recovered from unit E-3 of Member E of the Shungura Formation in Ethiopia, preserves most of an occipital, both parietals, and a small fragment of the left side of the frontal. Despite being a juvenile – one of the few belonging to *Paranthropus boisei/aethiopicus* – the L. 338y-6 calvaria already manifests many of the features seen in adult *Paranthropus boisei* crania (e.g., the heart-shaped foramen magnum, *striae parietalis*, temporal lines closest anteriorly, etc., although it lacks an occipito-marginal system of endocranial venous drainage). These features have provided important information about the ontogeny of *P. boisei*.

## labyrinth
*See* **bony labyrinth**; **inner ear**.

## La Chaise
The La Chaise cave complex, which is near Angoulême, France, has three main galleries: Abri Suard, Abri Bourgeois-Delaunay, and Grotte Duport (all named after the people who found them), plus the Grotte de la Tour, whose entrance is about 130 m west of Abri Suard. The Abri Suard and the Abri Bourgeois-Delaunay are the best-studied caves. Cranial and postcranial fragments assigned to *Homo neanderthalensis* have been recovered from the lower layers at the Abri Bourgeois-Delaunay (146–106 ka). The hominin fossils recovered from the Abri Suard, many of which come from juvenile individuals, are older (185–101 ka) and have been described as being either Neanderthal or pre-Neanderthal. (Location 45°40′12″N, 00°26′48″E, France.)

## La Chapelle-aux-Saints
The cave of Bouffia Bonneval near to the village of La Chapelle-aux-Saints, France, is best known for the nearly complete skeleton of *Homo neanderthalensis* recovered in 1908. This *c*.50 ka elderly male hominin skeleton (La Chapelle-aux-Saints 1) is significant because Marcellin Boule's original reconstruction of this fossil in 1911–12 portrayed this individual as a stooped and hunched. Along with his interpretation of La Ferrassie 1 and 2, this reconstruction was the basis of the stereotype of Neanderthals as substantially less evolved, both intellectually and physically, than modern humans. The reconstruction was corrected in the 1950s to a more

upright posture. Some researchers have argued that because this individual had widespread severe degenerative joint disease and had lost all of its molar teeth, it is evidence that Neanderthals had a complex social system that involved care for less able individuals. (Location 44°59′N, 01°43′E, France.)

## lactase persistence

Lactase persistence is the ability to metabolize lactose after infancy. All mammals can digest milk as infants because they produce the enzyme lactase that breaks down the sugar lactose into glucose and galactose, both of the which are readily absorbed into the bloodstream. However, this ability typically declines after the infant is weaned because the gene (called *LCT* in modern humans) that encodes the enzyme lactase is turned off. In post-weaning individuals who are lactose intolerant (i.e., their ability to produce lactase does not persist after weaning) lactose is metabolized by intestinal bacteria, which produce byproducts including hydrogen and methane (typically causing cramping, bloating, diarrhea, and nausea). In modern humans lactase persistence has evolved multiple times in populations with a history of herding and dairying.

## lactose intolerance

See **lactase persistence**.

## lacustrine

Anything that relates to lakes (e.g., "lacustrine sediments" are sediments formed in the bed of a lake). (L. *lacus* = lake and *inus* = relating to.)

## LAD

Acronym for **last appearance datum** (*which see*).

## Laetoli

(formerly Laetolil) The site now called Laetoli is where fossiliferous sediments known variously as the Garusi River Series, the Vogel River Series, and the Laetolil Beds are exposed along the headwaters of the Garusi River, Tanzania. The area was first explored by Louis Leakey and colleagues in 1935 and a hominin lower canine was recovered. In 1938–9, an expedition led by Ludwig Kohl-Larsen recovered three hominin fossils (Garusi 1, 2, and 4). In 1974, Mary Leakey located hitherto unexplored exposures from which her team recovered a hominin upper premolar and a juvenile mandible. The hominin-bearing Upper Laetolil Beds are exposed over an approximately 60 km²/23 sq. mile area and the discrete localities are numbered from 1 to 24. Fourteen Laetoli hominin fossils (LH 1–14) were collected between 1974 and 1975, and LH 15–19 and 21–30 were collected between 1975 and 1979; LH 20 is a monkey. An isolated upper molar (LH 31) was recovered by the Institute of Human Origins' expedition in 1987 and four more hominins (EP 1000/98, 162/00, 2400/00, and 1500/01) were collected by the Eyasi Plateau Paleontological Project between 1998 and 2001. Fossil footprints of small and large animals were first recognized in 1976, and the trails of at least three hominins (at site G in Locality 8) were first uncovered in 1978. The hominin fossils from the Upper Laetolil Beds (3.85–3.63 Ma) have been assigned to *Australopithecus afarensis*. They include the holotype of *Praeanthropus africanus* (Garusi 1), the holotype of *Au. afarensis* (LH 4), and an associated skeleton (LH 21). A fragment of the right side of a maxilla from the Upper Ndolanya Beds (2.66 Ma) has been assigned to *Paranthropus aethiopicus*; a contemporaneous proximal tibia may also belong to the same taxon. A partial cranium of an archaic *Homo sapiens* (LH 18) is known from the Late Pleistocene Upper Ngaloba Beds. Middle Stone Age and Later Stone Age artifacts are known from the Late Pleistocene Olpiro and Ngaloba Beds. (Location 03°13′27″S, 35°11′38″E, Tanzania.) *See also* **Australopithecus afarensis**; **Laetoli hominin footprints**; *Paranthropus aethiopicus*; *Praeanthropus africanus*.

## Laetoli hominin footprints

In 1976 Andrew Hill discovered animal prints and raindrop impressions preserved on an exposed airfall tuff. In the following year, a trail of five prints, potentially made by a hominin, was found at Footprint Site A and in 1978 the footprints of hominins and a variety of other animals were found at Footprint Site G. Both the A and G footprint trails were made in a series of layers of petrified volcanic airfall ash called the Footprint Tuff. The age of the footprint-bearing stratum is c.3.66 Ma. At Site G two of the hominin footprint trails (G-1, G-2/3) occur side by side, suggesting that two individuals may have walked together; the prints of the third individual (G3) were made in the prints of G2 and therefore some of the details of both trails are obscured. A total of 38 footprints were uncovered in the G-1 trail, and 31 in the G-2/3 trail. It is estimated that the hominins were moving at a walking pace of 0.4–0.45 m/s. Researchers continue to debate interpretations of the prints, but the consensus is that the Laetoli hominin prints represent evidence of hominins well adapted for bipedal walking with derived foot anatomy that includes relatively short toes and a hallux (big toe) that is more closely aligned with the other toes than is the case in the extant African apes. Prints of other animals found along with the hominin footprints include those of bovids, lagomorphs, giraffids (large and small), rhinoceratids, equids, suids, proboscideans, rodents, carnivores, and cercopithecids.

## La Ferrassie

La Ferrassie consists of a large and deep cave with a small rock-shelter on one side and a long rock-shelter on the other; both are hollowed out from a south-facing cliff that overlooks a valley in the Dordogne region of France. Its dating is problematic due to some mixing of the levels, but the oldest (Mousterian) levels have been correlated with the Ferrassie Mousterian at Combe Grenal (i.e., 74–68 ka and equivalent to the transition between Marine Isotope Stages 4 and 5). The remains of eight hominin individuals, two adults (La Ferrassie 1 and La Ferrassie 2), three juveniles, one neonate, and two fetuses, at least one of which was probably intentionally buried with a possible grave good, were recovered between 1909 and 1921. La Ferrassie 1 was one of the more complete Neanderthal skeletons (along with La Ferrassie 2 and La Chapelle-aux-Saints 1) used by Marcellin Boule to reconstruct the morphology and posture of Neanderthals. The small rock-shelter and the cave also contained archeological material from the Mousterian, Aurignacian, and Gravettian periods, and the talus slope below the longer rock-shelter contained Mousterian, Châtelperronian, Aurignacian, and Gravettian levels. (Location 44°57′N, 00°56′E, France.)

## Lagar Velho
A rock-shelter in the Lapedo gorge in central Portugal that is best known for the $c.30$ ka intentionally buried articulated postcranial skeleton of a 4–5-year-old child (Lagar Velho 1). The mandible, dentition, and fragments of most of the cranium were recovered from surrounding disturbed sediments and there was evidence of a ritual burial (e.g., burning of a Scots pine, body ornaments, a red-ochre-painted shroud, parts of a deer at the head and feet, and a juvenile rabbit skeleton across the legs). Although clearly modern human by most anatomical criteria, the Lagar Velho 1 child also has traits (e.g., low crural index, receding symphysis, semispinalis capitis fossae, $I_2$ shoveling, dental maturational pattern) associated with *Homo neanderthalensis*. The mosaic of anatomically modern human and archaic traits prompted the suggestion that Lagar Velho 1 provides evidence for admixture between Neanderthals and modern humans at the time of contact in Europe, but most assessments consider Lagar Velho 1 to be within the range of variation of *Homo sapiens*. (Location 39°45'20"N, 08°44'06"W, Portugal.)

## Lake Botsumtwi
This 74 m-deep lake in Ghana preserves a sedimentary record of paleoclimate for subtropical Africa stretching back $c.1$ Ma. Oxygen isotopes in carbonates preserve a detailed record of changes in African monsoon strength. Decadal- to centennial-length droughts are common over the past three millennia, suggesting that droughts similar to, or worse than, recent Sahelian droughts would have been a factor affecting modern human ancestors in the region. There is evidence for extreme aridity between 135 and 75 ka. *See also* **megadroughts**.

## Lake Mungo
Three gracile hominin skeletons (LM 1–3) were recovered in the core of an eroded lake-shore dune on the southern edge of the now dry Lake Mungo, in Australia. Two of the skeletons (LM 1 and LM 2) were cremated and buried in a conical pit, while the third was interred as a complete skeleton (LM 3). The most recent reliable dates suggest that the skeletons are $c.40$ ka. LM 1 and LM 3 are remarkably gracile, especially when compared to the later but much more robust individuals from Kow Swamp and Willandra Lakes. Two competing explanations have been offered regarding the transition from the earlier gracile to the later robust crania and postcranial skeletons in Pleistocene Australia. The first interprets the fossil evidence as consistent with a single evolving lineage that was initially relatively gracile and over time came to be more robust, perhaps as an adaptation to an increasingly harsh and arid environment. The second interprets the same fossil record as evidence for multiple migrations from Asia, with an early migration of gracile individuals from north Asia, and a later migration of more robust individuals from Southeast Asia. Mitochondrial DNA (mtDNA) isolated from LM 3 has a sequence unknown among modern Australian Aboriginals, but researchers who have re-examined the mtDNA evidence suggest that the "ancient mtDNA" from LM 3 was most likely either a contaminant and/or was the result of degradation. (Location 33°07'S, 143°10'E, Australia.)

## Lake Turkana
This lake, formerly known as Lake Rudolf, occupies the southern part of the Omo-Turkana Basin, in Kenya. The lake has fluctuated in size through time becoming larger when inflow exceeds any outflow and evaporation and smaller when it does not. The major fossiliferous

formations around Lake Turkana are the Koobi Fora, Mursi, Nachukui, Nkalabong, Shungura, and Usno Formations. The major fossil and archeological sites surrounding Lake Turkana are Koobi Fora, Lothagam, the Omo region, and West Turkana.

## Lake Turkana Basin
*See* **Turkana Basin**.

## Lamarckism
(aka neo-Lamarckism) Jean-Baptiste de Lamarck was one of the first to argue that species are not fixed entities and his theory of the "transmutation" of species was published in *Philosophie zoologique* in 1809. For Lamarck it was not species that changed, but individuals. Lamarck suggested that there was an inherent tendency for organisms to be transformed into "higher" (i.e., more advanced) forms, and that this tendency resulted in a kind of escalator effect where, over many generations, lower organisms would progressively be transformed, or transmuted, into more advanced taxonomic categories. Lamarck's transmutation theory remained an idea on the periphery of accepted science until the publication of Charles Darwin's *On the Origin of Species by Means of Natural Selection* in 1859. Many naturalists in France, Germany, and America accepted the fact of biological evolution, but relied upon Lamarck's notion of the inheritance of acquired characteristics to explain how evolution occurred. The neo-Lamarckian version was more acceptable than Darwin's because within the neo-Lamarckian framework it could be argued that evolution proceeded according to a natural, or divinely directed, plan that was still consistent with the history of life on Earth as recorded in the fossil record. It was also consistent with a "progressive" interpretation (i.e., orthogenesis) that saw the appearance of modern humans in the world as the ultimate goal of evolution. For example, whereas for Darwin the traits shared between the extant apes and modern humans were derived from a recent common ancestor, neo-Lamarckians were more likely to argue that these traits had evolved independently. Neo-Lamarckian ideas remained appealing and influential until the rise of the modern evolutionary synthesis in the 1940s. *See also* **modern evolutionary synthesis**.

## lamellar bone
*See* **bone**.

## landmark
Anatomy A place on a bone or tooth that can be defined precisely enough so that independent observers are likely to agree on its location (e.g., nasion – where the suture between the nasal bones meets the frontal bone and basion – the most posterior point on the anterior border of the foramen magnum in the sagittal plane). Geometric morphometrics Landmarks are loci that are homologous across individuals in a sample (i.e., the landmarks have the "same" location in every other member of the sample being investigated). The location of landmarks can be recorded in one, two, or three dimensions. *See also* **coordinate data; geometric morphometrics; semilandmark**.

## Langebaanweg
A site on the west coast of South Africa north of Cape Town known for its exceptionally well-preserved and diverse fauna. To date, over 230 vertebrate and invertebrate taxa have been described from deposits that date to the terminal Miocene/earliest Pliocene ($c.5$ Ma).

For many taxa, including carnivores, birds, and frogs, Langebaanweg is unique in terms of species number, specimens, and quality of preservation. Langebaanweg provides an important context for interpreting and dating hominin-bearing sites in other parts of Africa with comparable faunal assemblages. The open-air archeological locality known as the Anyskop Blowout is approximately 1 km/0.6 miles south of the Pliocene fossil beds. (Location 32°57′41″S, 18°06′38″E, South Africa.)

## language

Communication is ubiquitous among animals, but modern human language seems to be a unique form of communication. Phonology is the study of how specific sounds (or signs) encode meaning, semantics is the investigation of how this meaning is generated and constituted in language, grammar is the study of how the various elements of language are structured and organized into larger units, and pragmatics focuses on the way language is used socially. An important distinguishing feature of modern human language is recursive syntax, or recursion, which is stringing different sounds to generate new meaning (i.e., grammar) into a sentence. This feature has never been observed in any species other than modern humans. For example, while a dog may respond appropriately when asked to "fetch the ball" and when asked to "roll over," it will not understand the command "roll over the ball" even though such a command, odd though it may be, would be immediately understood by a 2-year-old modern human. Researchers who focus on the evolution of modern human language try to understand how and when our ancestors acquired the facility for recursion. It is still unclear if the capacity for recursion gradually evolved or if it appeared "fully formed" relatively recently in our evolutionary history. (ME *language* from OF *langage* from L. *lingua* = tongue or speech.) *See also* **syntax**.

## language, evolution of

Research on the evolutionary origins of modern human language has focused primarily on the origins of its distinctive recursive syntax. Unlike modern humans, nonhuman primate species that are otherwise very good at detecting patterns of speech syllables cannot detect patterns generated by recursive rules. Recursive syntax makes possible several important semantic features of modern human language. Modern humans can communicate about specific objects or events with which they can have had no acquaintance because they can use recursive syntax to generate descriptions of arbitrarily precise specificity. Few animal communication systems are capable of communicating information about spatiotemporally displaced referents, and none seem capable of specifying them with the precision that recursive syntax makes possible. Syntactic rules apply to parts of speech defined independently of their semantics (e.g., any noun phrase, referring to any object, combines with any verb phrase, referring to any action, to form a sentence). This makes possible two other powerful semantic properties: generality and domain independence. Generality enables modern humans to use language to represent situations that have never been perceived and perhaps never could be perceived (e.g., a can of tuna eating a cat). Domain independence is a byproduct of such generality, and explains the idea that noun phrases and verb phrases referring to members of disparate domains can be combined into sentences expressing domain-independent propositions. It is not clear how or why a group of hominins, not yet able to use a modern human-like language, would develop a capacity to think recursive, domain-independent, and fully general thoughts, while numerous species of primates have inhabited environments similar to those of our precursors without developing that capacity. This leads many to argue that structurally complex language evolved as a tool both for thinking *and* communication in the relatively recent past. Such views distinguish the faculty of language in the broad sense from the faculty of language in the narrow sense.

The unique component of the modern human faculty of language – in the narrow sense – is the capacity to detect and produce recursively structured signals. Others distinguish between "proto-language," in which unstructured combinations of two to three distinct vocalizations communicate task-specific information, and fully grammatical language. Some regard pidgin languages and the language of 2-year-olds as the linguistic equivalent of fossils (i.e., they are relics of proto-language). Fully grammatical language requires an appreciation of the various parts of speech and the recursive rules for combining them into sentences. Fully recursive, grammatical language appears to be a highly complex structure, with many components that make no independent contributions to functionality and which have no obvious independent function. Some argue that grammatical complexity arose when isolated hominin populations began interacting with each other and that the need for communication among individuals sharing little background knowledge or context drove the development of increasingly context-independent languages, capable of conveying clear, unambiguous information between interlocutors of vastly different backgrounds. Charles Darwin proposed that modern human language is descended from sexually selected mating songs and this proposal has received some support from evidence that sexually selected birdsongs approach the structural complexity of human language. The *FOXP2* gene, which is known from neuropathological evidence to play a role in the modern human capacity for structurally complex language, apparently plays a role in the capacity for sexually selected birdsongs in some species. The discovery of mirror neurons in primate brains, including those of macaques and modern humans, has been interpreted as evidence for the manual origins view of language evolution. To produce linguistic signals properly one must understand how they will be interpreted, and to interpret linguistic signals properly one must understand the intent with which they are produced. Mirror neurons, which fire when the same action is observed and executed, may be a mechanism for implementing this property of modern human linguistic communication.

## language, hard tissue evidence for

Hard-tissue evidence that may potentially be related to spoken language can be divided into evidence related to the generation of spoken language and evidence related to its perception. Two examples of hard-tissue evidence for speech are the size of the vertebral canal to predict the presence of the "extra" neurons in the thoracic spinal cord that allows for the fine control of the muscles of respiration, thus facilitating spoken language in modern humans, and the size of the hypoglossal vertebral canal to predict the presence of the "extra" neurons that facilitate the fine control of the muscles of the tongue. Neither has proved to be a reliable proxy for complex spoken language. Potential hard-tissue evidence related to the perception of spoken language involves the external and middle ears. The external ears of chimpanzees are longer and smaller in cross-section than those of modern humans and what is known of the size of the external auditory meati of early hominins prior to *Australopithecus afarensis* suggests that they were chimp-sized. Middle ear ossicles have been recovered at Swartkrans and Sterkfontein (an incus, SK 848, belonging to

*Paranthropus robustus*, from the former, and stapes, belonging to *Australopithecus africanus* and *Homo habilis*, from the latter). There is a marked similarity in size between the footplates of the early hominin stapes and the stapes belonging to the living great apes, and they are both substantially smaller than those of modern humans. But what, if anything, this implies about the speech perception capabilities of *Au. africanus* (and *H. habilis*) is unclear.

## language, soft tissue evidence for

With respect to the generation of spoken language only two categories of potential neuroanatomical evidence are relevant, namely evidence for the reorganization, or enlargement, of Broca's area and cerebral cortical petalias. Broca's area, which is located along the inferior frontal gyrus of the anterior prefrontal cortex, is enlarged in the left hemisphere of most modern humans and it is thought to play a role in the production of spoken language. The lateralization of Broca's area, though it is present in apes, occurs less frequently than it does in modern humans. A recent survey of extinct and extant hominins showed that while asymmetrical enlargement does occur in transitional and pre-modern *Homo* taxa (i.e., *Homo rudolfensis*, *Homo erectus*, *Homo heidelbergensis*, *Homo neanderthalensis*) it is absent in *Australopithecus afarensis*. However, doubt has been cast on the wisdom of using endocranial morphology as a proxy for identifying functional regions of the cortex. Petalias, or asymmetries in the cortical hemispheres of the brain, were once thought to be peculiar to modern humans (and thus a sound way of imputing functions such as spoken language), but they have now been identified in nonhuman higher primates.

## Lantian Chenjiawo

(蓝田陈家窝子) This *c.*650–500 ka site is the source of a generally well-preserved hominin mandible. Presumed to be an older adult female, the Chenjiawo mandible (PA 102) is the holotype of *Sinanthropus lantianensis*; it is now referred to *Homo erectus*. (Location 34°14′N, 109°14′E, eastern China.)

## Lantian Gongwangling

(蓝田公王岭) At *c.*1.15 Ma, Gongwangling is one of the oldest hominin fossil sites in mainland East Asia. Presumed to be a female of more than 30 years of age the Gongwangling partial cranium [PA 105 (1-6)] was, along with the mandible from Chenjiawo, assigned to "Lantian Man," or *Sinanthropus lantianensis*; it is now referred to *Homo erectus*. Quartzite and quartz artifacts (11 cores, five flakes, and four scrapers) were found in association with the fossil cranium. A large Acheulean-like handaxe was found at the nearby site of Pingliang. (Location 34°11′0.73″N, 109°29′38.94″E, eastern China.)

## Lantian Man

See **Lantian Chenjiawo**; **Lantian Gongwangling**.

## La Quina

There were two main archeological accumulations in a series of rock-shelters that overlook the floodplain of the Voultron River, France. The main one, Station Amont, contained several Mousterian layers and yielded many faunal remains, stone tools, and *Homo neanderthalensis* fossils. The smaller accumulation about 200 m to the southwest, called Station Aval, contained Mousterian, Châtelperronian, and Aurignacian levels. The upper Mousterian levels at Station Amont are c.48–43 ka; this suggests that the majority of the Station Amont is much older. Generally the site was deposited sometime during Marine Isotope Stages 3-4 (71–24 ka). One fairly complete Neanderthal skeleton, La Quina H5, was recovered and possibly represents an intentional burial. (Location 45°30′N, 00°17′E, France.)

## Laron syndrome
*See* **Homo floresiensis**.

## Lascaux
This cave site in France houses some of the most famous Upper Paleolithic cave paintings in Europe, with many scenes of animals, humans, and abstract designs. These 19–15 ka paintings have been the subject of many interpretations (e.g., astronomy, hunting magic, totem worship, and/or shamanism). The cave was once open to the public but it has been closed since 1963, although an exact replica of two of the galleries, Lascaux II, is available for public tours. Artifacts found at the site include lamps, decorated bones, and more than 300 stone tools. Some of the latter show evidence of having been used for engraving. (Location 45°03′15″N, 01°10′01″E, France.)

## laser ablation
High-energy lasers can transform a solid into plasma thus enabling its elemental composition to be measured by a mass spectrometer. The system can be used for determining stable isotope or trace element ratios. Laser ablation is a much more precise sampling strategy than bulk sampling as it helps avoid time-averaging stable isotopes within enamel and dentine; it can yield $^{13}C/^{12}C$ or Sr/Ca signals that sample days to weeks rather than years of time. This greater precision enables researchers to examine whether the $^{13}C/^{12}C$ ratio changes during the development of a tooth, as would be the case if the diet differed from one season to another, or to use the Sr/Ca ratio to detect when weaning occurs. However, laser ablation cannot eliminate time-averaging because of the dynamics of enamel mineralization; only about 15% of the mineral in enamel is laid down during enamel secretion, while the rest is deposited during enamel maturation and the timing and patterning of that process is poorly understood. Enamel maturation does not even start until the full crown thickness is attained, and it may take three times as long as the process of enamel secretion. *See also* **enamel development**; **stable isotopes**.

## laser ablation inductively coupled plasma mass spectrometry
(or LA-ICP-MS.) *See* **laser ablation**.

## last appearance datum
(or LAD) The term refers to the date of a taxon's last occurrence in the fossil record. For various reasons the LAD of a taxon is almost certainly earlier than when the taxon actually became extinct, or emigrated from that region. Just how much earlier the LAD is than its true extinction or emigration from that region is determined by two factors. The first is any error in the date and the second is the nature of the relevant fossil record after the LAD. Hominins are such a rare

component of the mammalian faunal record that researchers need to find a substantial number of nonhominin mammalian fossils (at least several hundred) without finding any evidence of a particular hominin taxon before it can be reasonably assumed that the hominin taxon in question was not part of the faunal assemblage being sampled. In such cases the LAD has an acceptably low 95% confidence interval, but in cases where there is a major unconformity (lack of a fossil record) spanning several hundred thousand years after the LAD of a taxon, then the LAD of that taxon has an unacceptably high 95% confidence interval. *See also* **first appearance datum**.

## last common ancestor
*See* **most recent common ancestor**.

## Last Glacial Maximum
(or LGM) The most recent interval in Earth's history when global ice sheets reached their maximum volume. Some ice sheets reached their maxima at 33 ka, all of them reached their maxima by 26.5 ka, and then they remained at these maxima until 20–19 ka. Sea levels, which are inversely correlated with glacial maxima, fell during Marine Isotope Stage 3 and reached their lowest level, called the Last Glacial Maximum low stand, between 26.5 and 19 ka.

## Late Pliocene Transition
(or LPT) The first of two periods (*c.*3.2 to *c.*2.7 Ma) during the Northern Hemisphere Glaciation (or NHG), when the rate of change in the $^{18}O/^{16}O$ ratio (a proxy for the intensity of glaciation) was particularly high. The second, later, period (*c.*1.2 to *c.*0.8 Ma) is called the Mid-Pleistocene transition (or MPT). The end of the LPT marks the beginning of the 41 ka-long obliquity-driven climate cycles and it is when ice-rafted debris began to appear in sediments in the North Atlantic.

## lateral enamel
*See* **imbricational enamel**.

## lateralization
The term used for the parts of the cerebral hemisphere whose functions are asymmetric. The classical example is that of Brodmann's areas 44 and 45 in the frontal lobe. These areas are involved with speech in the dominant hemisphere (the left in the majority of people, i.e., in those who are right-handed) and with spatial cognition on the nondominant side.

## lateral sulcus
The lateral sulcus is the *Terminologia Anatomica* term for the structure some comparative neuroscientists refer to as the Sylvian fissure. In primates it is located on the lateral surface of each cerebral hemisphere where it separates the temporal lobe from the frontal lobe and parietal lobe. Ontogenetically it is one of the earliest sulci to appear. Lateral sulcus length has been demonstrated as asymmetric (longer on the left than the right) in modern humans and in the great apes; a similar asymmetry of lateral fissure length has also been reported in several Old and New World monkey species. The greater length of the lateral fissure in the left hemisphere may be related to hemispheric specialization of auditory processing for vocal communication signals. (L. *sulcus* = groove or furrow.)

## Later Stone Age
The youngest period in the tripartite division of the African archeological lithic record formalized in 1929 by Goodwin and van Riet Lowe in their *The Stone Age Cultures of South Africa*. Although it was developed for sequences in southern Africa, the terminology has subsequently been applied to sites across sub-Saharan Africa. The Later Stone Age is a broad stage within a trend towards more sophisticated stone tools (i.e., a trend toward aspects of stone tool manufacture that are conceptually and technically more elaborate). The Later Stone Age is distinguished from the Middle Stone Age by the presence in the former of small microliths (with one blunted, or backed, edge), small circular scrapers, and in later assemblages the presence of ground stone tools and pottery. Although the term Later Stone Age is frequently used to describe stone-tool-equipped modern human foragers, the term may also encompass stone-tool-using pastoralist populations. Recognizing that first and last appearance dates vary locally and are subject to change as new sites are found, existing dating methods are improved, or when new methods are introduced, the present evidence suggests the Later Stone Age began $c.40$ ka. Its last appearance dates are contentious and some contemporary African populations still occasionally use stone tools.

## Late Stone Age
*See* **Later Stone Age**.

## lava
The term refers to the rocks formed when molten material (magma) extruded at the Earth's surface cools. The character of lavas varies widely with chemistry and cooling rate. The rate of cooling determines the texture of the lava, and the size of the crystals it contains. Rapid quenching of the molten material results in obsidian, whereas as slower cooling allows the growth of crystals. The exterior portions of a lava flow show a mix of fine crystals and a glassy groundmass, whereas the inner portions of a flow are more coarse-grained with large crystals. The most abundant types of lava relevant to paleoanthropology are basalt, andesite, rhyolite, trachyte, and phonolite. (L. *labi* = to fall.)

## Lazaret
A large cave on the Mediterranean coast of France on the western slope of Mount Boron in Nice. The site contains abundant archeological and faunal evidence as well as several fragmentary hominin remains. Biostratigraphy suggests an age of Marine Isotope Stages 5c and 5d ($c.115$–$92$ ka) for the different archeological units. The nine hominins fossils are considered to be pre-Neanderthal. The archeological layer, layer C, is divided into three units; CI and CII contain Acheulean artifacts and CIII contains Mousterian items of Acheulean tradition with many flake tools. (Location 43°41′38″N, 07°17′21″E, France.)

## LB1
An associated hominin skeleton from Liang Bua, a cave on the island of Flores, Indonesia. It includes evidence of the cranium, mandible, clavicle, humerus, ulna, part of the hand, vertebrae, ribs, sternum, partial pelvis, femora, tibiae, fibulae, and most of the foot. Two interpretations of this specimen have been put forward. The first concludes that its small overall size (the stature

of LB1 is estimated to be 106 cm and its body mass to be 25–30 kg), its especially small brain (approximately 417 cm$^3$) for an adult hominin, and its primitive, pre-modern *Homo* or transitional-hominin-like morphology, are most parsimoniously interpreted as evidence of a novel endemically dwarfed early *Homo* taxon. Accordingly it was made the holotype of *Homo floresiensis*. A second interpretation views LB1 as a pathological *Homo sapiens* individual afflicted by, variously, endocrine disorders or a range of syndromes that include microcephaly. According to this interpretation no new taxon needs to be erected, but researchers have concluded that what had been claimed to be "pathological" deformations cannot be explained without recourse to unrealistically exotic pathological explanations. *See also* **Homo floresiensis**; **Liang Bua**.

## learning
*See* **social learning**.

## least squares regression
*See* **ordinary least squares regression**.

## lectotype
If a holotype was not designated at the time of the original description of a taxon, one of the specimens (or syntypes) listed in that description can subsequently be designated as the lectotype. For example, Alexeev did not specifically name a holotype of *Pithecanthropus rudolfensis* so when it was transferred to *Homo* KNM-ER 1470 was proposed as the lectotype of *Homo rudolfensis*. (Gk *lectos* = chosen and *typus* = image.) *See also* **holotype**; **paratype**; **type specimen**.

## Ledi-Geraru study area
A study area between the Awash and Mille River basins in northeastern Ethiopia. It is directly east and somewhat north of Hadar, north of Dikika, and south of the Woranso-Mille study area. The deposits, which range from slightly less than 3.4 to 1.7 Ma, span much of the Hadar and Busidima Formations. Hominin cranial remains attributed to *Australopithecus afarensis* have been found as well as evidence of Oldowan, Acheulean, and Middle Stone Age technologies. (Location 11°20′N, 40°45′E, Ethiopia.)

## Le Moustier
Two rock-shelters along the Vézère river, in southwest France. The material recovered from the upper shelter led Gabriel de Mortillet to define the Mousterian technocomplex. The age range for the Mousterian levels is *c*.56–40 ka and the Châtelperronian level from the site is *c*.42 ka. Excavators found a deliberately buried associated skeleton of a young individual (Le Moustier 1) and a skull fragment (Le Moustier 3) in the much deeper deposit in the lower shelter, and later excavations found evidence of at least 11 different archeological layers plus the remains of an infant (Le Moustier 2). All specimens belong to *Homo neanderthalensis*. (Location 44°59′38″N, 01°03′35″E, France.)

## Lemuta Member
*See* **Olduvai Gorge**.

## leptomeninges
*See* **meninges**.

## Levallois
A way of manufacturing stone tools that involves the production of large, relatively thin, flakes (called Levallois flakes) from carefully prepared cores. Levallois cores have an upper surface from which Levallois flakes are removed; the lower surface is untouched. The two surfaces are separated by a striking platform that may extend around the full perimeter of the core. The "preparation" implied in the term "prepared core" consists of removing multiple small flakes to shape the various convexities on the upper surface of the core; it is these convexities that determine the size and shape of the resulting Levallois flakes. The shaping of the striking platform to ensure successful flake removal results in the characteristic facetted platforms see on Levallois flakes. The Levallois production strategy is also capable of producing blades or points. Levallois technology first appears in Acheulean sites in Africa and Eurasia, and it is seen commonly in many Middle Stone Age or Middle Paleolithic assemblages (e.g., the Kapthurin Formation). (etym. the Levallois industry takes its name from Levallois-Perret, a Paris suburb, the location of the archeological site where this method of stone tool manufacture was first recognized.)

## lever
A rigid bar with a fulcrum (aka pivot or point of rotation). A lever is a simple machine that can be used to multiply the mechanical force that can be applied to an object (i.e., the load). Power (aka force) arm length is the distance between the fulcrum and the point of application of the force; load arm length is the distance between the fulcrum and the load. For example, if the power arm is five times longer than the load arm, then the force applied to the load will be five times greater than the applied (power) force (i.e., the mechanical advantage is five). Levers are commonly found in musculoskeletal systems. For example, in the modern human foot, the calf muscles (aka triceps surae) contract and pull on the calcaneus via the Achilles tendon to provide force. This then rotates about the ankle joint (i.e., a fulcrum) to apply a force through the forefoot to the ground (i.e., a load).

## LGM
Acronym for **Last Glacial Maximum** (*which see*).

## LH 4
This 3.85–3.65 Ma well-preserved mandibular corpus from Laetoli, Tanzania, is the holotype of *Australopithecus afarensis*.

## LH 18
This *c.*120 ka specimen, which consists of a calvaria and a face with no contact between them, shows a mix of modern and archaic features. It was found at Laetoli, Tanzania. The modern ones include an expanded vault, rounded occipital, and low inion; the archaic features include a flat frontal region, supraorbital torus, small mastoid processes, an occipitomastoid crest, and a thick vault. The cranium has been likened to Omo I and II and to the Ndutu cranium. Until the

discovery of the Herto crania and the redating of Omo I and II from Omo-Kibish, many regarded this cranium as the best-preserved and earliest evidence of *Homo sapiens* in Africa.

## Liang Bua

This cave, which is 30 km/18.6 miles from the north coast of Flores, Indonesia, and 500 m above sea level, is a solution cavity. The entrance is 30 m wide and 25 m high and the cave extends for 40 m into Miocene limestone. Its north-facing entrance has been exposed by the Wae Racang River, which is now 30 m below and 200 m distant from the cave entrance. In 2003 the c.18 ka LB1 associated skeleton that became the holotype of *Homo floresiensis* was recovered. Thirty-two artifacts were found at the same level as LB1, but the greatest concentration of artifacts (5500 per m$^3$) is elsewhere in the site. Most are flakes made from volcanics and chert. Many artifacts are associated with the remains of a dwarf variety of *Stegodon* that is 30% smaller than *Stegodon florensis*. A formal analysis of the minimum number of *Homo floresiensis* individuals has yet to be presented, but estimates range between six and 12. The different estimates are due to uncertainties regarding particular aspects of the stratigraphy; further excavations may help resolve these issues. (Location 08°31′50.4″S, 120°26′36.9″E, Flores, Indonesia.) *See also* **Homo floresiensis**; **LB1**.

## life history

The relative rate at which members of a species proceed through important developmental, maturational and reproductive milestones. Life histories reflect the ways individuals within a taxon divide their energy between maturation, maintenance, and reproduction; the latter component is further subdivided between the production of offspring and their subsequent maintenance. An organism's life cycle is punctuated by trade-offs concerning the expenditure of time and energy. It is the sum of the average of these trade-offs that define the life history strategy of a taxon. These trade-offs include when to be born, when to be weaned, how many stages of development to pass through before reproductive maturity, when to procreate, and when to die. The life history of modern humans is divided into six recognized stages: infancy, childhood, juvenile, adolescence, adult, and old adult (specifically, postmenopausal women; men retain their ability to sire children nearly until death). On the basis of comparative physiological and behavioral data, it is proposed that childhood, adolescence, and grandmotherhood are unique to modern humans. Total life span in modern humans is relatively long and the intervals between developmental and reproductive milestones are also relatively long. Modern humans are also exceptional because they wean their infants early, their age at first birth is later than would be expected for a great ape of the same body mass, they have an absolutely long life span, and they decouple female fertility and mortality so that females have a long post-reproductive lifespan. Measures of life history include length of gestation, age at first molar eruption, age at weaning, age at sexual maturity, ages at first and last birth, interbirth interval, mean lifespan, and length of post-reproductive lifespan. *See also* **grandmother hypothesis**.

## ligaments

A band of connective tissue that connects one structure to another. The ligaments relevant to human evolution are the bands of collagen (e.g., ilio-femoral and sacro-iliac ligaments) that connect hard tissue structures. They protect joints by limiting the movement that can take place. (L. *ligamen* = a bandage, or "something that binds structures together.") *See also* **joint**.

## limbic cortex
*See* **limbic system**.

## limbic system
A group of interconnected structures in and beneath the cerebral cortex. The limbic system, which includes the amygdala, hippocampus, septum, basal ganglia, and the cingulate gyrus, is involved in memory, emotion, motivation, learning, and some homeostatic regulatory functions. (L. *limbus* = border or interface.)

## limestone
A sedimentary rock rich in calcium carbonate ($CaCO_3$) that is usually derived from the shells of microorganisms. It is easily dissolved by acidic water percolating through cracks to produce solution cavities, and when this happens over a long period of time these solution cavities develop into caves. If these caves connect with the surface, soil may be washed in. When this soil is mixed with stone fragments falling from the roof of the cave, it results in a type of rock called a breccia. Most of the early hominin cave sites in southern Africa are formed in a variety of limestone called dolomite, which is rich in magnesium. Many of these caves (e.g., Sterkfontein) were initially mined commercially for limestone that was used as either a building material or for the local manufacture of cement. (OE *lim* = birdlime.)

## Lincoln Cave
*See* **Sterkfontein**.

## LINE
Acronym for long interspersed nucleotide element, a DNA sequence of more than 5000 base pairs of which many are dispersed throughout the genome. LINEs are significant for studies of evolution because, like SINEs, they can be used to study population and species relationships. In addition, when LINEs copy themselves and insert the copy into the genome, neighboring DNA, which may contain regulatory elements, genes, or gene fragments, may also be copied and inserted. Such DNA may affect the regulation or sequence of a gene in the new location. *See also* **SINE**.

## lingual
The aspect of an upper and lower tooth crown that is closest to the tongue. Therefore, the lingual side of a tooth crown is the inner side of the crown. (L. *lingua* = tongue.)

## linkage
When two loci on a strand of DNA are physically near each other, the other alleles present at those loci are inherited together as a "linked" unit (i.e., they are not separated during recombination). The shorter the distance between the alleles the greater the linkage strength; thus linkage strength can be used to calculate the relative distance between loci, and from these data

genetic maps can be generated. Examples of linkage in modern humans include the Rhesus blood group and the enzyme 6-phosphogluconate dehydrogenase (an enzyme found in blood) on modern human chromosome 1. Before other polymorphic markers such as microsatellites were discovered the linkage between phenotypic traits was used to create a crude map of the modern human genome. *See also* **linkage disequilibrium**.

## linkage disequilibrium
The nonrandom association of alleles at two or more loci. The degree of association (aka linkage) between two loci is measured by the frequency with which recombination separates the two alleles (aka crossover frequency). The further apart two loci are the greater the chance they will be affected by recombination. Linkage disequilibrium occurs when a combination of alleles at different loci is found more often than expected from a random combination of alleles based on their frequencies in a population. Linkage disequilibrium is important in studies of modern humans because it can indicate the presence of positive selection in a portion of the genome. It can also be used to research population history since "young" populations show greater linkage disequilibrium in the genome than "older" populations. This is because recombination has had more time to break up linkage disequilibrium in "older" populations (e.g., sub-Saharan Africans typically have less linkage disequilibrium in their genomes than do Europeans).

## Linnaean binominal
The combination of the genus and species name (e.g., *Australopithecus afarensis*) that make up the proper scientific name of a species. The genus name is always capitalized and both names are always in a different font than the rest of the text. Usually they are italicized when typeset and underlined when hand-written. The genus can be abbreviated by reducing it to the first letter, or to the first few letters in cases where confusion may arise (e.g., *Au. afarensis*), but the species name must always be spelled out (e.g., *Homo sapiens* is abbreviated as *H. sapiens* and not as *Homo s.*).

## Linnaean taxonomy
A taxonomy based on the principles introduced by Carl Linnaeus in the mid-1700s. The two names (hence it is called a binominal) used for a species are a unique combination of a genus and a species name (e.g., *Homo sapiens*, *Pan troglodytes*, etc.).

## lithic
An adjective used to describe artifacts made of stone. Its use as a noun (e.g., "lithics") is common, but incorrect. Lithic artifacts are durable and the oldest known artifacts (presently from the site of Gona, Ethiopia) are made of stone. (Gk *lithos* = stone.)

## lithic analysis
Analyses of lithic assemblages undertaken to derive information about hominin behavior and site-formation process (i.e., collections of stone tools and their manufacturing debris). Depending on the research objectives, lithic analysis may include typological classification, metric and nominal trait analyses, quantification and/or description of flake scar patterns, refitting, microwear and residue analysis, and raw-material studies. The use of technological

types, such as flakes, cores, fragments, etc., to subdivide and describe lithic assemblages remains a fundamental step in virtually all lithic analyses. Nominal traits might include flake shape, platform type, and the occurrence of retouch. Artifact counts and metric and nominal data are used to make a wide range of inferences regarding knapping methods, techniques and skills, artifact transport patterns, and reduction intensity. Characteristic microwear "polishes" and macroscopic damage patterns identified through experimental replication are used to infer artifact function (e.g., the materials tools were used on, evidence of hafting and projectile use). Organic residues may also provide a direct indication of artifact function. Raw-material analyses are most commonly undertaken to identify the sources, transportation, and/or trade of lithic resources, but raw-material choice may also reflect the preference of the stone tool's maker.

## lithic assemblage
A collection of stone artifacts from some temporally and spatially bounded interval, typically a stratigraphic or excavation unit. This scale varies so that at some sites (e.g., Blombos Cave) archeologists use the term lithic assemblage to refer to the artifacts recovered from a single layer (e.g., CD) or a phase (e.g., the Still Bay from phase M1). (Gk *lithos* = stone.)

## lithostratigraphy
Recognition and correlation of sedimentary strata based on the appearance of the rock layers as seen through the naked eye, with a lens, or by using a microscope. (Gk *lithos* = stone, and see the etymology of stratigraphy.)

## "Little Foot"
Informal name for **Stw 573** (*which see*).

## Little Ice Age
*See* **Holocene**.

## locality
A location within a large fossil site where fossils have been found. At small sites and at excavated cave sites (e.g., Sima de los Huesos at Atapuerca, Spain) there is no need to do other than specify within which grid square and from what level in that square a fossil came from. But in large fossil sites (e.g., Koobi Fora, Middle Awash, Olduvai Gorge) the locations where one, or more, fossils have been found, are described as localities (e.g., FLKNN I and MNK I are two localities within Olduvai Gorge, and ARA-VP-1 is a fossiliferous locality within the Central Awash Complex of the Middle Awash study area).

## locomotion
The various ways animals use their anatomy to move from one place to another to acquire food, find mates, avoid predators, move to safe resting places, and interact socially with other members of a group. In a given day, virtually all extant primates use a repertoire of different types of locomotion to move from place to place. The locomotion practiced by a primate species has an impact on the architecture of most of the postcranial musculoskeletal system,

as well as on aspects of cranial anatomy. Thus, the anatomy of the skeleton, especially the postcranial skeleton, provides information about the locomotor adaptations and behaviors of fossil hominins. If an animal moves around with its trunk upright and supports itself exclusively on its hindlimbs, then its locomotion is described as bipedal. (L. *loco* = place and *motivus* = to cause motion.) *See also* **bipedal**; **bipedalism**.

## locus

The physical location of a gene or nucleotide in the genome. The locus of a gene is analogous to an individual's street address. For example, according to the March 2006 assembly of the modern human genome the gene for phenylthiocarbamide (PTC) tasting (*TAS2R38*) is on the long arm of chromosome 7 at position 141,318,900–141,320,042. (L. *locus* = place; pl. loci.)

## loess

A deposit of wind-blown silt that typically forms an extensive blanket over mid-latitude landscapes during a period of cold, dry climate. Eastern and central Europe and China, in particular, have significant loess deposits. Cycles of alternating loess and soils, indicative of climate fluctuation between cold/dry and warm/moist conditions, have produced a long stratigraphic record of paleoclimate back to *c.*2.5 Ma over most of the Loess Plateau of north-central China and as far back as *c.*22 Ma in the western part of this region. Connections between the climate history of the Loess Plateau and the Nihewan Basin have been implicated in the dispersal of early Pleistocene hominins to northeast Asia. Loess deposition in China during the Last Glacial Maximum extended as far south as the Bose Basin. (Ge. *loess* = loose.) *See also* **Bose Basin**; **Nihewan Basin**.

## Lokalalei

*See* **West Turkana**.

## Lomekwi

*See* **West Turkana**.

## long distance transport

*See* **carcass transport strategy**.

## long interspersed nucleotide element

*See* **LINE**.

## longitudinal arch

A structure oriented along the length of the foot, which allows the foot to take advantage of its intrinsic anatomical structures in order to stiffen it and to store and return elastic energy during the stance phase of walking. Among extant taxa this feature is unique to modern humans. The arch has two components, the medial and the lateral longitudinal arch. Both are formed by articulations between bones in the hind- and midfoot and both are reinforced by soft-tissue structures. Apes do not possess a longitudinal arch. Fossil footprints from Ileret, Kenya, indicate that the arch had evolved in its modern form by 1.53 Ma, and some suggest that the 3.66 Ma footprints from Laetoli provide evidence for a longitudinal arch much earlier in hominin evolution. *See also* **foot movements; Koobi Fora hominin footprints; Laetoli hominin footprints; walking cycle.**

## longitudinal data
Data derived from a series of observations on a single individual, or a group of individuals. Longitudinal data are preferable to cross-sectional data for studies that investigate how individuals change over time, but longitudinal studies take longer to conduct and longitudinal data are methodologically more difficult to collect. *See also* **cross-sectional data.**

## long-period incremental lines
Incremental features in tooth enamel or dentine that have an intrinsic temporal secretory rhythm of more than 1 day. The features are called striae of Retzius and perikymata in enamel and Andresen lines and periradicular bands in dentine. In primates, long-period incremental lines have been reported to range between 2 and 12 days, with a consistent periodicity within an individual. Counts and measurements of these features facilitate estimation of the crown formation time, root formation time, or the root extension rate. *See also* **dentine; enamel development; enamel microstructure.**

## lordosis
The term lordosis (and its antonym, kyphosis) refers to the angular relationship within or among sections of a compound structure (e.g., the vertebral column) or between two planes (e.g., the anterior and posterior components of the cranial base). Thus, the parts of the adult modern human vertebral column that are concave posteriorly (i.e., the cervical and lumbar regions) are referred to as being lordotic, whereas the parts that are concave anteriorly (i.e., the thoracic and sacral) are referred to as being kyphotic. Adult modern humans are distinctive among the extant great apes in normally having a pronounced lumbar lordosis. This is related to our habitual upright posture and it is particularly exaggerated in females in late pregnancy when the trunk's center of mass moves anteriorly. With respect to the cranial base, organisms or individuals with a larger, more obtuse (i.e., more chimpanzee-like) cranial base angle are described as having a more lordotic basicranium, whereas organisms or individuals with a smaller, more acute (i.e., more modern human-like) cranial base angle have a more kyphotic basicranium. (Gk *lordos* = bent backwards.)

## lordotic angle
*See* **lumbar lordosis.**

## Lothagam
A site in northern Kenya where the Nawata and Nachukui Formations are exposed. The lower member of the Nawata Formation is *c*.7.4–6.5 Ma, the upper member of the Nawata Formation is *c*.6.5–5.23 Ma, and the Apak Member, the oldest member in the Nachukui Formation, is between approximately 5.0 and 4.2 Ma. A hominin mandible (KNM-LT 329) was recovered in 1967 and between 1989 and 1993 researchers recovered around 1700 vertebrate and other tetrapod fossils. Two hominin teeth have been recovered from the upper member of the Nawata

Formation, a fragment of the right side of the mandibular corpus (KNM-LT 329) from the Apak Member, and four isolated teeth from the Kaiyumung Member of the Nachukui Formation. The four isolated teeth have been assigned to *Australopithecus* cf. *Australopithecus afarensis*; the other hominins are assigned to Hominidae gen. et sp. indet. (Location 02°54′N, 36°03′E, Kenya.)

## Lower Omo Basin

The depression through which the lower reaches of the Omo River runs. The Omo River rises in the Ethiopian Highlands and ends by draining into the northern end of Lake Turkana. The last part of its course and Lake Turkana occupy the low country between the Ethiopian and Kenyan domes to the north and south, respectively, and between the western and eastern walls of the East African Rift. The sediments in the Lower Omo Basin are divided into the Omo Group and the Lake Turkana Group (also called the Turkana Group). *See also* **Omo Group**; **Omo-Turkana Basin**.

## Lower Paleolithic

A stage of the European and Near Eastern Paleolithic defined predominantly by Plio-Pleistocene technocomplexes that typically do not include handaxes (e.g., Mode 1 and Mode 2). It spans a long period of human evolution (c.2.5 Ma to 200 ka) during which there were significant biological and behavioral shifts as well as several distinct technocomplexes. The equivalent term for African sites with similar technologies is the Early Stone Age. It is often directly associated with Oldowan technology.

## low-ranking prey

Within optimal foraging theory low-ranking prey are those taxa or food resources that provide below-average energetic returns (e.g., small mammals such as rabbits and hares are typically regarded as low-ranking prey). However, a growing body of ethnographic evidence suggests that other factors (e.g., prey mobility) may play an important role in explaining human foraging decisions and prey rankings. *See also* **high-ranking prey**.

## LPT

Acronym for the **Late Pliocene transition** (*which see*).

## LSA

Acronym for the Late Stone Age and Later Stone Age. *See* **Later Stone Age**.

## "Lucy"

Nickname for A.L. 288-1, the partial skeleton of *Australopithecus afarensis* discovered at Hadar, Ethiopia, in 1974. During a camp celebration of the discovery, The Beatles' album *Sgt. Pepper's Lonely Hearts Club Band* was playing and one of the researchers suggested calling the presumed female skeleton "Lucy" after the song "Lucy in the Sky with Diamonds." *See also* **A.L. 288-1**.

## Lukeino Formation

Lacustrine and fluvial Late Miocene (c.6.0 Ma to c.5.7 Ma) sediments that outcrop extensively in the eastern foothills of the Tugen Hills, west of Lake Baringo, Kenya. The Kabarnet Trachyte and Kaparaina Basalts Formations are, respectively, below and above the four members of the

Lukeino Formation. Most of the *Orrorin tugenensis* hypodigm was recovered from the Aragai, Cheboit, Kapcheberek, and Kapsomin localities within the penultimate Kapsomin Member of the Lukeino Formation.

## "Lukeino molar"

An isolated mandibular molar found in 1975 in the Lukeino Formation in the Tugen Hills. It is the holotype of *Orrorin tugenensis*. See also **KNM-LU 335**; ***Orrorin tugenensis***.

## lumbar lordosis

A posteriorly concave curvature of the small of the back that is caused by the wedge-shaped (anteriorly tall and posteriorly short) vertebral bodies and intervertebral discs in the lumbar region. This configuration of the lumbar spine is seen in modern humans, in all extinct pre-modern *Homo* taxa, and in individuals assigned to *Australopithecus africanus* (e.g., Sts 14 and Stw 431). The exaggerated lumbar lordosis seen in late pregnancy in modern humans helps compensate for the anterior migration of the center of mass (i.e., the added weight of the fetus, placenta, amniotic fluid, etc.). The lordotic angle (i.e., the angle between the plane of the superior endplate of the first sacral vertebra and the plane of the superior endplate of the fifth presacral vertebra) is usually less than 25° in the great apes and more than 25° in modern humans and in fossil hominins. (L. *lumbus* = loin and Gk *lordos* = bent backwards.)

## lumbar vertebral column

The region of the vertebral column between the rib-bearing thorax and the sacrum. A traditional definition of lumbar vertebrae is that they lack ribs and do not articulate with the sacrum. The modal number of traditionally defined lumbar vertebrae in the living great apes is four. The modal number of traditionally defined lumbar vertebrae in modern humans is five, but six functionally defined lumbar vertebrae are found in about 40% of some modern human populations (which results from the lowest thoracic vertebrae having no articulation with ribs). Lordosis, or posterior curvature, of the modern human lumbar region is an adaptation to bipedal positional behaviors, and is caused by the dorsal vertebral margins being shorter than the ventral. Modern humans are dimorphic in this respect, with females having three wedged vertebrae, males only two; these differences reflect the need to accommodate shifts in the position of the trunk's center of mass during pregnancy by increasing lumbar lordosis. Three early hominin specimens (two *Australopithecus africanus* and one *Homo erectus*) and several *Homo neanderthalensis* individuals are complete enough to reliably infer the number of lumbar vertebrae. All of the Neanderthal specimens have five lumbar vertebrae, and one *Au. africanus* specimen (Sts 14) has six functional lumbar vertebrae, the most cranial of which bears a unilateral rib. The lumbar counts of the second *Au. africanus* and the *H. erectus* specimen (i.e., whether there are five or six lumbar vertebrae) are disputed. (L. *lumbus* = loin.)

## luminescence dating

A radiogenic dating method that relies on either a thermal (thermoluminescence dating or TL) or an optical (optically stimulated luminescence dating or OSL) signal that reflects the numbers of electrons trapped within defects in the crystal lattice of the respective mineral(s) due to intrinsic and extrinsic nuclear radiation. In essence, these methods function like a counter that is set to zero by some kind of human action, which then gradually accumulates electrons at a constant rate over time. The number of electrons corresponds to the age since the counter reached zero. In thermoluminescence dating, the electron population is reset to zero when temperatures exceed 400 °C, such as might occur in a fire. In optically stimulated luminescence dating, trapped electrons are eliminated by prolonged exposure to daylight (i.e., when sediments are transported by wind or water, or if they lie for long periods on the surface). If these methods are to be of value for hominin evolution, the "resettings" must be brought about by events that are the result of hominin activity, and not the result of natural events. For example, when TL is used on burnt flint artifacts researchers assume the flints were burned in a fire not long after they were fashioned, so the TL date provides a conservative minimum date for the manufacture of the flint artifacts. Luminescence dating can be routinely applied to less than $c.100$ ka events involving minerals such as quartz, flint, and feldspars; in some circumstances they can be used for events that are more than 500 ka. Analytical problems with respect to OSL include inadequate exposure to sunshine so that previously acquired luminescence is retained.

## lumper
A researcher who favors fewer, more inclusive, taxa (e.g., some researchers advocate lumping all extinct *Homo* species into a single species, *Homo sapiens sensu lato*). (ME *lumpe* = to aggregate.) *See also* **alpha taxonomy**; **splitter**.

## lunate sulcus
A prominent sulcus found on the lateral surface of the occipital lobe of the cerebral hemispheres of some monkey and all extant ape taxa; it is also seen in a small percentage of modern humans. It is a rough approximation of the anterior boundary of the primary visual cortex. In the modern humans who have a lunate sulcus, its more posterior location relative to that of apes is thought to reflect the increased size of the human parietal association cortex and the consequent decreased size of the primary visual cortex. Endocasts of fossil hominins, both natural and those prepared by researchers, occasionally preserve evidence of the lunate sulcus and its location (either more anterior as in the extant apes, or more posterior as in modern humans) has been the main focus of debates about when in hominin evolution cortical reorganization occurred. For example, some researchers argue that *Australopithecus afarensis* (A.L. 162-28) and *Australopithecus africanus* (Taung 1 and Stw 505) have the lunate sulcus in a posterior, modern human-like position, but others claim the lunate sulci of Taung 1 and A.L. 162-28 are in a more anterior, extant ape-like, position. (L. *lunatus* = to bend like a crescent and *sulcus* = a deep furrow or groove.)

## Lydekker's Line
The eastern boundary of the biogeographical region known as Wallacea (i.e., the boundary between Wallacea and Sahul). *See also* **Wallacea**.

# M

### Ma
The abbreviation of an age determination expressed in millions of years. For example, "the approximate age of KNM-ER 3733 is 1.8 Ma." The terms "my" or "myr" should only be used to describe periods of time, as in "more than 2 myr have elapsed since *Homo* first appeared in Africa." *See also* **age estimate**.

### Maba
(马坝) Local farmers discovered hominin and nonhuman mammalian fossils while digging for natural fertilizers in this limestone cave site in China. A hominin calotte (PA 84) consisting of partial parietal bones, a frontal bone with fairly complete nasals, and most of the right orbit is that of an aged, probably male, adult. It is one of the few reliably dated ($c.129 \pm 10$ ka) hominins from China in this time period. (Location 24°40′28.51″N, 113°34′48.68″E, southern China.)

### macrodont
Literally having large teeth, but in paleoanthropology it usually refers to the relative size of the crowns of the postcanine teeth. Unlike the term megadont, which has been given a quantitative definition, macrodont is a qualitative term. (Gk *macros* = large and *odont* = tooth.) *See also* **megadont**.

### macroevolution
An inclusive term used to describe evolution above the level of the species. In essence, whereas microevolution is what can be learned about evolution from genetic experiments conducted in the laboratory and the field, macroevolution is what the fossil record can tell you about evolution on a geological scale. (Gk *macros* = large.)

### macrostructure
See **morphology**.

### Magdalenian
A European Upper Paleolithic technocomplex named after the site of La Madeleine, France. It is characterized by unretouched stone blades, bone needles, bone harpoons, and other bone tools, as well as figurative engravings of humans and animals, jewelry, and complex burials with many grave goods. It dates to between 17 and 11.5 ka.

---

*Wiley Blackwell Student Dictionary of Human Evolution*, First Edition. Edited by Bernard Wood.
© 2015 John Wiley & Sons, Ltd. Published 2015 by John Wiley & Sons, Ltd.

### magnetic anomaly time scale
*See* **geomagnetic polarity time scale**.

### magnetic polarity time scale
*See* **geomagnetic polarity time scale**.

### magnetic resonance imaging
(or MRI) A medical imaging method that uses magnetic fields and radiowaves to examine the anatomical structures in the body. The MRI scanner creates a strong magnetic field surrounding the area of interest, which excites hydrogen atoms in water present in the tissue. The excited hydrogen atoms emit a radio frequency signal that is detected by the scanner. Contrast between tissue types is obtained by varying the magnetic field (i.e., the magnetic coils on and off) to observe the rate at which the excited atoms return to their equilibrium state. Different tissue types have different concentrations of water, and their hydrogen atoms will thus return to equilibrium at different rates, allowing for tissue contrast. Different types of contrast can be detected in tissue depending on how the magnetic resonance signal is changed during scanning (e.g., T1-weighted, T2-weighted). *See also* **functional magnetic resonance imaging**.

### magnetochronology
*See* **geomagnetic polarity time scale**.

### magnetostratigraphy
The use of the magnetic polarity of a stratum, or a sequence of magnetic polarities of a longer section, to date sediments or igneous rocks. *See also* **geomagnetic polarity time scale**.

### magnetozone
*See* **geomagnetic anomaly**.

### Mahalanobis distance
A measure of multivariate distance between a point and the centroid of data set. Mahalanobis distance differs from Euclidean distance in that it takes into account the correlation between variables in the data set as well as the difference in standard deviations between variables; Euclidean distance only does the latter. This distinction is potentially important in calculating multivariate morphological distances. Mahalanobis distance requires knowledge of the standard deviations within each variable and correlations between variables, so the distance cannot be calculated between two individual points unless there is a reference group from which the standard deviations and correlations are calculated.

### Maiko Gully
*See* **Olduvai Gorge**.

### main cusp
(syn. primary cusp.) *See* **cusp**.

## Main Ethiopian rift
See **Ethiopian Rift System**.

## major axis regression
As with other methods of regression, this statistical analysis estimates the relationship between variables, specifically how one variable (the dependent variable) changes as a result of the other, independent variable. In major axis regression, the model parameters are fit by minimizing the sum of squared differences between observed values and the best-fit line as calculated by a perpendicular line from the best-fit line, thus taking into account error in both variables. The best-fit line produced by major axis regression is identical to the axis of the first principal component in a principal components analysis (using a correlation matrix, not a covariance matrix). Major axis regression should only be used when independent and dependent variables are measured in the same units; otherwise, reduced major axis regression should be used. Because major axis regression takes into account error in both variables and not just the independent variable when fitting model parameters, it is an example of a model II regression.

## Maka
This site is located on the eastern side of the Awash River in the Middle Awash study area in the Afar Rift System in present day Ethiopia; it is immediately south of Matabaietu and north of Belohdelie. All of the hominin remains are from the $c.3.4$ Ma Maka sands. All are attributed to *Australopithecus afarensis* and they provided the first substantial sample of *Au. afarensis* material outside of Hadar and Laetoli. (Location 10°30'40"N, 40°35'49"E, Ethiopia.)

## Makapansgat
(aka Makapansgat Limeworks) One of a number of cave complexes in the Makapan Valley in Limpopo Province of modern South Africa (formerly Northern Province). The cave was mined as the "Makapansgat Limeworks" (hence the prefix "MLD" for "Makapansgat Limeworks Deposits/Dumps") from around the beginning of the 20thC. The Makapansgat Formation is divided into five members (1–5), with Member 4 subdivided into two deposits (A and B). Subsequently, Member 4a became Member 4 and Member 4b was reclassified as the Central Debris Pile (CDP). All but two of the early hominins from Makapansgat come from Member 3; MLD 37/38 is one of two specimens from old Member 4a, new Member 4. All specimens are attributed to *Australopithecus africanus*. A combination of biostratigraphy and magnetostratigraphy suggest that Member 3 is between $c.2.85$ and 2.58 Ma and Member 4 is $c.2.58$ Ma. (Location 24°12'S, 29°12'E, South Africa.)

## Makapansgat Limeworks Deposit
See **Makapansgat**.

## Malapa
A cave site approximately 15 km/9 miles northeast of the concentration of sites in the Blauuwbank Valley, near Krugersdorp in Gauteng Province, South Africa, that includes Kromdraai, Sterkfontein, and Swartkrans. The cave filling comprises five facies, with A and B

below being separated by a layer of flowstone (flowstone 1) from C, D, and E above; the fossiliferous facies are D and E. The two specimens reported so far, MH1 and MH2, come from facies D, but it is suggested that there are additional hominin remains in facies E. The hominins in facies D and E are sandwiched between dated flowstones and because of this they can be given an unusually precise age of $1.977 \pm 0.003$ Ma. The published hominin remains consist of two associated skeletons: the juvenile MH1 and the adult MH2, plus evidence of *at least* one other individual. There is no archeological evidence at the site. (Location 25°53′39″S, 27°47′57″E, Gauteng Province, South Africa.)

## malar

The region of the face corresponding to the cheek is called the malar region. The external surface of the part of the maxillary bone that lies deep to the soft tissues of the cheek is called its malar surface. (L. *mal* = cheek.)

## Malema

A hominin fossil locality in the Chiwondo Beds in Malawi. The only hominin from Malema RC 11 is a maxillary fragment (HCRP RC 911) provisionally assigned to *Paranthropus boisei*. The significance of Malema is mainly biogeographical, for it extends the range of *P. boisei* southward by more than 1000 km/620 miles. (Location 10°01′18.59″S, 33°55′51.53″E, Malawi.)

## malleus

One of the three auditory ossicles in the middle ear. (L. *malleus* = hammer.) See **auditory ossicles**; **middle ear**.

## mammal size classes

*See* **bovid size classes**.

## mandible

The bone of the lower jaw. Each side of the mandible consists of a horizontal body and a vertical ramus (NB: it is incorrect to refer to a hominin fossil as a "right mandible"; it is the right side of a mandible). The body is formed from the fusion of the right and left corpus of the mandible at the mandibular symphysis early in development. Each corpus has an alveolar process superiorly, a base inferiorly, and external and internal surfaces. The alveolar process contains the sockets (aka alveoli) for the roots of the mandibular teeth. On the base either side of the midline are the digastric fossae. The mental foramen perforates the external surface of the corpus and the mylohyoid line divides the internal surface into the sublingual fossa superiorly and the submandibular fossa inferiorly. The symphysis has the alveolar process superiorly, the external or mental surface anteriorly, the internal or lingual surface posteriorly, and a base. The external surface of the corpus may have a mental protuberance but to qualify as a true chin there must be a depression, called an *incurvatio mandibulae*, superior to it. The internal surface of the symphysis may have tubercles and fossae for the genial muscles. It may describe a smooth curve, or be reinforced by transverse bony buttresses. These may be separate, as superior and inferior transverse tori, or in the form of a single transverse torus. When the internal surface of the symphysis between the alveoli and the torus is flat it is called a post-incisive planum. Each ramus has two surfaces (external and internal), four borders (superior, inferior, anterior and posterior), and two processes (condylar and coronoid) separated by the mandibular notch; the posteroinferior corner is called the angle. The anterior border of the ramus can be straight or sinuous. The internal surface of the ramus is perforated by the mandibular foramen (for the inferior alveolar nerve and vessels). (L. *mandere* = to chew and *mandibular* = lower jaw.) *See also* **true chin**.

## Mann–Whitney U test
A nonparametric statistical test used to determine whether the values of a sample of measurements are significantly larger or smaller than the values for a second sample. Used in the same situations as one would use a two-sample *t* test, the Mann–Whitney U test considers the relative rankings of the values in the two groups rather than the values themselves.

## manual digit
The manual digit is the distal component of a manual ray (the metacarpal is the proximal component). The manual digits or fingers of the primate, and therefore the hominin, hand are made up of three bones – the proximal, intermediate, and distal phalanges – but the thumb has just two phalanges.

## manual ray
A manual ray comprises a manual digit (i.e., the proximal, intermediate, and distal phalanges) plus its associated metacarpal. The thumb has just two phalanges. (L. *manus* = hand and *radius* = radiating line.)

## manuport
A stone found at an archeological site that is inferred to have been transported by hominins, perhaps for future use as a source of raw material for the production of stone tools. Manuports are usually identified as such by eliminating the possibility of other natural transport processes (e.g., deposition by streams). (L. *manus* = hand and *portare* = to carry.)

## Mapa
*See* **Maba**.

## Marine Isotope Stages
Periods of stability in the ratio of oxygen isotopes from marine sources, which are used to provide approximate ages for archeological and fossil sites. The initial crucial discovery that led to the use of oxygen isotopes for geochronology was the demonstration that the shells of foraminifera (aka forams) recovered from the sea floor contain two isotopic forms of oxygen, $^{16}O$ and $^{18}O$, and that the relative amounts of the "light" ($^{16}O$) and "heavy" ($^{18}O$) isotopes varied according to the temperature of the seawater. When the water is warm evaporation removes more of the lighter isotope, thus increasing the $^{18}O/^{16}O$ ratio, and when the water is cool the $^{18}O/^{16}O$ ratio declines. It was suggested that the $^{18}O/^{16}O$ ratio could also be used as a proxy for terrestrial glaciations because the $^{18}O/^{16}O$ ratio in the oceans would fall when $^{18}O$-rich water was locked up in expanded ice caps and enlarged glaciers. Researchers proposed that the maxima

and minima of the inferred paleotemperature curve be referred to as "stages" (Marine Isotope Stages or MIS, but also still referred to as Oxygen Isotope Stages or OIS) starting with MIS 1 for the most recent maximum. Thereafter odd numbers were used for peaks of maximum temperature and even numbers for troughs of minimum temperature. Older glacial/interglacial terminologies based on continental indicators of cold and warm climate phases have been replaced by the MIS numerical system, allowing more precise estimates of age. Thus the Würm glacial corresponds with MIS 2–4, the Riss with MIS 6, the Mindel with MIS 8, 10, 12, and 14, and Günz with MIS 16 (even-numbered MIS are cold periods). The recognition of MIS allowed the validation of the so-called astronomical theory of climate. Once it became clear that the orbital "Milankovitch cycles" were significant drivers of climate change, it became possible to use the mathematically generated time estimates to fine tune less-precise geological records (aka orbital tuning). The MIS system is increasingly used to indicate the age of fossil sites that cannot be dated precisely using methods that measure time in years. In some cases, layers of tephra at fossil sites are also found intercalated in the marine sediments sampled by deep sea cores. Whereas the latter sediments are deposited more or less continuously, some cycles of deposition may be missing from the nonmarine sections. Correlations between terrestrial and marine sequences have enabled the dates from the terrestrial sections to be orbitally tuned using the marine sequences. *See also* **astronomical theory**; **astronomical time scale**; **orbital geometry**; **orbital tuning**.

### marker bed

A layer or stratum that is physically or chemically distinctive enough to be identified independently of its context. Marker beds are used to link isolated blocks of sediments in the same fossil site or to provide a means of linking/correlating strata from one locality to another, or even from one site to another (e.g., the "grey tuff" and the "pumice tuff" in the Kapthurin Formation, Tugen Hills in Kenya). The ash spewed from each volcano has a unique mix of chemicals and even the ashes derived from different eruptions from the same volcano are chemically distinctive. Tephrostratigraphy uses the unique chemical profile of a tuff to trace it from one block of sediment to another and from one site to another.

### mastication

The processes involved when food is chewed by moving the lower teeth against the upper teeth in order to slice, fracture, or crush food into smaller pieces. This mechanical reduction, or comminution, of food in the mouth is a key feature of mammals and it was crucial for the relatively high energy consumption of mammals compared to reptiles of the same body mass. Mastication reduces the particle size of food, mixes it with saliva, and softens it so that it forms a bolus that can then be transferred to the esophagus by the process called swallowing. (Gk *mastikhan* = to grind the teeth.) *See also* **chewing**; **mastication, muscles of**.

### mastication, muscles of

An informal term used for the group of striated muscles that move the mandible during mastication. The main muscles of mastication are the masseter, temporalis, and the lateral and medial pterygoid. The main action of the masseter is to elevate the mandible, but its obliquely orientated fibers also play a minor role in side-to-side and protraction and retraction movements of the mandible. The anterior temporalis fibers are involved in elevating the mandible and the posterior fibers with retraction. The pterygoids act as an integrated unit. When the medial and lateral pterygoids of the same side contract the condyle on that side is drawn anteriorly so that movements then occur around an axis that passes through the opposite condyle. When the medial and lateral pterygoids of one side contract, followed by contraction of the medial and lateral pterygoids of the other side, the mandible is moved from side to side in a chewing movement. When the inferior fibers of the two lateral pterygoids contract together with the two medial pterygoids they draw the mandibular condyles and the articular discs anteriorly onto the articular eminence; this maximizes the gape. When the superior fibers of the lateral pterygoids are involved instead of the inferior fibers, together with medial pterygoids, the mandible is protruded. The motor branch of the trigeminal nerve supplies all the muscles mentioned above. Other muscles play an important role in mastication (e.g., the buccinator prevents masticated food collecting in the cheek) but they are not usually considered to be among the "muscles of mastication."

## masticatory apparatus

The hard (i.e., bones and teeth) and soft (i.e., muscles) parts of the skull involved in chewing food; only the hard tissues fossilize. The hard tissues include the upper jaw (aka maxilla), plus the upper (aka maxillary) teeth, and the lower jaw (aka mandible) and the lower (aka mandibular) teeth. The jaws and the teeth are among the more dense parts of the skeleton, and for this and other reasons they tend to resist post-mortem damage and survive long enough to be fossilized. This results in the hard tissues of the masticatory apparatus dominating the hominin fossil record. (Gk *mastikhan* = to grind the teeth, plus L. *ad* = to and *parare* = to make ready.)

## mastoid crest

*See* **mastoid process**.

## mastoid process

A bony prominence that projects inferiorly from the posterior part of the temporal bone. It is relatively large in modern humans (you can feel it just behind your ear). The shape and size of the mastoid process shows considerable inter-individual variation in all the higher primates, including extinct hominin taxa. However, there are trends in mastoid process size and shape that are reasonably consistent. In modern humans the mastoid has many air-filled cells (it is pneumatized), and it is typically the only substantially pneumatized part of the temporal bone, with a distinctive pyramidal shape. Among pre-modern *Homo* taxa, the mastoids of *Homo neanderthalensis* are distinctively small. In archaic hominins the mastoid is less distinct because it is just one part of a more general pneumatizion of the temporal bone. The sternocleidomastoid muscle and two deep neck muscles (the splenius and longissimus capitis) are attached to its lateral surface and the bony edge of their attachment is sometimes raised up as a mastoid crest. If the posterior-most fibers of the temporalis muscle are especially well developed, then the bone may be raised to form a supramastoid crest. In *Paranthropus* (and in most archaic hominin taxa assigned to *Australopithecus*) the mastoid normally projects further laterally than the supramastoid crest. In *Paranthropus* the supramastoid and mastoid crests are distinct and they are usually separated by a supramastoid sulcus. The posterior belly of the digastric muscle is attached to a groove (called the digastric notch) on the medial surface of the mastoid process. The suite of structures that can potentially be seen as you move medially from the tip of a mastoid process are the digastric fossa, the juxtamastoid eminence, a groove for the occipital artery, and the occipitomastoid crest. (Gk *mastos* = breast and *œides* = shape.)

## Matabaietu
See **Middle Awash study area**.

## Mata Menge
The best known of a group of Early and Middle Pleistocene sites in the Ola Bula Formation located in the Soa Basin of central Flores. More than 500 artifacts have been recovered from sediments that are *c*.800 ka. Aside from the intrinsic value of the archeological evidence Mata Menge is significant for two other reasons. First, because Flores is separated from the Sunda shelf by at least three deep sea channels, even when sea levels were lowered at times of global cooling hominins would have had to face a sea crossing of approximately 20 km/12.4 miles to reach Flores from the Sunda shelf. Second, the similarities between the artifacts at Mata Menge and those at Liang Bua (some 50 km/31 miles to the west) strengthen the case that the latter were manufactured by hominins other than modern humans. (Location 08°41'31"S, 121°05'43"E, Flores, Indonesia.)

## maternal effect
When an organism's phenotype is affected by the maternal genotype, environment, or condition. Sometimes an offspring will manifest aspects of the maternal phenotype due to maternal effects resulting from a supply of messenger RNA (mRNA) or proteins to the developing ovum. The maternal, or intrauterine, environment can affect offspring phenotype in ways such as their location in the uterus (e.g., female fetuses located between male fetuses can experience effects of androgens on morphology, behavior, and reproduction). The physiological condition of the mother can also affect her offspring [e.g., uncontrolled gestational diabetes mellitus may result in macrosomia (large size), delayed lung maturation, and problems with blood sugar regulation].

## matrix
Biology The combination of water and polymers that, along with cells, fibers, and fibrils, makes up connective tissues. The more polymerized the matrix the stiffer the connective tissue. Earth sciences The rock in which a fossil is embedded. If the matrix is very mineralized it may adhere to the surface of the bone. Extreme changes in humidity and temperature can produce cracks in fossils and when matrix enters these cracks they undergo further expansion. Matrix-filled cracks can artificially increase the size of a structure like the corpus of the mandible. Statistics When raw data, or reformulations of raw data, are presented in a series of rows and columns it is referred to as data matrix. More sophisticated matrices (e.g., correlation and covariance matrices) are used in multivariate analysis and in Euclidean distance matrix analysis. (L. *mater* = mother, in the sense of a substance from which something else originates.)

## Matuyama
See **geomagnetic polarity time scale**.

## Matuyama chron
*See* **geomagnetic polarity time scale**.

## MAU
Acronym for **minimal animal unit** (*which see*).

## Mauer
A commercial sandpit 16 km/10 miles southeast of Heidelberg, Germany. The stratified fluvial sands exposed there contain mammalian fossil fauna, but the only hominin from the site is a mandible discovered in 1907. When the mandible was found it was broken at the symphysis and some perisymphyseal bone has been lost. Most of the lower teeth are preserved in good condition. The left coronoid process is broken at the tip and the left mandibular condyle is deformed. The Mauer mandible, which dates from Marine Isotope Stage 15 (*c*.600 ka), is the holotype of *Homo heidelbergensis*. (Location 49°21′N, 08°48′E, Germany.) *See also* **Homo heidelbergensis**.

## Mauer mandible
*See* **Homo heidelbergensis**; **Mauer**.

## maxillary trigon
The name used to describe the triangular area of the midface of *Paranthropus robustus*. Its sides are the anterior pillars medially and the zygomaticomaxillary step laterally; the base of the triangle is the zygomaticoalveolar crest. There is often a hollowed area, the maxillary fossula, at the inferomedial corner of the maxillary trigon.

## maxillary visor
*See* **facial visor**.

## maximum life span
The greatest age to which a member of a group or species has lived. For modern humans, the maximum documented life span is the 122 years and 164 days recorded for Jeanne Calment. For nonhumans, the maximum life span is much harder to determine. Accurate life spans for zoo animals can typically only be obtained if the animals were born in captivity. To measure life span in the wild, an observer must be present for both the birth and death of an individual animal and must be able to identify the animal as a unique individual so that the birth and death can be linked. A further complication is that the maximum life span represents an extreme event, so it is sample-size-dependent. Given the difficulties both in determining ages at death for fossils and the widely varying sample sizes for various paleoanthropological taxa, any statements about the evolution of maximum life span are highly speculative.

## maximum likelihood
A method for estimating a statistical model's "most likely" parameters. This technique is often used to fit models of inferred phylogenetic relationships. Unlike cladistic analysis, the maximum likelihood method requires an explicit model of character evolution to be specified.

## Medieval Warm Period
See **Holocene**.

## Mediterranean climate
In Mediterranean climates, precipitation occurs mainly as winter-season rain, unlike the summer rains common in monsoonal climates. Mediterranean climates around the world (parts of California, coastal South Africa, and around the entire Mediterranean Sea) are characterized by high floral diversity and a large number of endemic species. The Mediterranean climate along the northern coast of Africa and the Near East is strongly affected by global climate patterns. Inter-annual to inter-decadal variability is influenced by the North Atlantic Oscillation and to a variable extent by the El Niño Southern Oscillation. Greenland ice core data indicate these patterns of variability persist for several hundred years.

## megadont
A term (which literally means having "very large teeth") that is used in paleoanthropology to refer to the large size of the crowns of the postcanine teeth (i.e., premolars plus molars) in *Paranthropus* and *Australopithecus*. The megadontia quotient (or MQ) was introduced to relate postcanine tooth area to estimated body mass. By analogy with the encephalization quotient, MQ = observed tooth area/$12.15 \times$ (body mass), where "observed tooth area" is the sum of the occlusal areas of the $P_4$, $M_1$, and $M_2$. Some researchers use megadont for the size of the postcanine teeth seen in *Paranthropus robustus* and hyper-megadont for the exceptionally large postcanine teeth of *Paranthropus boisei*, *Paranthropus aethiopicus*, and *Australopithecus garhi*. See also **macrodont**.

## megadontia
See **megadont**.

## megadontia quotient
See **megadont**.

## megadroughts
Droughts are periods of below-average rainfall and when such dry intervals last decades or more they are referred to as megadroughts. Drill cores from Lake Botsumtwi and Lake Malawi in East Africa indicate that periods of severe aridity between 135 and 75 ka were marked by 95% reduction in lake volume. This is notably drier than the conditions reconstructed for the Last Glacial Maximum, previously identified as a dry interval in the Late Pleistocene. The c.100 ka African megadrought would have restricted the total area available for year-round habitation by early modern humans, and may have caused a population bottleneck.

## megafauna

A term, which literally means "large animals," used for animals with an adult body mass of more than 44 kg (approximately the size of a large Labrador dog). Defined in this way, the megafauna in the Plio-Pleistocene faunas of Africa not only include elephants and rhinoceros, but also lions, Grant's gazelles, and warthogs. In the more temperate faunas of Pleistocene Europe megafauna include deer, boar, and bear. With an average body weight of $c.65$ kg, modern humans are also considered megafauna. This classification, however, illustrates the problem of this category in an evolutionary perspective; some of the earlier hominins almost certainly did not have a body mass over 44 kg, and neither would some populations of contemporary modern humans, and thus they would not qualify as megafauna. (Gk *megas* = large and L. *Fauna* = in Roman mythology the sister of the Faunus, the god of nature.)

## *Meganthropus*

A genus introduced and used informally in the early 1940s by Ralph von Koenigswald for the Sangiran D (1941) mandible. The genus name was used by other researchers (e.g., Franz Weidenreich), but the correct citation for the first use of this genus name is von Koenigswald (1950) for it was in this paper that he formally introduced the species name *Meganthropus palaeojavanicus*. *Meganthropus* fell into disuse because researchers judged it to be a junior synonym of either *Homo erectus* or *Paranthropus*. (Gk *megas* = great and *anthropos* = human being.)

## *Meganthropus africanus* Weinert, 1950

The binominal introduced by Hans Weinert for the taxon represented by the maxilla found at the site then called Garusi and now called Laetoli. Most researchers now regard *Meganthropus africanus* as a junior subjective synonym of *Australopithecus afarensis*. (Gk *megas* = great and *anthropos* = human being, and L. *africanus* = pertaining to Africa.)

## *Meganthropus palaeojavanicus*

See *Meganthropus*.

## melanin

A dark organic pigment that gives brown coloration to skin, hair, feathers, scales, eyes, or other tissues. In hominins, melanin helps protect the skin from damage by ultraviolet radiation. (L. *melan* from Gk *melas* = black.) *See also* **pigmentation**.

## Melka Kunturé

A complex of sites (e.g., Garba, Gomboré) along approximately 6 km/3.7 miles of outcrops exposed along either side of the Awash River in Ethiopia. The sites range from Early Stone Age (Oldowan) to Later Stone Age. The Melka Kunturé Formation at Gomboré is $c.1.4$–0.7 Ma, the section at Garba is >1.7–0.87 Ma, the Acheulean artifacts at Melka Garba are <1.26 Ma, and the bifaces and the hominins at Gomboré II are between $c.0.9$ and 0.7 Ma. Hominin remains from Gomboré I and II, Garba III, and Garba IV have been attributed to *Homo erectus*. The primary Oldowan localities are Gomboré I, Garba IV, Karre I, and Kella III; Gomboré I is considered a Developed Oldowan locality. Early Acheulean localities include Garba XII and XIII and Simbiro III; Gomboré II and Garba I are considered to be Middle and Late Acheulean, respectively. Garba III samples the Acheulean–Middle Stone Age transition. (Location 08°42′N, 38°34′E, Awash River, Ethiopia; alternate spellings include Melka Kunture, Melka-Kunturé, Melka Kontoure, and Melka Kontouré.)

### member
A physically distinct subdivision of a geological formation (e.g., the Okote Member of the Koobi Fora Formation).

### membranous labyrinth
*See* **bony labyrinth**.

### memory
A multifaceted faculty that includes different memory systems. There are at least two broad types of memory system: long-term memory and short-term memory. Long-term memory can be divided into explicit, or conscious, and implicit, or unconscious, memory. Explicit memory includes episodic or biographic memory, and semantic or factual memory systems. Implicit memory includes priming memory (i.e., a previous stimulus affecting the response to a later stimulus) and procedural memory (i.e., memory for performing actions). Although the term short-term memory is often used interchangeably with working memory, the two types of memory are different. Short-term memory is the storage of information for a short period of time, while working memory refers to processes used for manipulating and temporarily storing information. It is widely accepted that all mammals and many birds have both long- and short-term memory. Some have argued that certain memory subsystems such as episodic memory and working memory are unique to modern humans and are not shared with other mammals or other primates including the great apes, but this view is not widely accepted among comparative psychologists. (L. *memor* = to be mindful.)

### menarche
The time of onset of the first menstrual period in females, which is an important life history variable. Because it represents the beginning of near-monthly cycles of egg release from the ovary it also represents the beginning of fertility. The onset of menarche is influenced by a number of factors; nutritional status is thought to be particularly important, and social factors such as the presence of a father or adult male have also been proposed to play a role. Over the past 200 years, statistics show a trend toward decline in the average age at menarche in various modern human populations across the world. The mechanism behind this apparent trend is debated. It has been suggested that high levels of disease and infection associated with increased population sizes worked in the past to hold off the beginning of regular menstrual cycling and only recently, with improvements in nutrition and health care, modern human populations have started to move back to a younger age at menarche. Menarche generally occurs earlier in great apes and Old World monkeys than in modern humans, but there is some overlap among chimpanzees and modern humans (around 11 years of age). There is evidence to suggest that great apes and Old World monkeys, like modern humans, also pass through a phase of adolescent subfertility after menarche occurs. (Gk *men* = month and *arkhe* = beginning.)

## Mendelian inheritance
*See* **Mendelian laws**.

## Mendelian laws
Laws of inheritance established by Johann Gregor Mendel, an Augustinian monk who experimented with simple traits in pea plants. Mendel's work, which was initially published in 1865 and 1866, was largely ignored until the early 20thC. As a result of his experiments, Mendel came to two conclusions that are now known as Mendel's laws. The first, the law of segregation, states that during gamete formation the two alleles of a gene segregate so that each gamete receives only one allele. This law is true because of the mechanism of meiosis where the pairs of chromosomes are separated so that each gamete receives one of each type of chromosome. The second, the law of independent assortment, states that alleles from different gene loci assort independently during gamete formation. Mendel examined simple traits that were located on different chromosomes so the second law was true in all of his experiments. However, the second law is not true if two genes are located close together on the same chromosome. In this case, the alleles may not sort independently because of linkage. In general, genes on the same chromosome that are more than 50 centimorgans apart will not be linked, and thus the second law will apply to these genes. *See also* **linkage disequilibrium**.

## meningeal arteries
*See* **meningeal vessels**.

## meningeal vessels
The meningeal vessels (i.e., arteries and veins) run between the inner (aka endocranial) surface of the cranial vault and the dura mater that lines it. They either supply blood to the arteries or drain blood from the veins of both the cranial vault and the dura mater. They leave vascular impressions on the endocranial surface of the cranial vault and the pattern of these vascular markings has been explored for their taxonomic utility as well as for making inferences about the relative size of the components of the cerebral hemispheres. (Gk *meninx* = membrane.) *See also* **middle meningeal vessels**.

## meninges
The term refers individually or collectively to the three membranes (the pia mater, the arachnoid mater, and the dura mater) that cover the components of the central nervous system (i.e., the brain, the optic nerve, and the spinal cord). The innermost layer, the pia mater, adheres closely to the surface of the brain. The middle layer, the arachnoid mater, is separated from the pia by the subarachnoid space and from the outer dura by the subdural space. The outer dura mater has two layers: a fibrous inner meningeal layer and a more vascular and osteogenic outer (aka endosteal) layer. The inner meningeal layer of the dura mater has two prominent folds. In the sagittal plane, the falx cerebri and falx cerebelli separate the cerebral and cerebellar hemispheres, respectively. The approximately horizontal tentorium cerebelli separates the posterior cranial (also called the cerebellar) fossa from the rest of the cranial cavity. The dural venous sinuses (e.g., superior sagittal, transverse, sigmoid, occipital, and marginal) are

endothelium-lined spaces between the two layers of the dura. The endosteal layer of the dura is continuous at the foramina and sutures with the periosteum (aka pericranium) that covers the outer surface of the cranium. The pia and arachnoid are derived from ectoderm and are called the leptomeninges; the dura mater develops from mesoderm and is also called the pachymeninx. (Gk *meninx* = membrane.)

### menopause
The time when ovulation ceases; it marks the permanent end of female reproductive fertility. In modern human females this life history variable occurs, on average, around 50 years of age, but there is substantial within- and between-population variation. While other primates may also experience an end to fertility before death, modern humans differ because individuals outlive their fertility by a significant margin in the majority of populations. The extension of the life span beyond menopause in modern human women, compared to nonhuman primates and most other mammals, has been attributed to a number of factors (e.g., alloparenting). (Gk *men* = month and *pausis* = pause.) See also **grandmother hypothesis**.

### mental protuberance
A midline projection of bone near the base of the mandible on the external surface of the symphysis. For a mental protuberance to be classed as a true chin there must be a depression (aka *incurvatio mandibulae*) between it and the alveolar process superiorly. (L. *mentum* = chin.) See also **mandible**.

### mentum osseum
See **true chin**.

### meristic
Morphology that develops from craniocaudally or mesiodistally serially arranged segmental units. Meristic also applies to variation in the number or position of those segmental structures. The vertebral column and the dentition are examples of meristic structures. (Gk *meristos* = divided.)

### meristic variation
Variation in the number or nature of serial structures such as vertebrae or teeth. Additional vertebrae, sacralization of the fifth lumbar vertebra, additional teeth, etc., are all examples of meristic variation.

### Mesgid Dora
See **Woranso-Mille study area**.

### mesic
An environment or habitat that is moist. (Gk *mesos* = middle.)

### mesiobuccal cusp
The primary cusp on a maxillary (paracone) or mandibular (protoconid) molar.

## metacone

The main cusp distal to the paracone on a maxillary (upper) molar tooth crown. It is one of the components of the trigon. (Gk *meta* = behind or beyond and *konos* = pine cone).

## metaconid

The main mesial cusp on the lingual aspect of a mandibular (lower) (hence the suffix "-id") molar tooth crown or the lingual cusp of a bicuspid mandibular premolar. It is one of the components of the trigonid. (Gk *meta* = behind or beyond and *konos* = pine cone).

## meta-memory

The ability to reflect on one's own knowledge or memory. Studies with primates (rhesus macaques, capuchin monkeys, and the great apes) have suggested that meta-semantic-memory is shared and may be phylogenetically ancient. However, various authors have argued that meta-episodic-memory is unique to modern humans. This implies that other animals are unable to declaratively state or "re-experience" the past, an essential feature of all episodic memories.

## *Metridiochoerus*

A genus of African Suidae that underwent an adaptive radiation in the later Pliocene and Pleistocene, during which five species arose and then became extinct. A member of this genus eventually gave rise to the modern warthog, *Phacochoerus*. Because the *Metridiochoerus* radiation is relatively well dated at East African hominin sites, *Metridiochoerus* fossils have been used for biostratigraphy. (Gk *metridios* = fruitful and *choerus* = pig.)

## Mezmaiskaya Cave

A partial skeleton of a *Homo neanderthalensis* neonate less than 7 months old was found in the lowest Middle Paleolithic layer of this cave site in Russia; it included a damaged cranium and much of the upper body. This individual was the second Neanderthal to have its mitochondrial DNA examined. The size of the cranium allowed researchers to reconstruct Neanderthal life history as being as slow as that seen in modern humans. (Location 44°10′N, 40°00′E, Northern Caucasus, Russia.)

## MH1

This *c.*2.0 Ma associated skeleton of a juvenile male from Malapa in southern Africa is the holotype of *Australopithecus sediba*. The researchers who analyzed MH1 suggested that despite many primitive features [e.g., small brain size (*c.*420 cm$^3$), high brachial index, curved fingers, calcaneus more primitive than that of *Australopithecus afarensis*], it nonetheless shares cranial (e.g., more globular neurocranium, gracile face), mandibular (e.g., more vertical symphyseal profile, a weak *mentum osseum*), dental (e.g., simple canine crown and small tooth crowns), and postcranial (e.g., acetabulocristal buttress, expanded ilium, short ischium, possible evidence of a pedal arch) morphology with *Homo* taxa, both early and late. But the *Homo* features may reflect its immaturity.

## microcephaly

A spectrum of pathologies whose common denominator is that individuals have an unusually small brain for their body size. Some modern human microcephalic brains are relatively normally proportioned, but most have a disproportionately small cerebral cortex and hence a disproportionately large cerebellar cortex. Primary microcephaly has been linked with loss-of-function mutations at certain genetic loci (e.g., *ASPM* and *microcephalin* genes which show significant coding sequence changes in ape and human evolution). One interpretation of *Homo floresiensis* is that it samples a *Homo sapiens* population afflicted by a pathology that includes microcephaly. However, if *H. floresiensis* samples a relic population of archaic or transitional hominins that at one time or another had undergone insular dwarfing, then its small brain would not be considered to be pathological. (Gk *micro* = small and *kephale* = head.)

## micro-computed tomography

(or micro-CT or µCT) This method for imaging internal structures is based on the same principles as computed tomography (CT), but it uses dedicated equipment (e.g., a micro-focus X-ray tube) designed for the nondestructive testing of materials. The limit on sample size is roughly 7 cm in diameter, and the spatial resolution ranges between 5 and 50 µm. Micro-CT has been used to investigate the morphology of the enamel–dentine junction, enamel thickness, and enamel volume in single teeth. A special type of micro-CT, synchrotron radiation micro computed tomography (or SR-µCT) enables researchers to investigate microstructure of subsurface enamel. *See also* **computed tomography; synchrotron radiation micro computed tomography**.

## micro-CT

*See* **micro-computed tomography**.

## microfauna

In the context of paleoanthropology, microfauna can be used in several ways. It can refer to all small animals, all small vertebrates, or all small mammals (usually rodents and insectivores) found in a fossil faunal assemblage. Analyses of microfauna, principally small mammals (aka micromammals) that weigh less than 1000 g, have contributed significantly to studies of taphonomy, paleoenvironments, and biostratigraphy relevant to human evolution. (Gk *micro* = small and L. *Fauna* = in Roman mythology the sister of the Faunus, the god of nature.)

## microlith

Small blades or flakes that have been modified by removing even smaller flakes to form triangular or crescent-shaped pieces. One edge is typically blunted (or backed) for insertion into a prepared haft as part of a composite tool. Although abundant in some artifact assemblages over 40 ka (e.g., Enkapune ya Muto in Kenya), similarly shaped, but larger, forms dating to the Middle Pleistocene are seen in some African sites. (Gk *micro* = small and *lithos* = stone.)

## microsatellite

Microsatellites (aka short tandem repeats or STRs) are repeated sequences between 2 and 9 base pairs in length. Microsatellites occur throughout the genome and are often highly polymorphic. This makes them useful for studies of recent modern human population history, for linkage analyses, and for individual identification. (Gk *micro* = small and *satelles* = an attendant.)

## microstrain

(or µε) *See* **strain**.

## microstructure
See **enamel microstructure; incremental features**.

## microtomography
See **micro-computed tomography**.

## microwear
See **dental microwear; diet reconstruction; tooth wear**.

## midcarpal joint
See **wrist joint**.

## midden
An archeologically visible accumulation of household waste, which is usually a faithful record of daily activities. In many places, middens contain large numbers of shells, which make them alkaline and thus good preservers of organic remains.

## Middle Awash study area
The Middle Awash region is one of the three informal divisions of the part of the Awash River valley that runs through the Afar Triangle. It extends from Lake Yardi in the south towards the Hadar study area in the north. Major fossiliferous subregions within the Middle Awash include the Bouri peninsula, Central Awash complex, and Western Margin on the west side of the Awash River, and Bodo-Maka on the east side of the river.

## middle ear
The part of the ear between the external (aka outer) ear and the inner ear. It comprises a cavity and the hard- and soft-tissue structures contained within it. The cavity lies within the petrous part of the temporal bone between the medial end of the external ear (i.e., the tympanic membrane at medial end of the external auditory meatus) and the oval window of the bony labyrinth. It is about the size of three stacked 10 cent (US) or 1 pence (UK) pieces stood on end. The middle ear contains three small bones or auditory ossicles (from lateral to medial they are the malleus, incus, and stapes) that connect the tympanic membrane with the oval window, two small muscles that help dampen excessive movements of the auditory ossicles, blood vessels, and several nerves.

## Middle Ledi
See **Ledi-Geraru study area**.

## middle meningeal vessels
The middle meningeal vessels (i.e., arteries and veins) supply blood to (by way of the arteries) and drain blood from (by way of the veins) both the cranial vault and the dura mater of the middle cranial fossa. They leave vascular impressions on the endocranial surface of the cranial vault (the transmitted pulsations of the arteries are responsible for this, although it is the vein that is immediately adjacent to the bone of the cranial vault). The pattern of

these vascular markings has been explored for its taxonomic utility as well as to make inferences about the relative size of the components of the cerebral hemispheres. See also **meningeal vessels**.

## Middle Paleolithic

A stage of the European and Near Eastern Paleolithic that is defined by the dominant Mode 3 stone tool technology (i.e., the Mousterian and its regional variants). The Middle Paleolithic is usually associated with *Homo neanderthalensis* in Europe, but it has also been associated with some anatomically modern human fossils in the Near East (e.g., Skhul). The Middle Paleolithic has also been described at sites in India and the Far East, but it is not clear what hominin was responsible for the artifacts at these sites. The equivalent term for African sites with comparable technologies (but not necessarily similar ages) is the Middle Stone Age.

## Middle Pleistocene

An informal division of the Pleistocene epoch. The lower boundary of the Middle Pleistocene is the boundary between the Matuyama (C2) and Brunhes (C1) chrons at 781 ka. (Gk *pleistos* = most and *kainos* = new.)

## Middle Stone Age

(or MSA) The intermediate period in the tripartite division of the African archeological lithic record formalized by Goodwin and van Riet Lowe in their *The Stone Age Cultures of South Africa*. Although it was developed for sequences in southern Africa, the terminology has subsequently been applied to sites across sub-Saharan Africa. The MSA was initially defined by its technological and stratigraphic position between Early Stone Age and Later Stone Age sites. The distinctive features of a MSA archeological site are the absence of the handaxes and/or cleavers that distinguish Early Stone Age sites, the absence of the microliths that are typical of the Later Stone Age, and the presence of points, presumably used to make composite tools such as spears or similar hunting implements. Flake production is often by Levallois techniques, or comparable methods, and blades or elongated flakes are present in some assemblages, particularly in southern Africa. MSA artifacts are associated with the earliest fossil remains of *Homo sapiens* (e.g., Omo-Kibish). Recognizing that first and last appearance dates vary locally and are subject to change as new sites are found, existing dating methods are improved, or when new methods are introduced, the present evidence suggests the MSA began *c.*300 ka and ended *c.*30 ka.

## midfacial prognathism

See **prognathic**.

## midfoot break

Bending (or dorsiflexion) at the calcaneocuboid and cuboid-metatarsal joints during the push-off phase of walking, which is rare in modern humans but typical in nonhuman primates. This flexibility in the midfoot allows the forefoot to maintain its grip on arboreal supports while the hindfoot and other hindlimb elements propel the body during locomotion. Originally thought to occur solely at the transverse tarsal joint it is now recognized to be a more complicated foot motion involving both the transverse tarsal joint and the tarsometatarsal joints. Although a few modern humans have a midfoot break, it is usually more subtle than the motion pattern seen in nonhuman primates.

## Mid-Holocene Warm Event
See **Holocene**.

## Mid-Pleistocene transition
(or MPT) The second (*c*.1.2–0.8 Ma) of two periods during the Northern Hemisphere Glaciation (or NHG) when the rate of change in the $^{18}O/^{16}O$ ratio (a proxy for the intensity of glaciation) was particularly high. The end of the MPT (i.e., *c*.0.8 Ma) marks the beginning of the *c*.100 ka-long climate cycles.

## midtarsal break
See **midfoot break**.

## migration
See **gene flow**.

## Milankovitch cycles
See **astronomical theory**.

## Millennium Man
A term used by the media for ***Orrorin tugenensis*** (which see).

## Mindel
See **glacial cycles**.

## Mindel-Riss
See **glacial cycles**.

## mineralization
The process whereby minerals are added to a soft tissue, such as immature enamel, that results in an increase in its hardness so much so that it is classified as a hard tissue. During tooth formation, the primary organic matrix secreted by ameloblasts, cementoblasts, and odontoblasts is replaced by an inorganic crystalline lattice, which results in three of the body's hard tissues (enamel, dentine, or cementum). It is because of the highly mineralized nature of their component tissues that teeth are relatively common in the fossil record. In a sense an organism's teeth undergo the equivalent of fossilization during its lifetime. See also **calcification**; **dentine**; **enamel development**.

## minicolumn
A fundamental structural and functional unit within the cerebral cortex that is made up of a single vertical row of neurons with strong vertical interconnections among the layers. The core of a minicolumn contains the majority of the neurons, their apical dendrites, and both

myelinated and unmyelinated fibers, whereas the outer layer of each column has fewer neurons and more connections (e.g., dendrites, unmyelinated axons, and synapses). In modern humans there is bilateral asymmetry of minicolumn width in Wernicke's area (a brain area known to be important in the understanding of language that is also known as cytoarchitectonic area Tpt), with more space for neuronal interconnections in the left cerebral hemisphere. Similar analyses of minicolumns in area Tpt of chimpanzees and macaques have not revealed this asymmetry.

## minimal animal unit
(or MAU) A derived measurement used to quantify the frequencies of skeletal elements in a bone assemblage. The MAU is calculated as the minimum number of elements (or MNE) of a skeletal element divided by the number of times that element occurs in the complete skeleton (e.g., femur MAU = 0.5, rib MAU = 0.042). It is a useful measure for examining skeletal element survivorship, since elements that are rare in a skeleton may show disproportional frequencies within an archeological site. *See also* **minimum number of elements**.

## minimum number of elements
(or MNE) The minimum number of skeletal elements, or portions thereof, necessary to account for the observed specimens in a bone assemblage. When fossil bone assemblages are heavily fragmented, MNE counts are used to estimate the minimum number of bones originally deposited by hominins, carnivores, or any other depositional agent. MNE counts allow faunal analysts to compare the relative abundances of skeletal elements to those in a complete skeleton to address questions of skeletal element survivorship or differential transport by hominins.

## minimum number of individuals
(or MNI) The minimum number of individuals of a particular taxon necessary to account for the number of observed skeletal elements, or portions thereof, in a bone assemblage. It provides an estimate of the minimum number of individual animals originally deposited by any depositional agent. The MNI can be a useful measure when comparing taxonomic abundances in a faunal assemblage where the bone accumulator may have deposited some species largely intact and only certain parts of another. To some extent, MNI can also correct for differential fragmentation and for the fact that some species have a greater number of taxonomically diagnostic bones than others.

## minisatellite
A tandemly repeated sequence that is between 10 and 100 base pairs in length. Minisatellites, which are located throughout the genome, are particularly common at centromeres and telomeres. They are often polymorphic and have been used extensively for DNA fingerprinting (i.e., individual identification) and for linkage analysis. (It. *miniatura* = miniature illumination in a Medieval manuscript and L. *satelles* = an attendant.)

## Miocene
One of the five epochs that make up the Tertiary period, and the first of the two epochs that make up the Neogene period. Miocene refers to a unit of geological time (i.e., a geochronological unit) that begins 23.5 Ma at the end of the Oligocene and ends 5.2 Ma at the beginning of

the Pliocene. It is usually divided into three phases: Early, Middle, and Late. (Gk *meion* = less and *kainos* = new.)

## miombo woodland

Woodland dominated by trees belonging to the genus *Brachystegia*. Other trees growing in miombo woodland include members of the genus *Isoberlinia* and *Julbernardia*. Miombo woodlands range from areas where the trees are spaced so far apart that there are substantial areas of open tropical ($C_4$) grass between them, to patches of woodland where the canopy is more or less continuous. (Swa. *miombo* = term for the many types of trees in the genus *Brachystegia*.)

## mirror neurons

See **language, evolution of**.

## MIS

Abbreviation for Marine Isotope Stage. See **Marine Isotope Stages**.

## mismatch distribution

A histogram of the number of pairwise nucleotide differences within a sample of multiple individuals. A population that is expanding in size shows a smooth curve, while a population with a long-term constant population size shows a ragged distribution. As a population expands, the "wave" generated moves from left to right (i.e., as the differences between lineages increase with time) and the timing of the initial population expansion can be estimated. This method has been used in analyses of mitochondrial DNA as well as other loci and the results suggest that the population of anatomically modern humans expanded approximately 50 ka ago.

## missense mutation

See **nonsynonymous mutation**.

## mitochondria

The energy-producing organelles of the cell, located in the cytoplasm. Each cell typically contains several hundred mitochondria, and each mitochondrion has several copies of its own DNA. Similarities in the genetic code of mitochondria and an endosymbiotic type of Proteobacteria suggest that the former may be derived from the latter. (Gk *mitos* = thread and *khondrion* = small grain.) See also **mitochondrial DNA**.

## mitochondrial DNA

(or mtDNA) A circle of DNA about 16,570 base pairs in length. The mtDNA genetic code differs slightly from the nuclear DNA genetic code; for example, by having codons with different protein translations (e.g., in the mtDNA ATC codes for tryptophan while in nuclear DNA it is a stop codon). Also, unlike the nuclear DNA, most of the mtDNA genome is coding sequence except for some short sequences between genes and the displacement loop (also known as the D-loop or hypervariable region). The mitochondrial genome has been examined extensively to investigate modern human population history worldwide as well as nonhuman primate population history. Several important features of mtDNA make it useful for population genetic studies (e.g., it is typically maternally inherited, it does not undergo recombination, and it has a relatively fast mutation rate). Characteristic polymorphisms have been used to define mtDNA haplogroups (a group of similar mtDNA lineages). One limitation of mtDNA is that its lack of recombination means that it should be treated as a single locus (since everything is linked and passed from mother to child as a unit) in population genetic analyses. This reduces the statistical power of conclusions about population history that only use mtDNA. Although both men and women have mtDNA, only women usually pass it to their offspring, and therefore it is useful for examining female population history (including migration patterns and effective population sizes). Most of the early analyses of ancient DNA focused on mtDNA because it has a much higher copy number in each cell (>700) than nuclear DNA, making it much more likely to be recovered in samples where DNA quantity and quality are degraded. For example, several mtDNA hypervariable region sequences as well the complete mtDNA genome are available for Neanderthals. mtDNA has also been used extensively to generate primate phylogenies.

## "mitochondrial Eve"

(also called "African Eve") A concept that arose from research conducted in the 1980s. From studies of the mitochondrial DNA (or mtDNA) of current human populations, researchers concluded that all current humans can trace their matrilineal ancestry back to one common source, a modern human female who lived in Africa. Extrapolating from the mutation rate acquired from that research they estimated that the female mitochondrial common ancestor lived 290–140 ka, which they simplified to a figure of *c*.200 ka. This hypothetical female common ancestor was dubbed Eve, and the research that led to the notion of the mitochondrial Eve was immediately seen as support for the out-of-Africa hypothesis, and it was judged to be incompatible with the idea that *Homo sapiens* had evolved from regional populations of *Homo erectus*, as supporters of the multiregional hypothesis had argued. Subsequently, there has been debate over mutation rates in mtDNA and about other factors that could affect the original interpretation of the original research, but its general conclusion is still widely accepted. This research marked the beginning of a period of rapid growth in the application of molecular biology to the interpretation of recent human evolutionary history. *See also* **mitochondrial DNA**.

## mitochondrial genome

*See* **mitochondrial DNA**.

## MLD 1

This *c*.2.8 Ma partial calvaria from Makapansgat in southern Africa preserves most of the occipital bone and the posterior parts of the parietal bones (and perhaps also part of the face via MLD 6 and 23). It was made the holotype of *Australopithecus prometheus*, a taxon that is almost universally regarded as a junior synonym of *Australopithecus africanus*.

## MNE

Acronym for **minimum number of elements** (*which see*).

## MNI

Acronym for the **minimum number of individuals** (*which see*).

## MN zones
Acronym for mammal neogene. *See* **European mammal neogene**.

## mobile art
*See* **art**.

## mode
Archeology In his 1969 review of global world prehistory, Grahame Clark divided lithic technology into five categories, or modes, that characterized the dominant artifact form and the inferred underlying hominin behavior responsible for it. Modes emphasize only the most complex, or derived tool forms, and this feature has occasionally led to its use for comparing change in the archeological and hominin fossil records. Evolution George Gaylord Simpson's *Tempo and Mode in Evolution* helped push evolutionary biologists in the direction of recognizing two components of the evolutionary process: rate (or tempo) and pattern (or mode) of evolution. The various evolutionary modes include stasis, directional selection, and random drift. Researchers who have attempted to investigate the mode of evolution within the hominin clade suggest that stasis is the dominant signal. (L. *modus* = manner.) *See also individual modes*; **phyletic gradualism**; **punctuated equilibrium**; **tempo**.

## Mode 1
The first of the five technological modes defined by Grahame Clark in his review of lithic technology. Mode 1 artifacts were originally defined as "chopper-tools and flakes," but they are now widely equated with "least-effort flake production." Mode 1 assemblages are comparable to those of the Oldowan, characterized by the production of flakes from cobbles or similar naturally occurring rock forms by direct stone-on-stone percussion or the use of an anvil. Sharp flakes are the desired products, cores are generally thought to be byproducts, and there is no elaborate knapping plan.

## Mode 2
The second of the five technological modes defined by Grahame Clark in his review of lithic technology. Mode 2 artifacts were originally defined as "bifacially flaked handaxes," but it is generally considered to include cleavers as well. Mode 2 assemblages are comparable to those of the Acheulean, characterized by the production of handaxes and cleavers through bifacial flaking using stone or organic (e.g., wood or antler) hammers. Shaped cores are the desired product, and the flakes produced in the course of the shaping are considered waste. There is substantial variation, but Mode 2 forms should usually be worked on both sides (bifacial) and have a recognizable form.

## Mode 3
The third of the five technological modes defined by Grahame Clark in his review of lithic technology. Mode 3 artifacts were originally defined as "flake tools from prepared cores." The classic example is Levallois technology. Cores are carefully shaped (prepared) so that one or more standardized flakes of a predetermined shape may then be produced. Cores and the flakes produced in the course of the shaping are considered waste.

## Mode 4
The fourth of the five technological modes defined by Grahame Clark in his review of lithic technology. Mode 4 artifacts were originally defined as "punch-struck blades with steep retouch," but now it is often used to refer to blade production generally. Classically, a roughly cylindrical, conical, or wedge-shaped "prismatic" core is carefully prepared with long ridges around its circumference. Percussion above the ridges (often using a punch) produces multiple, uniform blades, which may then be retouched into a wide array of standardized tools.

## Mode 5
The fifth of the five technological modes defined by Grahame Clark in his review of lithic technology. Mode 5 artifacts were originally defined as the "microlithic components of composite artifacts." Clark associated Mode 5 with the Mesolithic, but microlithic technologies are now well known from the later Paleolithic. Microlithic tools may be produced using microblades detached from diminutive cores or snapped segments of larger blades that are retouched into standardized geometric shapes and hafted to make composite tools.

## modeling
<u>Biology, general</u> Modeling is routinely used in the experimental sciences. The aim is to build a model that best explains the observed phenomena. Well-designed experiments enable researchers to refine and validate their models by comparing what is observed with what the model predicted. The principles of modeling can also be used in the historical sciences, but there are difficulties in applying modeling in the absence of the ability to experiment. John Tooby and Irven DeVore distinguished between referential and conceptual modeling. In a referential model the real behavior of one, or more, living animal(s) is used to reconstruct the behavior of a fossil taxon (the referent). In contrast, in conceptual modeling the model is based on general principles that have been developed from observations of a wide range of animals, rather than only those closely related or analogous to the referent. In conceptual models individual taxa should be treated as sources of data points for comparative studies, not as models. For example, baboon behavior could be used as a literal referential model, or observations on baboons could be just one component of the information used to generate a conceptual model. A third category, strategic modeling, is based on the principle of uniformitarianism. The premise of strategic modeling is that species in the past were subject to the same fundamental evolutionary laws and ecological forces as species today, so that principles derived today are applicable throughout evolutionary history and although no present species will correspond precisely to any past species, the principles that produced the characteristics of living species will correspond exactly to the principles that produced the characteristics of past species; <u>Biology, bone</u> In bone biology modeling refers to any change in the mass and/or external shape of bone (e.g., during growth or as the result of excessive use). (L. *modus* = standard.)

## model I regression
*See* **regression**.

## model II regression
*See* **regression**.

## models of intermediate complexity
*See* **general circulation model; paleoclimate.**

## modern evolutionary synthesis
The modern evolutionary synthesis emerged as the dominant model of biological evolution during the 1930s and 1940s. Charles Darwin unveiled his theory of evolution through natural selection in 1859 and while most biologists came to accept the idea that species evolved over time there was disagreement over the mechanisms that caused evolution. By the beginning of the 20thC a substantial number of biologists and paleontologists supported versions of evolution that differed from Darwin's. These scientists tended to reject the idea of natural selection, and instead invoked either neo-Lamarckism, which stressed the role of the inheritance of acquired characteristics, or orthogenesis, which argued that biological properties internal to organisms caused certain groups to evolve along linear progressive lines. Reduced to its most critical components, the modern evolutionary synthesis argued that (a) evolution was the result of small genetic mutations accumulating in populations of organisms over long periods of time as well as the movement and recombination of genetic traits within populations, (b) natural selection was the primary mechanism by which species change over time, and (c) natural selection operating upon a population leads to adaptation to specific environmental conditions. Through their research, publications, and prominent institutional affiliations, Ernst Mayr, Theodosius Dobzhansky, George Gaylord Simpson, and Sewall Wright transformed modern thinking about the mechanisms underlying evolution. Members of this group began to apply the principles of the modern evolutionary synthesis to the problem of human evolution. The modern evolutionary synthesis was brought solidly into human origins research in 1950 at a symposium held at Cold Spring Harbor in New York. In a paper titled "Taxonomic categories in fossil hominids," Mayr sought to reform hominin taxonomy to reflect the new population thinking about species. At the same meeting George Gaylord Simpson criticized the persistence of orthogenesis in many interpretations of human evolution and explained how the hominin fossil record could be better interpreted from the perspective of the modern evolutionary synthesis. While there was some resistance among anthropologists and anatomists to the modern evolutionary synthesis (sometimes referred to as just "the synthesis"), by the 1980s the majority of human origins researchers had fully accepted this approach to the study of human evolution.

## modern human
When used as a noun (e.g., as in "the Herto crania more closely resemble the crania of modern humans than they do any other hominin taxon") the term modern human is equivalent to either all *Homo sapiens*, or to just extant *H. sapiens*. When used as an adjective (e.g., as in "modern human morphology") this term refers to the morphology seen in contemporary populations of *H. sapiens*. "Human morphology" (i.e., the morphology seen in taxa within the genus *Homo*) is sometimes used when the writer really means "modern human morphology." (L. *modernus* = now and *humanus* = people.) *See also* **Homo sapiens.**

## modern human behavior
The suite of behavioral patterns that characterize behaviorally modern *Homo sapiens* and that distinguish them from living primates and other members of the hominin lineage. There is intense debate concerning the nature, timing, and geographical origin of modern human

behavior. Disagreement stems in part from the difficulty of defining modern human behavior in the first place and then being able to recognize modern human behavior through the lens of the archeological record. Some behavioral patterns that are considered to be evidence of modern human behavior include symbolic behavior and personal adornment, effective exploitation of large mammals and dietary breadth, long-distance exchange networks, standardized lithic technology, and use of composite tools, among others. But there is a danger of historicism coming into play with respect to some of these lines of evidence, because archeologists tend to assume that the links between artifacts and behavior observed in the present (e.g., ochre and body adornment) were also operating in the past. As for timing, some researchers argue that modern human behavior emerged abruptly $c.50$ ka ago, creating a temporal lag between the origins of anatomically modern humans and the origin of behavioral modernity. Others suggest that the origin of modern human behavior is the result of the gradual accumulation of behavioral patterns associated with the appearance of the Middle Stone Age in Africa beginning $c.250$ ka, roughly coincident with fossil evidence for the appearance of anatomically modern humans, or *Homo sapiens*.

## Modjokerto
The old Indonesian spelling for the site now known as Mojokerto. Hence the fossils recovered from the site when the old spelling was in use retain that spelling (e.g., Modjokerto 1), as does the taxon *Homo modjokertensis*, now a junior synonym of *Homo erectus*. See also **Mojokerto**.

## Modjokerto 1
This $c.1.8$ Ma child's partial calvaria from Mojokerto is the holotype of *Homo modjokertensis*, now widely regarded as a junior synonym of *Homo erectus*. The specimen has been used to argue for a relatively primitive pattern of growth in endocranial volume in *H. erectus*.

## modularity
Modules in biology are entities within a system that are internally integrated, such as complex biological structures or pathways, that are comparatively independent from other surrounding modules with which they may interact. Some regard modules as abstract entities that reflect processes, or sets of connections within networks, while others regard them as physical entities within organisms. In the phenotype of complex organisms such as primates, there are typically many sets of integrated traits and such organisms are said to be modular. Modularization can help release evolutionary constraints, by allowing different parts of the phenotype to evolve independently. Thus, modularity is a determinant of evolvability. Modular organization also means that changes to one part of an organism or developmental system are less likely to produce deleterious effects elsewhere. (L. *modulus* = a measure in music.) See also **morphological integration**.

## module
An internally integrated unit, such as a complex biological structure or pathway, that is comparatively independent from other surrounding modules with which it may interact. Module is also used in paleoanthropology in dental metrics. See also **modularity**.

## Mojokerto
This exposed river-bed deposit in Indonesia (formerly spelled Modjokerto), 3 km/1.8 miles north of Perning, was the site of discovery of Modjokerto 1, a c.1.8 Ma child's partial calvaria that was made the holotype of *Homo modjokertensis* (now a junior synonym of *Homo erectus*). Computed tomography of the fossil's endocranial features indicates that its brain growth differed from that observed in modern humans. (Location 07°22'S, 112°38'E, Indonesia.)

## mokondo
A term used at the southern African cave sites for a sinkhole. This is an area of breccia that has been decalcified and then subsequently washed out, thus providing a cavity into which more recent breccia can accumulate. It is this type of sinkhole and new breccia accumulation that complicates the geology of sites such as Swartkrans. The cranium Stw 53 from Sterkfontein was recovered from breccia that was in a mokondo/sinkhole.

## molar
A type of postcanine tooth. In modern humans there are two deciduous and three permanent molars; the deciduous molars are replaced by the premolars. The permanent molars are the teeth distal to (i.e., behind) the premolars in the tooth row. Upper deciduous and permanent molars consist of a mesial trigon and a distal talon; the lower molar equivalents are the trigonid and the talonid. The upper deciduous molars are referred to as $dm^1$ and $dm^2$ (or $dp^1$ and $dp^2$) and the upper permanent molars as $M^1$, $M^2$, and $M^3$. The lower deciduous molars are referred to as $dm_1$ and $dm_2$ (or $dp_1$ and $dp_2$) and the lower permanent molars as $M_1$, $M_2$, and $M_3$. The molars are primarily used for chewing, crushing, and grinding food. (L. *mola* = mill stone.) See also **dental formula**; **dm**.

## molarized
A term introduced by John Robinson to describe mandibular premolars with such a well-developed talonid that they come to resemble mandibular molars. See also **talonid**.

## molecular anthropology
Developments in biochemistry and immunology during the first half of the 20thC allowed the focus of the search for better evidence about the nature of the relationships between modern humans and the great apes to be shifted from traditional, macroscopic, morphology to the morphology of molecules, particularly proteins. The earliest attempts to use the proteins of primates to determine the relationships among taxa were made just after the turn of the century, but the results of the first of a new generation of analyses were reported in the early 1960s, and Linus Pauling claimed it was he who coined the name molecular anthropology for this area of research. In 1960 Emile Zuckerkandl used enzymes to break up the hemoglobin (Hb) protein into its peptide components, and showed that when the peptides were separated (using starch gel electrophoresis) the patterns of the peptides in the gel for modern humans, gorilla, and chimpanzee were indistinguishable. Two years later, Morris Goodman used a process called immunodiffusion to study the affinities of the serum proteins of the apes, monkeys, and modern humans and he came to the conclusion that the patterns produced by the albumins of modern humans and the chimpanzee in the immunodiffusion gels were effectively identical.

Therefore it is likely that the structures of the albumins were also, for all intents and purposes, identical. In the 1970s Vince Sarich and Allan Wilson exploited these minor variations in protein structure to determine the evolutionary history of protein molecules and therefore, presumably, the evolutionary history of the taxa whose proteins had been sampled. They, too, concluded that modern humans and the African apes, in particular the chimpanzee, were very closely related. In a later paper Mary-Claire King and Allan Wilson suggested that 99% of the amino acid sequences of chimps and modern humans were identical.

The discovery, by James Watson and Francis Crick, of the structure of DNA, and the subsequent discovery by Crick and others of the genetic code, showed that it was the sequence of bases in the DNA molecule that determined the nature of the proteins manufactured within a cell. This meant that the affinities between organisms could be pursued at the level of the genome, thus potentially eliminating the need to rely on morphological proxies, be they traditional anatomy or the morphology of proteins, for information about evolutionary relatedness. The DNA within the cell is located either within the nucleus as nuclear DNA, or within the mitochondria as mitochondrial DNA (mtDNA). In the early stages of DNA research, DNA hybridization told researchers relatively little about a lot of DNA, whereas the sequencing method told them a lot about a little piece of DNA. Nowadays technological advances mean that whole genomes can be sequenced. Sequencing is favored because the knowledge about the type of differences between the base sequences provides some clues about the steps that are needed to produce the observed differences. This is because the base changes called transitions (A to G and T to C) readily switch back and forth, whereas less common transversions (A to C and T to G) are more stable and are thus more reliable indicators of genetic distance. Information from both nuclear and mtDNA suggests that modern humans and chimpanzees are more closely related to each other than either is to the gorilla. When these differences are calibrated using paleontological evidence for the split between the apes and the Old World monkeys, and if one assumes that the DNA differences are neutral, then this predicts that the hypothetical ancestor of modern humans and the chimpanzees and bonobos lived between about 5 and 8 Ma, and probably closer to 5 than to 8 Ma.

## molecular biology

The study of the molecules involved with the maintenance of life. Although the term suggests it should involve the biology of all molecules, in practice molecular biologists are mainly concerned with complex molecules (e.g., proteins). In particular they are interested in the roles played by a class of complex molecules called nucleic acids (e.g., nuclear DNA, mitochondrial DNA, RNA) that are involved in the maintenance of life.

## molecular clock

A term introduced by Emile Zuckerkandl and Linus Pauling in 1962 to refer to the use of differences between molecules to generate a tree of their relationships in which branch lengths are proportional to time. Such proportional trees can be calibrated using the fossil record and then used to estimate otherwise noncalibrated divergence times. Genetic differences between species can be estimated, among other approaches, using nucleotide sequence differences, DNA/DNA hybridization distances, amino acid sequence differences, restriction-enzyme pattern differences, and immunological distances. According to the neutral theory of molecular evolution some classes of mutations accumulate in a sufficiently clock-like manner over relevant time intervals as judged by the relative rate test, albeit in a manner that is modulated over longer time periods by factors such as generation time and effective population size. Thus, knowing the genetic distance between two species allows the estimation of the average time since the two genomes diverged. For example, paleontological evidence about the timing of the divergence between Old World and New World monkeys can be used to calibrate the molecular clock in primates and from this the divergence time of hominoids from Old World monkeys, or that of modern humans from chimpanzees, can be estimated. Given that population separation is recognized using derived features of one or both lineages, and that these features may not be present at the precise time of population separation, the fossil record will always underestimate the actual population separation time by some unknown amount.

The molecular clock is only valid if the locus evolves under selection. To use a molecular clock to estimate times of divergence, sufficient external data to calibrate the clock reliably (i.e., a well-dated fossil record) as well as relative rate tests to demonstrate rate constancy at the locus or loci being used are necessary. Recently, analyses of both the genetic data and the fossil record have frequently followed Bayesian approaches in which prior probabilities can be assigned both to a range of plausible fossil-based calibration times and to plausible variation in "local" rates of genetic change. Such approaches generate best-fit, most probable hypotheses given the data. It is now clear that rates of genetic change vary between and within tree branches and it is also rarely the case that the fossil record is dense enough to demonstrate with confidence that a divergence has not occurred, even though it can show that a divergence has occurred; hence caution needs to be used in interpreting any molecular clock estimates. *See also* **neutral theory of molecular evolution**.

## molecular evolution
Evolution involving changes in either DNA (due to one or more bases being substituted for another, or added, or deleted) or proteins (due to one or more amino acids being substituted for another).

## molecule
Two, or more, atoms held together by chemical bonds, or the smallest physical unit of a substance that demonstrates the properties of that substance (e.g., a single $H_2O$ molecule has all the properties of a larger volume of water). The term molecular biology suggests it should involve the biology of all molecules, but in practice molecular biologists are mainly concerned with complex molecules, such as proteins and nucleic acids, involved in the maintenance of life. (L. *moles* = mass.)

## moments of area
*See* **cross-sectional geometry**.

## monogenic trait
A trait caused by alleles at a single gene locus, following simple patterns of inheritance from the Mendelian laws of genetics. In modern humans monogenetic traits include wet or dry earwax and attached or unattached earlobes. Such phenotypic traits are easy to measure and

thus have been intensively studied; the single nucleotide polymorphism (or SNP) responsible for the variation in earwax phenotype was discovered in the *ABCC11* gene (NB: despite being cited as such in many textbooks, tongue rolling is *not* a monogenetic trait). (Gk *monos* = alone.)

### monophyletic
See **clade**.

### monophyletic group
See **clade**.

### monophyletic species concept
See **species**.

### monsoon
Monsoonal climates are characterized by summer-season precipitation. The West African monsoon is dominated by summer monsoonal precipitation associated with the northward migration of the intertropical convergence zone (ITCZ). Tropical Atlantic sea-surface temperatures are the primary variable controlling the strength of the West African monsoon on shorter time scales. Variations in continental heating driven by precession also influence the land/sea temperature contrast and the strength of the monsoon on longer time scales. The Indian monsoon indirectly affects climate in northeast Africa, causing drying. Prior to the uplift of the Himalayas and the Ethiopian Highlands, northeast Africa would also have had a substantial monsoon. Summer-season precipitation in monsoonal tropical climates creates favorable conditions for $C_4$ tropical grasses. (Ar. *mawsim* = seasonal, referring to the seasonal reversal of the winds.)

### Monte Carlo
A family of methods that allow for statistical tests using randomly drawn samples from a larger set of possible outcomes. The term comes from the use of random number generation in gambling. Monte Carlo methods are a subset of resampling analyses (as opposed to exact resampling analyses). (Fr. *Monte-Carlo* is one of the administrative areas of the state of Monaco, where Le Grand Casino is located.) See also **resampling**.

### Monte Christo Formation
The caves at Sterkfontein and Swartkrans are within this, the second oldest of the five geological formations recognized in the Malmani Subgroup of the dolomite uplands near to modern-day Krugersdorp in South Africa.

### Monte Circeo
A limestone promontory along the Tyrrhenian coast of Italy about 85 km/53 miles southeast of Rome that includes a number of caves known to contain Pleistocene deposits. There are over 30 caves within the mountain, but only three have produced hominin remains; from west to east they are Grotta Breuil, Grotta del Fossellone, and Grotta Guattari. In the past, the fossils found in the caves were named after the mountain itself (e.g., Circeo I, Circeo II), but more

recently workers have taken to naming the remains after the individual cave. Hence, Circeo I would now be known as Guattari I. (Location 41°14′N, 13°05′E, Italy; etym. from *Circe*, a sorceress from Homer's *Odyssey* who was said to dwell there.) *See also* **Grotta Breuil**; **Grotta Guattari**.

## Montmaurin

This site comprises a series of caves and chimneys in a limestone cliff overlooking the Seygouade River in the Pyrenees region of southern France. Several hominin dental remains were recovered, including a well-preserved mandible that has been variously interpreted as belonging to a pre-Neanderthal or *Homo heidelbergensis*, portions of a maxilla of uncertain taxonomic status, a juvenile mandibular fragment likely to be from *Homo neanderthalensis*, and a few isolated teeth. The site contains several levels, including late Acheulean assemblages associated with the earlier hominin, a typical Mousterian layer, and a Châtelperronian layer. Biostratigraphy and palynology suggest that the earliest layers (those containing the mandible) date to the Mindel-Riss interglacial (*c.*400 ka); most of the site ranges between Riss I (*c.*390 ka) and Würm I (*c.*100 ka). (Location 43°13′N, 00°36′E, France.)

## morphocline

The order of character states from the most primitive to the most derived. Morphoclines are usually established by two criteria: ontogenetic and outgroup. In the former, the earlier stages in the ontogeny of a character are assumed to be more primitive than the later stages, while in the latter closely related taxa are used as outgroups on the assumption that they are likely to display the primitive character state. For example, the ontogenetic criterion would suggest that the primitive condition for any tooth root is a single root, because all teeth with multiple roots start out as single-rooted. In contrast, the outgroup criterion suggests that the primitive condition for the root system of anterior mandibular premolar teeth within the hominin clade is the same as the most common root morphology for *Pan* (a mesiobuccal root and a plate-like distal root). Based on the outgroup analysis, there appear to be two derived morphoclines within the hominin clade. One trends towards root simplification with a single root at the derived end; the other trends towards complexity with teeth with two molar-like roots at the derived end.

## morphogenesis

The development of morphology or form during the embryonic stage. Along with pattern formation (i.e., the formation of discrete tissues along a spatiotemporal pattern) and growth, morphogenesis is one of the three major developmental processes. Morphogenesis is technically about more than just shape (i.e., the geometric features of an object that are left once location, orientation, and size have been removed). Morphogenesis is about the combination of size and shape (i.e., form) and it is thus not easily distinguishable from growth. Evolution, which has been neatly summarized as the influence of ecology on development, mainly involves minor adjustments to morphogenesis and to growth, which between them determine the shape and size of the bones and teeth that comprise the hominin fossil record. Genes that affect morphology do so via effects on the complex developmental processes and pathways subsumed within the term morphogenesis.

## morphogenetic fields

A region (or field) of cells in the developing embryo that will eventually lead to the development of an organ or structure (e.g., heart, limb). Such regions, which are under the influence of self-directed and self-contained regulation, have the potential to influence other cells they are brought into contact with so that they also develop according to the rest of the field. This phenomenon led to the idea that it was the field itself (rather than the individual cells that comprise it) that was responsible for the pattern of development leading to the eventual organ or structure. After the middle part of the 20thC the concept lost popularity largely because there were no direct tests of the concept and the burgeoning field of studying gene expression in embryos took hold. However, more sophisticated techniques have been able to demonstrate the molecular basis for the observations that led to the morphogenetic field concept. (Gk *morphe* = form and *genesis* = birth or origin.)

## morphological integration

The tendency for development to produce covariation among morphological traits. Under this definition, morphological integration is a property of the architecture of development and it is distinct from correlation or covariation observed in the phenotype. In its modern formulation, functional integration (or shared functions among traits) leads to genetic and developmental integration through natural selection. This, in turn, produces evolutionary integration, or the tendency for structures to exhibit coordinated evolutionary changes. Recent years have seen a surge in studies of integration in the literature on human evolution, where morphological integration is measured primarily via patterns of phenotypic correlation or covariation. For paleoanthropologists, understanding morphological integration is important at both practical and theoretical levels. Practically, it is necessary for structuring cladistic studies to avoid highly correlated (integrated) phenotypic traits, or for choosing extant models that share similar patterns of integration (and thereby presumably have evolved similarly) to a fossil group being studied. Theoretically, it provides important insights into the evolutionary process; the degree and pattern of integration can either constrain or facilitate the evolution of complex phenotypes, thereby having a profound impact on the manner in which morphological evolution proceeds. A simple way to think about this is through the following analogy: imagine three objects on your computer screen. You can "grab" each object with your cursor and move each independently; this is like selection acting independently on traits that are not integrated. But imagine using a grouping function to link the objects to each other, so that now they only move as a unit (i.e., they are now integrated). In grouping the objects, you have constrained the movement of each object such that they cannot move independently. So if one object becomes fixed, the others do as well. This would be like stabilizing selection acting on one trait, which would consequently constrain the evolution of the other, correlated traits. But at the same time, by selecting and moving just one of the three objects, you can move the entire group. In other words, directional selection acting to change a trait would facilitate the evolution of correlated traits, even though those traits are not themselves under the direct influence of selection. This is very different from the target and amount of selection necessary to move these three traits independently if they were not integrated, although the end product might appear the same. Obviously in a complex phenotype there are many sets of integrated traits, and in such a case an organism is said to be modular. Modularization can act to release evolutionary constraints, allowing different parts of the phenotype to evolve independently. (L. *integrare* = to make complete.) *See* **modularity**.

## morphology

The combination of the external and internal appearance of an animal, structure, or fossil, or the scientific study of form (e.g., as in "the morphological sciences"). It subsumes both gross (i.e., what is visible with the naked eye, also called macrostructure) and microscopic (what is visible only with a microscope, also called microstructure) morphology. There are two main systems for capturing and recording morphology. One uses measurements, the other records morphology by using presence/absence criteria, or by comparing a fossil's morphology with a series of standards. The former is referred to as metrical analysis, morphometrics, or just morphometry. The latter system (e.g., the numbers of cusps or roots a tooth has, or the presence or absence of a structure such as a foramen in the cranium) is called nonmetrical analysis because it relies on categorical methods for assessing the presence/absence or the degree of development of structures whose morphologies need to be compared. Morphometric methods traditionally use homologous, standardized locations called landmarks and record either the shortest distance between the landmarks (chord distance) or, if the surface between the landmarks is curved, the arc distance. The difference between the chord and arc summarizes the degree of curvature. Angles record the orientation of a structure relative to the sagittal or coronal planes, or to a reference plane such as the Frankfurt Horizontal or the orbital plane. Three-dimensional morphometric techniques can be used to record the size and shape of an object in three dimensions. The position of each reference point is recorded using a three-dimensional coordinate system, and the distances between pairs of recorded points can be recovered if needed. These types of three-dimensional data are used in a family of methods called geometric morphometrics that is being used with increasing frequency. Three-dimensional data can be captured using machines called "digitizers," or they can be captured from photographically, laser-, or computed tomography-generated images. Digitizers usually have a mobile arm with a fine, needle-like point at the end. Three-dimensional coordinate data can be manipulated using specialized software programs (e.g., Morphologika) and then converted it into a "virtual" solid object whose shape can be visualized and compared. These virtual surfaces are compared using a variety of analytical methods, some of which (e.g., thin-plate spline analysis) were developed by engineers for measuring and comparing the complex three-dimensional shapes of machinery components. (Gk *morphe* = form and *ology* = study of; syn. anatomy, phenotype.)

## morphometrics

The study of form (i.e., size plus shape) by measurement. Morphometric methods can be divided into traditional and geometric. Traditional morphometric methods typically apply bi- and multivariate analytical statistical techniques to a range of measurements, such as distances and distance ratios, angles, areas, and volumes. These methods were and still are valuable, but distance measurements do not preserve the geometric properties of the object, nor do they allow those geometric relationships to be reconstructed. Traditional morphometric data can be analyzed statistically, but if the research question is one that is purely based on geometry, then size must be removed. In the 1980s several innovations (e.g., coordinate-based analytical methods, the introduction of more sophisticated shape statistics, and the computational power to manipulate deformation grids) meant that researchers can now explore and visualize large high-dimensional data sets and take advantage of exact statistical tests based on resampling. This new approach to morphometry is referred to as geometric morphometrics because it preserves the original geometry of the measured objects during all phases of the analyses. Geometric approaches used in paleoanthropology include Procrustes analysis, Euclidean distance matrix analysis, and elliptic Fourier analysis. (Gk *morphe* = form and *metron* = measurement.) *See also* **geometric morphometrics**.

### morphospecies
A species group defined on the basis of morphology. Nearly all, or very nearly all, species founded upon fossils will of necessity be morphospecies. It is the type of species most paleoanthropologists try to recognize, but nevertheless it still leaves moot how much variation can be subsumed within a single morphospecies before a researcher should contemplate establishing another morphospecies to accommodate the "excess" variation. *See also* **species**.

### mortality
The number of deaths per unit of time in a population, scaled to the size of that population. Mortality rate is calculated as number of deaths per 1000 individuals in a given year. Low mortality rates relax the pressure on reaching reproductive age early and allow for ontogenies to proceed more slowly, especially during the juvenile period. Juvenile mortality and adult mortality each exert their own influence on fitness across a wide range of life history strategies.

### mosaic
The presence of cells with two different genotypes in the same individual. This can occur because a mutation during development is found in only a subset of the cells in the body, or when more than one fertilized zygote fuses early in embryonic development (chimerism). In primates, marmoset chimerism occurs because of the high frequency of twinning. Twins *in utero* can exchange stem cell lines by an interconnected lattice of blood vessels attached to the same placenta.

### mosaic evolution
A term introduced by Gavin de Beer to describe what William King Gregory and Robert Broom had earlier described as "palimpsest evolution." Organisms typically combine a mix of evolutionarily older and more recent traits that have evolved at different rates at different times. They thus present a combination, or mosaic, of "primitive" (plesiomorphic) and "advanced" or "derived" (apomorphic) characters. The mosaic contrasts often correspond to functionally distinct systems (masticatory, locomotor, nervous system, etc.), with those reflecting recent or current major adaptive shifts displaying the fastest rates of change and being the most derived conditions at any given time. The hominin fossil record has been cited as a particularly clear case of mosaic evolution. The term mosaic evolution is also applied to morphological changes that occur in stages as opposed to a synchronic package (e.g., different rates of brain, locomotor, or dental evolution).

## mosaic habitat
When a range of different habitat types are scattered across, or interspersed within, a given area. The paleohabitats of many Miocene, Pliocene, and Pleistocene hominin sites have been reconstructed as mosaics. These could be accurate representations of regional habitats (e.g., modern Maputaland and the Okavango Delta), but they could also be the result of "time-averaged" and "space-averaged" reconstructions that result in a false mosaic habitat signal. Modern evidence also shows that although habitats can be highly variable in a small area, they can also vary over even a short period of time, as has been seen for example at Amboseli, Kenya. (Gk *mouseion*=a picture or design made up of small components, and L. *habitare*=to inhabit.) *See also* **time-averaging**.

## most parsimonious tree
The tree (or cladogram) that requires the fewest number of evolutionary events (as recorded by character state changes, or steps) to account for the distribution of similarities among a group of taxa. The recovery of the most parsimonious tree (or trees) is the primary goal of cladistic analysis. *See also* **cladistic analysis**.

## most recent common ancestor
The species or occasionally the individual that gave rise to a group of species; the last ancestor shared among that group. Any group of species will share multiple ancestors that will have emerged at different times, but the most recent common ancestor is the one whose emergence is latest in time. Consider modern humans and orangutans. Even if we go no further back than the origin of the primates, modern humans and orangutans have at least three ancestors in common: the ancestor of all of the haplorhine primates, the ancestor of all of the hominoid primates, and the ancestor of all of the great apes. Of these ancestors, the most recent was the ancestor of all the great apes. Thus, the most recent common ancestor of modern humans and orangutans is the ancestor of all the great apes. *See also* **cladistic analysis; coalescent time**.

## motion analysis
The capture and analysis of quantitative information about movement, usually undertaken with the goal of studying functional morphology. *See also* **kinematics**.

## motor speech areas
*See* **Broca's area**.

## Mount Carmel
A mountainous region on the Mediterranean coast of modern-day Israel that contains several historic and prehistoric sites. One valley on its southwestern slope, called Wadi el-Mughara in Arabic and Nahal Me'arot in Hebrew (meaning "stream of caves" in both cases) is the location of the Middle and Late Pleistocene sites of el-Wad, Tabun, and Skhul. Another site on Mount Carmel, Kebara, has yielded an exceptionally complete postcranial skeleton of *Homo neanderthalensis*.

## Mousterian
A technocomplex characterized by the use of the Levallois technique and a high proportion of scrapers found in the European and Near Eastern record from sites dating to the Middle Paleolithic. It is associated with *Homo neanderthalensis* in Europe, but in the Near East it is

found in association with both Neanderthals (e.g., Tabun) and with anatomically modern human fossils (e.g., Skhul). François Bordes noted four distinctive variants, or facies, of the Mousterian: the Mousterian of Acheulean Tradition has distinctively small handaxes or backed knives; the Typical Mousterian is dominated by sidescrapers; the Denticulate Mousterian has a high proportion of notched or toothed flakes; and the Quina-Ferrassie (Charentian) Mousterian that was made predominantly using Levallois flakes. Bordes argued that these facies were distinct cultures within a large contemporary population. Lewis and Sally Binford argued that each tool type had a specific function, so that the different facies represented differences in behavioral demands at the various sites, not cultural differences. Harold Dibble and Nicholas Rolland take the view that the different tool types represent different points along the reduction sequence of tool making, from a raw flake to a heavily reused and retouched flake. They argue that the differences among the variants reflect differences in raw material and intensity of reduction.

## Movius' Line

A term introduced by Carleton Coon that is based on an observation made in 1948 by Hallam Movius in a seminal review entitled *The Lower Paleolithic Cultures of Southern and Eastern Asia*. Movius noted that there is a geographical demarcation between the "Hand-Axe Culture" seen in Africa, the Near East, and much of India, and the "Chopping-Tool Culture" seen in mainland and Southeast Asia. Evidence recovered since Movius' time has shown that hominins in eastern Asia could and occasionally did manufacture bifacially flaked, handaxe-like tools (e.g., Bose Basin in China and the Imjin/Hantan River Basins in Korea). However, the relative rarity of handaxes in eastern Asia is still broadly consistent with the geographic pattern noted by Movius.

## MRCA

Acronym for the most recent common ancestor of a group (e.g., molecules, taxa). *See* **coalescent time**; **most recent common ancestor**; **phylogeny**.

## MRI

Acronym for **magnetic resonance imaging** (*which see*).

## MSA

Acronym for the **Middle Stone Age** (*which see*).

## μCT

*See* **micro-computed tomography**.

## Mugharet el 'Aliya

(aka Dar el 'Aliya, The High Cave, and Tangier) A cave in Morocco containing Pleistocene sediments consisting of five beds of well-sorted windblown fine-to-medium sand derived from beaches formed beneath the cave. The site is notable for possessing stratified Aterian assemblages that date to 60–35 ka. The hominins recovered from the cave, which have been assigned to *Homo sapiens*, are of a similar age. (Location 35°45′N, 05°56′W, Morocco.)

## Mugharet el-Kebara
See **Kebara**.

## Mugharet el-Wad
See **Mount Carmel**.

## Mugharet el-Zuttiyeh
See **Zuttiyeh**.

## Mugharet es-Skhul
See **Skhul**.

## Mugharet et-Tabun
See **Tabun**.

## multipoint mapping
See **linkage**.

## multiregional hypothesis

An interpretation of the human fossil record formulated in the early 1980s by Milford Wolpoff and Alan Thorne that emphasized a phenomenon known as "regional continuity" in the fossil record of *Homo erectus* and *Homo sapiens*. The hypothesis began with their observations that a distinct morphological signature could be traced through the hominin fossil record in Australasia, but they later argued that a similar morphological continuity within geographical regions could be observed in the hominin fossil record over much of the Old World. The multiregional hypothesis argued that *H. erectus* was the first hominin to leave Africa, and that a million years ago these migrants dispersed to populate much of Asia and parts of Europe. It stressed that these regional populations of *H. erectus* encountered and adapted to quite different environments, and because of their relative geographic isolation, regional differences began to appear, yet there was sufficient gene exchange between adjacent groups to ensure that the populations continued to remain as one species. Thorne and Wolpoff's hypothesis is similar to Franz Weidenreich's polycentric theory of human evolution. According to Weidenreich, populations of early hominins living in relative geographical separation from other populations evolved into modern humans and because these populations were separated geographically racial variations emerged. Thus, there is a morphological continuity through time in any given geographical region, but Weidenreich was adamant that these populations did not evolve in complete isolation from one another because there was always an important amount of interbreeding between these populations that ensured all these geographical "races" still belonged to the same species. By the late 1990s most paleoanthropologists were interpreting the fossil and genetic evidence as incompatible with a "strong" version of the multiregional hypothesis. However, the results of recent analyses of the draft Neanderthal nuclear genome suggests that whereas modern humans from sub-Saharan Africa showed no evidence of any Neanderthal DNA, three modern humans from outside of Africa showed similar amounts (between 1 and 4%) of shared DNA with Neanderthals. These results are compatible with either a deep split within

Africa between the population that gave rise to modern humans and a second one that gave rise to present-day non-Africans plus Neanderthals, or with the hypothesis that there was hybridization between Neanderthals and modern humans soon after the latter left Africa. *See also* **assimilation model**; **candelabra model**; **out-of-Africa hypothesis**; **replacement with hybridization**.

## multivariate analysis
*See* **multivariate statistics**.

## multivariate size
Regardless of whether it is the size of a whole organism, or the size of a smaller subregion of an organism, size can be captured using one (univariate) or several (multivariate) variables. Multivariate measures of size typically reduce multiple size measurements to a single value. The four most common methods of doing so use (a) a geometric mean, (b) an equation for the size of a known geometric shape, (c) the first principal component of a principal component analysis (or PCA), or (d) centroid size. Multivariate measures of size generated through PCA are dependent on the specimens included in a sample and will change for a particular specimen depending on the other specimens included in the analysis. In contrast, measures of size using geometric means or size equations for known geometric shapes are sample-independent (i.e., the value for a particular specimen does not change, regardless of what other specimens are included in an analysis). Furthermore, the first principal component is almost always a mix of size *and* shape information. In geometric morphometrics analyses, size is usually calculated as centroid size (i.e., the square root of the sum of squared distances of a set of landmarks from their centroid). *See also* **size**.

## multivariate statistics
A type of statistical test that involves two or more variables. Multivariate statistics typically consider the interaction of multiple variables together as opposed to individual analyses of each variable. Commonly used multivariate statistical methods include multiple regression, principal components analysis, and discriminant function analysis.

## muscles of mastication
*See* **mastication, muscles of**.

## mustelid
The informal name for the Mustelidae (weasels and otters), one of the caniform families whose members are found at some hominin sites. *See also* **Carnivora**.

## mutation
One of the four forces of evolution and the only force that creates new variation. Mutations are caused by an error in DNA replication or through damage to the DNA. There are several types of mutation: (a) point mutations involve a change from one nucleotide to another, (b) inversions occur where a sequence is inverted so that the 5′ and 3′ ends are switched, (c) translocations or transpositions are the terms used when a sequence of nucleotides is moved to another location,

and (d) insertion/deletion (or indel) mutations are where one or more nucleotides are deleted from or added to a DNA sequence. In coding regions, types (b), (c), and (d) may cause a frameshift in the reading of the nucleotide sequence and frameshift mutations are typically highly deleterious. Point mutations can be transitions, where one purine (adenine or guanine) is replaced by another purine or where one pyrimidine (cytocine or thymine) is replaced by another pyrimidine. They can also be transversions, where a purine is replaced by a pyrimidine (or vice versa); in modern humans transitions are more common than transversions. Point mutations in coding regions may be synonymous (also called silent), when they do not cause a change in the protein sequence, or nonsynonymous. Nonsynonymous point mutations are called nonsense mutations when they involve a change to a stop codon, or missense mutations when a different amino acid is coded for. Another class of mutations involves the number of chromosomes. Mutations that result in changes in the numbers of the whole complement of chromosomes are not viable in humans, but other changes in chromosome number are possible. Chromosomal aberrations such as centric fusion can lead to the evolution of chromosome number and organization. For instance, sometime during hominin evolution, the chromosomes 2a and 2b found in the other great apes fused to form the single modern human chromosome 2.

### mutation rate
The occurrence of a mutation per generation or unit of time. Mutations are stochastic events but over longer periods of time they occur at a measurable rate. This rate differs among regions of the genome (for example, the mutation rate is faster in mitochondrial DNA than in nuclear DNA) and among species. Reasons for differences in rate can include the guanine/cytosine content of the DNA, differences in editing capability of the polymerase that replicates the DNA, and secondary structure of DNA in a specific region.

### m.y.
See **myr**.

### mylohyoid line
See **mandible**.

### myr
The abbreviation for a period of time expressed in millions of years, as in "5 myr have elapsed since hominins first appeared in Africa." The result of an age determination, or age estimate, should be expressed differently, as in "the approximate age of KNM-ER 3733 is 1.8 Ma." *See also* **age estimate**.

# N

## ¹⁵N/¹⁴N

The relative ratio of two stable isotopes of nitrogen ($^{15}N/^{14}N$ in the ‰ notation, or $\delta^{15}N$ values in parts per million) in fossil collagen reflects the foods an organism has consumed and where it consumed them. This is because nitrogen isotope ratios in animal tissues reflect those in the plants that they consume (e.g., $\delta^{15}N$ values in herbivores are about 3‰ higher than in the plants they consume, and $\delta^{15}N$ values in secondary carnivores are about 3‰ higher than those in herbivores: these differences reflect a phenomenon known as trophic enrichment). Meat intake should be reflected by higher values of bone collagen $\delta^{15}N$ than those found in plants and modern human diets that include marine foods have higher $\delta^{15}N$ values than those with only terrestrial foods. The $^{15}N/^{14}N$ and $^{13}C/^{12}C$ systems can also be used to investigate the timing of weaning. It has been suggested that the enriched levels of $^{15}N$ in *Homo neanderthalensis* indicate a diet that included a significant component of carnivory, but this type of profile is also consistent with plant foods.

## Nanjing Man
See **Tangshan Huludong**.

## Nariokotome
A complex of fossil localities in the Kaitio and Natoo Members of the Nachukui Formation in West Turkana, Kenya. An early hominin associated skeleton, KNM-WT 15000, attributed to *Homo erectus*, was recovered from locality Nariokotome III. See also **KNM-WT 15000**; **West Turkana**.

## "Nariokotome boy"
See **KNM-WT 15000**.

## Narmada
See **Hathnora**.

**narrow-sense heritability**

($h$ or $h^2$) Narrow-sense heritability is the portion of phenotypic variation that is additive (or allelic) (i.e., the potential influences of epistatis, dominance, or parental effects are ignored). It is the part of variation that natural selection can act on. See also **heritability**.

**nasoalveolar clivus**

The bone surface between the inferior margin of the anterior nasal (aka piriform) aperture and the alveolar process. In modern humans and most pre-modern *Homo* a distinct sill separates the nasal floor from the nasoalveolar clivus, but there is much less of a distinction within and among the great apes. The distinction also varies within and between archaic hominin taxa (e.g., in *Australopithecus africanus* the nasoalveolar clivus and the inferior part of the anterior pillars are in the same plane and form a unitary structure, the nasoalveolar triangular frame, which stands out from the rest of the face).

**nasoalveolar gutter**

The depressed upper part of the nasoalveolar clivus in *Paranthropus robustus* and *Paranthropus boisei*. In *P. robustus* the nasoalveolar clivus is scooped out and the anterior nasal (aka piriform) aperture is set below the surface of the face between the inferior ends of the anterior pillars. In *P. boisei* a similar appearance is due to the plate-like infraorbital region (maxillary visor) extending anteriorly ahead of the plane of the anterior nasal aperture. See also **facial visor**.

**nasoalveolar triangular frame**

See **nasoalveolar clivus**.

**nasopharynx**

See **pharynx**.

**natal group**

See **philopatry**.

**natural selection**

The process of differential survival and reproduction of genotypes. Natural selection acts on the phenotype, and in doing so affects the inherited alleles such that it maintains favorable alleles and genotypes in the population and removes deleterious alleles and genotypes. It promotes increased frequency of the most advantageous alleles and genotypes in a particular environment, and inevitably reduces the frequencies of other alleles and genotypes. It is one of the four forces of evolution (along with mutation, genetic drift, and gene flow). There are several different types of natural selection. These include directional selection (which can be positive or negative), balancing selection, diversifying selection, purifying selection, and stabilizing selection. Ultimately, all forms of natural selection result in differential birth and/or death rates for individuals with different fitness levels.

**Nazlet Khater**

A 40–35 ka underground chert-mining site in modern Egypt that has yielded one of the very few modern human skeletons from this time period in this region of Africa. (Location 26°47′N, 31°21′E, Egypt.)

## Ndutu

A c.400 ka site best known for the Ndutu 1 cranium. The latter provides a key morphological link in Africa between the crania of *Homo erectus* and those of *Homo heidelbergensis*. (Location 03°00'S, 35°00'E, Tanzania.) *See also* **Homo heidelbergensis**.

## $N_e$

*See* **effective population size**.

## Neandertal

*See* **Homo neanderthalensis**; **Kleine Feldhofer Grotte**; **Neanderthal**.

## Neanderthal

The taxon *Homo neanderthalensis*, or any specimens belonging to that taxon, should be referred to informally as "Neanderthal." Some researchers use "Neandertal" to reflect changes to the German language that included dropping the silent h so that "-thal" became "-tal." However, according to the International Code for Zoological Nomenclature the Linnaean binominal *Homo neanderthalensis* remains unchanged, so the old spelling should be used.

## Neanderthal 1

The holotype of *Homo neanderthalensis* (*which see*). *See also* **Kleine Feldhofer Grotte**.

## Neanderthaloid

An outdated term that was used in two distinct senses. A minority of researchers applied it to Middle Pleistocene European specimens (e.g., Mauer, Steinheim, Montmaurin, and Fontéchevade) that displayed one, or more, of the features characteristic of *Homo neanderthalensis* and were thus considered potential Neanderthal ancestors (i.e., "fossils resembling Neanderthals"). Used in this sense it is broadly equivalent to the terms "pre-Neanderthal" or "ante-Neanderthal" as used by some French workers. The second sense in which Neanderthaloid has been used is in connection with African and Asian fossil specimens (e.g., Jebel Irhoud, Kabwe, Elandsfontein, and Maba) that resembled, but were distinct from, *H. neanderthalensis*. Under the influence of the Neanderthal phase hypothesis "Neanderthaloid" acquired a phyletic connotation, implying that specimens so described represented the African and Asian *evolutionary* equivalents of European and Near-Eastern Neanderthals. "Neanderthaloid" was in common usage from the 1940s to the 1970s but recognition of the importance of focusing on shared, derived, characters when trying to delimit taxonomic groups has led to the realization that the term "Neanderthaloid" is nothing more than an ill-defined grade of fossil hominins and its use has now been largely abandoned. (Ge. *Neanderthal* = the valley of the river Neander and Gk *oeides* = resembling, as in having the shape or form of Neanderthals.)

## Neanderthal phase hypothesis

This hypothesis (aka the Neanderthal phase model) influenced interpretations of human evolution for much of the 20thC. It suggested that modern humans had evolved from an ape-like ancestor through several evolutionary stages, the most recent of which was represented by *Homo neanderthalensis*. Thus, supporters of a Neanderthal phase of human evolution believed that at least some Neanderthal populations evolved into modern *Homo sapiens*.

**nearly neutral theory of molecular evolution**
A modification of the neutral theory of molecular evolution that suggests mutations that are slightly disadvantageous or slightly advantageous may evolve like strictly neutral mutations. *See also* **molecular clock**; **neutral theory of molecular evolution**.

**negative allometry**
A relative size relationship in which part of an organism, or a variable that functions as a proxy for part of an organism, increases in size at a slower rate than the overall size of the organism or a variable that functions as a proxy for the whole of an organism (i.e., the variable becomes proportionally smaller as overall body size increases). *See also* **allometry**; **scaling**.

**negative assortment**
Negative assortative mating occurs when individuals mate with other individuals that are different from themselves at a higher frequency than would be expected by chance (i.e., when "like" mates with "unlike"). Red hair in modern humans shows moderate negative assortative mating, since it appears that red-haired people preferentially choose partners that do not have red hair.

**neocortex**
The phylogenetically newest part of the cerebral cortex. It comprises all of the cerebral cortex apart from the cortical areas that are linked with olfaction (piriform cortex) and memory and spatial navigation (hippocampus). Neocortex is present only in mammals, although homologous structures, which share certain features of molecular expression and connectivity with the neocortex, have been proposed to exist in reptiles and birds. The most striking and distinguishing architectural trait of the neocortex in almost all mammals is that its neurons are arranged in six well-defined layers. Although this six-layer structure is characteristic of the entire neocortex, the thickness of individual layers and the types of neurons within them vary in different functional regions of the cortex. The characteristic pattern of layering in different cortical areas (aka cytoarchitecture) was used by Korbinian Brodmann to organize the modern human cerebral cortex into about 50 numbered areas. The modern human neocortex, which is approximately four times larger than the neocortex of the living great apes, is larger than would be predicted by the observed allometric relationships within living hominoids. (Gk *neos* = new and L. *cortex* = bark; i.e., outer covering.)

**neocortex ratio**
The ratio of the volume of the neocortex to the volume of the rest of the brain. Researchers have proposed that this metric is correlated across primate species with variables such as social group size and grooming clique size, which might reflect the complexity of social cognition. They have also suggested that a major force driving the evolution of language was the need to track the status of the individuals in a relatively large social network. In this scenario, spoken language evolved as a replacement for grooming as a more efficient mechanism for maintaining social connections and exchanging information among a large numbers of individuals. *See also* **language, evolution of**.

## Neogene

The middle of the three periods that comprise the Cenozoic era. The Neogene period, which comprises two epochs (the Miocene and Pliocene), refers to a unit of geological time (i.e., a geochronological unit) between c.23.5 to either 2.58 Ma (Gelasian) or 1.8 Ma (Calabrian). Some geologists disagree with the decision to elevate the Quaternary to period status and thus they still include the Pleistocene and the Holocene epochs in the Neogene. In this case, the Neogene would span the interval of time from c.23.5 Ma to the present. (Gk *neos* = new and *genos* = birth.) *See also* **Plio-Pleistocene boundary**; **Tertiary**.

## neo-Lamarckism

*See* **Lamarckism**.

## neonatal line

An accentuated line in the enamel and dentine of deciduous teeth and in the first permanent molars of higher primates caused by the stress of birth interfering with enamel and dentine development. *See also* **enamel development**.

## neonate

Literally a "newborn," but in pediatrics the term is used for infants from birth to 4 weeks old. (Gk *neos* = new and L. *natus* = to be born.)

## neontological species

One of the many terms used to describe a contemporary species defined on the basis of criteria such as evidence of interbreeding or the lack of any evidence of interbreeding between individuals belonging to that species and individuals belonging to any other species. (Gk *neos* = new, *onto* = being, and *ology* = the study of, thus a "new," or contemporary, species; syn. **biological species**; **extant species**.) *See also* **species**.

## neoteny

A heterochronic process in which ancestral ontogenetic trajectories are dissociated so that shape change is retarded for a given size in the descendant relative to the ancestor resulting in a juvenilized or pedomorphic morphology. Neoteny has a history of being invoked as a particularly important heterochronic process involved in human evolution. Features such as large brains, short and flat faces, and hairlessness have been interpreted as the result of juvenilization via uncoupling of ancestral ontogenetic trajectories. However, modern human development is prolonged relative to that of apes so this runs counter to the requirements of neoteny. Neoteny does provide a viable explanation for the juvenilized skull and face in bonobos relative to common chimpanzees. (Gk *neos* = new plus *teinein* = to extend.) *See also* **acceleration**; **heterochrony**; **pedomorphosis**; **peramorphosis**.

## neurocranium

The part of the cranium that surrounds all but the anterior aspect of the brain, where the brain is in contact with the viscerocranium. It comprises the cranial vault (aka calvaria) that covers the top, back, and sides of the brain, and the basicranium that lies beneath the brain. (Gk *neuron* = nerve and *kranion* = brain case.)

## neuron

A class of cells in the nervous system. Neurons are specialized cells that transmit information to other cells. They have shorter-branching projections called dendrites that receive transmissions, and a long process called an axon that sends transmissions. Within the neuron, information is conducted through the cell with electrical signals (i.e., generated by movement of ions across the neuron cell membrane). At the synapse, or space between two neurons, the presynaptic neuron releases a neurotransmitter, or chemical messenger, into the synapse, which binds with receptors on the postsynaptic neuron's dendrites and begins the process of electrical conduction in the postsynaptic neuron. Some types of neuron in the cerebral cortex (e.g., pyramidal cells of the prefrontal cortex and the von Economo neurons of the anterior cingulate and frontoinsular cortex) have been linked with behavioral and cognitive specializations seen in the great apes and modern humans. (Gk *neuron* = a nerve cell with an appendage.)

## neutral allele

An allele that has a selection coefficient of zero ($s=0$) is neither advantageous nor disadvantageous to an individual's fitness. Any change in the frequency of such alleles is due to random genetic drift, gene flow, or mutation.

## neutral mutation

A mutation in the DNA code that has a selection coefficient of zero ($s=0$), which reflects that this change is neither advantageous nor disadvantageous to an individual's fitness. Any change in the frequency of such mutations is due to random genetic drift or gene flow.

## neutral polymorphism

*See* **neutral allele**.

## neutral theory of molecular evolution

A theory suggesting that the vast majority of molecular evolution involves the replacement of one neutral mutation by another neutral mutation. *See also* **molecular clock**.

## Ngandong

An open-air, river terrace site on the north bank of the Solo River in East Java best known for the hominin crania (more than 11) and two tibiae recovered from the site. The age of Ngandong has been, and still is, controversial. Startlingly recent ages (c.46–27 ka) have been met with considerable resistance from scholars who claim that the fossil hominins were originally from older deposits and, after sediment reworking, later came into close proximity with the more recently dated mammal teeth. They also claim that these young ages for the deposit do not correspond with the well-established faunal biostratigraphy of the region. The Ngandong hominins display more apparently derived features than does the collection of crania from the earlier Bapang-AG levels. Originally attributed to a new species, *Javanthropus soloensis*, the taxonomic status of these specimens is unclear. Some researchers consider them a subspecies of *Homo sapiens* while others suggest they should be subsumed into *Homo erectus*. (Location 07°18′S, 111°39′E, Indonesia; etym. named after a local village.) *See also* **Homo (Javanthropus) soloensis**.

### Ngrejeng

A fossil hominin site within the Sangiran Dome, Indonesia. Recent radiometric age estimates range between 1.51 and 1.24 Ma. The only hominin recovered, NG 8503, a juvenile mandibular corpus with an $M_1$ plus an unerupted $M_2$, has been referred to *Homo erectus*. (Location 07°20′S, 110°58′E, Indonesia.)

### Niah cave

Niah cave (or the Great Cave of Niah) is located on the coastal plain of Sarawak, northern Borneo, approximately 15 km/9.3 miles from the South China Sea. The so-called "Deep Skull," a modern human cranium found in the Late Pleistocene deposits, is between 46 and 34 ka. Its morphology is most similar to those of modern day Tasmanian and Negroid populations. An almost complete left femur, a right proximal tibia, and a talus likely belong to the same individual. (Location 03°49′09″N, 113°46′42″E, Sarawak, Borneo Island, Malaysia.)

### niche

The physically distinctive space, or habitat, occupied by a species. It can also mean the ecological role of a species. (L. *nidus* = nest.)

### niche construction

A term introduced to describe biotic environmental modification. Organisms adapt to their environment but they are also capable of modifying the environment (e.g., beavers regularly create ponds by damming streams). This organism-driven environmental modification is seldom as dramatic or substantial as the environmental modification that is brought about by tectonism or climate change, but it can be significant (e.g., modern humans significantly change their local habitats by agriculture, mining, construction, and other efforts). It is unclear at what stage in their evolution modern humans shifted from adapting to existing environments to being substantial modifiers of the environment.

### Nihewan Basin

The more than 12 Early and Middle Pleistocene sites identified in the eastern part of the Nihewan Basin are among the oldest evidence of hominins in northern China. In the western part of the basin the best-known Early Pleistocene sites are Goudi/Majuangou III (they are the same site), Xiaochangliang, and Donggutuo; Xujiayao is a major Middle–Late Pleistocene open-air site. Magnetostratigraphy suggests Goudi/Majuangou III is *c*.1.66 Ma, Xiaochangliang is *c*.1.36 Ma, and Donggutuo *c*.1.1 Ma. The more than 20,000 artifacts from sites in the Nihewan Basin are primarily small core and flake tools produced on locally abundant, diverse rock types (e.g., fine-grained silicified quartzite). One of the defining characteristics of the Nihewan Basin is the overall poor flaking quality of the stone raw materials. Many of the Nihewan core and flake tools indicate flaking techniques and capabilities similar to those in Oldowan assemblages in Africa. (Location approximately 41°13′N, 114°40′E, Shanxi and Hubei Provinces and Inner Mongolia Autonomous Region, northern China.)

### NISP

Acronym for **number of identified specimens** (*which see*).

## nomen
It refers to any scientific name at any level (e.g., Primates, Hominidae, Hominini, *Australopithecus, Australopithecus africanus*). It refers to the name itself, not to any taxon that may be designated by the name. (L. *nomen* = name; pl. nomina.)

## nomenclature
The principles and practice of providing names for taxa. The process of naming a new group is controlled by rules and recommendations set out in the International Code of Zoological Nomenclature, otherwise known as the "Code." The purpose of the rules and recommendations set out in the Code is to make sure the names given to new taxa do not duplicate those of existing taxa (i.e., homonymy) and to prevent two or more names being given to the same taxon (i.e., synonymy). It also describes procedures that are designed to avoid new taxa being given inappropriate names. (L. *nomenclatura* = calling by name; e.g., calling the role.) *See also* **International Code of Zoological Nomenclature**.

## nonindependence
*See* **phylogenetically independent contrasts**.

## nonmetrical traits
Some "difficult to measure" morphology is more amenable to nonmetrical analysis. Most skeletal nonmetrical traits (aka discrete variables, epigenetic traits) are craniodental, like many of the examples given below, but they can also be postcranial. Nonmetrical skeletal traits include additional bones called ossicles (e.g., bregmatic, asterionic), bony tubercles (e.g., marginal, pharyngeal), bony ridges or tori (e.g., angular, mandibular, palatine), bony processes (e.g., paracondylar process, third trochanter), bony connections called "bridges" (e.g., divided hypoglossal canal, mylohyoid bridge), fissures (e.g., squamotympanic, mastoid), sutures (e.g., infraorbital, metopic), foramina (e.g., mastoid, supraorbital), and additional cusps, ridges, and fissures on the enamel of tooth crowns. In some cases these traits can be recorded (also called scoring) as either present or absent, while in other cases they may be given a nominal value by comparing the morphology observed with pre-defined categories of expression of that feature (e.g., small/medium/large or well-expressed, expressed, weakly expressed, absent, etc.). *See also* **epigenetics**.

## nonparametric statistics
Statistical tests that do not assume data follow any particular distribution (e.g., the normal distribution) (e.g., Mann–Whitney U test, Wilcoxon signed rank test). Resampling techniques such as the bootstrap and randomization are also nonparametric. *See also* **parametric statistics**.

## nonrandom mating
When mating is affected by physical, genetic, or social preference or by a barrier (e.g., a geographic barrier such as a large river). Within a population, nonrandom (aka assortative) mating results in increased inbreeding and random genetic drift. In modern humans cultural factors (e.g., language, education, and religion) and physical characteristics (e.g., height and skin color) are known to affect mating patterns.

## nonrandom sampling
*See* **ascertainment bias; bias.**

## nonsense mutation
A mutation that changes a codon sequence from one that encodes an amino acid to one that encodes a stop codon and thus results in a truncated protein sequence. Such mutations are typically deleterious. *See also* **genetic code; mutation.**

## nonsynonymous mutation
A point mutation (i.e., change in one nucleotide) in a coding sequence that results in a change in the amino acid sequence of a protein. Most changes of this sort are deleterious and are subject to purifying selection, but a small fraction are adaptive and subject to positive selection. For example, in the modern human lineage there have been two nonsynonymous mutations in the *FOXP2* gene, which encodes FOXP2, a transcription factor. This new amino acid sequence is highly conserved in evolution and in modern humans changes in the *FOXP2* gene have been linked to problems with speech and grammar. Researchers have shown that this gene has been subject to recent positive selection. *See also* ***FOXP2;* genetic code; mutation.**

## nonsynonymous substitution
*See* **nonsynonymous mutation.**

## *norma basalis*
*See* *norma basilaris*.

## *norma basilaris*
One of the standard views of the cranium. It is the inferior view, the one that looks at the cranium from its underside, or base. In modern humans it mainly comprises the inferior surface of the hard palate and the undersides of the sphenoid, temporal, and occipital bones. (L. *normal* = made according to a pattern, or standard, and basilar is a neologism introduced by translators that presumably comes from *basis* = base or pedestal.)

## *norma frontalis*
One of the standard views of the cranium. It is the anterior view, the one that looks at the forehead and face. In modern humans it comprises the anterior surface of the face, and the anterior surfaces of the squamous part of the frontal bone and the parietal bones. (L. *normal* = made according to a pattern, or standard, and *frons* = brow or forehead.)

## normal
The state of the Earth's magnetic field when the needle of a magnet points to the North Pole. *See also* **geomagnetic polarity time scale.**

## *norma lateralis*
One of the standard views of the cranium. It is the view from the side. In modern humans it comprises the lateral surfaces of the frontal, temporal, parietal, and occipital bones. (L. *normal* = made according to a pattern, or standard, and *latus* = side or flank.)

## normal distribution

A probability distribution characterized by data falling symmetrically about the mean and with most data falling relatively close to the mean so that observations become less frequent as distance from the mean increases. When considering the dispersion of a normal distribution, about 68% of values fall within one standard deviation of the mean, about 95% of values fall within two standard deviations of the mean, and more than 99% of values fall within three standard deviations of the mean. In a normal distribution, the mean, median, and mode are equivalent. Most traditional statistical tests assume that data are normally distributed; these tests come from a branch of statistics referred to as parametric statistics. When data are not normally distributed, parametric statistical tests are at a greater risk of Type I error and nonparametric statistics should be used instead. (syn. Gaussian distribution, also called a "bell curve.") *See also* **nonparametric statistics**; **parametric statistics**.

## normal fault

*See* **fault**.

## normalizing selection

*See* **natural selection**.

## normal views

Standard views of the cranium (i.e., *norma verticalis*, superior view; *norma frontalis*, anterior view; *norma lateralis*, right or left lateral view; *norma occipitalis*, posterior view; *norma basalis*, inferior view). (L. *normal* = made according to a pattern, or standard.) *See also* **norma basilaris**; **norma frontalis**; **norma lateralis**; **norma occipitalis**; **norma verticalis**.

## norma occipitalis

One of the standard views of the cranium. It is the posterior view. In modern humans it comprises the posterior surfaces of the parietal and occipital bones. (L. *normal* = made according to a pattern, or standard, and *occipio* = to begin or commence.)

## norma verticalis

One of the standard views of the cranium. It is the view from above (i.e., looking down onto the vertex, or the highest point, of the cranium. In modern humans it comprises the superior surfaces of the frontal, parietal, and occipital bones. (L. *normal* = made according to a pattern, or standard, and *vertex* = highest point.)

## norm of reaction

*See* **reaction norm**.

## Northern Hemisphere Glaciation

(aka NHG, or the Late Pliocene transition) Refers to the period when major ice sheets covered many northern landmasses including North America, Greenland, and much of Asia. The onset of the Northern Hemisphere Glaciation was between 3.2 and 2.7 Ma.

## nuchal
The area at the back of the neck where you can feel the nuchal muscles. Used also for the nuchal area on the underside of the basicranium posterior to the foramen magnum where the nuchal muscles are attached. (ME *nucha* = nape, or back, of the neck.)

## nuchal crest
A bony crest caused by the attachment of the fascia covering the semispinalis capitis, a major nuchal muscle. Such crests are common in large male gorillas, but they are also seen in some hominin taxa (e.g., *Australopithecus afarensis* and in large, presumed male, hyper-megadont archaic hominins such as OH 5).

## nuchal muscles
The muscles (e.g., trapezius and sternocleidomastoid covering the semispinalis and splenius capitis) you can see and feel at the back of the neck.

## nuchal plane
The surface of the lower or inferior part of the squamous occipital bone. The superior nuchal lines demarcate the nuchal plane below and the occipital plane above, which is the upper part (aka scale) of the squamous occipital bone. The external occipital protuberance (aka inion) marks the midpoint between the two superior nuchal lines. (syn. lower scale of the occipital bone.)

## nuclear DNA
The DNA within the nucleus of eukaryotic cells. In most eukaryotes nuclear DNA, which in a modern human diploid cell consists of roughly 6.4 billion base pairs, forms the majority of the genome. The other component of the genome of eukaryotes is found within organelles in the cytoplasm called mitochondria (i.e., mitochondrial DNA). *See also* **DNA; mitochondrial DNA**.

## nucleotide
Components of DNA and RNA. They are composed of a nitrogenous base (either a purine or a pyrimidine), a five-carbon sugar (ribose or deoxyribose), and one or more phosphate groups.

## nucleus
Cell biology Nucleus refers to the membrane-bound organelle in a nucleated, or eukaryotic, cell that contains DNA (known as nuclear DNA). The transcription of DNA occurs *inside* the nucleus, whereas the process of translation occurs *outside* the nucleus in the cytoplasm. Neuroscience Nucleus refers to either a geographically localized group of neurons that share the same morphology and/or function (e.g., the caudate nucleus, or nuclei of the cranial nerves) or to a group of such nuclei at the base of the forebrain (i.e., the basal ganglia, aka basal nuclei). (L. *nux* = nut.)

## null hypothesis
A statement to be tested, which is typically framed as the absence of a difference between groups or the absence of a relationship between variables. The $P$ value is the probability that the observed pattern in the data could occur if the null hypothesis were true. If the $P$ value is

below some threshold (usually 0.05) then the test is said to show a statistically significant result and the null hypothesis is rejected. Most statistical tests rely on null hypothesis testing, but increasingly researchers use Bayesian methods to help choose among several competing hypotheses.

## number of identified specimens
(or NISP) A measure that reflects the number of identified specimens (i.e., a skeletal element, or portion thereof) of any taxon represented in a bone assemblage. The NISP serves as the baseline from which all other related quantitative units (e.g., minimum number of elements, minimal animal units, and minimum number of individuals) are derived. NISP counts are used to compare taxonomic abundances through time or among fossil localities. *See also* **minimal animal unit; minimum number of elements; minimum number of individuals**.

## numerical-age methods
*See* **geochronology**.

### $^{18}O/^{16}O$ ratio

Stable oxygen isotopes that can be measured in animal proteins (e.g., collagen) and in carbonate or phosphate present in shell, bone, or enamel apatite. The ratio of these oxygen isotopes reflects the isotopic ratio of the water ingested by an animal either as drinking water or in their diet. Oxygen isotope ratios in plants reflect their water source and the relative humidity of the environment in which they are growing. Plants growing in drier environments (and animals consuming these plants) have higher $^{18}O/^{16}O$ ratios (this is the ‰ notation; also commonly represented as $\delta^{18}O$ values in parts per million) than those growing in wetter environments, and leaves tend to have higher $\delta^{18}O$ values than fruits. As a result $\delta^{18}O$ can distinguish frugivores from folivores, and animals that drink infrequently (which have higher $\delta^{18}O$ values) from those that drink frequently (which have lower $\delta^{18}O$ values). Oxygen isotope ratios can also be used to interpret paleoclimate. The lighter $^{16}O$ evaporates more readily than $^{18}O$ from large bodies of water, thus rain and rain-derived terrestrial water sources have lower $\delta^{18}O$ values compared to ocean water. During colder periods the $^{16}O$-rich water that evaporates from oceans tends to be trapped on land as ice, leaving ocean and lake water relatively enriched in $^{18}O$. Thus, animals ingesting water during cooler glacial periods have greater $\delta^{18}O$ values than animals living during warmer interglacial periods. Global shifts in $\delta^{18}O$ values associated with periodic glacial/interglacial cycling were used to create the Marine Isotope Stages, which are used to calibrate sites in higher latitudes. See also **Marine Isotope Stages; stable isotopes**.

### objective synonym
See **synonym**.

### obligate bipedalism
See **bipedal**.

### obliquity
A measure of the angle between the Earth's rotational axis and a line that is perpendicular to its orbit. Obliquity varies on a *c.*41 ka cycle, and is one of the three rhythms or cycles that affect the Earth's orbital geometry and therefore its climate. It interacts with the *c.*23 ka precessional cycle to take the poles of the Earth either closer to or farther away from the sun. Higher angles of obliquity increase the effects of the seasons in polar and temperate regions, making the

differences between summer and winter much larger. Obliquity was the dominant effect on global climate between 3 and 1 Ma. *See also* **astronomical time scale; eccentricity; precession.**

## obsidian
A typically black, or dark-colored, silica-rich volcanic glass. Because it has no fixed planes of fracture the way that many other minerals do, it is possible to predict and control how obsidian fractures when it is hit, which makes it a valuable raw material for stone tool manufacture. When subjected to percussion it fractures in a conchoidal fashion (that is, its breakage pattern resembles a cone with the apex at the site of percussion). There is evidence from some sites (e.g., Mumba Shelter in Tanzania) that hominins traveled long distances to fetch obsidian to make stone tools. (etym. Pliny suggests it was given its name because obsidian is said to resemble a stone found in Ethiopia by Obsius.)

## obsidian hydration dating
(or OHD) A chronometric dating method based on the fact that obsidian (volcanic glass) weathers, or hydrates, and that the amount of this hydration increases with time. In principle, accurate measurements of the hydrated area ("hydration rim") and the rate of hydration allow researchers to calculate the time since the obsidian surface was first exposed (e.g., by flake removal) so the hydration present should be a measure of the time that has elapsed since the flake was made. There is substantial variability in hydration rates due to compositional variability within obsidian sources and individual artifacts; the latter is a consequence of substantial variation within local depositional settings (e.g., groundwater composition and past temperatures). Obsidian hydration dating remains accurate for coarse chronologies and some advocate its use primarily as a relative dating tool. Obsidian hydration dating, along with other methods, has been used to provide age estimates at Enkapune ya Muto and other sites in East Africa. The applicable time range is 200–100 ka.

## Occam's razor
William of Occam (or Ockham) was a 13th/14thC English philosopher who promulgated the rule (or "razor") that, all things being equal, researchers should favor the simplest hypothesis. In paleoanthopology Occam's razor is most frequently applied to phylogenetic reconstruction where the preference for the least complex explanation (i.e., the most parsimonious explanation) is a guiding principle.

## occipital
An unpaired cranial bone at the base and rear of the skull that develops from several components. From front to back they are a single basilar (or basioccipital) part, paired lateral (or condylar) parts, and most posteriorly a single, large, flat, or squamous part. All four components of the occipital contribute to the foramen magnum. The occipital bone forms the posterior part of the basicranium and contributes to the posterior part of the cranial vault. The squamous part of the occipital bone is divided into a lower part that forms the nuchal plane (aka lower scale) and an upper part that forms the occipital plane (aka upper scale). The superior nuchal lines separate inion is the midpoint of the midline division between the nuchal and occipital planes. The external occipital protuberance (aka inion)

marks the midpoint between the two superior nuchal lines. (L. *occipio* = to begin or commence, thus the part of the head that usually appears first during a normal birth; or from *ob* = against and *caput* = head, the occiput is the part of the head that rests on the ground when you sleep on your back.)

### occipital lobe
One of the four main subdivisions of the cerebral cortex of each cerebral hemisphere; it forms the posterior part of each cerebral hemisphere. The occipital lobe is primarily devoted to visual processing. The primary visual cortex, Brodmann's area 17 (also known as V1 and the striate cortex) is located within the calcarine sulcus, toward the rear of the occipital lobe. The most posterior point on the cerebral hemisphere, called the posterior pole, is formed by the occipital lobe.

### occipital sinus
One of the dural venous sinuses that make up the intracranial venous system. *See also* **cranial venous drainage; occipito-marginal system**.

### occipito-marginal system
A variant of the intracranial venous system. In most modern humans the deep cerebral veins drain the deep structures of the brain, and they themselves drain into the left transverse sinus, while the superior sagittal sinus drains into the right transverse sinus. Each transverse sinus drains into a sigmoid sinus that runs inferiorly in a sigmoid-shaped groove from the lateral end of the transverse sinus to the superior jugular bulb, which marks the beginning of the internal jugular vein. This pattern of dural venous sinuses is known as the transverse-sigmoid system. However, in just a few percent of modern humans and in the majority of the crania belonging to *Australopithecus afarensis*, *Paranthropus boisei*, and *Paranthropus robustus*, the venous blood from the brain takes a different route. Instead of draining into the transverse sinuses, the deep cerebral veins and the superior sagittal sinus drain into an enlarged midline occipital sinus, which runs down from the cruciate eminence towards the foramen magnum in a midline groove. The occipital sinus drains into uni- or bilateral marginal sinuses, which run(s) around the margin(s) of the posterior quadrants of the foramen magnum to drain into the superior jugular bulb and/or the vertebral venous plexuses. Individuals with a substantial occipital sinus and with marginal venous sinuses on both sides (e.g., OH 5) are said to show occipito-marginal system dominance. *See also* **cranial venous drainage**.

### occipitomastoid crest
A crest that may or may not follow the course of the occipitomastoid suture. It forms the bony lateral edge of the cranial attachment of the superior oblique, one of the short muscles of the base of the cranium. When the superior oblique muscle is confined to the occipital bone then the occipitomastoid crest and the occipitomastoid suture are likely to coincide. However, when that muscle is relatively large, or if its location is displaced laterally, then the occipitomastoid crest may run on the temporal bone instead of along the occipitomastoid suture. When the course of the occipital artery coincides with that of the occipitomastoid crest the latter may appear as a compound structure, but in effect it is a single crest divided by a groove for the

occipital artery. The occipitomastoid crest is especially prominent and pneumatized in *Australopithecus afarensis*, and in that taxon it forms the lateral face of the occipitomastoid suture. In *Homo neanderthalensis* the occipitomastoid crest region is located in a relatively inferior position, resulting in the impression of a less-projecting and less-robust mastoid process and the distinctive appearance of this anatomical region in Neanderthals. [L. *occipio* = to begin or commence, thus the part of the head that usually appears first during a normal birth; or *ob* = against and *caput* = head, the occiput is the part of the head that rests on the ground when you sleep on your back; *mastos* = breast and *oeides* = shape and L. *crista* = crest; literally "a crest along the junction (suture) between the occipital bone and the mastoid part of the temporal bone."]

## occlusal
The surfaces of the teeth that touch when the mouth is closed. They are the superiorly facing surfaces of the enamel cap of the mandibular teeth and the inferiorly facing surfaces of the enamel cap of the maxillary teeth. They occlude when the mandible is elevated during the chewing cycle. (L. *occludere* = to close.)

## occlusal morphology
Refers to the size and shape of the surfaces of the teeth that touch when the mouth is closed (i.e., the superiorly facing surfaces of the enamel cap of the mandibular teeth and the inferiorly facing surfaces of the enamel cap of the maxillary teeth). Occlusal morphology is usually described in terms of the number and relative size of main and accessory cusps, the pattern of primary and secondary fissures, and the presence/absence of depressions (aka foveae). (syn. cusp morphology.)

## occlusal wear
Loss of the volume of a tooth that occurs on its functional or occlusal surface (i.e., the surfaces of the upper and lower teeth that meet when the jaws are brought together as the mouth closes). Occlusal wear that can be seen with the naked eye is called gross dental wear (aka macrowear or mesowear). Occlusal wear that can only be seen with a microscope is called dental microwear. *See also* **tooth wear**.

## Oceania, peopling of
The large and small islands in the central, western, and southern Pacific. The large islands are Australia and New Zealand; the three major divisions of small islands that make up Oceania are Melanesia, Micronesia, and Polynesia. Anyone reaching them from Asia or the New World would have required significant advanced technologies, like shipbuilding and navigation, thus they were not populated by humans until relatively recently. Past theories on the human settlement of the Pacific Islands collectively known as Oceania have suggested that the immigrants originated in Island Melanesia, Southeast Asia, Japan, Micronesia, or even South America. Evidence from contemporary archeology, linguistics, and genetics now points to several major migrations into Oceania over a remarkably long time period, all of them coming from the west. The initial settlement by modern humans of Sahul (the ancient Australia/New Guinea continent) as well as neighboring islands to the east in the Bismarck Archipelago occurred

*c.*50 ka from areas immediately to the west, mostly now submerged as the Sunda shelf. This was most likely an extension of an early modern human expansion out of Africa that followed the shorelines of South Asia. As sea levels subsequently rose the inhabitants of Sahul and the nearby islands were isolated for the following 20–30 ka, but by the end of the Pleistocene there is evidence of increasing interaction with peoples to the west. Because of their complex settlement history, Polynesians form a complicated and diverse genetic, linguistic, and cultural phylogenetic entity, with the inhabitants of Melanesia being the most diverse. Evidence from mitochondrial DNA suggests that Melanesians have some very ancient genetic relationships with Australian Aborigines, as well as some relatively minor genetic similarities in particular regions with Polynesian and Micronesian groups. Some geneticists have recently suggested that there are indications of genetic affinities of Australo-Melanesians with pre-modern human groups, notably the Denisovans. *See also* **Helicobacter pylori**.

## Ochtendung

A few cranial fragments assigned to *Homo neanderthalensis* were discovered at the foot of an exposure of late Middle Pleistocene volcanic and loess deposits during quarrying operations in Germany in 1997. Three Middle Paleolithic stone tools were found associated with the hominin remains. Stratigraphic correlation suggests a date of Marine Isotope Stage 6 (*c.*186–128 ka). (Location 50°21′N, 07°23′E, Germany.)

## ochre

A family of minerals identified by the red-, yellow-, or orange-colored streak they produce, and the compounds (e.g., hematite [$Fe_2O_3$] and goethite [$\alpha\text{-}Fe^{3+}O(OH)$]) from which those colors derive. When ground into powder or shaped into "crayon" form, ochres can be used as pigment to decorate the body and other surfaces. Notable archeological examples include the *c.*40 ka Mungo III burial from Lake Mungo in Australia which was discovered covered in powdered ochre, the *c.*77 ka specimens of symbolically incised ochre from Blombos Cave, South Africa, and the assemblage containing multiple types of ochre from between 270 and 170 ka at Twin Rivers, Zambia. Other known applications include, but are not limited to, animal hide tanning, medicine, vegetable preservative, and as a component of the mastic used for hafting tools. (Gk *okhros* = pale yellow.)

## Ockham's razor

*See* **Occam's razor**.

## odontoblasts

*See* **dentine**.

## odontogenesis

*See* **dentine**.

## OH 5

A well-preserved *c.*1.84–1.83 Ma cranium from locality FLK in Bed I at Olduvai Gorge that preserves all of the upper dentition, and much of the face, cranial base, and cranial vault. It is undistorted and the surfaces of the bones and teeth are remarkably well preserved. Its endocranial volume is approximately 500 cm³. It is the holotype of *Zinjanthropus boisei*, a taxon now usually included in *Australopithecus* or *Paranthropus*.

## OH 7

This 1.85–1.825 Ma specimen from FLK NN in Bed I at Olduvai Gorge consists of a juvenile skull and 13 hand bones. Its endocranial volume is 687 cm³. It is the holotype of *H. habilis*. Most researchers regard *H. habilis* as the earliest representative of the genus *Homo*, but some are less convinced that the morphology *H. habilis* shares enough features of the *Homo* clade and grade to justify its inclusion in *Homo*. Functional morphological interpretations of the hand fossils vary from those that emphasize their similarity to modern humans, to those that stress their similarities to *Australopithecus*, including *Australopithecus afarensis*. The pollical distal phalanx of OH 7 has no ungual spines, nor is there a ridge for the insertion of a flexor pollicis longus tendon. Researchers have suggested that the lack of a distinctive ungual fossa and ungual spines implies limited precision-grip capability.

## OH 8

This 1.85–1.825 Ma left foot from FLK NN in Bed I at Olduvai Gorge is complete except for the phalanges, the distal ends of the metatarsals, and the calcaneal tuberosity. There are also some dental fragments. The specimen is one of the paratypes of *Homo habilis*. The conclusions of functional morphological interpretations of the whole foot and of the component bones range from those that emphasize similarities to modern humans to those that stress similarities to *Australopithecus*. Some researchers claim that the OH 8 foot and the OH 35 distal tibia and fibula belong to the same individual, and it has been claimed that crocodile tooth marks on both sets of bones match up. But questions remain over whether they can be associated geologically (the OH 8 foot and the OH 35 distal tibia come from different levels in the Olduvai sequence). Furthermore, an investigation of the extent of the congruence between the reciprocal joint surfaces on OH 8 and OH 35 showed that they were unlikely to belong to the same individual, and a more detailed examination of the bite marks suggested they were not made by the same animal. Although OH 8 was allocated to *H. habilis* there are no taxonomically diagnostic cranial remains associated with it.

## OH 9

This *c*.1.7–1.2 Ma calvaria from LLK in Bed II at Olduvai Gorge lacks much of the vault, but it does preserve the supraorbital region. Its endocranial volume is 1067 cm³. It is the holotype of *Homo leakeyi*, but the OH 9 calvaria is seen by many as an early example of *Homo erectus sensu stricto* in Africa and it exemplifies the proposition that the distinctive features of *H. erectus* are best expressed in larger specimens.

## OH 12

This >0.78 Ma skull from VEK in Bed IV at Olduvai Gorge consists of the posterior vault, parietal, and facial fragments plus a few fragments belonging to the coronoid process of the mandible. Its endocranial volume of around 730 cm³ must be regarded as a relatively crude estimate. OH 12 is one of the youngest African fossils attributed to *Homo erectus*, yet its estimated endocranial volume is small. It was the first evidence that *H. erectus*-like hominins subsumed a substantial range of variation in absolute size and endocranial volume.

## OH 13

This c.1.7 Ma skull of a subadult hominin from MNK in Bed II at Olduvai Gorge consists of much of the vault, the greater part of the mandible and both maxillae, all of the mandibular teeth, and some of the maxillary postcanine teeth. Its endocranial volume is around 650 cm$^3$. A paratype of *Homo habilis*, OH 13 provides some of the best dental and mandibular evidence for that taxon. Although its postcanine teeth are absolutely small, when they are scaled to its relatively small body mass it is still megadont. The OH 13 skull has some "derived" features and because it is one of the youngest specimens in the *H. habilis* hypodigm it has been suggested that there was a temporal trend in *H. habilis* morphology, but the antiquity of OH 24 casts doubt on that hypothesis.

## OH 16

A <1.80 Ma fragmented skull of a subadult hominin from FLK in Bed II at Olduvai Gorge. The best-preserved part of the calvaria is the supraorbital region and the region of the mandibular fossa. Fifteen maxillary and 13 mandibular permanent teeth are preserved; the $M_3$s were erupted, but the $M^3$s were not. The endocranial volume of the existing, poor, reconstruction is about 640 cm$^3$. Though originally only "provisionally referred" to *Homo habilis*, OH 16 is now generally thought to belong to that taxon.

## OH 24

This c.2.0–1.85 Ma cranium from locality DK in Bed I of Olduvai Gorge is distorted and very fragmented, but it does preserve substantial parts of the vault, face, palate, and base. No anterior teeth are preserved, but both upper premolars and the $M^1$ and $M^3$ are preserved. Its endocranial volume is approximately 600 cm$^3$. The OH 24 cranium is one of the temporally oldest specimens attributed to *H. habilis* yet it has some of the more "derived" features also seen in OH 13.

## OH 28

A <780 ka left femur (lacking the head, greater trochanter, and the condyles) plus much of the ischium and the ilium (the crest and the anterior superior iliac spines are lacking) of a left pelvic bone from locality WK in Bed IV of Olduvai Gorge. The apparently derived features shared among the femora recovered from Zhoukoudian and the OH 28 femur (e.g., anteroposterior flattening of the shaft, convex medial border, and the low position of the minimum breadth of the shaft) link OH 28 with *Homo erectus*, and provide support for the hypothesis that while *H. erectus* lower limb morphology (e.g., iliac pillars and the impressions for the ilio-femoral ligament) is consistent with that taxon being a habitual, or obligate biped, it shows consistent departures from the lower limb morphology of modern humans. Prior to the discovery of OH 28 there was no reliable information about the pelvis of *H. erectus*.

## OH 62

This c.1.85–1.83 Ma fragmentary associated skeleton is from locality FLK in Bed I of Olduvai Gorge. Although the skull and all of the major long bones are represented in the OH 62 associated skeleton, it is in about 300 fragments. The palate is the most complete part of the skull, followed by the face and the mandible. The tooth crowns that are preserved are very worn. The long bones of the upper limb are better preserved than those of the lower limb, but even the relatively well-preserved humerus cannot provide a thoroughly reliable estimate of its undamaged length. In spite of its very fragmentary condition, OH 62 is a significant specimen for at least two reasons. First, it provides the only unambiguous evidence about the postcranial skeleton of *Homo habilis*, for none of the other postcranial elements linked with *H. habilis* have associated taxonomically diagnostic cranial elements. Second, despite the fact that the shafts of the long bones are incomplete, most researchers consider enough is preserved to suggest the limb-length proportions of the OH 62 *H. habilis* individual are not significantly different from those of *Australopithecus afarensis*, but they *are* significantly different from those of modern humans. This inference is supported by an analysis of the cross-sectional geometry, demonstrating that OH 62 has a degree of humeral strength relative to femoral strength comparable to that of *Au. afarensis* and unlike that seen in *Homo erectus* and modern humans.

## OH 65
A 1.84–1.79 Ma complete upper dentition in a well-preserved palate and lower face from locality 64 in Bed I of Olduvai Gorge. Some researchers argue that OH 65 is consistent with the hypothesis that variation within the "early *Homo*" fossils from East Africa was best explained as sexual dimorphism within a single species, thus undermining the case for a second species such as *Homo rudolfensis*. Other observers have concluded that OH 65 has more in common with the facial morphology of crania such as KNM-ER 1813, and thus supports the idea of two distinct species. It is also possible that it belongs to *Homo ergaster*, or early African *Homo erectus*.

## OHD
*See* **obsidian hydration dating**.

## OIS
Abbreviation of Oxygen Isotope Stages. *See* **Marine Isotope Stages**.

## Okladnik'ov Cave
This cave site in Siberia is best known for being the being the most easterly of the sites containing evidence of *Homo neanderthalensis*. Several hominin fossils have been recovered but their taxonomic status is much debated, with some arguing they are *H. neanderthalensis*, others saying they evolved from Neanderthals, and still others suggesting similarities between these fossils and Asian *Homo sapiens* and *Homo erectus*. Recent studies of mitochondrial DNA from one *c.*24 ka adult and a *c.*38–30 ka juvenile suggest that the juvenile is a Neanderthal, while the adult likely represents another taxon. (Location 54°45′50″N, 84°02′34″E, Altai Mountains, Siberia, Russia.)

## Ola Bula
*See* **Soa Basin**.

## Old World monkeys
*See* **Cercopithecoidea**.

## Oldowan

The Oldowan, which is currently the oldest recognized stone tool industry, is known from sites that date from c.2.6 to 1.6 Ma, primarily in Africa. Oldowan stone artifacts are of three types: tools, utilized pieces, and débitage. An artifact is only considered a "tool" if it shows signs of being intentionally retouched or flaked. Tools were further subdivided into light-duty (e.g., small, retouched flakes) and heavy-duty (e.g., choppers, discoids, polyhedrons) categories. Utilized pieces consist of hammerstones, anvils, or flakes that were damaged through use. Debitage consists of unmodified flakes, flake fragments, and other knapping debris. The term manuport is used for natural stones that had been transported and discarded without being modified. There is debate about whether the flakes, not the cores, were the desired end product. Oldowan artifacts were used to process animal and plant foods, and wood. Recent analyses emphasize that the hominins responsible for Oldowan artifacts displayed a keen understanding of the physical properties of different raw materials and preferred to flake good-quality rock (i.e., tractable and not readily prone to shattering). Older Oldowan occurrences include those in the Omo Shungura Formation, the Gona Paleoanthropological study area, and the Hadar study area in Ethiopia, and West Turkana and Kanjera in Kenya. More recent (i.e., between 2.0 and 1.6 Ma) Oldowan sites have been found in North (e.g., Ain Hanech), southern (e.g., Sterkfontein), and East (e.g., Koobi Fora, Olduvai Gorge) Africa. The recovery of a comparable technology to the Oldowan at Dmanisi, Georgia, at c.1.8 Ma, and from Majuangou and Yuanmou, China, at about the same time, suggests that the earliest travelers out of Africa took with them the capacity to make the Oldowan tool kit.

## Olduvai Gorge

An approximately 40 km-/25 mile-long dry river valley incised into the western margin of the eastern limb of the East African Rift System in Tanzania that exposes sediments dating from c.2 Ma to the recent past. The hominin fossil and archeological localities at Olduvai Gorge, which are located in sedimentary rocks in either the Main Gorge or in a branch called the Side Gorge, all lie above the Naabi Ignimbrite. They are divided into seven beds (aka Olduvai Beds). The four beds immediately above the basalt are identified using roman numerals (Beds I–IV, with Bed I at the bottom); the three beds above them were given local names (Masek immediately above Bed IV, Ndutu, and Naisiusiu). Beds I and II can be readily distinguished, but there are places where Beds III and IV cannot be distinguished; some propose merging them into a single unit called Beds III–IV. The sediments of Beds I and II contain several tuff layers (Bed I, Tuffs IA–IF; Bed II, Tuffs IIA–IID) and Olduvai was the first fossil hominin site to be dated using potassium-argon dating methods, which provide dates for the volcanic eruptions that created the tuffs. Within Bed II, just above the Lemuta Member, there is an abrupt change in the fauna, which is referred to as the "faunal break." The sediments exposed in the gorges at Olduvai were laid down in and around a paleolake, called Lake Olduvai. The paleolake varied in size, sometimes being small and highly saline and at other times larger with fresher water. The localities at Olduvai are identified using a letter and number code. The letters refer to the names used for the small valleys, or korongos (aka karongas) made by the streams that drained into the Main and Side Gorges, and to the situation of the locality within that small stream valley; the roman numeral numbers refer to the bed the locality is in. Most of the korongos were named after the Leakey's relatives and colleagues, with additional letters to denote how adjacent sites are geographically related to each other. So FLKN I is a fossil locality in Bed I in the Frida Leakey korongo that is north of locality FLK I, and FLK NN I is a fossil locality in Bed I in the Frida Leakey korongo that is north of locality FLKN I. Each locality has a geological locality number and an archeological number associated with it (e.g., FLK NN I is geological locality 45 and archeological site no. 31). Argon-argon, geomagnetic, and potassium-argon dating suggest the sediments at Olduvai Gorge span the period between $c.2$ Ma to $c.1300$ years BP. The Naabi Ignimbrite at the base of the sediments is securely dated at $c.2.0$ Ma. More than 70 fossil hominins have been recovered from Olduvai Gorge including OH 5, the type specimen of *Paranthropus boisei*, OH 7, the type specimen of *Homo habilis*, OH 9, a *Homo erectus* cranium, and OH 62, a partial skeleton of *H. habilis*. (Location 02°56–59′S, 35°15–22′E, Tanzania.) *See also individual OH fossils.*

## Olduvai subchron
One of the first events (now called subchrons) to be recognized in the geomagnetic polarity time scale (GPTS). It is an episode of normal magnetic direction within the Matuyama reversed epoch (now called the Matuyama chron) from $c.1.9$ to $c.1.785$ Ma that was recognized in the sediments of Olduvai Gorge. *See also* **geomagnetic polarity time scale**.

## Olorgesailie
A large site complex located in the southern Kenyan part of the East African Rift System with numerous archeological and paleontological sites near the margins of a fluctuating lake. It is best known for dense concentrations of handaxes and other Acheulean implements, but the upper layers contain substantial evidence of the Middle Stone Age. Sediments of the Olorgesailie basin are divided into two geological formations. The Early and Middle Pleistocene-aged Olorgesailie Formation has been studied in detail; the Middle and Later Pleistocene Oltulelei Formation has only been recently defined. Argon-argon dating of tephra and magnetostratigraphic analyses of sediments provide age estimates of $c.990$ ka for the middle of Member 1 of the Olorgesailie Formation and $c.490$ ka for Member 14; the Brunhes-Matuyama boundary occurs near the base of Member 8. The diminutive partial cranium KNM-OG 45500 was recovered in 2003 from 970–900 ka sediments. The >34,000 recovered stone artifacts demonstrate substantial technological and typological variability within the Acheulean behavioral system. Middle Stone Age sites in the overlying Oltulelei Formation are reported to date from $c.500$ to 220 ka but have not been formally published. (Location 01°35′S, 36°37′E, Kenya.)

## omnivore
*See* **eclectic feeder**; **omnivory**.

## omnivory
The ability to eat a wide range of plants and animals. Several primates, including vervets, baboons, chimpanzees, and modern humans, are described as omnivores, but the ability of modern humans to process foods (e.g., by cooking) means they can incorporate a particularly wide range of items into their diets. (L. *omni*, from *omnis* = all, every, plus *vorous*, from *vorare* = to devour.) *See also* **eclectic feeder**.

### Omo 18-1967-18

(or Omo 18-18) A *c.*2.6 Ma adult mandibular corpus that extends from a fracture through the $M_3$ alveolus on the right side through to a fracture just mesial to the left $M_3$ alveolus. It is the holotype of *Paraustralopithecus aethiopicus*. Almost all researchers regard *Paraustralopithecus* as a junior synonym of the genus *Paranthropus*. Researchers wanting to maintain a distinction between the pre- and post-2.3 Ma parts of the East African hypodigm of *Paranthropus* do this by recognizing the species *Paranthropus aethiopicus*. See also **Paraustralopithecus aethiopicus**.

### Omo I

This *c.*195 ka associated skeleton from Omo-Kibish, in Ethiopia, comprises a skull and parts of the axial skeleton, the shoulder girdle, pelvic girdle, and the upper and lower limbs. Its endocranial volume is approximately 1435 cm³. The cranial vault of Omo I is high and globular, with a nearly vertical frontal bone, rounded occipital and pronounced parietal bosses; the mandible bears a slight chin. The cranium and postcranium are modern human in overall morphology, but they retain some primitive features. The revised age of 195 ka makes this skeleton among the earliest evidence of a modern human-like morphology.

### Omo II

This *c.*195 ka well-preserved older adult calvaria from Omo-Kibish, in Ethiopia, preserves all of the vault bones except portions of the frontal and the posterior and lateral parts of the basicranium. Its endocranial volume is around 1435 cm³. The vault is high and the frontal is moderately receding, but closer to vertical than in Kabwe or Saldanha. The supraorbital torus is modern human-like, but the occipital bone is strongly angled, the maximum vault breadth is across the supramastoid tubercles, the mastoids are mediolaterally thick, and the tympanic is robust, suggesting affinities with *Homo erectus* or *Homo heidelbergensis*. Whereas Omo I is evidently anatomically modern, the angled occipital and robust temporal bone of Omo II align it with pre-modern *Homo*. Geologic evidence indicates the two are coeval, raising questions of whether they sample a single, morphologically variable population or two morphologically distinct populations, or whether the Omo II calvaria might have eroded from an older context.

### Omo Basin

*See* **Lower Omo Basin**.

### Omo Group

The older of the two inclusive stratigraphic sequences (the younger one is called the Turkana Group) exposed within the region known as the Lower Omo Basin (also called the Lower Omo Valley) and the Turkana Basin. It consists of the Mursi, Nkalabong, Shungura, and Usno Formations to the north, the Koobi Fora Formation to the east, and the Nachukui Formation to the west. *See also* **Omo-Turkana Basin**.

### Omo-Kibish

A complex of sites in the Kibish Formation, part of the Turkana Group, that outcrops along the lower Omo River in Ethiopia, where a series of hominins (Omo I, II, and III) were recovered. The site has been dated by correlation to sapropels in the Mediterranean. Member I correlates with the *c*.197 ka S7, the KHS Tuff within Member II correlates with S6, Member III with the *c*.104 ka S4, and Member IV with the *c*.8 ka S1. Members I–III contain scattered concentrations of small Middle Stone Age artifacts. Most of the tools are not retouched, but some leaf-shaped bifacial points as well as sporadic, larger handaxes, picks, or lanceolates (which resemble Acheulean artifacts) were recovered. Member IV contains Later Stone Age artifacts and bone harpoons. (Location approximately 5°23′N, 35°56′E, Ethiopia.)

## Omo Rift Zone

Part of the East African Rift System, a series of river valleys and basins that extends from the Afar Rift System in the northeast, via the Main Ethiopian Rift System to the Omo Rift Zone in the southwest. Hominin fossil and archeological sites in the Omo Rift Zone include Fejej, the Usno and Shungura Formations, and the site complexes around Lake Turkana (e.g., Koobi Fora, West Turkana, Lothagam, and Kanapoi).

## Omo-Shungura

*See* **Lower Omo Basin**.

## Omo-Turkana Basin

The Lower Omo Basin and the Turkana Basin are effectively two components of an area of low country between the Ethiopian and Kenyan domes (north and south) and between the walls of the East African Rift System (east and west). The Omo-Turkana Basin extends for about 500 km/310 miles from north to south and up to about 100 km/62 miles from east to west. For the past few million years or so the larger southern component of the basin has been occupied by Lake Turkana. The latter has fluctuated in size, becoming larger when inflow exceeds any outflow and evaporation and smaller when it does not. The sediments in the Omo-Turkana Basin are divided into the Omo Group and the Turkana Group. The major fossiliferous formations within the Omo-Turkana Basin are the Koobi Fora, Mursi, Nachukui, Nkalabong, Shungura, and Usno Formations. The major hominin fossil and archeological sites within the Omo-Turkana Basin are Koobi Fora, Lothagam, the Omo region, and West Turkana.

## ontogenetic

Any criterion, explanation, or test that involves ontogeny. For example, the ontogenetic criterion is the proposal that the morphology of a character early in its ontogeny is likely to be the primitive condition of that character. *See also* **ontogeny**.

## ontogenetic allometry

*See* **allometry**; **scaling**.

## ontogenetic criterion

*See* **ontogenetic**.

## ontogenetic scaling

*See* **allometry**; **scaling**.

## ontogeny

All of the phases of life history that precede the cessation of growth. It is also used to refer to the unique developmental history of each individual organism. (Gk *onto*=being and *gena*=to give birth to.)

## operational taxonomic unit

(or OTU) Any taxon used in a cladistic analysis or phylogenetic analysis. In phylogenetic comparative analyses an operational taxonomic unit refers to an actual extant or fossil taxon for which data have been measured as opposed to an internal branching node within a phylogeny. To avoid circular reasoning, hypotheses about the nature of operational taxonomic units must be generated using phenetic and not cladistic methods. *See also* **phylogenetic comparative analysis**.

## opposable thumb

The ability to bring the palmar pulp surface of the thumb (aka first manual digit) squarely into contact with – or opposite to – the palmar pads of one or more of the other digits. The thumb abducts and flexes (aka conjunct rotation) across the palm while the finger(s) flex to meet the thumb. A truly opposable thumb is found only in Old World monkeys, apes, and modern humans. In modern humans, opposition is facilitated by a broad area of contact between the pulps of the finger(s) and thumb, making the modern human hand especially well adapted to manipulate objects. In apes, opposability is hampered by the relatively short thumb compared to long fingers. In contrast, terrestrial monkeys, such as baboons and mandrills, have more enhanced manipulative skills because the thumb-to-finger length ratio is more similar to that in modern humans. Pseudo-opposability refers to the ability to move the thumb close to the fingers, but without allowing pulp-to-pulp contact between the thumb and finger(s). *See also* **power grip**.

## optically stimulated luminescence dating

*See* **luminescence dating**.

## optimal foraging theory

A theory from evolutionary ecology which is concerned with the foraging behavior of an organism. It looks at the diet of a predator in terms of energy. Specifically, the model compares the net amount of energy gained from the prey relative to the amount of energy expended in acquiring it (i.e., energetic returns). The model predicts how the number of prey types included in a predator's diet (aka diet breadth) will respond to changes in its environment. The fundamental assumption of optimal foraging theory is that all foragers behave optimally, with the objective of maximizing energetic returns (i.e., obtaining more calories with minimal effort). Early archeological applications of optimal foraging theory date to the 1970s and focused largely on changes in subsistence behavior. Today, optimal foraging theory is increasingly used to examine subsistence change through time in response to environmental change, resource depression, demographic shifts, or technological innovation.

## orbital geometry

Three aspects of the Earth's relationship to the sun, including the tilt of the Earth's axis, the stability of that axis, and the shape of the Earth's orbit around the sun. Changes in orbital geometry are predictable over the Neogene, but predictions break down further back in time.

Predictable changes in (a) the tilt of the Earth's axis (called obliquity) follow a $c.41$ ka cycle, (b) the extent to which the Earth's axis wobbles (called precession) follow a $c.23$ ka cycle, and (c) the shape of the Earth's orbit round the sun (called eccentricity) follow a $c.100$ ka cycle. These changes in orbital geometry alter the latitudinal and seasonal distribution of insolation (i.e., incoming sunlight) with demonstrated influences on Earth's climate. The regularity of these cycles during the Neogene is such that astrochronology can be used to calibrate, or "tune" (as in "orbital tuning"), paleoclimate cycles to the astronomical time scale. *See also* **astronomical theory**; **astronomical time scale**.

## orbital tuning

The process of matching a stratigraphic record to the calculated fluctuations in insolation (i.e., incoming sunlight) based upon astronomical theory. Orbitally tuned age models are dependent on absolute dating boundaries (e.g., the geomagnetic polarity time scale). Within those age constraints cyclic variations in a paleoclimate proxy may be tied to a dominant orbital frequency. The presence of other orbital frequencies provides independent corroboration of whether or not the "tuning" is valid. This approach has been very powerful for improving the age-model resolution for the Pliocene and Pleistocene using the oxygen isotope stratigraphy of benthic foraminifera. *See also* **astronomical theory**; **astronomical time scale**; **Marine Isotope Stages**.

## Orce region

Located in a geographical depression of exposed Lower Pleistocene sediments in the southeastern Spanish province of Andalucía, this region has been reported to have one of the best Plio-Pleistocene mammalian fossil records in Europe. There are three main sites: Venta Micena, Fuentenueva, and Barranco León. Venta Micena is best known for the debate over whether a cranial bone from the site is from a hominin. It is almost certainly not a hominin. Mode 1 artifacts have been recovered from Barranco León, as has a hominin deciduous molar that is possibly the oldest hominin in Europe. (Location 37°43′N, 02°28′W, Spain; etym. after a nearby town.)

## ordinary least squares regression

As with other methods of regression, this statistical analysis estimates the relationship between variables, specifically how one variable (the dependent variable) changes as a result of the other, independent variable. In ordinary least squares regression, model parameters are fit by minimizing the sum of the squared differences between observed and predicted values of the dependent variable. Least squares techniques may be applied to simple (bivariate) regression or multiple regression. It is an example of Model I regression. *See also* **regression**.

## ordination
*See* **seriation**.

## oropharynx
*See* **pharynx**.

## Orrorin tugenensis Senut et al., 2001

The genus and species established in 2001 by Senut et al. to accommodate cranial and postcranial remains recovered from the c.6.0 Ma Lukeino Formation at Aragai, Cheboit, Kapcheberek, and Kapsomin, Baringo District, Kenya. The femoral morphology has been interpreted as suggesting that *Orrorin tugenensis* was at least a facultative biped, but other researchers interpret the radiographs and computed tomography scans of the femoral neck as indicating a mix of bipedal and nonbipedal locomotion. Otherwise, the discoverers admit that much of the critical dental morphology is "ape-like." *O. tugenensis* may prove to be a hominin, but it is equally and perhaps more likely that it belongs to another part of the adaptive radiation that included the common ancestor of panins and hominins. If the morphological differences between *Ardipithecus ramidus* and *O. tugenensis* do not justify their being assigned to different genera, then the latter taxon would be transferred to the genus with priority (i.e., *Ardipithecus*) as *Ardipithecus tugenensis*. The holotype of *O. tugenensis* is BAR 1000′00. (etym. *Orrorin* = original man in the Tugen language; *tugen* = Tugen Hills.) See also **BAR 1000′00**.

## orthogenesis

A model popular in the early 20thC that assumed biological evolution proceeded in straight lines instead of the branching pattern of evolution promoted by Charles Darwin. By the end of the 19thC most biologists had accepted the principle of biological evolution, yet many of them rejected Darwin's mechanism of natural selection in favor of neo-Lamarckian models of evolution consistent with orthogenesis. Supporters of orthogenesis believed that it was an inherent biological property of all organisms to evolve along predetermined lines of development, thus selection and adaptation play little, or no, role in orthogenetic models of evolution. Orthogenesis finally fell out of favor in the 1950s in response to the rise of the modern evolutionary synthesis and to the general acceptance of models of evolution that were closer to that set out by Darwin. (Gk *orthos* = straight and *genes* = to be born.)

## orthognathic

A term that literally refers to jaws that do not project, but technically it refers to skulls in which no part of the face or mandible (except the chin) projects forwards. It is usually quantified by measurements of the distance between a landmark on the base of the cranium (e.g., basion, porion) and a landmark on the front of the face (e.g., subnasale, alveolare). *Paranthropus boisei* and *Paranthropus robustus* are examples of fossil hominin taxa that have relatively flat or orthognathic faces. (Gk *orthos* = straight and *gnathos* = jaw; ant. prognathic.) See also **prognathic**.

## orthograde

See **posture**.

## os

In the Latin version of anatomical terminology "os" is used as a prefix for many of the bones. Some researchers still use the Latin form (e.g., "os coxae" instead of the official English-language term "hip bone" or "pelvic bone"), but to be consistent those researchers should refer to *all* bones in their Latin form (e.g., the scaphoid as the "os scaphoideum" and "caput ulnae" instead of the "head of the ulna"). It is more appropriate to use the English-language terms listed in the *Terminologia Anatomica* than to continue to use terminology that is arcane. However, for some bones "os" is still used as a prefix of their formal name (e.g., "os centrale" is both the Latin *and* the English-language term for a cartilaginous nodule that in modern humans usually fuses with the scaphoid, but which in the non-African apes more often persists as a separate carpal bone). (L. *os* = bone, as in *osseus* = bone or bone-like.) *See also* **anatomical terminology**.

## os centrale
A cartilaginous nodule that in modern humans usually fuses with the scaphoid, but which in the non-African apes more often persists as a separate carpal bone.

## os coxae
*See* **os; pelvis**.

## OSL
Abbreviation of optically stimulated luminescence dating. *See* **luminescence dating**.

## ossification
The process of bone formation (aka osteogenesis). In both types of ossification – intramembranous and endochondral – bone formation begins at a focus called an ossification center. Bone formation by intramembranous ossification (e.g., in the flat bones of the cranial vault) begins at centers of ossification within flat sheets or "membranes" (hence the term intramembranous) of collagen fibers. Endochrondral (i.e., "within-cartilage") ossification is a two-stage process. First, cells called chondroblasts form a hyaline cartilage model of the future bone. Chondrocytes then swell (hypertrophy) and die, leaving only a mineralized shell of cartilage matrix. This mineralized cartilage is resorbed by osteoclasts (strictly speaking, chondroclasts) and osteoblasts start forming osteoid matrix on the walls of the spaces. This happens first at the primary center of ossification, which is soon surrounded by a sleeve of subperiosteal bone that forms beneath the periosteum. In long bones the bone formed from the primary center of ossification is called the diaphysis. Primary centers usually appear between 7 weeks and 4 months of intrauterine life. Some bones (e.g., auditory ossicles, zygomatic, carpals, and tarsals) form from a single primary center; others (e.g., sphenoid, temporal) form from multiple primary centers. Secondary centers of ossification, or epiphyses, form in complex bones (e.g., vertebrae), at the ends of long bones, and at the sites of bony processes (e.g., trochanters of the femur and the tubercles of the humerus). Long bones can continue to grow in length because new cartilage is laid down in epiphyseal plates between the primary center in the shaft of the bone and the epiphyses at either end. Bones stop growing in length when the primary and secondary centers of ossification fuse into a single continuous bone. Epiphyseal plates are sometimes visible in radiographs as a dark line between the diaphysis and the epiphysis. (L. *os* = bone and *facere* = to make.) *See also* **bone remodeling; modeling**.

## ossification center
*See* **ossification**.

### osteoblasts
See **ossification**.

### osteoclasts
See **ossification**.

### osteocytes
See **ossification**.

### osteogenesis
See **ossification**.

### ostrich egg shell
The fossilized eggs, or more often fragments of eggs, of ostrich-related species are the most common avian fossils in African hominin sites, especially in open-air sites. They are useful for dating and for paleoenvironmental reconstruction, for ostrich-related birds are indicative of open, dry savanna-like habitats. See also **amino acid racemization dating**.

### ostrich egg shell dating
See **amino acid racemization dating**.

### OTU
Acronym for **operational taxonomic unit** (*which see*). See also **cladistic analysis**; **phylogenetic comparative analysis**.

### Oulad Hamida 1
See **Thomas Quarry**.

### outer table
See **cranial vault**.

### outgroup
See **cladistic analysis**.

### outgroup criterion
One of the criteria used to determine the polarity (i.e., which is the most primitive state and which the most derived) of the states of a character. The principle is that the character states seen in closely related, preferably primitive, sister taxa of the group being studied are likely to manifest the primitive states for that character. See also **cladistic analysis**.

### out-of-Africa hypothesis
A hypothesis about modern human origins that emerged during the 1980s. It drew upon paleontological, archeological, and genetic evidence to support the idea that modern humans had evolved within Africa, and then had migrated in one, or very likely more than one, dispersal

out of Africa to populate the Old World and later the New World. Thus, according to the hypothesis, all modern humans originate from a group of *Homo sapiens* that left Africa relatively recently. Importantly, the out-of-Africa hypothesis rejects the idea that *Homo erectus* populations existing in Asia and elsewhere had evolved into *H. sapiens*, or had given rise to any current modern human populations, as is argued by the supporters of the multiregional hypothesis. The out-of-Africa hypothesis for the origin of modern humans is now sometimes referred to as "Out of Africa 2" in recognition of the fact that the first hominin dispersal out of Africa was undertaken much earlier by *H. erectus*. The most widely accepted version of the out-of-Africa hypothesis is called the "recent African origin" or "replacement model." It suggests that modern *H. sapiens* evolved from a *Homo heidelbergensis* or *Homo rhodesiensis* ancestor in Africa *c.*200 ka and that the hominin groups existing outside of Africa did not contribute in any significant way to the ancestry of current modern human populations. Some of these early *H. sapiens* migrated out of Africa *c.*100 ka and spread throughout the Old World where they replaced existing populations such as *Homo neanderthalensis*. Support for this hypothesis came from research that concluded that all living human populations shared a most recent female common ancestor who lived in Africa *c.*200 ka. While there is evidence of gene exchange between Neanderthals and anatomically modern humans, as well as between Denisovans and anatomically modern humans, the primary ancestral contribution to modern humans in Eurasia clearly derived from the out-of-Africa migrations of early modern people. Thus, whereas the strongest form of the out-of-Africa hypothesis has apparently been falsified, in its weaker form it has survived refutation. However, there is little evidence that would allow one to falsify the weak forms of either the out-of-Africa or the multiregional hypothesis. *See also* **assimilation model**; **candelabra model**; **multiregional hypothesis**; **replacement with hybridization**.

## owl pellet
The regurgitated, indigestible remains of food ingested by owls (aka owl pellets) are considered to be the main way microfaunal remains enter the fossil record at archeological and paleontological sites. Owls are specialized predators of microfauna. The activities of owls are thus important for the preservation of an important part of the paleontological and paleoenvironmental record. *See also* **microfauna**.

## oxygen
An element commonly used in paleoclimate reconstructions and stable-isotope biogeochemistry. Oxygen has two common isotopes ($^{18}O$ and $^{16}O$) and one rare isotope ($^{17}O$); all three isotopes are stable and do not decay. *See also* **$^{18}O/^{16}O$**; **paleoclimate**; **stable isotopes**.

## Oxygen Isotope Stages
*See* **Marine Isotope Stages**.

## paedomorphosis
See **pedomorphosis**.

## Pakefield
An archeological site between Pakefield and Kessingland on the coast of eastern England that has been exposed by a fall in the level of the North Sea. It consists of sediments of the Cromer Forest-bed Formation (CF-bF) that probably belong to Marine Isotope Stage 17 (but they could be as old as Marine Isotope Stage 19). Thirty-two worked flints, including a simply flaked core and a crudely retouched flake, were found *in situ*. (Location 52°25.9′N, 01°43.8′E, England.)

## Paleoanthropological Inventory of Ethiopia
One of the few attempts to locate potential hominin fossil sites on a large, regional, scale. It used images from two space-based systems, Landsat thematic mapping (TM) and large-format camera (LFC). The two sets of data were combined to help identify promising sedimentary basins, and then researchers explored the basins by vehicle and on foot to verify the presence of potential sites. Several sources of hominin fossils in the Ethiopian Rift Valley were identified in this way, including the site complex within the Kesem-Kebena basin in the Awash River system upstream of the Hadar and Middle Awash study areas and the site of Fejej between the River Omo to the west and Lake Chew Bahir to the east. *See also* **satellite imagery**.

## paleoanthropological terminology
The terminology used to describe hominin fossils is a mixture of the terminology used by modern human anatomists (i.e., anatomical terminology) and terms introduced by anthropologists and paleoanthropologists when they have encountered structures in nonhuman primates and in the hominin fossil record that are not ordinarily seen in modern humans. Some of these terms were initially described using the Latin form (e.g., *crista occipitomastoidea, planum alveolare*) in which case it is conventional to use italics for these terms. However, there is no need to italicize terms that were initially described in English (e.g., nasoalveolar gutter, preglenoidal plane). *See also* **anatomical terminology**.

---

*Wiley Blackwell Student Dictionary of Human Evolution*, First Edition. Edited by Bernard Wood.
© 2015 John Wiley & Sons, Ltd. Published 2015 by John Wiley & Sons, Ltd.

## paleoanthropology

A term interpreted differently in Europe and North America. In Europe it is synonymous with human, or hominin, paleontology, whereas in North America it is interpreted more inclusively to include archeology as well as hominin paleontology. By the end of the 19thC Europe's leading anthropologists and archeologists recognized paleoanthropology as a science that dealt specifically with human paleontology. The most up-to-date interpretation of paleoanthropology used in the North American, more inclusive, sense can be found on the website of the Paleoanthropology Society and suggests that "paleoanthropology is multidisciplinary in nature and the organization's central goal is to bring together physical anthropologists, archeologists, paleontologists, geologists and a range of other researchers whose work has the potential to shed light on hominid behavioral and biological evolution." (Gk *palaios* = ancient, *anthropos* = human being, and *ology* = study of; literally the study of ancient humans.)

## paleoclimate

Reconstructions of climate on a global or regional scale made for time periods when it is not possible to take direct measurements of climate. Information about past climates comes from biotic or chemical proxies for the four climate variables (i.e., temperature, precipitation, wind speed and wind direction). The biotic proxies use productivity (e.g., in warmer phases the oceans have a greater biomass and tree rings are wider) and evidence of the types of plants and animals that were living at the time. The chemical proxies include changes in oxygen isotope ratios recorded in foraminifera and in changes in air bubbles in ice. (Gk *palaios* = ancient and *klima* = which refers to the slope of the Earth's surface, is the origin of L. *clima* = referring to climate, or latitude.) *See also* **climate**; **climate forcing**.

## paleoclimate reconstruction

*See* **paleoclimate**.

## paleocommunity deme

*See* **deme**.

## paleodeme

*See* **deme**.

## paleodemography

The demography of past populations or samples from which no records of age or sex are available. These populations or samples may be modern humans from the recent past, or they may be extinct taxa.

## paleoecology

Interactions between an organism and its environment in the past. Paleoecology is of interest in hominin paleontology because it enables researchers to investigate the factors that might have affected past human behavior and evolution from a holistic perspective. Studies of paleoecology are focused on either the interactions that center on a particular species (paleoautecology), or reconstructions of community ecology at a particular site or during a well-defined time interval. (Gk *palaios* = ancient, *oikos* = house, and *logia* = study.)

## paleoenvironment

Information about the climate, landscape, and the vegetation of a fossil locality in the remote geological past (i.e., it is a more inclusive concept than paleoclimate). (Gk *palaios* = ancient and OF *environ* = round about.) *See also* **environment**; **paleoenvironmental reconstruction**.

## paleoenvironmental reconstruction

The process of deducing past environmental conditions from the physical and biological evidence left behind in the geological record. Paleoenvironmental reconstruction relies on the principles of uniformitarianism; in other words, by studying present processes we will understand what occurred in the past. Evidence, which can be abiotic or biotic, can take many forms (e.g., sedimentology, soil chemistry, paleobotany, and vertebrate and invertebrate paleontology). Knowledge of the abiotic environment can be gained either directly from evidence of natural processes in the preserved sediments or indirectly from a study of the biotic components and their inferred ecological preferences (e.g., the presence of a perennial river could be inferred from cross bedding in the sedimentary deposits, or from fossil evidence of riverine fish, reptiles and mammals, or from preserved freshwater oyster reefs). Paleoenvironmental reconstructions of fossil localities rely on a multiproxy approach that attempts to reconcile the different lines of evidence recovered from a site to create a coherent picture of the past. Multidisciplinary paleoanthropological projects now routinely take this approach, since past environments are seen as a key aspect of understanding evolutionary processes.

## paleohabitat

A past habitat. *See also* **habitat**.

## Paleolithic

The "old stone age," defined by John Lubbock in 1865 as the time when humans coexisted with extinct fossils and when stone implements were rough-hewn. It is in contrast to the Neolithic, or "new stone age," when stone implements were generally polished. The term now refers both to a time period (the sum of human history from the invention of stone tools until *c*.10 ka) and to a distinct economic, social, and technological complex, usually defined by reliance on flaked stone tools, low population sizes, and foraging. The Paleolithic is customarily divided into three stages (Lower, Middle, and Upper), each having unique features. In Africa, the equivalent terms are Early, Middle, and Later Stone Age.

## paleomagnetic dating

The use of the magnetic polarity of a stratum, or a sequence of magnetic polarities of a longer section, to date sediments or igneous rocks. (Gk *palaios* = ancient and *magnes* = magnet.) *See* **geomagnetic polarity time scale**.

## paleomagnetic record
See **geomagnetic polarity time scale**.

## paleosol
A fossil soil. Paleosols are distinguished on the basis of their structure, color, and mineral segregation. The nature of paleosols reflects the factors that control soil formation (i.e., climate, vegetation, relief, parent material, and time) and thus paleosols can be used to reconstruct these variables in the geologic past. (Gk *palaios* = ancient and L. *solum* = ground.)

## palimpsest
The term originally referred to a parchment, the surface of which had been scraped in places to allow for reuse, and so by extension the text written on it was likely produced by several authors on different occasions. In paleoanthropology a palimpsest refers to an archeological or fossil assemblage formed of elements deposited at different times and perhaps by different causes or agents. It is useful to distinguish between "depositional" and "behavioral" causes of palimpsests. For example, the abraded *Homo erectus* fossils (calotte, femur, molar tooth) recovered some distance apart from a sandbank on a bend of the Solo River at Trinil, central Java, Indonesia, are very likely a "depositional" palimpsest in the sense that the bones were doubtless redeposited on the sandbank at different times. Archeological assemblages can also be "behavioral" palimpsests, indicating that the site was visited by one or several different individuals, possibly at several times for multiple different tasks (e.g, FLKN at Olduvai Gorge). Any site that has a sample size of artifacts or fauna large enough to be detected in the archeological or fossil record is likely to be a palimpsest, whether formed over days, months, years, decades, or thousands of years. (Gk *palimpseston* = scrape again.)

## palimpsest evolution
A term used to describe what is usually referred to as "mosaic evolution." See also **mosaic evolution**.

## palynology
The study of pollen and other organic microfossils recovered from sedimentary rocks and sediments. Palynology includes the use of pollen (a) to reconstruct past environments and climates, (b) to investigate of the effects of the activity of modern humans and extinct hominins (e.g., logging and burning) on the flora, and (c) in the creation of plant species lists to assist in reconstructing the diet of extinct hominins. (A neologism created in 1944 by Hyde and Williams, its roots are Gk *paluno* = to sprinkle and *pale* = dust, and the suffix *ology* = study of.) See also **pollen**.

## *Pan*
The genus that includes the common chimpanzees (*Pan troglodytes*) and the bonobo (*Pan paniscus*). Molecular evidence suggests that they split around 2 Ma ago, with *P. paniscus* presently confined to the region in the Democratic Republic of the Congo that lies to the south of the Congo River. The geographical range of *P. troglodytes* extends from Senegal and Guinea in the west to Uganda and Tanzania in the east, but it is not continuous between these two

longitudinal extremes for there are no chimpanzees in the relatively arid Dahomey Gap. DNA evidence suggests that chimpanzees from western Nigeria and from localities along the Nigeria/Cameroon border are distinct from the main population of Western chimpanzees, *P. troglodytes verus*. For reasons of priority, these populations are called *Pan troglodytes ellioti*. The range of the central chimpanzee (*Pan troglodytes troglodytes*) extends from its northern boundary with *P. troglodytes ellioti* in northern Cameroon (this approximates to the Sanaga River) to the Oubangui River in the east and the Congo River to the east and south. The Oubangui River separates the range of *P. troglodytes troglodytes* from that of the eastern chimpanzee, *Pan troglodytes schweinfurthii*, and the Congo River separates it from the range of the bonobo (see above). Molecular and morphological evidence are consistent in suggesting that *P. troglodytes verus* is the most distinctive subspecies of *P. troglodytes*.

## *Pan* fossils
See **KNM-TH 45519–45522 and 46437–46440**.

## panin
The informal or vernacular term for a specimen that belongs to a taxon within the tribe Panini (e.g., *Pan paniscus* is a panin taxon and KNM-TH 45519 is a panin fossil).

## Panini
The tribe that includes chimpanzees and bonobos and all the fossil taxa more closely related to chimpanzees and bonobos than to any other living taxon.

## Panina
If the tribe Hominini is interpreted to include the clades that contain modern humans and extant chimpanzees/bonobos, then some researchers discriminate between the two clades at the level of the subtribe. In this case the clade that contains extant chimpanzees/bonobos would be called the Panina and the clade that contains modern humans would be called the Hominina.

## paninan
The informal term for individuals or taxa within the subtribe **Panina** (*which see*).

## *Pan paniscus*
A species of extant chimpanzee known as the bonobo. It was previously referred to as the "pygmy chimpanzee," but given similar body masses and the extensive overlap between the limb dimensions of *Pan paniscus* and *Pan troglodytes* the epithet "pygmy" is inappropriate. In addition to its importance as one of the living taxa most closely related to modern humans some researchers take the view that among the extant *Pan* taxa *P. paniscus* may be the best model for the hypothetical common ancestor of chimpanzees/bonobos and modern humans. See also **Pan**.

## papionin
The informal name for the tribe **Papionini** (*which see*).

## Papionini
One of two tribes within the subfamily Cercopithecinae (the other is the Cercopithecini, or guenons), the family Cercopithecidae, and the superfamily Cercopithecoidea. The papionini comprises baboons and their allies. The extant genera included in the Papionini are *Papio* (common baboons), *Mandrillus* (drills and mandrills), *Theropithecus* (geladas), *Macaca* (macaques), *Cercocebus*, *Rungwecebus*, and *Lophocebus* (mangabeys). Fossil genera within the Papionini include *Parapapio*, *Dinopithecus*, and *Gorgopithecus*. Papionins tend to be relatively large-bodied and they are mainly terrestrially adapted (except for several macaque and mangabey species that are arboreal), whereas cercopithecins are smaller and generally more arboreal. The papionins are the best studied of the groups within the Old World monkeys, and a good deal is known about their evolution and past diversity, in part because their fossils are relatively common at many archeological and hominin paleontological sites. *Parapapio* and *Theropithecus* were the two dominant genera in the Plio-Pleistocene. *Parapapio* was extinct by the early Pleistocene. *Theropithecus*, which had a wider geographic distribution and a longer temporal span than *Parapapio*, has only one surviving member, the gelada, which is confined to the highlands of Ethiopia and Eritrea. The most abundant and widespread modern sub-Saharan African papionin genus, *Papio*, was poorly represented in the East African Plio-Pleistocene record, although it was more prevalent in southern Africa. Its present-day diversification apparently began when the modern chacma baboon, *Papio ursinus*, diverged in southern Africa c.1.8 Ma and then dispersed out and speciated during the Pleistocene. (L. *papio* = baboon, which gives rise to *Papio*, the name for the genus that contains the baboons.)

## paracone
The main cusp beside (i.e., on the buccal aspect of) the protocone on a maxillary (upper) molar tooth crown, or the main buccal cusp of a bicuspid upper premolar. It is part of the trigon component of an upper molar crown (Gk *para* = beside or by and *konos* = pine cone; syn. eocone, mesiobuccal cusp, tuberculum anterium externum.)

## parallel evolution
The independent origin of similar traits in closely related taxa (e.g., if the coronally rotated petrous bones seen in *Homo* and *Paranthropus* were not inherited from their most recent common ancestor, or if bipedalism evolved independently in *Australopithecus afarensis* and *Australopithecus africanus*). Convergent evolution is the independent origin of similar traits in distantly related taxa (e.g., the acquisition of thick-enameled teeth in orangutans and cebus monkeys). Convergent and parallel evolution are examples of homoplasy, for both concern the appearance of a similar morphology in two taxa that is not seen in their most recent common ancestor. *See also* **convergent evolution**; **evolution**; **homoplasy**.

## paralogous genes
Genes formed by a gene-duplication event. Paralogous genes often have similar functions and they may form a series of related genes known as a gene family (e.g., the *HOX* gene family and the globin gene family). Homology between paralogous genes of the same species can be used to establish gene phylogenies. (Gk *para* = beside and *logos* = reason.)

## paralogue
See **paralogous genes**.

## parameter
The true (and usually unknowable) population values that researchers try to estimate with sample statistics. Most statistical analyses involve collecting data on a sample that is part of a larger population. The statistics calculated for the sample (e.g., mean, standard deviation, variance, or range) are estimates, determined with some error, of the corresponding population parameters. Even if a researcher measures every known example of a trait (e.g., the mesiodistal length of all of the well-preserved mandibular first molars belonging to an early hominin taxon such as *Australopithecus afarensis*) the value determined is still a sample statistic for the researcher is using the fossil record to make an inference about the mean of the trait for the entire species. Many treat the terms parameter and variable as synonyms, but they are not. For example, the mesiodistal length of mandibular first molars is a variable, the mean and standard deviation of the mesiodistal length of mandibular first molars for all of the members of the species *Pan troglodytes* are parameters, and the mean and standard deviation of the mesiodistal length of mandibular first molars of the sample of *P. troglodytes* in the Powell-Cotton Collection are sample statistics (i.e., they are estimates of the parameters of the population of *P. troglodytes* from which the sample in the Powell-Cotton Collection is drawn). (Gk *para* = alongside and *meter* = measure.)

## parametric statistics
Statistical tests that assume data follow particular probability distributions (e.g., normal distribution, or a bell curve). Most common statistical tests (e.g., $t$ test, $F$ test, analysis of variance, etc.) are parametric. Inappropriate use of parametric tests when data do not meet the assumptions of parametric tests will result in outcomes being subject to increased Type I error. In such cases one should use nonparametric statistics. *See also* **nonparametric statistics**.

## *Paranthropus*
Hominin genus established in 1938 to accommodate the type species *Paranthropus robustus*. Many researchers considered the differences between the hypodigms of *Australopithecus* and *Paranthropus* insufficient to justify a second genus for the southern African archaic hominins, so it became a junior synonym of *Australopithecus*. However, the genus *Paranthropus* has been revived by researchers who consider that *P. robustus*, *Paranthropus boisei*, and *Paranthropus aethiopicus* belong to a separate hominin clade. Species in the genus *Paranthropus*, which are often informally called robust australopiths, share postcanine megadontia, thick enamel caps on their molars and premolars, a distinctive occlusal morphology in which the cusp tips of the postcanine teeth are located towards the center of the crown, and a number of highly derived features in the cranium (especially in the facial skeleton) generally thought to be functionally related to the generation and dissipation of high masticatory loads. (Gk *para* = beside and *anthropos* = human being.)

## Paranthropus aethiopicus (Arambourg and Coppens, 1968) Wood and Chamberlain, 1987

A new name combination used by researchers who do not recognize *Paraustralopithecus* as a separate genus, but who do consider >2.3 Ma hyper-megadont hominins from the Turkana Basin belong to a species that antedates and is distinct from *Paranthropus boisei*. The hypodigm of such a species would include a well-preserved adult cranium from West Turkana (KNM-WT 17000) together with mandibles (e.g., KNM-WT 16005) and isolated teeth from the Shungura Formation; some also assign L. 338y-6 to this taxon. The only postcranial fossil that is considered to be part of the hypodigm of *Paranthropus aethiopicus* is a proximal tibia from Laetoli. The fossil evidence for *P. aethiopicus* is similar to that for *Paranthropus boisei* except that the face of the former taxon is more prognathic, the cranial base is less flexed, the incisors are larger, and the postcanine teeth are not so large or morphologically specialized. When *P. aethiopicus* was introduced in 1968 it was the only megadont hominin in this time range, but with the discovery of *Australopithecus garhi* it is apparent that robust mandibles with long premolar and molar tooth rows are being associated with what are claimed to be two distinct forms of cranial morphology. (Gk *para* = beside, *anthropos* = human being, L. *australis* = southern, Gk *pithekos* = ape, and *aethiopicus* = Ethiopia.) *See also* **Laetoli**; ***Paraustralopithecus aethiopicus***.

## Paranthropus boisei (Leakey, 1959) Robinson, 1960

For researchers who support the hypothesis that *Australopithecus robustus* and *Australopithecus boisei* are sister taxa it makes sense to revive the genus *Paranthropus* as a way of recognizing that clade. In this case *Zinjanthropus boisei* that became *Au. boisei* (Leakey, 1959) would become *Paranthropus boisei* (Leakey, 1959), a name combination first used by John Robinson. *P. boisei* has a comprehensive craniodental fossil record. There is evidence of both large- and small-bodied individuals and the range of the size difference suggests a substantial degree of body-size sexual dimorphism. It is the only hominin to combine a massive, wide, flat, face with massive premolars and molars, small anterior teeth, and a modest endocranial volume (around 480 cm$^3$). The face of *P. boisei* is larger and wider than that of *Paranthropus robustus*, yet their brain volumes are similar. The mandible of *P. boisei* has a larger and wider body or corpus than any other known hominin, and the tooth crowns apparently grew at a faster rate than has been recorded for any other early hominin. There is, unfortunately, no postcranial evidence that can with certainty be attributed to *P. boisei*, but some of the postcranial fossils from Bed I at Olduvai Gorge currently attributed to *Homo habilis* may belong to *P. boisei*. The fossil record of *P. boisei* extends across about 1 million years of time (c.2.3–c.1.3 Ma) during which there is little evidence of any substantial change in the size or shape of the components of the cranium, mandible, and dentition. The holotype is OH 5 and fossils assigned to *P. boisei* come from Chesowanja, Konso, Koobi Fora, Malema, Olduvai Gorge, Peninj (Natron), Shungura Formation, and West Turkana. (Gk *para* = beside and *anthropos* = human being, and *boisei* to recognize the help provided to Louis and Mary Leakey by Charles Boise.) *See also* ***Zinjanthropus boisei***.

## Paranthropus crassidens Broom, 1949

Hominin species introduced in 1949 to accommodate SK 6, an adult mandible recovered at Swartkrans from breccia now believed to be derived from Member 1. Broom opted for the new species name because he considered the Swartkrans teeth to differ from those of *Paranthropus robustus* from Kromdraai. However, most researchers do not make a specific distinction between the hypodigms, so *P. crassidens* is now regarded as a junior synonym of *Paranthropus robustus*. (Gk *para* = beside, *anthropos* = human being, and L. *crassus* = dense.) See also **Paranthropus robustus**.

### *Paranthropus robustus* Broom, 1938

Hominin species established in 1938 for fossils recovered from what was then referred to as the "Phase II Breccia" (now called Member 3) at Kromdraai. The hypodigm comes from Cooper's, Drimolen, Gondolin, Kromdraai (Mb 3), and Swartkrans (Mbs 1, 2, and 3). The dentition is well represented and some partial crania are well preserved, but most of the mandibles are crushed or distorted. The postcranial fossil record is more meager. The brain, face, and chewing teeth of *Paranthropus robustus* are larger than those of *Australopithecus africanus*, yet the incisors and canines are smaller. The crania of *P. robustus* include specimens that have ectocranial crests, whereas there are no *Au. africanus* crania with unambiguous crests. What little is known about the postcranial skeleton of *P. robustus* suggests that the morphology of the pelvis and the hip joint is much like that of *Au. africanus*. It was most likely capable of bipedal walking, but most researchers subscribe to the view that *P. robustus* was not an obligate biped. It has been suggested that the thumb of *P. robustus* would have been capable of the type of grip necessary for stone tool manufacture, but this claim is not accepted by all researchers. Most researchers subsume *Paranthropus crassidens*, recovered from Swartkrans, into *P. robustus*. The holotype is TM 1517 and the taxon spans the period between c.2.0–1.5 Ma. See also **Paranthropus**.

### parapatric speciation

A mode of speciation in which a new species evolves when part of a species population colonizes a new ecological niche in a geographical zone adjacent to that of the "parent" species. Unlike allopatric speciation, there is no physical boundary separating the parent and "daughter" populations, but because these two populations are subject to different selective pressures the hybrid individuals that result from mating between the populations experience reduced fitness. Over time, this reduces gene flow between the populations and favors mating within each population. This effect, in combination with natural selection affecting each population in different ways, eventually leads to the two populations becoming reproductively isolated.

### parapatry

Two organisms with adjacent but not significantly overlapping geographic ranges are described as being parapatric (e.g., most extant baboon species are parapatric). In parapatric speciation there is usually no physical barrier between populations and new species arise from contiguous (adjacent) populations. It may also occur when a subpopulation of the parent population moves into an area with a different environment and thus different selective pressures, causing a selective advantage for traits that differ from those seen in the majority of the parent population. This can cause clinal variation and eventually divergence and speciation. Interbreeding between populations occurs at hybrid zones where the populations meet, but because there is no significant gene flow between the populations, each population maintains its unique characteristics. It has been suggested that baboon-like hybrid zones were important in creating diversity and hence speciation within *Homo*. (Gk para = beside and *patris* = fatherland.) *See also* **parapatric speciation**; **reticulate evolution**; **speciation**.

## paraphyletic
An adjective used to describe a taxon that is part of a paraphyletic group. *See also* **paraphyletic group**.

## paraphyletic group
A taxonomic grouping that includes only part of a clade or monophyletic group. For example, if the term "great apes" is used to refer to a group that includes chimpanzees, gorillas, and orangutans, then the term refers to a paraphyletic group. This is because there is sound evidence that humans are most closely related to chimpanzees, and this group is next most related to gorillas and then orangutans. The clade structure of those taxa is (((*Homo*, *Pan*) *Gorilla*) *Pongo*); there is no clade made up of *Pan*, *Gorilla*, and *Pongo* that does not include *Homo*. (Gk para = beside and *phulon* = a class.)

## paraphyly
The state of a taxonomic group that is paraphyletic. *See also* **paraphyletic group**.

## parasagittal crests
A term usually used for two sharp crests of bone running either side of the sagittal (aka inter-parietal) suture. The bony crests form when the medial borders of the fascia covering the temporalis muscles are raised up. In extant hominoids the interval between the crests is occupied by fat and small blood vessels. (Gk *para* = beside, L. *sagitta* = an arrow, and L. *crista* = cock's comb, or a tuft of feathers, in the midline of a bird's head.) *See also* **sagittal crest**; **sagittal keel**.

## paratype
Specimens other than the holotype listed in the original description of a taxon. (Gk *para* = beside and *typus* = image.) *See also* **holotype**; **lectotype**; **type specimen**.

## *Paraustralopithecus* Arambourg and Coppens, 1968
Hominin genus established to accommodate the type species *Paraustralopithecus aethiopicus*. The genus *Paraustralopithecus* is now regarded as a junior synonym of either *Australopithecus* or *Paranthropus*. (Gk *para* = beside, L. *australis* = southern, and Gk *pithekos* = ape.) *See also* *Paraustralopithecus aethiopicus*.

## *Paraustralopithecus aethiopicus* Arambourg and Coppens, 1968
Hominin species established to accommodate a mandible Omo 18-1967-18, the holotype, from the Shungura Formation, Omo region, Ethiopia. Some sink *Paraustralopithecus aethiopicus* into *Australopithecus boisei* (or *Paranthropus boisei*), but others have assigned >2.3 Ma hyper-megadont fossils from the Shungura Formation and West Turkana into either *Australopithecus aethiopicus* or *Paranthropus aethiopicus*. When *P. aethiopicus* was found it was the only hyper-megadont hominin in this time range, but with the discovery

of *Australopithecus garhi* it is apparent that robust mandibles with large premolars are being associated with two distinct forms of cranial morphology. (Gk *para* = beside, L. *australis* = southern, Gk *pithekos* = ape, and *aethiopicus* = Ethiopia.) *See also* **Omo 18-1967-18**.

## parcellation
A term used in neuroscience to refer to the subdivision of the cerebral cortex into anatomically distinct regions based on differences in cytoarchitecture, myeloarchitecture, chemoarchitecture, or patterns of connectivity. Korbinian Brodmann used an early version of parcellation to divide the cerebral cortex into the cytoarchitectonically distinctive Brodmann's areas. *See also* **Brodmann's areas**.

## parental effects
*See* **genomic imprinting; heritability**.

## parental investment
Any investment by a parent in an individual offspring that increases the offspring's chance of survival (and hence reproductive success) at the cost of the parent's ability to invest in other offspring. Parental investment is usually unbalanced between the sexes, and the sex that invests the most in offspring is thought to be the more discriminating sex when it comes to finding a mate. Aspects of parental investment in modern humans and nonhuman primates include gestation, lactation, carrying of offspring, food provisioning, and protection. In most primates, females are the higher-investing sex, although callitrichines and some modern human populations show significant male parental investment.

## parietal art
*See* **art**.

## parietal lobe
One of the four main subdivisions of the cerebral cortex of each cerebral hemisphere. It is located between the frontal lobe anteriorly and the occipital lobe posteriorly. The parietal lobe is involved in processing information about somatic sensation, body position, attention, visual and spatial relations, and language. The parietal lobe contains the primary somatosensory cortex, which lies on the postcentral gyrus.

## parietal-occipital plane
In some crania (e.g., OH 5) the external aspect of the squamous part of the occipital bone and the external aspect of the posterior part of the parietal bones form a continuous flat surface that is referred to as the parietal-occipital plane, or planum.

## parsimony
The principle that the most economical explanation is the one that should be adopted in the first instance. It is the equivalent of Occam's razor. For example, in phylogenetic systematics the parsimony principle means that the cladogram that involves the fewest evolutionary

changes (i.e., the fewest character state changes) is the one that should be preferred. (L. *parsimonia* = frugality.) *See also* **Occam's razor**.

**passive scavenging**
A carcass transport strategy that involves culling small amounts of meat or marrow from animal carcasses that have been heavily ravaged and abandoned by their initial predators. Opportunistic passive scavenging has been documented ethnographically in some hunter–gatherer groups (e.g., the Hadza), but obligate passive scavenging of terrestrial mammals is considered to be outside the range of modern human behavior. *See also* **scavenging**.

**pastoralist**
Mobile societies that rely on livestock as a food source and which are driven to move seasonally in search of forage and water. Pastoral systems based on the herding of domesticated or partially domesticated animals arose 12,000–10,000 years BP, approximately at the same time as crop domestication. Worldwide, the most common animals herded by pastoralists are sheep, goats, cattle, horses, donkeys, camels, reindeer, and yaks. Economically, pastoralism is viable in regions too arid for the planting of crops, and pastoralists frequently control their ecosystems by using fire to limit the growth of trees and to rejuvenate pasture. Milk, blood, meat, skin, and dung are common products of pastoral societies, but wool or hair can be important as well. Agriculturalists and pastoralists frequently develop extensive regional trade networks to exchange goods, and historically competition for land has often led to conflict between pastoral and agricultural groups. (L. *pastor* = a shepherd.)

**pattern**
*See* **mode**.

**Paviland**
*See* **Homo sapiens**; **radiocarbon dating**.

**PCA**
*See* **principal components analysis** (*which see*).

**PCR**
Abbreviation of **polymerase chain reaction** (*which see*).

**p-deme**
*See* **deme**.

**Pech de l'Azé**
A site in France, consisting of four adjacent localities (I–IV), that contains evidence of a *Homo neanderthalensis* child associated with a Mousterian of Acheulean Tradition (or MTA) assemblage. Pech IV may be as old as 100 ka; the Mousterian levels are c.51–41 ka. (Location 44°51′29″N, 01°15′14″E, France; etym. Occitan *pech* = hill; *azé* = donkey.)

## Pech Merle

This extensive cave site near the village of Cabarets, France, is known for its cave paintings. Stylistic features of the artwork suggest the site dates to the Gravettian period and radiocarbon dating of the pigments has provided an age of c.25 ka. (Location 44°40′27″N, 01°38′40″E, France; etym. Occitan *pech* = hill; *merle* is unknown.)

## pectoral girdle

The bones by which the proximal part of the upper limb is attached to the axial skeleton. The pectoral girdle, which comprises the scapula and clavicle, is part of the postcranial skeleton. (L. *pectoralis* from *pectus* = breast and ME *girdle* = sash.)

## pedal digit

See **digit**; **foot**.

## pedal ray

See **foot**; **ray**.

## pedogenic

Refers to the structure, color, mineralogy, and horizontal zonation characteristics of forming soils. Pedogenic carbonate, which is a mineral precipitate formed in the subsurface (or B) horizon of a soil, is most commonly found in arid to semi-arid conditions. Nodules and rhizoliths preserve the carbon and oxygen stable-isotopic signatures during the time of soil formation and thus in ancient soils they can be used as a proxies to reconstruct paleovegetation, paleoclimate, and paleoenvironment. (Gk *pedon* = Earth and *genos* = birth.) See also **$C_3$** and **$C_4$**.

## pedomorphic

See **pedomorphosis**.

## pedomorphosis

The state in which adults of one species retain features seen only in the juveniles of their ancestors. Pedomorphic morphology results from three different kinds of processes that change the timing of development (time hypomorphosis, rate hypomorphosis, or neoteny). All of these processes produce, in varying ways, descendants with ancestral juvenile characteristics or shapes retained during later stages of ontogeny. Since the juvenile features are shifted to the adult, there are no novel pedomorphic morphologies. (Gk *paedo* = child and *morphe* = form.) See also **acceleration**; **heterochrony**; **neoteny**; **peramorphosis**.

## Peking Man

The informal name given to the taxon represented by the fossil hominin remains recovered from Locality 1 in the Zhoukoudian Lower Cave, a site not far from Beijing (formerly Peking), China. The remains were initially assigned to *Sinanthropus pekinensis*, but early in 1940 they were transferred to *Pithecanthropus erectus* and later that year Franz Weidenreich formally suggested that taxon be a junior synonym of *Homo erectus*. See also **Homo erectus**; **Zhoukoudian**.

## pelvic girdle

The pelvic girdle, which is part of the postcranial skeleton, comprises the bones that form the pelvis [i.e., the two pelvic (aka hip) bones and the sacrum]. (L. *pelvis* = basin and ME *girdle* = sash.)

## pelvis

The pelvis surrounds the birth canal and provides attachments for the propulsive muscles of the hindlimb, the muscles of the anterior abdominal wall, as well as the muscles that move and support the spine. Its shape reflects the overall shape of the torso. It is formed by the sacrum and two pelvic, or hip, bones; the three components together are called the pelvic girdle. Three primordia – the ilium, ischium, and pubis – fuse during early childhood to form a single pelvic bone. The ilium is the broad, cranial portion, the pubis the ventral portion that meets in the midline, and the ischium the dorsal and caudal portion; all three contribute to the acetabulum or hip joint socket. Apes, and especially great apes, have craniocaudally elongated pelves consistent with their shortened lumbar vertebral column and stiff torso, which provides stability for hoisting the body up during arm-hanging and climbing. Apes have a broad rib cage and shoulder joints facing laterally and correspondingly reoriented, dorsally rotated, iliac blades. The modern human pelvis is shorter and broader than in any other extant primate. The wide modern human sacrum accommodates the wide set of the vertebral zygapophyses in the lumbosacral region that is necessary for upright posture. The expanded iliac blades of the modern human pelvis are rotated so that the external surface faces partly laterally; it is on this surface that the lesser gluteal musculature (gluteus minimus and medius) originates. These muscles abduct the femur at the hip joint and keep the pelvis level during the single support phase of modern human bipedal gait. Among the great apes, only modern humans have this abductor mechanism. This iliac form also places the abdominal oblique muscles in a position to wrap around the abdomen and assist in the lateral pelvic tilt that is part of the bipedal gait of modern humans. The iliac blades of *Australopithecus* tend to be more horizontal and more widely flared than those of *Homo*.

With the expansion of the size of the pelvic inlet in *Homo* to accommodate larger-brained neonates, the acetabulae are more widely set and the ilia are more vertical. The acetabulocristal buttress becomes more pronounced and the portion of the ilium anterior to this buttress also expands. The part of the ilium dorsal to the sacroiliac joints is expanded in modern humans and provides a large area of attachment for the erector spinae musculature. Hominins have short, retroflexed ischia, which allow the hamstring muscles the leverage to assist in extending the thigh in bipedal postures; this pattern contrasts with the ischia of great apes that are elongated and dorsally directed, thus providing maximum leverage when the thigh is flexed during quadrupedal locomotion and in climbing postures. In modern humans, because of the tight relationship between fetal head size and the pelvic inlet, male and female pelves are sexually dimorphic. As such, the sex of a skeleton can be determined with a fair degree of accuracy from the form of the pelvic bone. Because of the wider set of the hip joints and the wider pelvic inlet in modern human females, the angle between the caudal rami of the pubes is wider than in males, and their internal margins are everted, forming a lip that flares ventrally. In modern humans, the pelvic inlet is broader mediolaterally than anteroposteriorly (i.e., it is platypelloid), whereas the opposite proportions are found at the outlet. As a consequence, in modern humans the neonatal head rotates during the birth process; *Australopithecus* and early *Homo* pelves appear to be platypelloid at both the inlet and the outlet and would not have had a rotational birth mechanism. Climate influences pelvic shape in modern humans because body breadth heavily influences surface-area-to-volume ratios, thus affecting a body's capacity for radiating heat. Tropical populations tend to have narrower pelves than those in more northern latitudes, with *Homo neanderthalensis* having the broadest pelves of all known hominins. (L. *pelvis* = basin.) *See also* **lumbar vertebral column**; **scapula, evolution in hominins**; **sexual dimorphism**.

## Peninj
A site located west of Lake Natron near the boundary between Kenya and Tanzania. Several stratigraphic intervals are represented in the Peninj exposures, but the Peninj 1 hominin mandible and the artifacts come from the 1.7–1.3 Ma Humbu Formation. Peninj 1, which can be confidently assigned to *Paranthropus boisei*, is between c.1.5 and 1.3 Ma. (Location 02°00–50'S, 35°40'–36°20'E, Tanzania.)

## peramorphosis
A category of morphologies that result from three different kinds of processes that change the timing of development (time hypermorphosis, rate hypermorphosis, or acceleration). Peramorphosis, unlike pedomorphosis, introduces novel morphology because it is the result of extending the ancestral ontogenetic trajectory beyond its original endpoint. (Skt *para* = beyond plus Gk *morphe* = form.) *See also* **acceleration**; **heterochrony**; **neoteny**; **pedomorphosis**.

## percussion
The action of one object striking another. Percussion is used in the study of stone tools to refer to the technique by which rock is fractured. In direct percussion, the stone, bone, antler, or wood hammer strikes the rock directly, whereas with indirect percussion an intermediate object, typically a chisel-like piece of bone, wood, or antler is placed on the rock and struck, directing the impact force for the more reliable production of elongated flakes and blades.

## perforator
A flake or blade with one (or rarely multiple) small, pointed tip(s) shaped by finely controlled retouch. Sometimes referred to as a "borer," perforators are inferred to have been used in the scraping or drilling of holes in materials such as wood, bone, shell, and hide. (L. *perforare* = to bore.)

## Perigordian
A term that refers to early Upper Paleolithic technologies predominantly found at sites in central and southwestern France, and named after the Périgord region. The results of more recent excavations suggest that what was originally called the Perigordian I should be subsumed within the Châtelperronian, and the Perigoridan II and III should be subsumed within the Aurignacian. The Upper Perigordian (Perigoridian IV, V, and VI) is still seen as a separate technology, characterized by backed points made on blades, rare bone tools, carved, stylized figurines of women (called Venus figurines), parietal art, and use of shelters, and dated to

28–21 ka. Similar technocomplexes found at sites outside of this area of France are usually called Gravettian.

## perikymata

External ridges that encircle the crowns of all permanent teeth; they are especially pronounced on the labial surface of the crowns of the anterior teeth. The troughs between the ridges correspond with striae of Retzius. Perikymata counts can be used to estimate the time it takes to form the imbricational enamel (or lateral enamel), but to do this you also need to know the periodicity of the perikymata (i.e., how much time has elapsed between each perikymata). In modern humans the perikymata become more closely packed as you move from the cusp apex to the cervix, but this is not the case in some hominin taxa (e.g., *Paranthropus*). (Gk *peri* = around and *kymata* = waves.) See also **enamel development**; **striae of Retzius**.

## period

A unit of geological time (i.e., a geochronologic unit). It is a subdivision of an era, and it is itself subdivided into epochs. For example, the Neogene period is within the Cenozoic era and is itself divided into the Miocene and Pliocene epochs. The corresponding chronostratigraphic unit equivalent of a period is a system.

## periodicity

The number of daily cross-striations between striae of Retzius in enamel, and/or the number of von Ebner's lines between Andresen lines in dentine. The periodicity of these long-period incremental lines is consistent within an individual, but it varies among individuals of a taxon. The modal value for the periodicity of long-period incremental lines in fossil hominins is 7 days in australopiths and 8 days in *Homo*. The presence of incremental lines in enamel allows researchers to determine the rate and duration of crown formation, and in some cases it is possible to infer the age at death of individual specimens. (Gk *periodos* = interval of time.) See also **Andresen lines**; **cross-striations**; **dentine**; **striae of Retzius**.

## periodontal ligament

A fibrous connective tissue ligament that exists in the 300–500 μm-wide space between the surface of the tooth root and the inner aspect of the alveolar bone socket. The periodontal ligament (aka PDL) is rich in blood vessels and proprioceptive nerve endings that relay information about bite force to the brain and trigger protective reflexes. In addition to attaching a tooth to the bony walls of its alveolus, the PDL also functions as a hydrodynamic damping mechanism that helps absorb and distribute the forces generated when a tooth is loaded during biting. (Gk *peri* = around *odonto* = pertaining to dentine.) See also **chewing**.

## peripatric speciation

A mode of speciation in which a new species evolves as a consequence of a small subpopulation of the original species population colonizing a new geographic area (e.g., animals trapped on a large mat of vegetation that separates from a much larger island). Because these peripheral populations are small and geographically isolated from the parent population, genetic drift causes the new population to become genetically distinct and, eventually, reproductively isolated from the parent population. This process is also known as the founder effect. Peripatric speciation can be thought of as a special case of allopatric speciation insofar as reproductive isolation (and, hence, a new species) evolves in geographically isolated populations. *See also* **founder effect**; ***Homo floresiensis***.

## Perissodactyla

The mammalian order in which the axis of the leg passes through a single-hoofed toe (unlike the Artiodactyla in which the axis passes between two hoofed toes). Perissodactyls underwent an adaptive radiation earlier in the Cenozoic, particularly in North America, but during the Neogene this order was relatively impoverished, with low species diversity. The perissodactyl taxa found at early hominin sites include members of the Rhinocerotidae (rhinoceros) and the Equidae (zebras, horses, and their allies). (Gk *perisso* = odd-numbered and *daktulos* = toe.)

## permanent dentition

In modern humans and other catarrhine primates the permanent dentition consists of the two incisors (I), one canine (C), two premolars (P), and three molars (M) in each quadrant of the jaw. The permanent teeth either replace the deciduous dentition [i.e., the permanent incisors (I) replace the deciduous incisors (i), the permanent canines (C) replace the deciduous canines (c), and the premolars (P) replace the deciduous molars (dm)] or they are formed in the alveolar process distal to the deciduous dentition (i.e., the permanent molars, M). *See also* **deciduous dentition; teeth**.

## permutation test

*See* **randomization**.

## Perning

*See* **Mojokerto**.

## Peştera Cioclovina Uscata

Excavations at Cioclovina Cave, which is part of a large karstic system, have produced Mousterian and Aurignacian lithic material and a hominin calotte. (Location 45°35′N, 23°07′E, Romania; etym. in Romanian literally the "dry Cioclovina Cave.")

## Peştera cu Oase

This site consists of previously sealed galleries that form part of a karstic complex in southwestern Romania. The remains of two *Homo sapiens* individuals, Oase 1 and 2, have been directly dated using AMS radiocarbon dating to *c.*35 ka and *c.*29 ka, respectively, making them the oldest reliably dated modern human remains in Europe. (Location 45°01′N, 21°50′E, Romania.)

## PET

Acronym for positron emission tomography, a neuroimaging modality. *See* **positron emission tomography**.

## petalia

A cerebral petalia is a modest relative expansion of one part of one side of the cerebral hemisphere. Petalias are considered by some to be a manifestation of a more general anatomical asymmetry of the brain. The most typical configuration in modern humans is the relative expansion of the right frontal and the left occipital lobes. Some authors have argued that this pattern (known as a torque) is unique to modern humans and is linked with language ability, but some earlier hominin species show similar asymmetrical cerebral petalias and recent comparative studies using measurements from magnetic resonance imaging scans suggest that left-occipital and right-frontal petalias are also seen in chimpanzees. There is no evidence of the left–right petalia torque in any primate species outside of the great apes. (Gk *petalos* = leaf, outspread.)

## Petralona

The Petralona Cave in Greece is best known for the discovery of a remarkably complete fossil hominin cranium that has most recently been assigned to *Homo heidelbergensis*. The cranium, which was found on a ledge, has neither sedimentological nor paleontological context so its age is unknown. (Location 40°22′08″N, 23°09′33″E, Greece.)

## petro-median angle

The angle between long axis of the petrous part of the temporal bone as seen from below (i.e., in *norma basilaris*) and the median, or sagittal, axis of the cranium. The petro-median angle is more acute (i.e., smaller) in *Pan*, whereas in later *Homo* and in *Paranthropus boisei* it is closer to the coronal plane and is thus more obtuse (i.e., larger).

## petrous

Refers to the dense and hard (hence "rock-like") wedge-shaped part of the temporal bone. (L. *petrous* = rock-like; the same root as "petroleum" which is literally "rock oil.")

## pharynx

The hollow soft-tissue structure that connects the mouth and nose with the esophagus and the larynx. The pharynx consists of three parts – nasopharynx, oropharynx, and laryngopharynx – which are posterior to the nasal cavity, oral cavity, and larynx, respectively. The pharynx, which is made exclusively of soft tissues so there is no direct evidence of it in the fossil record, is one of the structures that differ substantially in shape in chimpanzees and modern humans. The basicranium and the pharynx in the modern human neonate are much the same as they are in an adult chimpanzee; this configuration allows modern human neonates to suckle and breathe at the same time. However, once the cranial base flexes, the soft palate no longer seals off the mouth from the respiratory tract. Instead, in adult modern humans the soft palate seals off the nose and nasopharynx from the mouth and oropharynx. (Gk *pharunx* = a chasm.)

## phenetic analysis

A form of analysis that uses information about the phenotype. Contrast this with a cladistic analysis that just focuses on shared/derived, or apomorphic, aspects of the phenotype. Phenetic methods should be used to identify taxa (also called alpha taxonomy) and cladistic methods should be used to investigate the relationships among taxa. (Gk *phainen* = to show.) *See also* **alpha taxonomy; morphology; phenotype.**

## phenetic species concept
See **species**.

## phenotype
The term used for the observable characteristics of a living organism, from its molecular structure up to its overall size and shape. The phenotype is determined by complex interactions between the genotype and the environment. (Gk *phainen* = to show and *typus* = image.) See also **epigenetics**; **phenotypic plasticity**.

## phenotypic plasticity
The tendency for the same genotype or genetic program for development to generate a range of phenotypes (called reaction norms) in response to different environmental settings. The phenotypic differences, which can be behavioral as well as morphological, may be initiated early in development or may take place later during adulthood. Ultimately the capacity for plasticity and the degree to which an organism's phenotype is malleable is dependent on the fitness benefit of the plasticity. In modern humans the relationship between body condition and reproductive capacity in women, whereby very low body fat resulting from undernutrition or extreme physical exercise results in irregularities or cessation in menstrual cycling, is an example of phenotypic plasticity. (Gk *phainen* = to show, *typus* = image, and *plaissen* = to mold.)

## phenotypic variance
See **heritability**.

## philopatry
The term used when an individual remains within their natal group (i.e., the group they were born into). Primate social groups tend to be structured so that once either males or females reach sexual maturity they move away from the group in which they were born, presumably to minimize inbreeding. In Old World monkeys it is common for the males to move away from their natal group, but females stay in theirs; hence this is known as female philopatry. This contrasts with chimpanzees in which it is the males that tend to remain in their natal group (i.e., male philopatry). (Gk *philo* = loving and *patris* = fatherland.)

## phoneme
See **spoken language**.

## phonolite
See **lava**.

## phyletic gradualism
The phenomenon in which an ancestral species is gradually transformed through many incremental stages into a new descendant species. This mode of evolution, which is typically associated with anagenesis (but can also be associated with cladogenesis), is the one envisioned by Charles Darwin to have been the end result of natural selection. There are a growing number of paleoanthropologists who believe that *Australopithecus anamensis* was

transformed into *Australopithecus afarensis* through this process, and many researchers believe that *Homo neanderthalensis* evolved from earlier even more archaic pre-modern *Homo* populations using the same mode. A taxonomic implication of phyletic gradualism is that the boundaries of species are likely to be difficult to locate and define and because of this its adherents are likely to recognize a smaller rather than a larger number of species in the fossil record. *See also* **anagenesis; lumper.**

## phylogenetic
Adjective formed from the noun phylogeny. For example, a phylogenetic tree is a branching diagram that depicts a hypothesis about the shape of part of the tree of life. A phylogenetic analysis (also known as cladistic analysis) is a form of analysis that is designed to recover information about relationships that are the product of phylogeny. Confusingly phylogenetic analysis results in a branching diagram called a cladogram and not a phylogeny or a phylogenetic tree. A phylogenetic tree is a more complex hypothesis than a cladogram for it includes time and specifies ancestors and descendants whereas a cladogram does not. (Gk *phylon* = race, *genesis* = birth or origin, and *etikos* = from.)

## phylogenetically independent contrasts
A type of phylogenetic comparative analysis that takes into account phylogenetic relatedness when comparing continuous variables among three or more taxa. Most statistical analyses assume statistical independence of data points, but in phylogenetic comparative analyses species are not treated as statistically independent data points. This is done because if taxa share a recent common ancestry then they are more likely to share similar values for a variable because they have been phylogenetically independent from each other for a shorter period of time (i.e., less time has elapsed since they were a single species). Phylogenetically independent contrasts works by converting species data into standardized differences between pairs of species (contrasts between two operational taxonomic units), between a species and an internal node within the phylogeny (contrasts between an operational taxonomic unit and a hypothetical taxonomic unit), or between pairs of internal nodes (contrasts between two hypothetical taxonomic units). The proper name for the method is phylogenetically independent contrasts, but it is often referred to in its shortened form, "independent contrasts." *See also* **phylogenetic generalized least squares.**

## phylogenetic analysis
*See* **cladistic analysis.**

## phylogenetic comparative analysis
A type of analysis recognizing that data from each of multiple species are not statistically independent. It is a method of analysis that takes into account phylogenetic information when identifying relationships across multiple taxa. The two most commonly used types of phylogenetic comparative analysis are phylogenetically independent contrasts and phylogenetic generalized least squares. *See also* **phylogenetically independent contrasts; phylogenetic generalized least squares.**

## phylogenetic constraint

A cause of phylogenetic inertia, this is the persistence of a trait despite changes in aspects of the environment that can be expected to be selectively important for that trait. The most commonly invoked phylogenetic constraints are limited genetic variation, pleiotropy, and functional interdependency. *See also* **phylogenetic inertia**.

## phylogenetic generalized least squares

A type of phylogenetic comparative analysis that takes into account phylogenetic relatedness when comparing continuous variables among three or more taxa. *See also* **phylogenetically independent contrasts**.

## phylogenetic inertia

The persistence of a trait despite changes in aspects of the environment that can be expected to be important for selection for that trait. Thus, phylogenetic inertia applies to two situations. One is where a trait persists after the selective force that produced and/or maintained it stops operating. The other is where a trait is unaffected by new environmental conditions that should have resulted in selection acting on the trait. In the latter case, the lack of change is usually attributed some phylogenetic constraint. In paleoanthropology phylogenetic inertia has played an important role in the debate about the locomotor behavior of the australopiths. Some researchers who contend that australopiths were striding bipeds invoke phylogenetic inertia to explain why these australopiths retain primitive, ape-like, traits (e.g., their curved fingers and toes). (syn. phylogenetic effect, phylogenetic lag.)

## phylogenetic species concept

Under this concept species are defined as the smallest aggregation of populations diagnosable by a unique combination of character states. *See also* **species**.

## phylogenetic systematics

An integrated approach to phylogeny reconstruction and classification devised by Willi Hennig. The phylogeny reconstruction aspect of phylogenetic systematics, which is commonly called cladistic analysis, is discussed in that entry. Under the principles of phylogenetic systematics, classifications should only reflect information about phylogeny. Thus taxonomic groups should always be monophyletic. This emphasis on monophyly distinguishes phylogenetic systematics from the other major approach to classification, evolutionary taxonomy, which allows taxa to be paraphyletic if members of a clade exhibit significant adaptive differences.

## phylogenetic tree

A phylogenetic tree is a branching diagram that tries to capture as accurately as possible the evolutionary history of a group of taxa. The evolutionary history of a taxon is the path taken as one traces its ancestors (initially recent, but subsequently increasingly remote) back into the tree of life. It is a more complex hypothesis than a cladogram because it specifies ancestors and descendants. Several different phylogenetic trees may be compatible with a single cladogram. In a phylogenetic tree, time is on the vertical axis and morphology on the horizontal axis.

## phylogeny

The phylogeny of a taxon is the same as its evolutionary history. The evolutionary history of a taxon is the path taken as one traces ancestors (initially recent, but subsequently more remote) back into the tree of life. Evolutionary history is usually represented visually as a branching diagram called a phylogenetic tree, also called simply a phylogeny. A phylogeny is a more complex hypothesis than the hypothesis about relationships set out in a cladogram; the former includes specific hypotheses about ancestors and descendants, whereas the latter does not. Thus, a single cladogram may be consistent with several different phylogenetic trees. Strictly speaking, only taxa have phylogenies; individual organisms have an ontogeny, but not a phylogeny. (Gk *phulon* = tribe or race and *gena* = to give birth to; syn. ancestry, genealogy.)

## phylogeny reconstruction

The generation of an hypothesis or set of hypotheses concerning the phylogenetic relationships among a group of taxa. Phylogeny reconstruction can be carried out in a number of different ways. In paleoanthropology the main methods are phenetic and cladistic analysis *See also* **cladistic analysis; phenetic analysis**.

## phylogram

*See* **branch length**.

## phytolith

One of several kinds of plant microfossil, phytoliths are small noncrystalline silica bodies found within the tissues of plants. They function as a form of structural support and/or as a physical defense mechanism against herbivory. They are thought to be one of the major causes of microscopic marks on teeth. They can be recovered from archeological sediments, stone tools, ceramics, and dental calculus, and can provide information on which plants were present in the environment or in the diet of individuals and groups. *See also* **dental microwear**.

## pig

Informal inclusive term to describe the taxa within the family Suidae. (syn. suid, swine.) *See* **Suidae**.

## pigmentation

Colored material in plant or animal cells that is not due to structural color (such as iridescence). Examples of pigmentation variation in modern humans and other primates include the color of the skin, hair, and eyes. Skin color is determined by the relative amounts of melanin, blood, and keratin. Over 100 genes have been identified as being involved in pigmentation in mammals. Many of these are not involved in normal pigment variation within a species, but were discovered because they are mutated in a genetic disease or condition (e.g., albinism). Recent analyses have identified close to 20 genes involved in normal skin, hair, and eye color variation in different modern human populations. It has been suggested that the first hominins, like chimpanzees, possessed fair skin covered by protective fur. One theory suggests that the ancestors of modern humans evolved dark skin to reduce damage from solar radiation around the time ($c.1.2$ Ma) they lost their body hair. Several hypotheses (not necessarily mutually exclusive) exist to explain normal human pigmentation variation. These include sexual selection, purifying selection because of ultraviolet intensity and selection for sufficient vitamin D production, and genetic drift.

## Piltdown

In 1908 a laborer found "an unusually thick human parietal bone" in the grounds of Barkham Manor near to the village of Piltdown in Kent, England. Three years later Charles Dawson, a local solicitor, claimed he found "a larger piece of the same skull that included a portion of the left supra-orbital border" and a year later he found a third hominin cranial fragment, the right side of a mandible, three more pieces of parietal bone, and a piece of occipital. These fragments were used to reconstruct the Piltdown I skull and the mammal fossils that were supposedly found with the hominin "fossils" suggested the remains were very old. Sir Arthur Smith Woodard's interpretation was that the cranium and jaw belonged to a single individual, and that while "the skull is essentially human"…"the mandible appears to be that of an ape, with nothing human except the molar teeth." In 1913 a pair of nasal bones and a canine tooth were recovered. Two years later fragments of a second cranium, Piltdown II, were recovered. Charles Dawson became sick towards the end of 1915 and he died on August 10, 1916. Thereafter, although Smith Woodward visited the sites of Piltdown I and II many times he found nothing. In the meantime, discoveries in China and southern Africa of hominins that had more modern-human-like teeth and smaller brains suggested that there were two sorts of early hominin; one, like Piltdown, with a large brain and ape-like teeth and jaws and one with the opposite combination. The enigma of Piltdown was cleared up in the 1950s when researchers applied the fluorine dating method to the hominin and the nonhominin fossils. Whereas the large-mammal fossils had fluorine levels of 1.6–3%, the levels in the alleged Piltdown I hominin fossil (around 0.2%) were similar to levels found in contemporary bones. Closer inspection of the Piltdown "fossils" showed that the teeth had been filed down to simulate tooth wear and radiography showed that the roots of the molars in the jaw were ape-like. Subsequent fluorine analyses showed (a) that the cranial bones and the mandible were not the same age, and (b) that the mandible was almost certainly modern. Tests also showed that the "paleolithic" flint had been deliberately stained with chromium. All this was more than sufficient evidence to say with confidence that the Piltdown hominin fossils were modern and not ancient, and that their association with the Piltdown site was the result of an elaborate hoax. The chief suspect is Charles Dawson, but numerous other people besides Dawson were, and continue to be, suspected as the perpetrator(s) and the authorship of the Piltdown fraud has still not definitively been resolved.

## pink breccia
See **Swartkrans**.

## Pinnacle Point
A series of at least 15 caves and rockshelters exposed in quartzite cliffs near Mossel Bay, South Africa. The most extensively studied cave is one of the few Middle Stone Age (MSA) archeological sites that date to Marine Isotope Stage 6 ($c$.190–130 ka). Two hominin fragments were found in disturbed sediments that likely derive from MSA strata. The archeological evidence

includes multiple hearths, abundant shellfish remains, Levallois and other flakes, cores, and bladelets made predominantly of quartzite, and more than 50 pieces of red ochre, many of them ground or scraped. Faunal remains from the MSA deposits include tortoise and a variety of bovids. (Location 34°12'S, 22°05'E, South Africa.)

## piriform aperture
The opening into the nose from the face. The size, shape, and especially the form of the floor of the opening (e.g., smooth or sharp) have been used by paleoanthropologists as taxonomic and functional indicators. (L. *pirium* = pear.)

## *Pithecanthropus* Haeckel, 1868
The genus name *Pithecanthropus* was introduced by Ernst Haeckel for the penultimate (29th) stage in his human evolutionary scheme. The first researcher to use the genus *Pithecanthropus* was Eugène Dubois, who in 1894 used it to accommodate the taxon just one year earlier he had named *Anthropopithecus erectus*. The genus name was used until relatively recently for new species (e.g., *Pithecanthropus rudolfensis*) but in 1940 Franz Weidenreich was the first among many to recommend that *Pithecanthropus* (and *Sinanthropus*) should be transferred to *Homo*, so *Pithecanthropus* is now one of the many junior synonyms of *Homo*. (Gk *pithekos* = ape and *anthropos* = human being.)

## *Pithecanthropus dubius* von Koenigswald, 1950
*See* **Homo erectus**; **Sangiran Dome**.

## *Pithecanthropus erectus* (Dubois, 1892), Dubois, 1894
A new combination used by Eugène Dubois for the hominin species a year earlier he had named *Anthropopithecus erectus*. Its holotype is the Trinil 2 calotte. *See also* **Homo erectus**; **Pithecanthropus**; **Trinil 2**.

## *Pithecanthropus rudolfensis* Alexeev, 1986
The initial binominal of the species that was later transferred to *Homo* as *Homo rudolfensis*. *See also* **Homo rudolfensis**.

## -pithecus
Postfix meaning ape, or "ape-like." (Gk *pithekos* = ape.)

## plain film radiography
Also known as conventional radiography, this is the technology clinicians use to take standard radiographs of the chest or fractured bones. Because the image is made from a single source of X rays, all you see on a conventional radiographic image is an outline of the densest structure between the source and the X ray film. The first radiograph of a fossil hominin was made in 1904, but Franz Weidenreich was the first paleoanthropologists to make intensive use of plain film radiography as evidenced by the many radiographic images (he refers to them as skiagrams) in his monographs on the Zhoukoudian Lower Cave fossils.

## plane of interest
See **computed tomography**; **confocal microscopy**.

## planktonic foraminifera
See **foraminifera**.

## plantar aponeurosis
See **foot function**; **longitudinal arch**; **push-off**; **windlass effect**.

## plant microremains
Plant microremains are discrete plant parts that cannot be seen with the naked eye. They consist of phytoliths, pollen, starch grains, calcium oxalate crystals, and other, less diagnostic forms. The first three types have been identified at archeological sites, and have been used to reconstruct diet and/or the paleoenvironment. (Gk *micros* = small.) See also **phytolith**; **pollen**; **starch grains**.

## planum nuchale
See **nuchal plane**.

## planum temporale
A feature of the surface of the superior part of the temporal lobe of the cerebral hemisphere. In line with the predominant left-sided language dominance seen in modern humans, the surface area of the planum temporale is greater on the left than on the right side in modern humans. However, the planum temporale also displays left-hemisphere dominance in great ape brains. Because the presence and size of the planum temporale is not accessible using natural or prepared endocasts, we have no evidence about its evolution within the hominin clade. (L. *planus* = flat and *tempus* = time, thus it is literally "the flat part of the surface of the brain that lies beneath the temple"; NB: the part of the hair that tends to go gray first.)

## plastic deformation
Bones can be distorted by pressure either before they are fossilized or before fossilization has substantially hardened and stiffened them. These distortions result in permanent plastic deformation. This is difficult to correct, but researchers have recently used computer software to generate information about the average shape of other fossils belonging to the same fossil taxon, or in the case of possible early hominins the average shape of chimpanzee and modern human crania, in order to determine the most likely shapes of deformed fossil crania. (Gk *plastein* = to mold and L. *deformare* = undo form.)

## plasticity
See **phenotypic plasticity**.

## platform
The portion of a flake that includes the part of the striking platform removed from the core when a flake is detached. (syn. butt.) See also **facetted platform**; **striking platform**.

## pleiotropy

When a single gene influences multiple phenotypic traits. For example, high levels of testosterone in young adult modern human males may contribute to reproductive success, but later in life they are correlated with an increased risk of prostate cancer. Pleiotropy may be one of the reasons why phenotypic characters are found to co-vary for reasons other than shared evolutionary history. (Gk *plein* = more and *tropos* = towards.)

## Pleistocene

A term introduced by Charles Lyell to refer to the first of the two epochs that comprise the Quaternary period. Pleistocene refers to a unit of geological time (i.e., a geochronologic unit). The dates of the period are currently under debate. From 1948 to 2009, the beginning of the Pleistocene was 1.8 Ma, but in 2009 the International Union of Geological Sciences (IUGS) lowered this boundary to 2.58 Ma; the end of the Pleistocene remains 11.5 ka. For various reasons the 2009 IUGS decision is contentious; therefore, in this Dictionary Pleistocene is used in the sense meant before the change (i.e., 1.8 Ma–11.5 ka; *see* **Plio-Pleistocene boundary** for details). (Gk *pleistos* = most and *kainos* = new.)

## *Plesianthropus* Broom, 1938

A hominin genus established to accommodate the early hominin species previously referred to as *Australopithecus transvaalensis*. *Plesianthropus* is now almost universally regarded as a junior synonym of *Australopithecus*. (Gk *plesios* = near to and *anthropos* = human being.) *See also* **Australopithecus**; **Australopithecus africanus**.

## *Plesianthropus transvaalensis* (Broom, 1936) Broom, 1938

A hominin species, originally called *Australopithecus transvaalensis*, that was established to accommodate hominin fossils recovered from Sterkfontein. Most researchers consider *Plesianthropus transvaalensis* to be a junior synonym of *Australopithecus africanus*. The holotype is TM 1511 from Sterkfontein. [Gk *plesios* = near to and *anthropos* = human being and *transvaal* = refers to the old name for the province (now called Gauteng) where Sterkfontein is located.] *See also* **Australopithecus africanus**; **Australopithecus transvaalensis**.

## plesiomorphic

The primitive condition, or state, of a character used in a phylogenetic or cladistic analysis. (Gk *plesio* = near and *morphe* = form.) *See also* **cladistic analysis**.

## plesiomorphy

The state of a character in the hypothetical most recent common ancestor of a clade, or the state of a character in an outgroup (syn. **symplesiomorphy**). *See also* **cladistic analysis**.

## Pliocene

The second epoch of the Neogene period. Pliocene refers to a unit of geological time (i.e., it is a geochronologic unit). It previously spanned the interval between *c.*5.3 and 1.8 Ma (its end coincided with the beginning of the Calabrian stage), but in 2009 the International Union of Geological Sciences allocated the Gelasian stage to the Pleistocene so until and unless their recommendation is overturned the Pliocene now spans the interval of time between c.5.3 and 2.58 Ma (its end now coincides with the beginning of the Gelasian stage). In this Dictionary Pliocene is used in the sense meant before the 2009 International Union of Geological Sciences decision. The Pliocene is typically subdivided into Early and Late, with the boundary at c.3.6 Ma. (Gk pleion = more and kainos = new.) See **Plio-Pleistocene boundary**.

## Plio-Pleistocene

The period of geological time that consists of both the Pliocene and Pleistocene epochs and includes the time period more recent than c.5 Ma until c.12 ka. It almost certainly includes the origin and subsequent evolution of *Homo*. Some authors use Plio-Pleistocene because it approximates the period of time during which the northern hemisphere has been glaciated. (Gk pleion = more, pleistos = most, and kainos = new.) See also **Northern Hemisphere Glaciation**; **Plio-Pleistocene boundary**; **Pleistocene**; **Pliocene**.

## Plio-Pleistocene boundary

In 2009 the International Union of Geological Sciences (or IUGS) ratified a proposal that the boundary between the Pliocene and Pleistocene epochs be lowered from the Calabrian-Gelasian boundary at 1.8 Ma (near the base of the Olduvai subchron) to the Gelasian-Piacenzian at 2.58 Ma (near the base of the Matuyama chron). In this Dictionary Pleistocene is used in the sense meant before the 2009 IUGS decision. Proponents of the 2009 decision cite the lack of faunal turnover at the Calabrian-Gelasian boundary and stress that the latest glacial episode may not be fundamentally different from the rest of the Neogene. The resulting 44% expansion of the Pleistocene is opposed by a large body of researchers who work on late Cenozoic (post-Miocene) subjects on the grounds that this radical shift was not adequately justified under chronostratigraphic guidelines and that representatives of the affected disciplines were not consulted. *See also* **Pleistocene**.

## plunge

In structural geology, plunge refers to the dip and dip direction of linear features such as the axial trace of a fold or the orientation of the long axis of sedimentary clastic rocks or fossils. Nonrandom plunge values (i.e., orientation) can be taken as evidence of reworking of fossils by the action of streams, rivers, or slope wash.

## PM

Abbreviation for the posterior maxillary plane of Enlow. *See* **posterior maxillary plane**.

## PM plane

*See* **posterior maxillary plane**.

## point

Artifacts that are pointed in shape. Typically (but not always) points are inferred to represent the tips of hunting weapons such as spears or arrows. Stone points are characteristic of many Middle Stone Age sites in Africa (e.g., Blombos Cave), and they also occur at some similarly aged Middle Paleolithic sites in Eurasia. Ethnographic, historic, and archeological data demonstrate that points were also made of other materials such as bone and wood, although for obvious reasons these rarely preserve in the archeological record. (OF *pointe* = sharp end.)

## polarity
Earth sciences Refers to each end of an axis passing through a sphere and in that sense it is used to refer to the direction, either normal or reversed, of the Earth's magnetic field. *See also* **geomagnetic polarity time scale**. Systematics In cladistic analysis it is used to refer to the alignment of a sequence of character states, with the most primitive character state at one end and the most derived at the other. (Gk *polus* = axis.) *See also* **cladistic analysis; morphocline**.

## polarity chronozone
*See* **chron**.

## polarity reversal
*See* **geomagnetic polarity time scale**.

## polarization
Earth sciences Refers to rocks that have magnetized particles so that at the time of deposition of sediments, or when a lava cools, the direction of the Earth's magnetic field is preserved. At present the Earth's field is polarized so that the needle of a compass points to the north; this is called the normal direction. This has not always been the case and the Earth's history has seen a series of polarity reversals related to changes in pattern of flow in the fluid Earth core. *See also* **geomagnetic polarity time scale**. Systematics In systematics, polarization is the process of deciding which state of a character is primitive and which is derived. Several techniques have been developed to facilitate polarization. These include the stratigraphic criterion, the ontogenetic criterion, and communality analysis, but the most widely used technique is outgroup analysis. *See also* **cladistic analysis**.

## pollen
One of several kinds of plant microremains (the main others are phytoliths and starch grains). Pollen is the male gamete of flowering plants and it is either wind-borne or carried by insects. The shapes and sizes of pollen grains are usually unique to the plant taxon (e.g., at the level of the family, tribe, or genus and sometimes even at the level of the species) that produced it, and can be used to identify the presence of these plants in the archeological record. Pollen is particularly hardy and fossilizes well, but because pollen can travel long distances it is more useful for reconstructing the paleoenvironment than it is for reconstructing diet. The study of pollen is called palynology. (L. *pollen* = fine flour, mill dust.) *See also* **palynology; paleoenvironmental reconstruction**.

## polygenic trait
A phenotypic trait that results from the combined effect (action of) alleles at more than one locus and the environment. Because they depend on the simultaneous presence of several alleles as well as an environmental influence, polygenic traits such as skin color, hair color, height, weight, and blood pressure have more complex hereditary patterns than simple monogenic

traits, and do not follow simple Mendelian patterns of inheritance. The variation in a polygenic trait in a population is expected to follow a normal distribution (i.e., a bell curve).

### polymerase chain reaction
(or PCR) This is a method used to copy, or amplify, a fragment of DNA. The process has three steps: (a) heating the sample to break apart the two strands of DNA, (b) cooling the sample in the presence of short fragments of complementary DNA (called primers) which therefore bind to the area of interest on the sample DNA, and (c) warming the sample to increase the function of an enzyme that causes the new double strand of DNA to grow. This process is repeated many times, so that although initially only a small amount of sample DNA is present, at the end of the PCR many millions of copies are produced.

### polymorphic
See **polymorphism**.

### polymorphism
A morphological or genetic feature or character with alternative specifiable states within a biological population. For the genotype, for example, polymorphism includes having more than one type of allele at a locus (e.g., the blood type system has three possible alleles, A, B, and O). Examples from morphology relevant to human evolution include the several discrete forms the root system of a tooth may take, or the sex-specific variation seen in the occurrence of sagittal crests. Not to be confused with polytypism, which describes a species with multiple phenotypically and/or genotypically distinct populations that may be classified as subspecies (such as seen in chimpanzees and orangutans). (Gk *poly* = much and *morphe* = form, so literally "many forms.")

### polypeptide
A string of amino acids that forms all, or part of, a protein. (Gk *poly* = much and peptide, a compound containing two or more amino acids linked by a peptide bond.)

### polyphyletic
An adjective used to describe a taxon that is part of a **polyphyletic group** (*which* see).

### polyphyletic group
A taxonomic grouping that includes taxa from more than one clade or monophyletic group. The "baboon" group, if it includes savanna (*Papio*), forest (*Mandrillus*), and gelada (*Theropithecus*) baboons, is an example of a polyphyletic group because there is sound molecular evidence that mandrills belong in a separate clade. Recent analyses suggest that the morphological similarities between mandrills on the one hand, and savanna baboons and gelada baboons on the other, are the result of convergent evolution or parallel evolution and are therefore examples of homoplasy rather than homology. (Gk *poly* = many and *phylon* = races.)

### polyphyly
See **polyphyletic group**.

## polytypic
See **polytypism**.

## polytypism
A species with multiple phenotypically and/or genotypically distinct populations that may be classified as subspecies (e.g., chimpanzees or orangutans). Not to be confused with polymorphism, which refers to a feature or character of the phenotype or genotype with alternative specifiable states (e.g., number of cusps on a molar tooth or the pattern of intracranial venous sinuses) within a biological population. (Gk *poly*=much and *tupos*=impression.)

## ponderal index
Ratio of body mass relative to stature, calculated as $[(body\ mass)^{1/3}/stature] \times 100$. Modern humans have a low ponderal index compared to the extant great apes. Archaic hominins such as *Australopithecus afarensis* appear to have had a high, relatively ape-like, ponderal index. (L. *pondus*=weight.)

## *Pongo*
The genus that includes living orangutans. The present consensus is that the genus includes two species: *Pongo pygmaeus*, the Bornean orangutan, and *Pongo abelii*, the Sumatran orangutan. The ancestors of orangutans most likely migrated from the mainland to Sumatra and from there to Borneo. Some researchers are in favor of a deep (*c*.2 Ma) divergence and suggest that the Bornean and Sumatran orangutans are in an early stage of speciation, but others estimate the split time to be *c*.1 Ma. Some researchers recognize three subspecies of the Bornean orangutan: *Pongo pygmaeus pygmaeus*, the northwest Bornean orangutan known from Sarawak (Malaysia) and northwest Kalimantan (Indonesia), *Pongo pygmaeus wurmbii*, the Central Bornean orangutan, known from southwest and central Kalimantan (Indonesia), and *Pongo pygmaeus morio*, the northeast Bornean orangutan, known from East Kalimantan (Indonesia) and Sabah (Malaysia).

## Pontnnewydd
A cave site in North Wales where researchers found an Acheulean assemblage and several hominin remains, including a well-dated tooth that is the second-oldest hominin fossil in Britain (*c*.200 ka; Swanscombe is the oldest at *c*.400 ka), and similar to the fossils from Krapina, Croatia. (Location 53°13′N, 03°28′W, Wales, UK)

## population
Evolution The group of interest for a study. A population may include the entire species, a subset, or a small sample taken from a species. Statistics The full set of possible values from which a smaller set of values, known as a sample, is drawn for use in statistical tests. For example, a sample of cranial capacities can be drawn from the population of all chimpanzee crania which have ever existed. Statistics based on a sample are used to make inferences regarding the population from which the sample is drawn. See also **sample**.

## population bottleneck
See **bottleneck**.

## population genetics
The study of allele frequencies in populations. Specifically, it focuses on the effects of the four forces of evolution (natural selection, genetic drift, gene flow, and mutation).

## population size
Population size is literally the number of individuals alive at any one time for a given population. Typically, in modern human demography the population size is specified at mid-year, although if the population is stationary the population size is unchanging and could be measured at any time. Population size is used synonymously with census size, as the number of individuals alive at any one time. Population size is usually greater than effective population size, which is the size of an idealized population that would produce the same amount of genetic drift as observed in the actual population.

## population structure
A population is described as structured if mating is not random. For example, in modern humans, geography (including distance or barriers), language, and other cultural factors influence mate choice and thus gene flow.

## Porc-Épic Cave
A cave in the Ethiopian Highlands formed in Mesozoic limestone. Obsidian hydration and AMS radiocarbon dating suggest a Late Pleistocene age ($c.77–33$ ka). The only hominin is a mandibular fragment including two premolars and three molars. The Middle Stone Age artifacts are primarily made of locally available chert. Small Levallois, discoidal, and single and double platform cores produced flakes, blades, and bladelets; many are retouched into a wide range of point forms (about 40% of the formal tools) and scrapers. The well-preserved faunal assemblage is dominated by bovids, lagomorphs, and hyrax, and cutmarks and bone breakage patterns suggest that it is primarily hominin-accumulated. The material recovered within the Middle Stone Age layers includes 419 complete perforated opercula of the terrestrial gastropod *Revoilia guillainnopsis*. Their spatial distribution, evidence of polish within the perforations, and the lack of any evidence that they were used as food suggests a possible symbolic use, perhaps as beads. (Location 09°34′N, 41°53′E, Ethiopia.)

## porcupines
The informal name for part of the suborder Hystricomorpha of the order Rodentia. Porcupines are of interest from a taphonomic perspective as they are known to accumulate bone assemblages that can include remains of small (e.g., rabbit-sized) and large (e.g., wildebeest-sized) mammals. Such assemblages are characterized by high frequencies (60–100%) of specimens with evidence of porcupine gnawing. *See also* **rodents**.

## positive allometry
This term refers to a relative size relationship in which part of an organism, or a variable that functions as a proxy for part of an organism, increases in size at a faster rate than the overall size of the organism, or a variable that functions as a proxy for the whole of an organism (i.e., the variable becomes proportionally larger as overall body size increases). *See also* **allometry**; **scaling**.

## positive assortment
In simple terms, this occurs when individuals choose to mate with those who are similar to themselves (i.e., when "like" mates with "like"). Examples of positive assortative mating in modern humans include matings between individuals with similar religious beliefs, education, height, and skin color. When positive assortative mating is based on heritable traits, it tends to reduce overall variation.

## positive selection
See **natural selection**.

## positron emission tomography
(or PET) A relatively noninvasive neuroimaging modality that enables researchers to visualize and measure specific biochemical reactions *in vivo* (i.e., the subjects are alive and conscious). It involves the injection or consumption of radioactive compounds (called radiotracers) that are differentially distributed. Their concentration is captured by detectors that respond to photons generated by radioactive decay. Oxygen and glucose are commonly used as radiotracers because they accumulate in tissues that are metabolically active. The radiotracers accumulate in cells proportional to the latter's rate of metabolism. When the radioactive material decays, it emits a positron that collides with an electron and in the process two photons are emitted. The image generated by detecting the emitted photons shows the distribution of radioactivity in the brain, which is a reflection of regional metabolism during the period when the tracer was taken up. Applications of this method include the visualization of the brain areas that are activated during communicative and cognitive tasks. [etym. a positron, which is the antimatter equivalent of an electron, is emitted by radioactive material; emission refers to emitted positrons and photons and Gk *tomos* = a cutting (from *temnein* = to cut) and *graphein* = to write.]

## post-incisive planum
A feature found when the mandibular symphysis and the perisymphyseal part of the mandibular corpus are thick. In that event the bone immediately posterior to the incisors is not vertically oriented as it is in modern humans, but instead it slopes more horizontally and posteriorly.

## post-reproductive life span
See **grandmother hypothesis**; **life history**.

## postcanine teeth
The teeth distal to the canine. In the deciduous dentition it refers to the first and second deciduous molars, and in the permanent dentition to the two premolars and the three permanent molars. The rest of the teeth in each quadrant are called the anterior teeth.

## postcanine tooth row
In the deciduous dentition it refers to the first and second deciduous molars, and in the permanent dentition to the two premolars and the three permanent molars.

### posterior maxillary plane

(or PM) One of several measurements used to describe and compare crania. It is called a plane, but it was defined as a line. One terminus of this line is the location in the mid-sagittal plane of a line joining the most inferior and posterior points on both maxillary tuberosities (called the pterygomaxillary point), while the other terminus is the location in the mid-sagittal plane of a line joining the most anterior points on the lamina of the greater wings of the sphenoid (called the posterior maxillary point). The superior terminus of the PM plane marks the boundary between the middle and anterior cranial fossae and the inferior termini mark the posterior extent of the face. It has been suggested that one of the architectural constraints in the cranium is that the angle between the posterior maxillary plane and the neutral horizontal axis of the orbit (i.e., a line joining the midpoint of the orbital opening with the midpoint of the optic canal) is 90°, and because of this some researchers have suggested that the PM plane should be used more widely as a reference for functional studies of the cranium. It was used to reconstruct the face of KNM-ER 1470 as being more prognathic than in the original reconstruction. *See also* **basicranium**.

### posterior parietal cortex

The region of the parietal lobe that lies posterior to the primary somatosensory cortex (aka postcentral gyrus). It is involved in the coordination of somatic sensation, auditory information, and vision. It is a crucial region for integrating sensory perception of the spatial location of objects in the external world and information about body position. By integrating information about the state of the animal with that of potential targets for behavior, areas within the posterior parietal cortex are thought to create a context or frame of reference for guiding movement. Because this region is densely connected with the frontal lobe, it may contribute to the ability to voluntarily inhibit behavior, as well as adapt movements according to novel sensory stimuli. The posterior parietal cortex is functionally lateralized, such that the posterior parietal cortex of the left hemisphere is specialized for processing linguistic information, whereas in the right hemisphere it is specialized for spatial information. Some functional regions within the posterior parietal cortex of modern humans are activated in positron emission tomography (PET) imaging studies of modern humans learning how to fashion Oldowan-style stone tools. In the absence of comparable data from apes, however, it is not clear whether these posterior parietal areas are specific to *Homo sapiens* or whether they are shared with our close relatives. Some have argued that the posterior shift of the lunate sulcus and the relative reduction of the adjacent primary visual cortex in the modern human brain indicate that areas of the posterior parietal cortex have disproportionately expanded within the hominin clade. *See also* **parietal lobe**.

### posterior teeth

*See* **postcanine teeth**.

### postural feeding hypothesis

One of several hypotheses put forward to explain the origin of upright posture and bipedal locomotion in the hominin clade. Like the seed-eating hypothesis, it is based on a set of careful observations of both primate behavior and anatomy. The hypothesis suggests that among chimpanzees and baboons, bipedal postures are employed during both arboreal and terrestrial feeding, and typically involve reaching into higher branches to gather fruits. While standing bipedally, chimpanzees frequently use one of their upper limbs to steady themselves by simultaneously hanging from a branch. The supporters of the hypothesis point out that the postcranial skeleton of australopiths retains several ape-like features associated with arm-hanging, but it lacks features that would have made bipedal locomotion energetically efficient. The hypothesis suggests that the earliest hominins may have been dependent on arboreal fruit resources, and evolved anatomical adaptations to bipedal posture as a result of habitually employing an upright posture during feeding. These adaptations allowed facultative bipedal locomotion, but selection favoring the evolution of an obligate pattern of energetically efficient, or endurance, bipedalism did not occur until later in hominin evolution. *See also* **bipedal**.

## posture

The position, or attitude, of the whole body or of a part of the body. When the long axis of the trunk is oriented vertically this is referred to as an upright or orthograde posture; when the long axis of the trunk is oriented horizontally this is referred to as a horizontal or pronograde posture. Musculoskeletal adaptations for posture and locomotion are not always the same. For example, in chimpanzees, suspensory arm-hanging under a branch is the only behavior requiring complete abduction of the arm. This suggests that some morphological features related to shoulder mobility (e.g., the superiorly narrow "cone-shaped" rib cage) may be adaptations to an arm-hanging posture rather than to a climbing form of locomotion.

## potassium-argon dating

(or $^{40}$K/$^{39}$Ar dating) A radioisotopic dating method that uses potassium-bearing minerals, especially those from layers of volcanic tephra. These are usually in the form of ash, in which case they are called tuffs. The method compares the amounts of $^{40}$K and its daughter product, $^{39}$Ar. Because the potassium is measured as a solid and the argon daughter product as a gas, these measurements are made on two different samples of rock and from many mineral crystals from each sample. The error introduced tends to limit the application of potassium-argon dating for hominin evolution to sediments that are more than 1 Ma. Because relatively large amounts of material are needed the method is only suitable for samples that are not likely to be contaminated with older or younger crystals. The potassium-argon technique had early and important implications for the understanding of the time depth of human evolution, particularly in East Africa. In 1965 Evernden and Curtis published potassium-argon ages for the Olduvai Gorge sequence that placed the base of Bed I at 1.85 Ma and the top of Bed II at more than 500 ka. These had important implications especially for the age of what was then called *Zinjanthropus boisei* (OH 5), which was associated with the 1.85 Ma date, as well as for *Homo habilis* (OH 7) and pre-Chellean man (OH 9), which came from Bed II. Subsequent work has shifted the ages of the Olduvai stratigraphy to earlier dates, especially for Bed II. Nonetheless, this first set of dates provided sound evidence of a much greater antiquity for hominins than had been previously accepted. Potassium-argon dating has been largely superseded by argon-argon dating. *See also* **Olduvai Gorge**.

## power grip

A grip in which the object is held by the fingers and palm, with the thumb as a buttress. A distinctive modern human form of the power grip is the squeeze grip of cylindrical tools (e.g., hammer handles) that are held obliquely in the palm. This type of grip is facilitated in modern humans by having shorter fourth and fifth metacarpals compared to the rest of the palm, increased mobility in the ulnar direction of both the wrist and the fingers, and enhanced robusticity of the ulnar side of the hand. Chimpanzees and other apes are not capable of using this grip, and instead grasp the object transversely or diagonally across the fingers, usually without active involvement of the palm or the (relatively short) thumb. *See also* **opposable thumb**.

## power scavenging

*See* **confrontational scavenging; scavenging**.

## power stroke

Refers to the slow close phase of the chewing cycle that ends at minimum gape. It used to be thought that the power stroke consisted of two phases, with phase I being the last part of slow close phase and phase II being the start of the slow open phase. However, it is now clear that among primates little or no bite force is generated after minimum gape, so that the power stroke is equivalent only to phase I. *See also* **chewing**.

## *Praeanthropus*

The genus name introduced to accommodate the maxilla found at Garusi (now called Laetoli). *Praeanthropus* is the genus name that should be used by those who support removing *Australopithecus afarensis* from *Australopithecus*. (L. *prae* = before and Gk *anthropos* = human being.) *See also* **Laetoli;** ***Praeanthropus afarensis***.

## *Praeanthropus afarensis* (Johanson, 1978)

A new name combination suggested by researchers who concluded that *Australopithecus afarensis* was the sister taxon of *Australopithecus africanus* and other later hominins. They argued that if the same genus name was used for *Au. afarensis* and *Au. africanus*, *Australopithecus* would be a paraphyletic taxon, and so they suggested that the hypodigm of *Au. afarensis* should be transferred to a different genus. They argued correctly that the genus name *Praeanthropus* was available, but if *Praeanthropus* was used as the genus name, then the species name *afarensis* no longer had priority, for *Meganthropus africanus* Weinert, 1950 obviously had priority over *Au. afarensis* (Johanson et al., 1978). However, this latter change in nomenclature would result in two hominin species called "*africanus*": *Australopithecus africanus* and *Praeanthropus africanus*. Application was made to the International Commission on Zoological Nomenclature to have the specific name "*africanus* Weinert, 1950" suppressed, and the Commission upheld the application. So, if the *Au. afarensis* hypodigm is to be removed from *Australopithecus*, then the taxon should be referred to formally as *Praeanthropus afarensis* (Johanson, 1978). (L. *prae* = before, Gk *anthropos* = human being, and *afarensis* recognizes the contributions of the local Afar people.)

### *Praeanthropus africanus* Weinert, 1950
See ***Praeanthropus afarensis***

### preadaptation
A feature or trait that is designed to perform a functional role it does not currently perform. Preadaptations may be co-opted to perform the new functional role, and in the process become exaptations (also called co-options). For example, the lesser gluteal (buttock) muscles are used by apes as extensors during climbing, but when their attachments are rearranged as part of the skeletal adaptation to an upright posture and bipedal walking, they function as abductors to prevent the contralateral side of the pelvis from dipping down during the stance phase of walking. *See also* **exaptation**.

### precentral gyrus
The precentral gyrus is the part of the cerebral cortex located in the frontal lobe, anterior to the central sulcus and posterior to the precentral sulcus. It contains the primary motor cortex and portions of the premotor cortex, which correspond to Brodmann's areas 4 and 6, respectively. The primary motor cortex contains a map (aka topographic representation) of the body that is not to scale (e.g., regions used in tasks requiring precision and fine control, such as the face and hands, are disproportionately large in modern humans and other primates). The size of the primary motor cortex remains relatively constant across primates in proportion to overall body size. (Gk *gyros* = circle, ring.)

### precession
A change in the orientation of the Earth's axis, which is one of the three cycles that influences the Earth's orbital geometry and therefore its climate. Precession is the result of the gravitational pulls exerted on the Earth by the sun and moon, which together cause the Earth's axis to "wobble" in a predictable way in a *c.*23 ka cycle. Today the North Pole points towards Polaris, the North Star, but 12 ka it pointed towards Vega. Until 3 Ma the precessional cycle was the dominant influence on global climate. (L. *praecedere* = to go before.) *See also* **astronomical time scale**; **eccentricity**; **obliquity**.

### precision
The quality of being able to do something within well-defined limits; a precise value is one that has very little error. Precision is not the same as accuracy. Imagine a tight cluster of three darts on a dartboard far from the bull's-eye, the intended target. The placement of these darts is precise, but not accurate. (L. *praecision* = to cut off.)

### precision grip
Any grip that uses the thumb and one or more of the fingers. There are several types of precision grip, for example between the thumb and index finger (two-jaw), and between the thumb, index finger, and middle finger (three-jaw). Two-jaw grips include pad-to-pad (e.g., pinching a coin), tip-to-tip (e.g., threading a needle), or pad-to-side (e.g., using a key). Nonhuman primates are not capable of many of the precision grips used by modern humans. Chimpanzees more often use a pad-to-side grip between the thumb and the side of the index finger. Among extant primates, baboons come closest to being able to use the pad-to-pad grip because their hand proportions are most similar to those of modern humans.

## precocial
The offspring of precocial organisms are born at a relatively advanced stage of development (e.g., with eyes and ears open and able to move about on their own shortly after birth). While modern human neonates are born with their eyes open and with hair, they fall short in many of the criteria when compared to truly precocial animals. Thus they are described as secondarily altricial or semi-precocial. (L. *praecox* = early ripening.) *See also* **altricial**.

## predation
Predation is the process by which one organism – the predator – hunts, kills, and eats another organism – the prey. The discovery that hominins were prey animals (e.g., damage in the orbits and on the cranial vault of the Taung 1 cranium are compatible with it being killed by a raptor and leopard tooth marks have been found on hominin fossils from Swartkrans) contradicted the views of some early paleoanthropologists who believed that hominins were "bloodthirsty killers." Observations of modern primate behavior plus taphonomic and archeological evidence suggest that as far as early hominins are concerned there was no clear division between "the hunter" and "the hunted." (L. *praedari* = to rob or plunder.)

## predator
*See* **predation**.

## Předmostí
This 29–27 ka open-air site, which is located in northeastern Moravia, is one of the earliest sites where modern human fossils were recovered. Mousterian, Aurignacian, and Pavlovian (Gravettian) artifacts have been recovered from different levels within the site. The hominin remains come from the Gravettian levels. (Location 49°30′N, 17°25′E, Czech Republic.)

## pre-eminent bipedalism
*See* **bipedal; bipedalism**.

## prefrontal cortex
The part of the cerebral cortex that lies anterior to the premotor cortex, which, in turn, is anterior to the primary motor cortex. The prefrontal cortex is important for decision-making, moderating social behavior, and the planning of complex sequences of actions. The prefrontal cortex of modern humans comprises a relatively larger proportion of the frontal lobe than in great apes, but there is debate about whether the entire prefrontal cortex or components of it have undergone disproportionate expansion in modern humans beyond what would be predicted by the known allometric relationships within higher primates. *See also* **allometry**.

## prehensile
A hand (or foot, or lips) capable of isolating, pinching, and lifting an object. (L. *prehensus* = to grasp.) *See also* **hand, function; precision grip**.

## pre-Neanderthal hypothesis
A hypothesis that asserted both modern humans and Neanderthals evolved from populations that succeeded *Homo erectus*, but pre-dated the emergence of "classic" Neanderthals. Also known as the pre-Neanderthal theory, it was accepted by a number of leading human

paleontologists during the mid-to-late 20thC. Theoretically, the pre-Neanderthal hypothesis can be seen as the intellectual precursor of the out-of-Africa hypothesis for modern human origins in which modern humans are seen as evolving from Middle Pleistocene ancestors outside of Europe. The term pre-Neanderthal (Fr. *anté-Néandertalien*) is also used to refer to hominin remains from Europe that date from about Marine Isotope Stages 8–13 ($c.530$–$300$ ka).

## prepared core

A stone core that shows signs that the striking platform or flake release surface has been abraded or chipped to increase the likelihood of successful flake removal. For some, but not all, researchers, "prepared core" and "Levallois core" are synonymous. *See also* **Levallois**.

## pre-sapiens hypothesis

A model of human evolution advocated by a number of prominent scientists during the early 20thC. According to the pre-sapiens hypothesis (aka the pre-sapiens theory) fossil hominins morphologically similar to modern humans had existed since at least the early Pleistocene, if not earlier. This meant that other more recent forms of hominin (e.g., *Homo neanderthalensis* in Europe, *Sinanthropus* in China, and *Pithecanthropus* in Java, Indonesia) could not be ancestral to modern humans. Most of the lines of evidence that buttressed the pre-sapiens hypothesis proved to be unreliable and discoveries in southern Africa of australopiths strengthened the view that modern humans evolved from an ape-like ancestor through an australopith stage followed by a "pithecanthropine" stage. The popularity of the pre-sapiens hypothesis helps to explain why during the early 20thC so many scientists rejected the australopiths and the Neanderthals as modern human ancestors, and why there was so much influential support for the fraudulent Piltdown fossils.

## pre-sapiens theory

*See* **pre-sapiens hypothesis**.

## pressure pad

Devices (aka pressure-distribution systems) that use sensors to measure forces. They are typically arranged in a matrix within a pad (rigid or flexible) that allows normal forces (those perpendicular to the surface) to be measured across an area over time. Pressure pads have been used most commonly to examine the forces exerted by the foot from heel contact through toe-off, but they are also used to measure the force distributions during hand grips, across seats or saddles, as well as during other interactions between the body and a substrate.

## prey-choice model

A model derived from optimal foraging theory designed to predict whether a forager will pursue or ignore a prey item it encounters. The prey-choice model (aka encounter-contingent prey-choice model) assumes that the forager (a) wants to maximize energetic returns, (b) is aware of the energetic returns associated with different types of prey, and (c) searches for prey in a patch where encounter rates with different prey types are random. The model suggests that high-ranking prey will always be preferred because they provide higher-than-average energetic returns, and that energetically less profitable low-ranking prey will only be added to the diet when encounter rates with high-ranking prey decline and overall foraging efficiency declines. *See also* **high-ranking prey**; **low-ranking prey**; **optimal foraging theory**.

### prey rankings
*See* **high-ranking prey**; **low-ranking prey**.

### pre-*Zinjanthropus*
The term used to refer to the hominin remains that were subsequently included in *Homo habilis*. *See* ***Homo habilis***.

### primary cartilaginous joint
*See* **joint**.

### primary centers of ossification
*See* **ossification**.

### primary consumers
*See* **trophic level**.

### primary cusp
(syn. main cusp.) *See* **cusp**.

### primary producers
*See* **trophic level**.

### primary visual cortex
*See* **striate cortex**.

### primate archeology
The name used to describe the study of the "past and present material record of all members of the order Primates." Early reports of tool use in nonhuman primates focused on the way chimpanzees made and used tools, thus terms like "cultural panthropology" and "chimpanzee archeology" were used for the study of this phenomenon. However, tool manufacture and tool use in primates is emphatically not confined to chimpanzees so a more inclusive term was needed. The immediate goals of primate archeology are to investigate the spatial, environmental, and other contexts in which modern humans, early hominins, and nonhuman primates use probing, pounding, and digging tools, in order to better understand how evidence of such behaviors may be preserved in the archeological and fossil records.

### primer
Primers are short single-stranded fragments of DNA (usually 15–25 base pairs long) complementary to a portion of the DNA sequence of interest that can be used to specify the segment of DNA to be amplified. *See also* **polymerase chain reaction**.

## primitive
A morphological feature or character present in an outgroup or in a hypothetical common ancestor. A primitive (aka symplesiomorphic) character state is at the primitive end (aka pole) of a morphocline (syn. primitive condition). (L. *primitivus* = first of its kind.)

## principal components analysis
(or PCA) A form of multivariate analysis of standardized data in which the output takes the form of linear transformations of the original variables. The first of these linear transformations (aka the first principal component) describes the most variation in the data set (i.e., it has the highest eigenvalue), the second contains the most variation that is independent of the first variable, the third contains the most variation that is independent of the first two variables, and so on. Because they represent linear transformations of the original variables, each principal component axis (i.e., each eigenvector) is described by loadings or coefficients that reflect the relative contribution of each of the standardized variables to the variance explained by that principal component. High loadings reflect a strong contribution of the original variable, and positive loadings indicate that an increase in the original variable will increase the value of the principal component. These loadings can be used to identify the relative contributions of the original variables to the overall variation present in a data set. When examining PCAs of morphological data, some researchers consider the first principal component to be a general measure of multivariate size, but because it also incorporates shape differences that are correlated with size it is not a measure of size alone.

## priority
The principle used in nomenclature which suggests that the taxon name with the earliest publication date should be regarded as the senior name and thus should have priority over any other name for that taxon published thereafter. (L. *prior* = first.) See also **synonym**.

## prism
See **enamel**; **enamel microstructure**.

## problem-solving
The ability to overcome various kinds of obstacles to achieve a desired goal (e.g., the manufacture of a stone tool requires problem-solving skills). Researchers have demonstrated that both chimpanzees and orangutans have elementary problem-solving skills.

## Proboscidea
The order of hoofless mammals with elongated trunks that includes the extant African (*Loxodonta*) and Asian (*Elephas*) elephants, both of which are subsumed within the only extant family of proboscideans, the Elephantidae. During the early stages of human evolution the Proboscidea were a more diverse group than they are today. Members of the family Deinotheriidae (proboscideans with downward curving tusks) are known from African early hominin sites, and the family Gomphotheriidae (proboscideans with shovel-shaped tusks) is known from South American paleoindian sites. Numerous other families of proboscideans were extant during the course of earlier hominoid and hominin evolution. Most

proboscideans are regarded as megafauna, although some very early and dwarfed lineages are exceptions to this. (Gk *proboscis* = elephant's trunk, from *pro* = forward and *boskein* = to nourish or feed.)

## proboscidean
The informal term for the order that includes hoofless mammals with elongated trunks. *See also* **Proboscidea**.

## Procrustes analysis
A method, based on the least squares principle, for comparing the shape of objects. First, the landmarks of all of the objects are translated so that they share the same centroid. Second, they are scaled so they all have the same centroid size. Third, they are rotated around the centroid until the sum of the squared Euclidean distances between the homologous landmarks is minimized. The resulting coordinates are called Procrustes shape coordinates; the differences between any individual and the average shape are called Procrustes residuals. The Euclidean distance (aka Procrustes distance) (i.e., the square root of summed squared coordinate-wise differences) between two sets of Procrustes shape coordinates is a summary of the similarity or dissimilarity in shape between two landmark configurations. (etym. in Greek mythology, Procrustes, a son of Poseidon, was an innkeeper who insisted his guests were an exact fit for the only bed provided for them. Short guests were stretched and tall ones had their legs amputated!)

## prognathic
Literally the degree to which the jaws project forwards, but it is also used for projection of the mid- and upper face. Prognathism can be midline, lateral, or both, and can affect the upper, middle, lower, or all of the face. It is usually quantified by measurements of the distance between a landmark on the base of the cranium (e.g., basion, porion) and a landmark on the front of the face (e.g., subnasale, alveolare), but it can also be measured using landmarks confined to the face (e.g., sellion, prosthion). Prognathism usually increases during ontogeny and it is usually more pronounced in males than in females. As a generalization, there is a long-term trend within hominins towards the reduction of prognathism. Flat or weakly projecting faces are called orthognathic. (L. *pro* = forward and *gnathos* = jaws; ant. orthognathic.) *See also* **orthognathic**.

## prognathism
The tendency for a face to be prognathic. *See also* **prognathic**.

## projectile point
A pointed stone or bone artifact that is thought to have been attached to the tip of a flight weapon (e.g., an arrow). The functions of projectile points are inferred from wear-pattern analysis, residues, and comparisons with ethnohistoric and experimental projectile points. Current evidence suggests projectile point technology developed in Africa and then spread to Eurasia after 50 ka.

## promoter

The part of the gene where transcription is initiated and where regulation of gene expression occurs. Promoters are typically located at the 5′ flanking end (i.e., upstream) of the gene. (L. *promotus* = to move forward, or advance.) *See also* **transcription**.

## pronation

Upper limb Internally rotating the forearm so that it moves from the anatomical position (i.e., palm facing anteriorly) to the pronated position (i.e., palm facing posteriorly). Lower limb The foot is pronated during the part of the walking cycle when the leg is internally rotated as the body moves forward over the foot. (L. *pronare* = to turn face downwards.)

## pronograde

*See* **posture**.

## protein

A large molecule made up of one, or more, chains of amino acids. Proteins are fundamental components and products of cells (e.g., enzymes, hormones, and antibodies are all proteins) that are specified by the triplet DNA code. DNA is transcribed into messenger RNA (mRNA), and then the nucleotide sequence of the RNA is translated into sequences of amino acids. (Gk *protos* = first.) *See also* **amino acid; deoxyribonucleic acid**.

## protein clock

*See* **molecular clock**.

## protein domain

A portion of a protein with a specific structure or function. The same domain may be present in multiple members of a protein family or related groups of proteins. For example, the *HOX* genes all contain a protein domain known as the homeodomain that can bind DNA. It is encoded by a 180 base pair DNA sequence known as the homeobox. (L. *domininus* = a lord, hence the territory under the control of a lord.) *See also* ***homeobox*** **genes**.

## protocone

One of the main cups of a mammalian upper postcanine tooth. It is the mesial cusp on the lingual aspect of a maxillary (upper) molar tooth crown, and the lingual cusp of a bicuspid upper premolar. It is part of the trigon component of an upper molar crown. (Gk *protos* = first and *konos* = pine cone; syn. mesiolingual cusp.)

## protoconid

One of the main cups of a mammalian lower postcanine tooth. It is the mesial cusp on the buccal aspect of a mandibular (lower) (hence the suffix "-id") molar tooth crown, and the buccal cusp of a bicuspid mandibular premolar. It is part of the trigonid component of a mandibular postcanine tooth crown. (Gk *proto* = first and *konos* = pine cone; syn. mesiobuccal cusp.)

## proto-language
See **language, evolution of**.

## protostylid
An accessory cusp on the buccal face of the protoconid on the crown of a mandibular molar. It ranges in its expression from a pit to a well-developed cuspulid; it is sometimes associated with the buccal groove. (syn. paramolar tubercle, distoconid.) *See also* **protoconid**.

## provenance
The provenance of a fossil or an artifact is the horizon from which it was excavated or, if it is a surface find, the horizon from which it originated. Provenance includes the exact context in which an object was discovered (e.g., spatial location, lithostratigraphic, and chronostratigraphic contexts) and for artifacts it may also include sourcing an object to its point of origin (e.g., the source for the obsidian used to make tools). If a fossil or an artifact is excavated (i.e., discovered *in situ*) then its provenance is not in doubt, but most fossils and some artifacts are found exposed on the surface of the landscape (i.e., *ex situ*). In these cases geologists try to trace the horizon of origin by looking nearby for evidence of likely overlying fossiliferous strata, or by matching the rock matrix adhering to the fossil to the appearance (aka lithology) of any overlying strata exposed nearby. In some circumstances the provenance of a surface find can be determined precisely; in others the provenance, and thus inferences about the age or the paleoenvironment of a fossil, are much less precise. The term provenience is sometimes used synonymously with provenance, particularly by North American researchers. (syn. provenience.)

## provenience
See **provenance**.

## provisioning hypothesis
This hypothesis, which is one of several devised to explain the origin of upright posture and bipedal locomotion within the hominin clade, models human origins within a demographic and evolutionary socioecological framework. Mothers must balance the need to care for their infants with the need to care for themselves, and accidents when infants accompany mothers during maternal foraging are a major source of infant mortality. To offset accidental infant death, one solution is for males to provision both mother and child. Provisioning behavior, which should favor the evolution of bipedalism because this mode of locomotion frees the hand for carrying, also reduces the need for female mobility and in turn should minimize the incidence of accidental infant death. But the hypothesis argues that males will only adopt this behavior if (a) they can be assured that they are the biological father of the infant and (b) they have exclusive reproductive privileges with the female in question. It is difficult to evaluate the provisioning hypothesis because it makes few predictions that can be tested using the fossil record. *See also* **bipedal**.

## proximal
In the direction of the root of a limb (i.e., where it is attached to the body). Thus, the shoulder joint is proximal to the elbow joint and the ankle joint is proximal to the hallucial metatarsophalangeal joint.

## proxy

Proxy used to refer to a person authorized to act for someone else, but it now refers to any sort of substitute. So a variable (e.g., $^{18}O/^{16}O$ ratio) that tracks climate is called a proxy, as is a variable (e.g., femoral head size or orbital area) that is correlated with body size or body mass. (L. *procurare* = to take care of.)

## pseudogene

A gene without a function (i.e., genes that have arisen by duplications, frameshift mutations, or deletions). Pseudogenes are of interest to evolutionary biologists because they share ancestry with functional genes (for example, the beta-globin gene cluster includes several pseudogenes) and they can be used to understand the evolution of a gene family. Once a gene becomes nonfunctional it evolves neutrally, so pseudogenes can be used to reconstruct gene phylogeny. (Gk *pseudes* = false and *genes* = to be born.)

## pulp chamber

Enlarged end of the pulp cavity (root canal) within the dentine. It has been argued that pulp chamber proportions may distinguish Neanderthals from modern humans, with the former having larger (taurodont) molar pulp chambers. However, because the pulp chamber decreases in size with advancing age (due to secondary and tertiary dentine development) it is important that comparisons are made between teeth of approximately the same age and degree of wear. (L. *pulpa* = fleshy part of an animal or fruit.) *See also* **taurodont**.

## pumice

Pumices, which are found in the tephra that are ejected from volcanoes, are formed by the rapid cooling (or quenching) of molten rock (magma) that has a high proportion of volatile material (gas). The gas gives pumice its characteristic buoyancy. In addition to the vesicles, and the glass that forms the vesicle walls, pumices commonly contain primary volcanic minerals including feldspar, pyroxene, quartz, and zircon. Pumices range in size from small size particles (i.e., ash and lapilli) to large-size boulders. Potassium-rich feldspars extracted from pumice clasts have long been a mainstay of isotopic dating efforts because they can be unequivocally related to the eruptive event that generated the tephra deposit.

## punctuated equilibrium

When an ancestral species persists more or less unchanged over long stretches of geological time before suddenly speciating into one or more descendant species. The term was coined as an alternative to the paradigm of phyletic gradualism that had previously been used to explain the evolutionary transformation of one species into another. Proponents of punctuated equilibrium argued that the periods of morphological stasis commonly observed in fossil species over stretches of geological time were, in fact, meaningful data and not merely an artifact of poor sampling in the fossil record. They further argued that these long periods of stasis (or equilibrium) were periodically punctuated by short periods of rapid evolutionary change, hence the name. They claimed that punctuated equilibrium was a logical consequence of evolutionary theory, insofar as species originate when a small-to-medium sized subpopulation becomes isolated from a much larger parent population. In such modest-sized populations, natural selection and genetic drift can result in relatively rapid evolutionary change. The two studies that have looked at the fossil records of hominin taxa across substantial periods of geological time (i.e., *Paranthropus boisei* and *Australopithecus afarensis*) both concluded that the data were consistent with the punctuated equilibrium model.

## push-off

The propulsive phase of the modern human walking cycle in which the center of mass travels over the forefoot and the foot pushes against the ground to generate forward momentum. Push-off (aka toe-off), which takes place just after the midpoint of stance phase in the modern human walking cycle, is initiated by activity of the soleus and gastrocnemius plantarflexing the foot at the ankle joint. During the push-off phase inversion of the hindfoot locks the calcaneocuboid joint, thus limiting mobility in the midtarsal region. As the heel lifts, and the metatarsophalangeal joints dorsiflex, the plantar aponeurosis (fascia) tightens in what is called the "windlass effect," causing the foot to stiffen and thus be an effective lever. *See also* **foot arches**; **walking cycle**; **windlass effect**.

## *P* value

Associated with a statistical test, the probability that the observed pattern in a data set could occur if the null hypothesis were true. If the *P* value is below some threshold (usually 0.05) then the test is said to show a statistically significant result and the null hypothesis is rejected.

## "pygmy chimpanzee"

See *Pan paniscus*.

## pyriform aperture

See **piriform aperture**.

## pyroclastic

A term used to describe rock fragments that have been ejected explosively from a volcanic source. Ash is an example of a pyroclastic material. (Gk *puro* = fire and *klastos* = broken.)

## pyrotechnology

The skills used in making, curating, and using controlled fire. Fire residues rarely preserve well, so the earliest evidence of the hominin control of fire almost certainly postdates the acquisition of the skills needed to make and curate fire. The first sound evidence for controlled fire consists of burned bones, ashed plant material, and results from spectroscopic investigation of sediments from *c*.1 Ma at Wonderwerk Cave in South Africa. The next-oldest evidence for fire comes from burned flints at Gesher Benot Ya'akov at *c*.760 ka, and fire use appears not to have been systematic until *c*.400–300 ka. Some authors even argue that during the Middle Paleolithic in Europe fire use was still intermittent. Thus, they suggest it would be unwise to assume that all groups of hominins post-1 Ma understood how to make and curate fire. Earlier claims for controlled fire prior to 1 Ma are difficult to substantiate because of the potential to confuse a burned-out tree stump with a hearth, and attempts to use reconstructed maximum temperature to discriminate between controlled fires and natural fires have yielded inconclusive results. Most of the explanations for the use of controlled fire have focused on its utility for cooking, warmth, and protection, but researchers have suggested that as early as 164 ka at Pinnacle Point fire was being used to pre-treat silcrete to make it easier to flake.

# Q

### Qafzeh
(alternate spellings: Djebel Qafzeh, Jebel Qafzeh) A cave site about 2.5 km/1.5 miles southeast of Nazareth, Israel, in the Wadi el-Hadj, where the remains of 14 individual hominins have been recovered. Two individuals (Qafzeh 1 and 2) are from the Upper Paleolithic levels and the rest (Qafzeh 3–18) are from the Middle Paleolithic levels. Thermoluminescence and electron spin resonance spectroscopy dating suggest that the Middle Paleolithic levels are $c.$100–90 ka. The Middle Paleolithic Qafzeh hominins are generally interpreted as sampling a robust population of anatomically modern humans, and many have commented on the affinities between the Qafzeh hominin remains and those from Skhul. Because the Skhul and Qafzeh remains represent the earliest known modern human remains outside of Africa, they have played a crucial role in understanding the radiation of modern humans from Africa. There are very few artifacts in the Upper Paleolithic levels, and this makes difficult a precise identification of the industry that is represented. The assemblage in the Middle Paleolithic levels is broadly defined as Levantine Mousterian. (Location 32°41′N, 35°18′E, Israel; etym. Ar. الجبل *djebel* = mountain and قفزه *qafzeh* = jump or leap.)

### Qesem Cave
Discovered during construction work in 2000, this cave site preserves a well-dated sequence of Acheulo-Yabrudian (terminal Lower Paleolithic) technology. The uranium-series dates on speleothems (382–207 ka) suggest that in the Near East the Acheulo-Yabrudian represents a distinct and quite long cultural phase between the Acheulean and Mousterian technocomplexes. (Location 32°11′N, 34°98′E, Israel.)

### QTL
Abbreviation for **quantitative trait locus** (*which see*).

### qualitative variants
See **nonmetrical traits**.

### quantitative genetics
A statistical subsection of genetics that helps researchers understand the evolutionary dynamics that led to phenotypic differences. When applied to longer time scales (e.g., to between-species differences) it is referred to as "evolutionary" quantitative genetics.

## quantitative trait locus

(or QTL) A genetic variant that segregates with a quantitatively measured and normally distributed phenotype (e.g., height or levels of low-density lipoproteins). QTLs are typically regions of DNA that show significant variation within and between populations, but which do not themselves necessarily have an effect on the phenotype. Their statistical association with phenotypic variation indicates that a genomic region close to that locus (e.g., a known gene or one that has yet to be identified) significantly influences the phenotype. Quantitative traits (e.g., height) are often referred to as "complex" traits because they can be influenced by more than one gene, and also by environment, sex, and age.

## quantitative traits

See **polygenic trait**; **quantitative trait locus**.

## quarry

Any area repeatedly visited by hominins to obtain stone raw material. Stone raw material may be freely available on the surface, but in other instances quarries provide evidence that raw material was extracted after hominins used simple lever systems to remove subsurface boulders. Quarry sites typically include large amounts of debris associated with stone tool production (particularly in the early stages of their manufacture) as well as evidence of the tools (e.g., hammers) needed to make stone tools. The MNK "chert factory site" at Olduvai Gorge is a good example of a quarry site.

## quartz

Crystalline silica ($SiO_2$), which is one of the most common rock-forming minerals in the continental crust, is found in many different types of rock. Quartz is widely used in the production of stone tools yet it is prone to shatter, thus making it difficult to manufacture large tools. Chalcedony (e.g., agate, chert, and flint) has better flaking characteristics and is typically preferred for making stone tools. *See also* **conchoidal fracture**; **percussion**.

## quartzite

A typically coarse granular to crystalline rock, sedimentary or metamorphic in origin, composed primarily of quartz. Although its material properties are dependent upon the size of the individual quartz grains, percussion generally results in a conchoidal fracture and quartzite artifacts retain a durable sharp edge. It appears to have been widely used at certain archeological sites where it was readily available (e.g., Olduvai Gorge and Klasies River).

## Quaternary

A unit of geological time (i.e., a geochronologic unit) that includes major glaciation events and the appearance of modern humans. As of 2009, the International Union of Geological Sciences redefined the Quaternary as the most recent of the three periods that comprise the Cenozoic era. Thus, it spans the time interval from 2.58 Ma (the earliest onset of major glaciation) to the present and is divided into the Pleistocene (2.58 Ma–11.7 ka) and Holocene (11.7 ka–present) epochs. (L. *quarter* = four times.)

## Quina Mousterian

See **La Quina**.

# R

### race
In the field of genetics, "race" has been used historically to describe a strain, group, or subspecie within a species. With respect to modern humans, the term has become tied to political and sociocultural concepts that do not correspond to any meaningful biological units or biological realities. Race is not appropriate as a biological term or as a classificatory unit for modern humans.

### radiocarbon dating
A radioisotope-based dating method that is based on the decay of the radiocarbon isotope $^{14}C$ (aka radiocarbon) to $^{14}N$. The method can be applied to organic material such as bones, teeth, seeds, and charred wood/charcoal, and to inorganic precipitates (i.e., carbonates) that contain sufficient amounts of $^{14}C$. All living organisms incorporate some $^{14}C$ into their body tissues, either during photosynthesis (plants) or from dietary carbon (animal consumers), but when an organism dies it no longer does so. Because $^{14}C$ decays to $^{14}N$ at a constant rate, the relative amount of $^{14}C$ left in organic material can be used to estimate the age of the material. However, because the half-life of the breakdown of $^{14}C$ to $^{14}N$ is $5730 \pm 40$ years, the method can only be applied to material that is less than c.50–60 ka (i.e., less than 1% of the temporal span of the hominin fossil record) because samples older than this no longer contain enough $^{14}C$ to date. Conventional radiocarbon dating measures the radioactive decay of $^{14}C$ via the emission of beta particles (electrons). Accelerator mass spectrometry (or AMS) radiocarbon dating uses a particle accelerator to help remove the ions that interfere with the counting of $^{14}C$. This allows the accurate measurement of very low concentrations of $^{14}C$ and, because it also directly measures individual $^{14}C$ atoms, much smaller samples can be processed. Because AMS radiocarbon dating can routinely date samples of 1 mg of carbon this means that previously undateable samples (e.g., single hominin teeth and individual grains of domesticated cereals) can now be dated. Typical starting weights required for AMS (e.g., 10–20 mg of seed/charcoal/wood, 500 mg of bone, 10–20 mg of shell carbonate) are about 1000 times less than the weights required by conventional systems. The AMS method also allows for more thorough chemical pretreatment of samples, which is particularly important for older samples (>25 ka BP) where small amounts of modern carbon contamination may have a large effect on the measured $^{14}C$ fraction and hence the date. Differences in the atmospheric production of $^{14}C$ through time, due to changes in the Earth's geomagnetic field and solar output, mean that $^{14}C$ ages (also known as uncalibrated radiocarbon ages) need to be calibrated to correct for these potentially

---

*Wiley Blackwell Student Dictionary of Human Evolution*, First Edition. Edited by Bernard Wood.
© 2015 John Wiley & Sons, Ltd. Published 2015 by John Wiley & Sons, Ltd.

substantial errors. The discrepancies between the raw and calibrated radiocarbon ages can be substantial, and average error in the age estimate tends to increase with the age of the specimen. For specimens around 6–8 ka the uncalibrated $^{14}$C ages are 750–1000 years too young, for c.20 ka specimens the uncalibrated $^{14}$C ages may be c.3 ka too young, and for those between 35–40 ka the discrepancy can be c.5 ka. Radiocarbon dates are reported using the notation "BP" which stands for "before present" and indicates the age of an object in years prior to 1950 (so that an object that is dated to 1300 BP is from AD 650). Calibrated radiocarbon dates are given the prefix "cal." See also **radiometric dating**.

## radiogenic dating
See **geochronology**; **radiometric dating**.

## radioisotopic dating
See **geochronology**; **radiometric dating**.

## radiometric dating
A family of dating methods based on measuring radiation. To use these methods, one measures the amount of parent material that remains and/or the amount of daughter product accumulated, or the amount of radiation received (e.g., optically stimulated luminescence dating and electron spin resonance spectroscopy dating). One or more of these two measurements, in combination with the relevant decay constant, allow scientists to determine how long the decay process has been running. Some radiometric methods (e.g., potassium-argon dating, argon-argon dating, radiocarbon dating, fission track dating) work independently of local levels of cosmic and other background radiation. Other radiometric methods are sensitive to local burial conditions. To be useful for dating fossil hominins, a radiometric system must (a) have an appropriate length half-life, which must be long enough for there to be more than a small number of atoms to measure, and short enough that there are perceptible changes in the ratio of the parent to the daughter atoms within the past 5–8 Ma; (b) have a stable daughter product; (c) be a "closed" system, which means that there is no other source of daughter product and the accumulated daughter product cannot be removed; and (d) be such that the beginning of the decay process of the parent atom must coincide approximately with either the death of the animal or the formation of the rock in which the animal is buried. (L. *radiare* = to emit beams, from *radius* = ray, and *metricus* = relating to measurement.)

## rain forest
A forested region that receives more than 250 cm of rain a year, and is characterized by high species diversity and dense tree cover. It is likely that some hominids and perhaps even hominins (particularly those in the Miocene and Pliocene) existed in areas that contained tropical rain forest. Tropical forest is seen as a less favorable habitat for hominins than woodland because of the inaccessibility of fruits in the higher canopy and the cryptic nature of forest animals. Nonetheless, there is increasing archeological evidence for forest exploitation by hominins in Africa during the Pleistocene and archeological and paleontological data also suggest that modern humans colonizing Southeast Asia c.46–34 ka exploited a range of habitats, including rain forest.

**random assortment**
When alleles of different genes assort independently during gamete formation. During cell division in eukaryotic organisms, pairs of homologous chromosomes align randomly at the metaphase plate (i.e., which of a pair of chromosomes is from the male parent, and which is from the female, is due to chance). This random assortment process creates an unpredictable mixture of maternally and paternally inherited alleles in the newly formed gamete. This phenomenon, along with chromosomal crossing over, helps increase genetic diversity by producing new genetic combinations. *See also* **Mendelian laws**.

**random drift**
*See* **genetic drift**.

**randomization**
(aka permutation test) A type of analysis used in hypothesis testing that uses an iterative process of sampling without replacement to generate values for a test statistic. If a researcher wants to know whether a difference between two samples of different sizes is significant, then a randomization procedure can be developed to do this. Exact randomization looks at all possible ways to split the data, but usually a large number (typically $n \geq 5000$ or more) of randomly selected possibilities are used as a proxy for the full set of possible combinations. *See also* **resampling**.

**random mating**
A mating system in which every individual has an equal chance of interbreeding with every other individual of the opposite sex, regardless of genetic, physical, or social preferences. This type of mating system contrasts with situations such as those described for positive assortment. *See also* **positive assortment**.

**range expansion**
A type of dispersal that occurs when the area in which a population is found increases but still encompasses the original ancestral area. For example, towards the end of the Pliocene a trend toward environmental aridity led to an expansion of grasslands in Africa. Many grazing mammalian species, and the carnivores that preyed on them, experienced range expansion during this time period.

**range shift**
A type of dispersal that occurs when the area in which a population is found changes its location. For example, when aridification in the southern hemisphere meant that dry (aka mesic) vegetational zones shifted towards the equator, mammals in the southern hemisphere that were adapted to mesic conditions shifted their ranges northward.

**rarefaction**
A statistical technique developed by ecologists that is used by paleontologists and zooarcheologists to compare the number of species (aka species richness) between samples with unequal sizes. Because the number of species observed in an assemblage is influenced by the number of

specimens that are sampled, rarefaction is used to correct for sampling effects. It can also be used to determine how many species one would expect to observe in a fossil assemblage and to determine whether a faunal community has been sufficiently sampled.

## ray
A manual digit plus its metacarpal, or a pedal digit plus its metatarsal. Rays in the hand or foot are numbered (e.g., second manual ray or fourth pedal ray), but manual digits are normally named (e.g., ring finger, thumb). (L. *radius* = radiating line.)

## reaction norm
The range of possible phenotypes that can develop in response to a range of environments. A reaction norm is usually illustrated by plotting the curvilinear relationship between a measure of the phenotype and a measure of the environment (e.g., body size and average annual rainfall). This method is used to show the contribution of environmental variation to phenotypic variation, or how differing genotypes within a species respond to varying environments. A narrow distribution indicates a highly predictable phenotype from any given genotype (little to no contribution from the environment), whereas a broad distribution indicates that phenotypic variation is mostly a consequence of variation in the environment. (Ge. *reaktionsnorm*.) See also **phenotypic plasticity**.

## reassembly
Bones and teeth that are fragmented but otherwise undamaged and undistorted, can be reassembled. The process of reassembly is painstaking work and formerly researchers had to sit down with the original pieces to try to fit them together by hand. However, researchers have more recently developed and used computer software to create "virtual fossils," the fragments of which can be moved and rotated on the screen. This means that researchers can work at reassembling hominin fossils without risking damage to the originals.

## recent African origin of modern humans
See **out-of-Africa hypothesis**.

## recognition species concept
(aka RSC) A "process" definition of a species that tries to overcome the major problem of the biological species concept, which is that the biological species concept requires comparison of one species to another (i.e., it is relational, not free-standing). The RSC is defined as the most inclusive population of organisms that share a common fertilization system. In other words, species are composed of animals that share a specific mate-recognition system (aka SMRS). See also **species**.

## recombination
The exchange of DNA between individual chromosomes in a pair during meiosis. Recombination (aka crossing over) shuffles genetic material between homologous chromosomes so that alleles at different loci located far apart will not be linked. The frequency of recombination varies across physical distance on a chromosome, and there are recombination hotspots that are short

regions (just a few thousand base pairs) of the genome at which recombination is substantially elevated relative to the genome average. The frequency of recombination between two loci can be used to calculate relative distance in the genome (aka genetic map) in centimorgans, with one centimorgan being the distance between two loci for which one recombination occurs out of 100 meioses. *See also* **linkage**.

## reconstruction

If only part of an undistorted fossil bone or tooth has been preserved, the missing parts can be reconstructed. This may involve duplication if the missing piece is a bilateral structure and the other side (aka the antimere) is preserved, or it may involve extrapolation if there is no antimere or if only part of a structure is preserved. In general, the more complete a tooth or bone is, the more reliable the reconstruction. The most difficult problems occur when a fossil has been deformed. If the deformation has affected only one side of a bilateral structure then the undeformed side can be used to reconstruct the deformed side. If both sides are deformed, the only recourse for researchers is to use computed tomography (or CT) to digitally visualize the fossil, and then use software programs based on D'Arcy Thompson-type transformation grids to estimate the undeformed shape of the fossil. There are at least two problems with these methods. First, if fragmentary fossils are reconstructed on the basis of more complete ones, the sample tends to regress towards the "mean" (i.e., towards the more complete fossils). Second, it is difficult to determine which parts of a virtual fossil are "real" and which are reconstructed.

## recursion

*See* **language**.

## reduced major axis regression

(aka RMA) As with other methods of regression, this statistical analyses estimates the relationship between variables, specifically how one variable (the dependent variable) changes as a result of the other, independent variable. RMA minimizes the sum of products of differences between observed values and the best-fit line in both the independent and dependent variables. The principles of RMA are similar to those of major axis regression, but RMA is preferred in cases where the independent and dependent variables are not measured in the same units. Because RMA takes into account error in all variables, it is an example of a model II regression. *See also* **regression**.

## reduction sequence

*See* **chaîne opératoire**.

## Reduncini

A tribe of antelopes within the family Bovidae that comprises the reedbucks (*Redunca*), kobs (*Kobus*), and their allies. Their modern distribution is limited to Africa but they originated in Eurasia during the late Miocene. Modern examples prefer well-watered environments and many live close to permanent sources of water. Fossil reduncines are frequently found at hominin sites, where they are an indicator of the local availability of marshy habitats. (L. *reduncus* = curved back.)

## reference digit

The digit of the hand or foot that determines the names that are given to the sideways movements of the other digits and thus the names of the muscles responsible for those movements. The middle finger is the reference digit of the hand, and the second toe is the reference digit of the foot. The movement of any other digit away from the reference digit is called abduction. Likewise, the movement of any other digit towards the reference digit is called adduction. So, for example, the act of moving the index finger away from the middle finger is called abduction, and any muscle that moves the index finger away from it is called an abductor.

## referential modeling

In referential modeling, the real behavior of one, or more, living animal(s) is used as a model to reconstruct the behavior of a fossil taxon (the referent). For example, referential modeling can be used to infer social structure in fossil hominins based on the anatomy and behavior seen in closely related extant species. For example, if the extant species that exhibit substantial sexual dimorphism in canine crown height always have a multi-male social structure, then the principles of referential modeling suggest that a fossil hominin with similar levels of such sexual dimorphism in canine crown height most likely also had a multi-male social structure. (L. *referre* = to carry and *modus* = standard.) See also **modeling**.

## refit

An analysis that restores the integrity of stone tools or reconstructs the manufacturing (aka reduction) sequence of stone tools by fitting flakes and flake fragments back together. Refitted breaks conjoin broken artifacts, and reduction refits conjoin two or more artifacts within a reduction sequence. The extent of refitted breaks can be important in revealing post-depositional processes such as trampling, whereas reduction refits allow the precise reconstruction of how ancient hominins knapped stone, literally in a "blow-by-blow" fashion. Reduction refits of long sequences of removals have been crucial for demonstrating the many ways in which a single block of stone may be modified over the course of its reduction (e.g., the production of Levallois points as well as blades). See also **chaîne opératoire**.

## refugium

The discrete location of a small residual population of a species that was once much more widespread. Refugial populations can arise when inhabited areas get cut off from the rest of the range occupied by a species. For example, during the Pleistocene glaciations the land bordering the Mediterranean acted as a refugium for hominins and other animals that were not well adapted to the extremely cold conditions that prevailed in the higher latitudes. It was the populations in these refugia that then re-expanded into higher latitudes during interglacial periods. Any isolated population separated from their parent population in refugia offers the potential for allopatric speciation. Refuge populations can also lead to endemism (i.e., a species that is unique to a particular, often small, area). (L. *refugium* = a place of safety; pl. refugia.)

## Regourdou

A collapsed cave site located along the Vézère River in the Dordogne area of France that is on the same hillside as Lascaux cave. Comparisons with other sites suggest an age of late Marine Isotope Stage (MIS) 5 or early MIS 4 (*c.*70 ka) for the layer containing an intentional burial consisting of at least two *Homo neanderthalensis* individuals. The two main archeological layers both contain Mousterian stone tools. (Location 45°03′N, 01°11′E, France.) *See also* **Lascaux**.

## regression

Earth science The contraction of a body of water (lake or sea) to expose an area that used to be part of the lake or sea bed as land surface. Statistics Quantification of the relationship between two or more variables in which values for one variable (the dependent variable) are predicted based on the values of one or more other variables (the independent variables). In simple regression, a line is fitted to a plot of two sets of continuous measurements; the slope, the intercept, and the scatter about the line are all parameters that can be used to describe a data set. Multiple regression predicts values of the dependent variable based on multiple independent variables, with the associated parameters identifying the effect size and significance of each independent variable. Model I regression methods minimize errors in the dependent variable, whereas model II regression methods simultaneously minimize errors in both the dependent and independent variables. (L *regress* = to step back.)

## Reilingen

A hominin partial calvaria was recovered from this commercial gravel pit in Germany. The mammalian fauna from the site ranges in age from the Holstein interglacial (*c*.300–200 ka, possibly Marine Isotope Stage 11) to the Late Würm glacial (Marine Isotope Stages 2–4). The researchers who published the hominin fossil prefer the older date, especially because the fossil shares many affinities with crania from Sima de los Huesos, Steinheim, Swanscombe, and other archaic Neanderthals. Researchers who interpret *Homo neanderthalensis* as an evolving lineage would interpret the Reilingen calvaria as belonging to the second, or pre-Neanderthal, stage in this lineage. (Location 49°17′N, 08°33′E, Germany.)

## relative dating

A dating method that relates a horizon, or an assembly of fossils or artifacts, to an externally validated sequence of change of some kind (e.g., climate, the direction of the Earth's magnetic field, morphology, artifact design). Fossils or artifacts incorporated within a horizon are used to link that horizon to an absolutely dated horizon elsewhere (e.g., mammalian fossils within the caves in southern Africa are used to date their parent horizon by matching them with the same type of fossils found in absolutely dated horizons at East African sites). (L. *relatus* = to relate or compare with something else.)

## relative rate test

*See* **molecular clock**.

## relative taxonomic abundance

In a fossil assemblage, relative taxonomic abundance quantifies the numerical abundance of a given taxon relative to the total number of taxonomic groups in that assemblage. Relative taxonomic abundances are typically quantified using the number of identified specimens (or NISP) or minimum number of individuals (or MNI). Relative taxonomic abundances are particularly useful for making paleoenvironmental reconstructions (e.g., if the relative taxonomic abundance of alcelaphine bovids increases through time at a paleontological locality one might reasonably infer an expansion of grassland environments through time). In zooarcheology relative taxonomic abundance is used to examine how human subsistence strategies varied through time and space, particularly in response to factors such as environmental change, the development of new technologies, or demographic shifts. *See also* **minimum number of individuals**; **number of identified specimens**.

## relative tooth size
The absolute sizes of tooth crowns in relation to body mass or to variables that serve as a proxy for body mass (e.g., the megadontia quotient, or MQ). It also refers to the relative size of individual tooth crowns, or of one tooth type (e.g., incisors, canine, premolars, and molars) relative to another.

## relative warps
*See* **geometric morphometrics**.

## remodeling
*See* **bone remodeling**.

## Rensch's Rule
*See* **sexual dimorphism**.

## replacement with hybridization
Replacement with hybridization (aka the Afro-European sapiens hypothesis) is a variant of the out-of-Africa hypothesis. It suggests that the earliest evidence of modern human morphology appeared in East and southern Africa and that members of these populations migrated out of Africa to populate Eurasia, eventually replacing archaic hominins in those regions. These early African populations of *Homo sapiens* would thus be the direct ancestors of all modern human populations, but unlike the strong version of the out-of-Africa hypothesis it is also accepted that these early modern humans may have interbred with pre-modern *Homo* groups such as *Homo neanderthalensis*, aboriginal East Asians, presumably *Homo erectus*, and the Denisovans. *See also* **assimilation model**; **candelabra model**; **multiregional hypothesis**; **out-of-Africa hypothesis**.

## reproductive effort
The proportion of energy or materials an organism devotes to reproduction rather than to growth and maintenance. Reproductive effort in female modern humans and nonhuman primates is generally higher than in males because of anisogamy (unequal sizes of the gametes), internal gestation, and lactation. Relative to nonhuman primates, it might be expected that total lifetime reproductive effort in modern humans is comparatively high due to our life history (e.g., cooperative foraging and breeding, long juvenile periods), but the comparative evidence suggests that human lifetime reproductive effort is not unusually high. *See also* **life history**.

## reproductive investment
The energy invested by females or males in sexual maturation, sexual behavior (including competition, copulation, and mate guarding), gestation, and postnatal support of offspring. In most mammals, including primates, where reproduction is a costly event, reproductive investment is

higher per reproductive event for females compared to males. Life history theory predicts that reproductive investment should be high when high fitness returns are expected. This, combined with the cost/benefit calculation regarding current versus future reproduction (which is affected by offspring number, sex ratio, etc.) should determine an individual's level of reproductive investment throughout their reproductive lifetime. *See also* **life history**.

## resampling

A process of iteratively sampling from a data set to produce confidence intervals for a sample parameter (e.g., mean, median, standard deviation, etc.), to conduct significance tests for some test statistic (e.g., difference in the mean between two groups or for differences in a correlation coefficient from some pre-specified value such as zero, etc.), or to validate predictive models (e.g., to assess the accuracy and precision of predictions from regression or discriminant function analysis). Resampling techniques are preferred over more traditional techniques because they can find confidence intervals and *P* values for test statistics that do not have standard analytical solutions. Resampling techniques used in studies of human evolution include the bootstrap, jackknife, randomization (also known as permutation tests), and cross-validation. In each of these techniques a subset of data is sampled to generate a parameter, a test statistic, or a predictive model and then this procedure is repeated multiple times to generate a confidence interval, *P* value, or measure of model validation. If all possible subsets of the data are used then it is an exact test (this is required for the jackknife and is usually the case in cross-validation), but in many cases there are more possible ways to sample the data than can feasibly be performed (as is often the case in bootstrapping and randomization). In these cases a large number (e.g., 10,000) of randomly selected subsets of data are used; these versions are known as Monte Carlo methods. *See also* **randomization**.

## residue analysis

The study of biotic and abiotic residues left on the surfaces of stone, bone, or metal tools, or on pottery, to understand what the tool was used for. These residues can be identified by microscopy if directly visible (e.g., in the case of large particles, like fur, feathers, or plant residues), by spectroscopy (e.g., infrared spectroscopy) if they leave identifiable chemical signatures (e.g., for more amorphous or invisible residues, like amino acids, peptides, and lipids), or by a variety of other tests designed to identify specific residues (e.g., immunology assays to identify blood types). Residue analysis must take into account the possibility that the residue did not result from an intentional contact with the substance, but rather by precipitation, secondary deposition, or other forms of contamination.

## resorption

The removal of bone tissue that occurs when bone is remodeled during growth or after a fracture, or when the overall volume of a bone is reduced because it no longer has a function (e.g., the alveolar process of the mandible resorbs after the teeth have been lost). Bone resorption is carried out by osteoclasts and their activity leaves distinctive areas of localized erosion called Howship's lacunae that have been used to identify taxonomically distinctive areas of bone resorption on the surface of the face. Resorption also refers to the process whereby the roots of deciduous teeth are cleared out of the path of the developing permanent teeth during the passage of the latter through the mandible or maxilla. *See also* **bone remodeling; eruption; ossification**.

## resource defense model

A model of Oldowan site formation based on the hypothesis that the species responsible for bone assemblages were part of the predator guild. The model suggests that meat from hunted and scavenged carcasses was transported to places with fixed, defendable resources (e.g., trees, water, plant foods, sleeping sites). Group defense allowed these focal sites to be used regularly for activities that would lead to the gradual accumulation of archeological debris, and thus their eventual recognition as archeological sites. While the resource defense model resembles the home base hypothesis (e.g., delayed consumption, food transport to a central place, and potentially extensive food sharing), it lacks the latter's emphasis on a sexual division of labor.

## resource depression

A reduction in the abundance or availability of prey that results directly from the foraging behavior of the predator. In archeological contexts resource depression is associated with diminished access to high-ranking prey and may reflect an expansion of diet breadth and increased utilization of low-ranking prey. There is much interest in understanding prehistoric resource depression in an effort to understand how hominin foragers impacted the ecosystems they inhabited.

## retention index

A measure of the goodness of fit between a cladogram and a character state data matrix. In contrast to the consistency index, the retention index (or RI) is not sensitive to the number of taxa or the number of characters included in a character state data matrix. In principle, RI values can range between 1 and 0. An RI of 1 indicates a cladogram that requires no homoplasy, with the level of homoplasy increasing as the index approaches 0. *See also* **cladistic analysis; consistency index**.

## reticulate evolution

In reticulate evolution species form by the hybridization of two existing species rather than by bifurcation. This model of evolution is close to how some researchers interpret evolution in geographically widespread groups such as contemporary baboons. In this interpretation, peaks of morphological distinctiveness are equivalent to what are interpreted as species differences in other taxa. Yet at the boundaries between these peaks of morphological distinctiveness there are evident hybrid zones where baboons are less morphologically distinct. In the reticulate evolution model the location and nature of the peaks of morphological distinctiveness and of the hybrid zones change over time. (L. *reticulum*, dim. of *rete* = net.)

## retouch

The removal of one or more small flakes to shape a stone or bone tool. The amount of retouch and the number and shape of edges that have been retouched are key elements in many formal stone tool classification schemes. (Fr. *retoucher* = to alter or touch up.)

## Retzius' line
See **striae of Retzius**.

## reverse fault
See **fault**.

## rhinocerotid
Informal name for the family Rhinocerotidae, a member of the order Perissodactyla. Rhinocerotids include the rhinoceroses and their allies, whose members are found at some hominin sites. *See also* **Perissodactyla**.

## rhyolite
A term used for fine-grained volcanic rocks, often rich in phenocrysts of alkali feldspar and quartz. When available to fossil hominins, it appears that rhyolite was often used for stone tool production.

## riparian
Refers to the bank of a river or lake, so a "riparian" paleoenvironment would be one that is related to the bank of a stream, river, or lake (NB: "riparian rights" are what you pay for when you buy a fishing license).

## Rising Star
(aka Dinaledi Chamber) A hominin-bearing paleontological site, located in the Rising Star cave system near the well-known sites of Swartkrans and Sterkfontein, which is within the Cradle of Humankind World Heritage Site in South Africa. The site was discovered in September 2013 by exploration teams from the University of the Witwatersrand. Fossils were collected during two expeditions, one in November 2013 and a second in March 2014. The hominin fossils come from the Dinaledi Chamber, which is more than 30 m underground and extremely difficult to access. The remains of what are estimated to be more than a dozen archaic hominin individuals have been recovered, and those who have visited the Dinaledi Chamber estimated that many more fragmentary hominin fossils remain to be recovered. The geological age of the hominin fossils is presently not known. (Location 26°1′13″S, 27°42′43″E, Gauteng Province, South Africa.)

## Riss
See **glacial cycles**.

## riverine
Refers to a river, or to the bank of a river. A "riverine" paleoenvironment would be one that focused on the river itself, if it was aquatic, or on the bank of the river, if it was terrestrial. A terrestrial riverine paleoenvironment is one where the flora and fauna are like those found on the banks of contemporary ephemeral or permanent rivers. In the case of a permanent river this usually means a forest-like environment.

## r/K-selection theory
A hypothesis that there is a systematic relationship between the type of environment experienced by a taxon, and the type of reproductive strategy it adopts to maximize its likely success. Unstable or unpredictable environments tend to favor $r$-selected species, in which the individuals are small-bodied, mature early, tend to have many widely dispersed offspring, and have short generation times and lifespans. This contrasts with stable, predictable environments, which favor $K$-selected species, in which the individuals are larger-bodied, mature later, tend to have fewer offspring that require intensive investment, and have relatively long generation times and lifespans. Modern humans are usually categorized as a $K$-selected species.

## robust
Morphology A robust morphological variant is stronger than other variants. In the mandible this means mandibular bodies (corpora) that are relatively wide, and in the long bones it means a relatively thicker layer of cortical bone that results in wider shafts. Nomenclature Robust is used to describe a subset of archaic hominin taxa assigned to either *Australopithecus* or to *Paranthropus* (i.e., *Au.* or *P. robustus*, *Au.* or *P. boisei*, and *Au.* or *P. aethiopicus*) that have large postcanine teeth (aka postcanine megadontia or hyper-megadontia) and thick (i.e., robust) mandibular bodies. (L. *robustus* = strength, from *robus* = oak.) *See also* **robust australopith**.

## robust australopith
Informal term for archaic hominin species with large postcanine teeth (aka postcanine megadontia or hyper-megadontia) and thick mandibular bodies. The term, which was originally used for *Paranthropus robustus*, has subsequently been applied to *Paranthropus boisei* and *Paranthropus aethiopicus*. Many researchers now avoid the term because several archaic hominin taxa exhibit postcanine megadontia and relatively robust mandibular bodies, but lack other aspects of the derived morphology of *Paranthropus*. Similarly, some refer to *Australopithecus* taxa that do not have such large postcanine teeth and thick mandibular bodies (e.g., *Australopithecus africanus* and *Australopithecus afarensis*) as gracile australopiths.

## robusticity
The tendency to be robust. In mandibles this means having a high value for the mandibular robusticity index (corpus width/corpus height) and for long bones this means a high value for an index that relates a long bone's width to its length. (L. *robustus* = strength, from *robus* = oak.)

## Roc de Marsal
A cave that overlooks a tributary of the Vézère River in southwest France in which a nearly complete skeleton of a *Homo neanderthalensis* child, isolated teeth of additional individuals, and Middle Paleolithic artifacts were found. (Location 44°54′21″N, 00°58′45″E, France.)

## rock
A rock is a naturally occurring solid composed of one or more mineral phases, native elements, or organic accumulations. There are three fundamental types of rock: igneous (solidified from a molten phase, or melt), metamorphic (formed by solid-state recrystallization,

usually under high pressure and/or temperature), and sedimentary (otherwise formed under Earth-surface conditions). With very few exceptions hominin fossils are found in sedimentary rocks.

### rodents
A diverse group of mammals with continuously growing central incisors. The relatively species-specific habitat preferences and diets of rodents make them useful as paleoecological indicators. Rodents make up much of the microfaunal evidence found at early hominin sites. (L. *rodere* = to gnaw, or eat, away.) *See also* **microfauna**.

### root
Dentition The part of a tooth embedded in the bony alveolus that is made up of an inner core of dentine and the enclosed pulp; it is covered with cementum that attaches it to the wall of the alveolus. The boundary between the crown and the root on the outside of the tooth is called the cemento-enamel junction (or CEJ) or the cervix, and the boundary between the crown and the root within the tooth is called the enamel–dentine junction (or EDJ). Incisors and canines usually have a single root, premolars may have one or more roots, and molars usually have two or more roots. Osteology Describes the place where one structure is attached to another (e.g., the root of the zygomatic process of the maxilla). Cladistics Refers to the base of a cladogram. (OE *rot* = root.)

### rostral
The rostral direction is in the direction of the beak (or its equivalent), and is the opposite of caudal (toward the tail). When applied to the human body (or other bipedal species), the term rostral is equivalent to cranial (i.e., superior). Within the brain, rostral–caudal specifically refers to anterior–posterior, whereas dorsal–ventral refers to superior–inferior. (L. *rostrum* = beak, muzzle, or snout.)

### rostrum
In vertebrate anatomy the rostrum is an informal term for the snout or muzzle. As a practical matter, in primates the rostrum refers to the premaxilla and the portion of the maxilla that projects anterior to the orbits and to the roots of the zygomatic arches. In hominins, and particularly in *Homo*, the trend towards flatter faces (aka facial orthognathism) has resulted in extremely reduced rostra. (L. *rostrum* = beak, muzzle, or snout.)

### routed foraging
A model for Oldowan site formation that proposes that hominins were recurrently drawn to fixed resources (e.g., stone outcrops, stands of trees acting as midday resting sites, water sources). At such locations carcass parts would have accumulated over time and because of these accumulations the location would subsequently have become visible as an archeological site.

### *r*-selection
*See* **r/K-selection theory**.

**running**
See **endurance running hypothesis; gait.**

**Ruscinian**
See **biochronology; European mammal neogene.**

# S

### Saccopastore
Two *Homo neanderthalensis* crania and some Mousterian artifacts were found in this c.130–100 ka gravel quarry on the left bank of the Aniene River, a tributary of the Tiber, within the boundaries of modern Rome. (Location 41°57′N, 12°32′E, Italy.)

### sagittal
A plane that corresponds to the midline of the body. It takes its name from the arrow-shaped inter-parietal cranial suture in the cranium of a neonate/infant that runs in that plane from bregma (the tip of the arrow) to lambda (the feathers of the arrow). (L. *sagitta* = an arrow.)

### sagittal crest
A sharp crest of bone running along the course of the inter-parietal, or sagittal, suture that forms when the medial borders of the fascia covering the two temporalis muscles fuse. If the fusion is complete there is just one crest, but if it is incomplete it is referred to as a compound sagittal crest (aka parasagittal crests). A sagittal crest is seen in male gorillas and orangutans and in larger-bodied, presumed male, *Paranthropus aethiopicus* and *Paranthropus boisei* crania such as KNM-ER 406. (L. *sagitta* = an arrow and *crista* = cock's comb, or a tuft of feathers, in the midline of a bird's head.) *See also* **parasagittal crest**.

### sagittal keel
A blunt bony prominence running longitudinally along the course of the inter-parietal, or sagittal, suture. Most, but not all, *Homo erectus* crania have a sagittal keel. In general sagittal keels are better expressed in larger, presumed male, crania. (L. *sagitta* = an arrow and ME *kele* = the main longitudinal timber in a wooden boat, now used for the underside of a boat, or any projection designed to prevent a boat from "keeling over.")

### Sahel
Semiarid region of north-central Africa to the south of the Sahara, which is the transitional area between the desert and savanna biomes. The fossil evidence for *Sahelanthropus tchadensis* comes from sites in the Sahel, but c.7 Ma ago the climate in this region was much less arid. The environment where *S. tchadensis* lived probably resembled those in East and southern Africa where other fossil hominins are found. (etym. Ar. *sahil* = sea coast, shore, border of the desert.)

---

*Wiley Blackwell Student Dictionary of Human Evolution*, First Edition. Edited by Bernard Wood.
© 2015 John Wiley & Sons, Ltd. Published 2015 by John Wiley & Sons, Ltd.

## *Sahelanthropus* Brunet et al., 2002

A genus established in 2002 to accommodate a new species that was introduced for cranial remains recovered from the *c.*7 Ma Anthracotheriid unit at locality TM 266 at Toros-Menalla in the Chad Basin, in Chad, Central Africa. The authors claimed that various features of the cranium either distinguished the new fossils from *Pan*, or could not be accommodated in any of the existing hominin genera. Some researchers suggest that the morphological differences between species assigned to *Ardipithecus* and to *Sahelanthropus* do not justify their being assigned to different genera. If that is the case, the former genus would have priority, and *Sahelanthropus* would be the junior synonym of *Ardipithecus*. The type species is *Sahelanthropus tchadensis* Brunet *et al.*, 2002. (etym. Ar. *sahil* = sea coast, shore, border of the desert; Gk *anthropos* = human being.)

## *Sahelanthropus tchadensis* Brunet et al., 2002

Hominin species established to accommodate cranial remains recovered from the Anthracotheriid unit at Toros-Menalla in the Chad Basin, in Chad, Central Africa; the former has been dated by biostratigraphy and cosmogenic nuclide dating to *c.*7 Ma. The authors believe that, despite primitive features in the cranium and dentition, this taxon is a primitive hominin and not a panin because of its small, apically worn canines and the intermediate thickness of the postcanine enamel. If the hominin classification is upheld, and the age of the specimen is confirmed, then *S. tchadensis* would be the oldest known hominin and it would be a refutation of the hypothesis that hominins were confined to the area either within, or to the east of, the East African Rift System. There are no published postcranial remains. The holotype is the TM 266-01-060-1 cranium. (etym. Ar. *sahil* = sea coast, shore, border of the desert; Gk *anthropos* = human being; *tchadensis* = Chad.) See also **TM 266-01-060-1**.

## Sahul

The continental landmass (aka Australasia) that comprises present-day Australia, Tasmania, and New Guinea. During particularly cold periods when substantial volumes of water were locked up in ice accumulations at high latitudes, sea levels would have fallen low enough for Sahul to have been one continuous land mass. However, even during the coldest periods hominins would have had to make a deep-water crossing, accidental or deliberate, to reach Sahul from Sunda (the landmass that comprises the parts of the Asian continental shelf that were connected to the mainland, including present-day Sumatra, Borneo, and Java). The northwestern margin of Sahul marks the eastern boundary of Wallacea. [etym. Ballard (1993) suggests that it may be a word used by the Macassan people for a sandbank or shoal on the northwestern continental shelf of Australia to the southwest of the Aru Islands and separated from Timor by the deep waters of the Timor Trough.] See also **Sunda**.

## Saint-Acheul

This *c.*400 ka open-air site in the suburbs of Amiens, France, is the type site of the Acheulean technocomplex. Handaxes were found in the gravel pits in the mid 19thC and as unequivocally hominin-made artifacts they helped demonstrate the antiquity of hominins (contrary to the biblical view prevalent at the time). (Location 49°52′N, 02°19′E, France.) See also **Acheulean**.

## Saint-Césaire
A fragmentary and compressed *Homo neanderthalensis* skeleton was recovered in the *c.*40 ka upper Châtelperronian layer at Saint-Césaire, France. Seventeen archeological layers document the transition from the Mousterian through the Châtelperronian to the Aurignacian. The hominin fossil was found associated with *Dentalium* shells; some have interpreted these as grave goods associated with an intentional burial. (Location 45°44′57″N, 00°30′11″E, France.)

## Saldanha
*See* **Elandsfontein**.

## Salé
An open-air site located 5 km/3 miles northeast of Salé city on the Atlantic coast of Morocco that revealed a *c.*400 ka partial hominin cranium assigned to *Homo erectus* or "archaic *Homo sapiens*." (Location 34°02′N, 06°48′W, Morocco.)

## Sambungmacan
A site on the south bank of the Solo River in Central Java. In the Bapang Formation at this site, researchers found three hominin cranial fossils and a partial tibial shaft attributed to *Homo erectus*. Electron spin resonance spectroscopy and uranium-series dating of the deposits suggest a relatively young age of 53–27 ka for the hominins. (Location 07°21′S, 111°07′E, Indonesia.)

## sample
A subset of a larger population of possible measurements. For example, a sample of cranial capacities can be drawn from the population of all chimpanzee crania that have ever existed. Statistics based on samples are used to make inferences about the populations from which the samples are drawn. *See also* **population**.

## sample statistics
*See* **parameter**.

## sampling with replacement
In statistics it is possible to sample with or without replacement. If we had a bag of nine tiles, marked 1–9, and wished to choose three tiles from that bag, we could do it two ways. First, we could draw a tile, record the number, and return the tile to the bag. Alternately, we could draw a tile, record the number, and set it aside. The first case is sampling with replacement, and it is possible to draw the same tile twice. The latter case is sampling without replacement, and it is not possible to draw the same tile twice. These two methods have implications for the total number of possible combinations of three tiles that can be selected from the bag, which is important for resampling methods, like bootstrapping and randomization. Bootstrapping uses sampling with replacement, whereas randomization uses sampling without replacement.

## San
Hunter–gatherer people who live in Botswana, Zambia, Zimbabwe, and South Africa. All San speak Khoisan language variants, and as recently as 1990 some practiced an entirely foraging-based economy. Today almost all of the San rely heavily or entirely on farming. The San have been

the subject of many ethnographic studies designed to better understand subsistence strategies and to test hypotheses derived from behavioral ecology. See also **forager**; **hunter–gatherer**.

## sand
Fine (i.e., a mean diameter between 0.063 and 2 mm) sedimentary particles. When they are consolidated or cemented (lithified) they are called sandstone. Two characteristics of sand and sandstone make them a common medium for the preservation of fossils. First, the physical processes responsible for accumulation of sands are likely to associate bones and teeth in the same deposits; second, the porosity inherent in sands facilitates mineralization of the fossil material and hence its long-term preservation (e.g., the rich Plio-Pleistocene fossil assemblages of the Turkana Basin are derived primarily from the sands of the ancient river channels).

## sandstone
See **sand**.

## Sangiran Dome
A relatively low-lying massif surrounded by large volcanoes in Central Java, Indonesia. Almost 80 hominin fossils, including 10 well-preserved crania, 14 mandibular and maxillary specimens, and dozens of isolated teeth, have been recovered from sites in the Sangiran Dome. The fossil hominins come from either the Sangiran Formation (formerly the Pucangan Formation) or the Bapang Formation (formerly the Kabuh Formation) that are separated by a fossil-rich layer formerly known as the Grenzbank Zone. The oldest hominin specimens derive from the upper levels of the Sangiran Formation, but most of the fossils are derived from the Bapang Formation. The lower layers of the Bapang were characterized by high-energy deposition (e.g., in a fast-flowing river), so many of the mammalian fossils, including the hominins, are abraded and fragmentary. Higher in the Bapang Formation the sediments are finer-grained (indicating lower-energy deposition, e.g., a slower flowing river or backwater) and the fossils from those strata are better preserved and more complete. The Sangiran Formation has argon-argon dates of 1.92–1.58 Ma, but only the uppermost section of the formation is hominin-bearing. The Bapang Formation has argon-argon dates of 1.58–1.0 Ma. The Notopuro Formation, which caps the section but has yet to yield hominins, has an argon-argon date of 780 ka. The fossil hominins from Sangiran are all >1.0 Ma. Hominin fossils recovered from the Sangiran Dome include the holotypes of *Meganthropus palaeojavanicus* and *Pithecanthropus dubius* and the Sangiran 17 cranium, the most complete fossil hominin cranium to be discovered in Indonesia. All of these specimens have since been referred to *Homo erectus*. Stone tools, including utilized large flakes and bolas stones, were reported from Ngebung, a locality in the Bapang Formation in the northwestern part of the Sangiran Dome. (Location 07°20′S, 110°58′E, Indonesia.)

## Sangoan
An artifact industry found across Central Africa and portions of East Africa. The Sangoan is defined typologically by the presence of heavy-duty tools such as picks, choppers, and core-axes as well as a number of light-duty tools such as scrapers. At Kalambo Falls, Zambia, Sangoan assemblages fall between Acheulean and Middle Stone Age strata.

### sanidine

A common potassium-bearing mineral of the feldspar group found in pumices. Sanidine crystals are especially suitable for potassium-argon dating.

### sapropel

A sedimentary rock that contains more than 5% organic material. Sapropels occur in lacustrine and marine deposits. Sapropels in the eastern Mediterranean Sea provide proxy records of monsoonal rainfall in East Africa (particularly in the Blue Nile headwaters in the Ethiopian Highlands) and they are also used for the orbital tuning of the associated marine sequences. (Gk *sapros* = rotten and *pelagos* = sea.)

### satellite imagery

The traditional ways of identifying potential hominin fossil sites are to use conventional geological maps or aerial photographs, or to comb the records of local geological or natural history societies for records of fossil finds by amateurs. However, satellite images (e.g., NASA's Advanced Spaceborne Thermal Emission and Reflection radiometer, ASTER) can now be used to locate previously unrecognized locations whose geological structures and reflectance match those of existing hominin fossil sites. The first time satellite imagery was used on a large scale to locate sites was as part of the systematic inventory of paleontological and archeological sites in Ethiopia in the 1980s. This project resulted in the identification of productive new localities such as Aramis, Bouri, Fejej, Kesem-Kebena, and Konso-Gardula. *See also* **Paleoanthropological Inventory of Ethiopia**.

### savanna

An inclusive term that covers a large number of African biomes including woodland, bushland, shrubland, and grassland. Savanna biomes are generally characterized as a landscape with grass and with either scattered trees or open canopy trees. Savannas tend to occur on land surfaces of little relief, the soils tend to be low in nutrients, they have distinct wet and dry seasons, and they are often associated with precipitation ranging between 500 and 2000 mm per year. (Taino *zabana* = grassland.)

### scaling

The relationship between a variable and the overall size of an organism. Scaling subsumes isometry (no change in a variable with changes in size) and allometry (changes in a variable accompanying changes in size). For example, brain size scales with negative allometry relative to body mass. Thus, although larger-bodied taxa generally have larger brains than smaller-bodied taxa, because their brain size increases at a slower rate than their body mass, their brains are relatively smaller when scaled to body mass than those of smaller taxa. Scaling relationships are typically examined between species as interspecific allometry, within species as intraspecific allometry, or in a growth series as ontogenetic allometry. Scaling relationships can also be used for functional analyses, and may reflect functional equivalence or adaptation. Functional equivalence refers to underlying structural relationships between a variable and size. For example, long bones must be wider in large taxa to maintain the same strength relationship relative to body size, because mass is a function of the volume (i.e., length cubed) of an organism whereas bone strength is a function of the cross-sectional area (i.e., length squared). Thus, a species that is twice the body length of another (geometrically similar) species may have eight times ($2^3$) the mass, and would need bones with greater cross-sectional area than expected by isometry (twice the width of the small species) to maintain the functionally equivalent relationship. (L. *scalae* = ladder.) *See also* **allometry**.

## scapula, evolution in hominins

The scapula, which is a complex, flat bone that develops from multiple ossification centers, lies against the dorsolateral portion of the ribcage. In modern humans it spans ribs 1–7 and the glenoid fossa faces laterally, whereas in the great apes the fossa is situated more superiorly, producing a permanent "shrugged" shoulder appearance that provides strength and stability for the arm in suspensory climbing behaviors. Modern humans have a mediolaterally broad scapula; *Gorilla* and *Pongo* approach the modern human condition, whereas the scapulae of *Pan* and *Hylobates* are narrower. Most apes have a markedly oblique scapular spine, with *Hylobates* having the most oblique orientation (its spine is nearly parallel to the axillary border), whereas modern humans have a more transversely oriented spine. These differences are a reflection of the relative size of the fossae for the muscles that attach on the dorsal surface of the scapula. The hominin scapular fossil record is sparse and most of the specimens consist of glenoid fragments with some portions of the scapular spine and axillary border attached. Most scapular specimens attributed to *Australopithecus* (e.g., A.L. 288-1 and DIK-1-1) display the ape-like condition of a relatively cranially oriented glenoid fossa, while the glenoid fossa of an early *Homo erectus/ergaster* scapula, KNM-WT 15000, is more laterally facing.

## scavenger
*See* **scavenging**.

## scavenging

The acquisition of meat or marrow from the carcasses of animals that were killed by predators, or from an animal that died due to other natural causes. Perhaps the best-known contemporary scavengers are vultures, but some carnivores that primarily hunt (e.g., lions, spotted hyenas) are known to drive off the hunter (aka confrontational scavenging) or scavenge on an opportunistic basis (aka passive scavenging). The extent to which early hominins hunted versus scavenged is debated, but in no extant modern human groups is scavenging a major component of their subsistence strategy. (ME *skawager* = a collector of tolls, from the OF *escauwer* = to inspect; thus, someone who lives off another.) *See also* **confrontational scavenging**; **passive scavenging**.

## Schöningen

A site exposed in a brown coal mine in Germany that is best known for the recovery of wooden spears from the middle of the sequence. Two kinds of wooden spears were recovered. The older spears from Marine Isotope Stage (MIS) 11 (*c.*478–424 ka) levels are merely pointed, but were found in close association with flint tools and numerous horse bones that show signs of butchery. The more recent spears from MIS 10 (*c.*424–374 ka) levels show grooves that might have been slots for flint spearheads. At present the spears from Schöningen are the oldest known evidence for both spears and composite tools. (Location 52°08′N, 10°59′E, Germany.)

## sclerotome
*See* **somite**.

## scraper
A stone tool with retouched edges that meet at an angle of approximately 60–90°. The function of scrapers for use in processing hides and plant material has been inferred by ethnographic analogy, actualistic studies, and subsequent use-wear analysis. Although scrapers are reported from Oldowan sites in Africa, they are particularly common at Middle Paleolithic sites in Western Europe.

## seasonality
Regular changes in weather pattern (in both tropical and temperate regions) and day length (in temperate regions) within an annual cycle. These changes are driven by fluctuations in the orbital geometry of the Earth that affect the amount of sunlight reaching the Earth's surface (aka insolation) during the year. At higher latitudes weather is determined primarily by temperature, whereas in the tropics rainfall is an important determinant of weather, with wet and dry seasons within the annual cycle. Seasonality influences habitat and the availability of food. Primates cope with seasonal shifts in resources and habitat quality by switching to fallback foods. Seasonality was almost certainly an important influence on evolution within the hominin clade and it likewise probably affected hominin behavior. It is also possible that some characteristics of modern humans were adaptations to highly seasonal environments. For example, our propensity to gain body fat very quickly may have allowed early humans to gain weight during seasons of plenty in order to provide a buffer in times of scarcity. Modern humans use an array of cultural tools (e.g., food storage) to buffer against seasonality, which allows them to inhabit regions normally inaccessible to primates. The development of technologies and skills that facilitated the acquisition of meat is likely to have helped hominins exploit high latitudes because meat is available all year round whereas many plant foods are only available seasonally.

## seasons
*See* **seasonality**.

## secondarily altricial
*See* **altricial**.

## secondary cartilaginous joints
*See* **joint**.

## secondary centers of ossification
*See* **ossification**.

## section
Earth sciences A place where strata are exposed naturally or deliberately when researchers dig a trench to see the strata more clearly. Anatomical sciences A section refers to the surface of a bone or tooth that has been exposed by either a natural break or a saw cut made deliberately to provide information about the internal structure of bones and teeth. *See also* **cross-section**.

## sectorial
When the upper canine and the anterior lower premolar form a honing complex, the blade-like edge of the latter tooth is referred to as a sectorial premolar. In primates, a large, sharp canine has a social rather than dietary function. The loss of a honing complex and sectorial lower premolar in hominins is assumed to be related to a shift in behavior so that males no longer need to display or fight with their teeth to get access to mates. (L. *secare* = to cut.) *See also* **honing**; **sexual selection**.

## sedge
The common name for the Cyperaceae, a family of monocotyledonous flowering plants that superficially resemble grasses or rushes. Most sedges are found in wetlands, but some are adapted to seasonally dry environments. Many sedges have large corms (underground storage organs) which are often eaten by baboons, and which may have been important foods for early hominins. Several types of sedge have developed a photosynthetic system that initially fixes carbon dioxide in molecules containing four carbon atoms (i.e., they are $C_4$ plants). Some researchers have suggested that sedges may have been a source of forage for the early hominins that have a significant $C_4$ signature. *See also* $C_3$ **and** $C_4$; $C_4$ **foods**.

## sediment
Material that settles in the bottom of a liquid. The rocks in which most hominin fossils are preserved are rocks that have formed from sediment. The sediment can be coarse (e.g., gravel), finer (e.g., sand), or much finer (e.g., silt). (L. *sedimentum* = the act of settling.)

## sedimentary rock
Rocks formed from sediment. There are two categories of sedimentary rock. Clastic sedimentary rocks are formed from the fragments of other rocks that are transported from their source and then deposited in water. The second category includes all the rocks formed by precipitation (e.g., limestone is formed by the precipitation of calcium carbonate that comes from marine organisms).

## seed-eating hypothesis
One of several hypotheses put forward to explain the origin of an upright trunk in the hominin clade. The seed-eating hypothesis rests on the observation that australopiths share several derived characteristics (e.g., manual dexterity, efficient and powerful mastication, a reduced reliance on incision, and an upright trunk) with gelada baboons (aka *Theropithecus* or geladas). These shared features were evidently not inherited from the most recent common ancestor of these taxa, but rather represent examples of convergent evolution. In geladas this suite of features is related to seed-eating, and the seed-eating hypothesis suggests that the earliest hominins evolved in edaphic (wet) grasslands in which grain seeds would have been plentiful. *See also* **bipedal; tool-use hypothesis**.

## selection
*See* **natural selection**.

## selection coefficient

A measure of the fitness of a given phenotype relative to another fitter phenotype. The selection coefficient is equivalent to the Darwinian fitness of a phenotype. *See also* **fitness**.

## selective sweep

The relatively rapid fixation (i.e., an increase to an incidence of 100% in a given population) of an allele. When a selective sweep occurs, the reduction of nucleotide diversity around the selected allele is useful for identifying regions of the genome that have been recently subjected to directional selection. For example, in modern Europeans there is evidence of a selective sweep around the lactase gene because of the spread of an allele conferring lactose tolerance into adulthood.

## semantic memory

*See* **memory**.

## semicircular canals

The three semicircular canals comprise one of the three main subdivisions of the bony and membranous labyrinths of the inner ear, and they provide information about movement and orientation. Each canal comprises two-thirds of a circle. The anterior and posterior canals are at right angles to each other and each is oriented approximately 45° to the sagittal plane. The separate lateral canal is in the horizontal plane, whereas the conjoined anterior and posterior canals are in the vertical plane. The canals are arranged so that the anterior canal on one side and the posterior on the opposite side are "coplanar" (i.e., the long axis of the anterior canal on one side is parallel to the long axis of the posterior canal on the opposite side). Motion is detected by the relative movement that occurs between the hairs of the hair cells embedded in the walls of the part of the membranous labyrinth that lines the swelling (aka ampulla) at one, or both ends of the canals. Only the bony labyrinth is preserved in fossils, but its size and shape are a good proxy for the size and shape of the membranous semicircular canals. The relative size of the arcs of the canals differs in modern humans and the great apes, likely due to differences in posture. The modern human-like semicircular canal morphology is first seen in crania assigned to early African *Homo erectus* or *Homo ergaster*; it is claimed that the canals of *Homo neanderthalensis* are smaller than those of modern humans. *See also* **bony labyrinth**; **inner ear**.

## semilandmark

For some anatomical regions and structures (e.g., cranial vault, tooth crowns) there are only a limited number of possible traditional landmarks (e.g., suture intersections, cusp tips) and substantial amounts of morphology between the landmarks would be lost unless there was some other way of capturing it. Semilandmarks (aka sliding landmarks) were introduced as a way of extending landmark-based statistics to smooth curves and surfaces. Semilandmarks were initially applied to two-dimensional outlines, but researchers have subsequently developed the algebra needed to capture curves and surfaces in three dimensions. The same number of semilandmarks is placed in roughly homologous locations on every specimen of the sample; the number used depends on the complexity of the curves or surfaces to be captured and on the level of detail being investigated. The semilandmarks are then allowed to slide along the curves and surfaces so as to remove the effect of the arbitrary initial placement. *See also* **geometric morphometrics; landmark.**

## senior synonym
The name for a taxon that has historical priority [e.g., *Homo erectus* (1893) has historical priority over *Homo habilis* (1964)]. *See also* **synonym.**

## sensu lato
A more inclusive (i.e., less restrictive) interpretation of a taxon. For example, *Homo habilis sensu lato* includes specimens assigned to both *Homo habilis sensu stricto* and *Homo rudolfensis*, whereas *Paranthropus boisei sensu lato* includes specimens assigned to both *Paranthropus boisei sensu stricto* and *Paranthropus aethiopicus*. (L. *sensu* = sense and *lato* = lax.)

## sensu stricto
A less inclusive (i.e., more restrictive) interpretation of a taxon. For example, *Homo habilis sensu stricto* does not include specimens assigned to *Homo rudolfensis*, and *Paranthropus boisei sensu stricto* does not include specimens assigned to *Paranthropus aethiopicus*. (L. *sensu* = sense and *stricto* = strict.)

## seriation
The process by which archeological or fossil evidence is assembled into an ordered series (e.g., hominin fossil crania could be seriated on the basis of geological age, completeness, cranial capacity, etc.). (syn. ordination.)

## sexual dimorphism
Sexual dimorphism describes difference(s) between males and females of the same species. Sexual dimorphism among primate species has been attributed to a wide range of influences in addition to sexual selection (e.g., different factors leading to reduced female body mass, differences between male and female diet, and substrate use). Evidence from fossil hominins suggests a general decrease in sexual dimorphism of body and canine size through time, with some evidence of a decrease occurring during the time of *Homo erectus* that was likely brought about by an increase in female body size. However, direct inferences of social structure and mating behavior based on sexual dimorphism in the fossil record are complicated by the fact that multiple types of group-living social structures overlap in the degree of sexual dimorphism produced in living primates. In addition, archaic hominins, and especially hyper-megadont archaic hominins (e.g., *Paranthropus boisei*) are unusual in their pattern of sexual dimorphism in that they exhibit high levels of body size dimorphism, but low levels of canine size dimorphism; in most other primates levels of sexual dimorphism are usually similar for body mass and canine size. The unusual hominin pattern may reflect the loss of the canine-premolar honing complex and a change in canine shape in the hominin lineage, shifting the canine away from its use as a weapon and thus removing the canine as a target of sexual selection. (Gk *dis* = twice plus *morphe* = form.)

## sexual selection
A type of selection in which individual selection is influenced by mate choice (i.e., some individuals reproduce more because the traits they possess are more attractive to members of the opposite sex and/or they are successful in competing with individuals of the same sex for mates). Charles Darwin considered that differential access to mates constituted an evolutionary mechanism separate from natural selection.

## Shanidar
A cave approximately 645 km/400 miles north of Baghdad in the Zagros Mountains of Iraqi Kurdistan. The remains of approximately 10 *Homo neanderthalensis* individuals have been recovered and identified, at least four of which appear to have been intentionally buried. The 50–45 ka remains are generally grouped with those from Amud, Kebara, and Tabun into a Near-Eastern "late archaic" Neanderthal population. One individual (Shanidar 4) was found buried in soil that included highly elevated pollen counts, which was interpreted as the remains of flowers that were interred as grave goods, although others have argued the pollen is the result of digging by rodents. Others individuals (Shanidar 1 and 3) had suffered trauma and interpersonal injury. (Location 36°50′N, 44°13′E, Iraq.)

## shape
The form of an object is the sum of its shape and size. Shape is a spatial property of objects that is independent of location (translation and rotation) and size. At its most basic level, shape can be thought of as the relationship between two or more size variables (e.g., humerus length expressed as a percentage of radius length, or a ratio of first molar occlusal area to total postcanine occlusal area). Shape has been measured in a variety of ways in human evolution studies. For example, shape is often quantified as a measurement which has been adjusted using another measure of size, either through a criterion of subtraction using a regression analysis, or by dividing it by a proxy for size (e.g., body mass, geometric mean of multiple measurements). In geometric morphometrics, shape is described using a collection of landmarks that have been adjusted for size by being divided by centroid size. The study of size-related shape change (aka scaling) involves the principles of isometry (no change in the relationship between size and shape) and allometry (changes in shape as size changes). Shape is used in analyses of functional morphology (e.g., relating limb proportions to locomotion, relative tooth size to diet), alpha taxonomy, phylogenetic reconstruction, and paleoecological reconstruction.

## shape variables
Variables describing an object that are independent of the overall scale, orientation, and location of the object. For example, the distances between the apices of a triangle (i.e., interlandmark distances) describe the form of that triangle (i.e., those distances are invariant to rotation and translation) and if they are scaled to overall size they become shape variables.

## shared-derived character
*See* **autapomorphy**; **synapomorphy**.

## short-period incremental lines

Features in tooth enamel or dentine that have an intrinsic secretory rhythm equal to, or less than, 1 day. Cross-striations in enamel and von Ebner's lines in dentine have been shown to have a circadian (about 24 hour) rhythm. Counts and measurements of circadian features help estimate the daily enamel secretion rate, crown formation time, dentine formation time, and root extension rate. Short-period lines in enamel are much easier to visualize than those in dentine. *See also* **dentine; enamel development; enamel microstructure.**

## shoulder girdle

*See* **pectoral girdle**.

## sidereal

Anything that relates to stars or constellations, but particularly to measurements based on events related to stars [e.g., their daily motion, the time it takes for the moon to revolve around the Earth (approximately 27 days), or the Earth around the sun (approximately 365 days)]. Sidereal dating methods (e.g., tree ring dating, or dendrochronology) use such events for dating. (L. *sidereus* from *sidus* = star.) *See also* **geochronology**.

## Sidi Abderrahman

One of several sites discovered as a result of limestone quarrying during harbor construction near Casablanca, Morocco. Current age estimates for Sidi Abderrahman suggest a Middle to Late Pleistocene age, with the earliest deposits likely forming before Marine Isotope Stage 11 (i.e., 424–374 ka). Two hominin mandibular fragments are attributed to *Homo erectus* or *Homo erectus mauritanicus* and the Sidi Abderrahman sequence preserves handaxes, cleavers, and other Acheulean implements including Levallois cores. (Location 33°35′N, 07°40′W, Morocco.)

## Sidrón

*See* **El Sidrón**.

## Sierra de Atapuerca

*See* **Atapuerca**.

## sigmoid sinus

*See* **cranial venous drainage**.

## Silberberg Grotto

Part of the Sterkfontein cave complex that contains exposures of Sterkfontein Members 1, 2, and 3. Hominins recovered from the cave include the exceptionally well-preserved Stw 573 associated skeleton, which may be either *Australopithecus africanus* or another yet-unnamed species. A combination of uranium-series dating, electron spin resonance dating, and magnetostratigraphy suggests that the breccia exposed in the Silberberg Grotto is between 2.6 and 2.2 Ma. *See also* **Sterkfontein**.

## silent substitution

*See* **mutation**.

## silt
Very fine (i.e., a mean diameter between 0.063 mm and 4 µm) sedimentary particles. Consolidated or cemented (aka lithified) silt is referred to as siltstone. Silts accumulate on levees when flow energy is lost during overbank flooding. Silt is the dominant particle in loess, the aeolian deposit associated with glacial activity.

## siltstone
See **silt**.

## Sima de las Palomas
A *c.*35 ka karstic shaft site in Spain, in which remains of at least 63 hominin individuals have been recovered. Although the hominin fossils have been interpreted as possessing some "modern" features, the overall morphological evidence is consistent with them being assigned to *Homo neanderthalensis*. (Location 37°47′59″N, 00°53′45″W, Murcia, Spain.)

## Sima del Elefante
One of the breccia-filled caves that make up the Cueva Mayor-Cueva del Silo within a range of limestone hills called the Sierra de Atapuerca, near Burgos in northern Spain. Although it contains evidence of only one hominin individual, its early date (*c.*1.2–1.1 Ma) means that it is among the earliest evidence for fossil hominins in Europe. The only hominin recovered from the Sima del Elefante, a partial mandible, was provisionally assigned to *Homo antecessor*, but a more exhaustive assessment suggests that the mandible should be referred to *Homo* sp. indet. (Location 42°20′60″N, 03°31′09″W, Spain.)

## Sima de los Huesos
One of the breccia-filled cave systems that make up the Cueva Mayor-Cueva del Silo within a range of limestone hills called the Sierra de Atapuerca, near Burgos in northern Spain. To date, more than 6500 hominin specimens belonging to at least 28 individuals of diverse ages and both sexes have been recovered; the site has yielded at least one example of every bone in the body. The sediments are overlain by a speleothem which has yielded a minimum uranium-series dating age of less than 450 ka for the hominin-bearing layers. The hominins from the Sima de los Huesos have been assigned to *Homo heidelbergensis* and the researchers involved suggest that the Sima de los Huesos hominins sample the population that was the direct ancestor of *Homo neanderthalensis*. Other researchers think that the Sima de los Huesos hominins sample the early stages of *H. neanderthalensis* and they would include them in that taxon. The only archeological evidence is a single Acheulean handaxe. (Location 42°20′57″N, 03°30′51″W, Spain.)

## simple regression
See **regression**.

## *Sinanthropus* Black, 1927
Hominin genus introduced by Davidson Black in 1927. The type species is *Sinanthropus pekinensis* Black, 1927. See also **Sinanthropus pekinensis**.

## *Sinanthropus pekinensis* Black, 1927

Hominin species established by Davidson Black in 1927 to accommodate three fossil hominin teeth recovered in 1921, 1923, and 1927 at Choukoutien (now called Zhoukoudian) in China. In 1940 *Sinanthropus pekinensis* was transferred to *Pithecanthropus* as either *Pithecanthropus erectus* or *Pithecanthropus pekinensis*, and later that year it was formally transferred to *Homo* as *Homo pekinensis*. Most commonly, these specimens are viewed as part of the larger *Homo erectus* sample. However, some researchers have recently suggested that the differences between the *H. erectus* remains from China and those from Java, Indonesia (at Ngandong and Bapang-AG), deserve taxonomic recognition. In this case, the Chinese sample would be renamed *Homo pekinensis* while the Java sample would retain the name *Homo erectus*. The holotype of *S. pekinensis* is Ckn. A 1. (Gk *Sinai* = Chinese and *anthropos* = human being.) See also **Homo erectus**.

## SINE

Acronym for short interspersed nuclear element. These are short sequences (i.e., fewer than 500 base pairs) of DNA that are dispersed throughout the genome. They make up a class of mobile elements or transposons that are derived from reverse-transcribed RNA molecules. In modern humans the most common SINEs are the Alu elements, which comprise almost 13% of our genome. They are useful for phylogenetic analyses and for studies of population history because each SINE serves as a single character for tracing relatedness. *See also* **LINE**.

## single-crystal argon-argon dating

*See* **argon-argon dating**.

## single-crystal laser fusion $^{40}Ar/^{39}Ar$ dating

*See* **argon-argon dating**.

## single nucleotide polymorphism

(or SNP) A difference between two individuals at a single nucleotide (e.g., whether an individual has dry or wet earwax is due to a single nucleotide difference). (syn. point mutation.) *See also* **mutation**.

## single species hypothesis

(or SSH) The single species hypothesis argues that because human culture is such a specialized ecological niche no more than one culture-bearing hominid could have existed at one time. Thus, once culture had been acquired there could only ever have been one synchronic hominin species. Proponents of the single species hypothesis argued that the differences between *Austalopithecus africanus* and *Paranthropus robustus* (aka the gracile and robust australopiths) were probably due to sexual dimorphism in a single species rather than the existence of two separate species. However, the two taxa were found in separate sites and were separated by approximately half a million years. By dint of persistent advocacy the single species hypothesis survived for nearly another decade, but its demise came with the demonstration that an undoubted early African *Homo erectus* cranium (KNM-ER 3733) and an undoubted

*Paranthropus boisei* cranium (KNM-ER 406) had both been found *in situ* in the Upper Member of the Koobi Fora Formation. *See also* **competitive exclusion**.

### sinkhole
*See* **mokondo**.

### sister taxa
Two taxa that share either synapomorphies or a unique combination of symplesiomorphies. A pair of sister taxa is the minimum size for a clade, or monophyletic group. Pairs of sister taxa are the basic units of clades.

### site
A location, large or small, where fossils and archeological remains are found. It can vary from a cave the size of a room (e.g., Tabun) to an area of hundreds of square kilometers (e.g., Hadar study area, Koobi Fora, Middle Awash study area). Large sites may be subdivided into areas, localities, or collecting regions identified with names or numbers (e.g., Area 103, A.L. 333, Aramis), but if the site is small there is no need to specify separate localities. (L. *situs* = location.)

### size
Size is a spatial property of an object that is independent of its location (translation and rotation) and shape. More generally, size refers to the extent of a physical object's presence in one or more spatial dimensions. Size variables have two distinct properties: dimension and scope (e.g., part or all of the body). Size measures may be one-dimensional (linear), two-dimensional (area), or three-dimensional (volume or mass). With regard to scope, measurements may reflect overall organismal size (e.g., body mass, a geometric mean of measurements from throughout the skeleton), the size of an organ or organ system (e.g., brain mass, gut surface area, molar crown area, testicular volume), or the size of a particular anatomical region of interest (e.g., a geometric mean of measurements from the proximal femur). Size can be measured as a single variable or calculated from multiple variables. For example, the volume of a brain can be measured directly by placing it in a beaker of liquid and measuring the volume of liquid displaced, or it can be estimated using formulae based on the height, width, and length of the brain. Size is important in studies of alpha taxonomy, phylogeny reconstruction, and sexual dimorphism. Body size in particular is related to life history variables, biogeography, and substrate use. *See also* **multivariate size**.

### size classes
*See* **bovid size classes**.

### SK 6
A c.2.0–1.5 Ma adolescent mandible from Member 1 at Swartkrans, South Africa. It is the holotype of *Paranthropus crassidens*, a taxon that most researchers have since subsumed into *Paranthropus robustus*. *See also* **Swartkrans**.

## SK 15
A *c*.1.5–1.0 Ma adult mandible from the boundary between the Member 1 Hanging Remnant and Member 3 from Swartkrans, South Africa. It is the holotype of *Telanthropus capensis*, a taxon that has since been subsumed into *Homo erectus*. See also **Swartkrans**.

## SK 847
(plus SK 80 and 846b; aka "SK 847 composite cranium") In 1969 it was realized that these three specimens from the *c*.2.0–1.5 Ma Member 1 Hanging Remnant pink breccia at Swartkrans, South Africa, formed the partial cranium of a single adult individual. While there is near unanimity that SK 847 is not part of the *Paranthopus robustus* hypodigm, there is much less agreement about what taxon SK 847 does sample. The superficial similarities between KNM-ER 3733 and SK 847 have led some researchers to support an allocation to *Homo erectus sensu lato*, but other analyses suggest that SK 847 has similarities with, but also certain differences from, *Homo habilis sensu stricto*. See also **Swartkrans**.

## skeletal element survivorship
The probability that a particular skeletal element will survive the destructive taphonomic processes (e.g., carnivore ravaging, trampling, dissolution, etc.) that can alter a bone assemblage. High-survival elements are mostly those composed primarily of dense cortical bone with little or no cancellous bone (e.g., long bones, crania, and mandibles). These are the elements that are best for making hominin behavioral inferences because their high survivorship provides the most accurate representation of the bones originally discarded by hominins. The low-survival elements have thin cortical walls and are mostly cancellous bone (e.g., vertebrae, ribs, pelvis, scapula, long-bone epiphyses, etc.). Phalanges, carpals, and tarsals are considered low-survival elements because they tend to be swallowed whole by carnivores.

## skeletal part frequency
The frequency of different skeletal elements in a fossil bone assemblage is typically quantified using indices (e.g., minimal animal unit, minimum number of elements, minimum number of individuals, number of identified specimens). In zooarcheological contexts, observed skeletal part frequencies are the result of a complex interaction between the hominin utilization and transportation of animal carcasses (i.e., carcass transport strategies) and destructive taphonomic processes operating following discard (e.g., carnivore ravaging).

## skeleton
In vertebrates it refers to the hard tissue (bone and cartilage) elements that make up the endoskeleton that protects and supports the soft tissues. It is divided into the axial skeleton (comprising the skull, vertebral column, and thorax) and the appendicular skeleton (limb girdles and limbs). (Gk *skeletos* = to dry up.)

## Skhul
A cave on Mount Carmel, Israel, where the remains of 10 individual hominins (including two infants) and 16 isolated fragments attributed to *Homo sapiens* have been recovered. Layer B at the site also includes Mousterian artifacts. The most recent uranium-series and electron spin

resonance spectroscopy dating of hominin and faunal remains suggest a date of 130–100 ka. Skhul and Qafzeh are the sites of the earliest known modern human remains outside of Africa. (Location 32°40′N, 34°58′E, Israel.)

## Skildergat
See **Fish Hoek**.

## skull
The bony component of the head made up of 28 bones, some of which are unpaired (e.g., the frontal and the occipital), but most of which are in symmetrical pairs (e.g., the parietal and temporal). The skull comprises the cranium plus the mandible. (ME *skulle* = skull.) *See also* **cranium; mandible**.

## skullcap
See **calotte**.

## sliding landmark
See **semilandmark**.

## SMRS
Acronym for the specific mate-recognition system, which has been proposed as the mechanism for species defined on the basis of the individuals (usually females) identifying potential mates. *See also* **recognition species concept; species**.

## SNP
Abbreviation of **single nucleotide polymorphism** (*which see*.)

## Soa Basin
Several Early and Middle Pleistocene sites, of which Mata Menge is the best known, are located in the Soa Basin in central Flores, Indonesia.

## social enhancement
The phenomenon that an innate, species-typical behavior is more likely to occur if conspecifics are engaged in the same behavior. Examples of social enhancement (aka social facilitation or contagion) in modern humans include reactions to other people laughing, crying, or yawning. Social enhancement can resemble behavior matching or imitation because the presence of another individual has the potential to lead to synchronization of behaviors over space and time. Animal behaviorists believe that social enhancement is crucial in group-living species for group cohesion, behavioral coordination, foraging efficiency, and predator avoidance. While social facilitation alone does not lead to complex behavior matching or imitation, it has the potential to increase the opportunities in which observational learning can occur, and observational learning has been implicated in the development of the ability to manufacture tools.

## social facilitation
*See* **social enhancement**.

## social intelligence
Encompasses a range of skills including pro-social behaviors such as social learning, imitation, teaching, communication, theory of mind, and cooperation. Some have argued that social intelligence is a requirement of living in groups when important resources (e.g., food and potential mates) are limited. The social intelligence hypothesis proposes that group living in primates and other animals has favored social manipulation by adding to cognitive power through natural selection. It has received support from the positive correlation in primates between group size and neocortex ratio (i.e., the volume of neocortex relative to that of the other parts of brain) and the superior performance of transitive inference in species with larger group size as compared to that of closely related counterparts with smaller group size.

## social intelligence hypothesis
*See* **social intelligence**.

## social learning
Modern human development is very much dependent on learning that is influenced by the actions of others who provide cues and signals that guide an observer's attention to specific behaviors (e.g., social relationships) or to objects (e.g., food). The learning that results from observation or interactions with others falls under the general category of social learning. It is likely that learning to make stone and other types of tools involves social learning. Social learning is different from (a) individual learning, which is learning achieved through trial and error, (b) the learning that arises in isolation without the benefit of interacting with a conspecific or a model, and (c) instinctual behaviors. Using this broad definition, social learning is widespread in the animal kingdom. In fact, for mammals and birds, social learning has been shown to play a critical role in the development and maintenance of species-specific behaviors (e.g., the acceptance or avoidance of novel foods, mate choices, and predator-avoidance strategies). Among primates, social learning may require reasoning about other individuals, or even reasoning about one's own behavior. Examples of socially learned primate behaviors include specialized grooming, communicative signals, and tool-use techniques. Imitation represents a specific type of social learning, initially considered one of the few traits that animals had in common with modern humans, and many languages even have terms that involve the words "ape" or "monkey" to describe imitation (e.g., "aping" the professor). However, there is debate over whether or not monkeys, apes, or any animals other than modern humans are able to truly imitate. For this reason, the literature on social learning distinguishes terms for the different ways in which learning can take place, such as social facilitation, contagion, stimulus enhancement, local enhancement, emulation, imitation, and teaching. (L. *socius* = companion and OE *lernen* = to gain knowledge.) *See also* **theory of mind**.

## soil carbonates
*See* **paleosol**.

### Solo
See **Ngandong**.

### Solo Man
See **Ngandong**.

### solution cavities
See **limestone**.

### Solutrean
An Upper Paleolithic technocomplex that dates to the Last Glacial Maximum (*c*.21–17 ka). It is characterized by large, flat, and thin leaf-shaped points and it is associated with an increased site density, greater prevalence of art (particularly parietal art like cave paintings and engravings), and the hunting of reindeer. (etym. after the site of Solutré.)

### somite
Bodies develop in the form of several discrete, repeated segments, the earliest expression of which are the somites. Somites develop into myotomes (aka muscles) and sclerotomes (aka vertebrae). The first four or so somites are incorporated into the basicranium. The rest contribute to the development of the vertebrae, the intervertebral discs, and ribs. Each vertebra is formed from the caudal half of its equivalent somite, plus the cranial half of the somite below it (e.g., the C4 vertebra is formed from the caudal half of the C4 somite and the cranial half the C5 somite). (Gk *soma* = body.) See also **vertebral number**.

### Southern Afar Rift System
The term used for the southern part of the Afar Rift System (aka Afar Depression). This section of the Ethiopian Rift System includes the Dikika, Gona, Hadar, Middle Awash, and the Woranso-Mille study areas.

### South Turkwel
A site between Lodwar and Lothagam on the west side of Lake Turkana in Kenya. The two hominin fossils from South Turkwel are KNM-WT 22936, part of a right mandibular corpus, and KNM-WT 22944, an associated skeleton that includes a piece of weathered mandible and tooth crowns, four hand bones, and a proximal foot phalanx. A reasonable maximum age for the two hominin fossils is the age of the Lokochot Tuff (i.e., 3.58 Ma) with a minimum age of *c*.3.2 Ma. Morphologically the two hominin specimens are consistent with an attribution to *Australopithecus afarensis*, but the researchers who found them attribute them to *Australopithecus* sp. indet. (Location 02°19′N, 36°04′E, Kenya.)

### spatial packing hypothesis
An attempt to explain variation in the overall shape of the cranium in terms of physical interactions between the growth and development of the brain, the basicranium, and the face. It is based on the premise that the craniofacial complex consists of independent and partially independent modules that interact as the cranium develops because they are physically adjacent or

have developmental connections. In particular, the hypothesis holds that the growth of the brain relative to the growth of the basicranium is a major determinant of the cranial base angle and thus of the position of the face relative to the brain case. Thus, flexion of the basicranium enables an increase in brain size to be accommodated without any significant changes in its width and/or length. *See also* **basicranium**.

## spear

A weapon consisting of a long shaft and a sharp tip that can be thrust into, or thrown at, an animal. Many stone points recovered in the archeological record are inferred to have been manufactured in order to be mounted onto a wooden shaft as the tips of spears, but sharpened wooden tips are also effective. Chance preservation of some wooden spears suggests that the latter types date to more than 400 ka.

## specialist

*See* **stenotopy**.

## speciation

The process by which descendant species evolve from ancestral ones. Speciation is often discussed in terms of the ecological and/or biogeographic factors that may have led to the evolution of new species and a terminology has been devised to distinguish among these various factors (e.g., allopatric speciation, peripatric speciation, parapatric speciation, sympatric speciation). Ultimately, however, although it is almost certainly true that at least some hominin species evolved as a result of cladogenesis, it is very difficult to assert with any confidence the precise type of speciation involved with the emergence of any particular hominin taxon. Hominin species may also have arisen by hybridization.

## species

The next-to-least inclusive category in the Linnaean taxonomic system. There has been consistent confusion between how species are conceptualized and they are recognized in the fossil record. One scheme divides species concepts into two categories: one that emphasizes the processes involved in the generation and maintenance of species (process-based) and one that emphasizes the methods used for recognizing them (pattern-based). The main concepts in the process category are the biological species concept (or BSC), the evolutionary species concept (or ESC), and the recognition species concept (or RSC); the main pattern-based species concepts are the phenetic species concept (or PeSC), the phylogenetic species concept (or PySC), and the monophyletic species concept (or MSC). The three pattern-based concepts emphasize different aspects of an organism's morphology, so they are all variations on the morphospecies concept. In practice most researchers of human evolution use some version of the PySC. They search for the smallest cluster of individual organisms that is "diagnosable" on the basis of the preserved morphology. Most diagnoses of early hominin taxa inevitably emphasize craniodental morphology because these are the most often preserved elements in the fossil record. (L. *specere* = to look; NB: *spectaculum*, *specimen*, and *spectio*, the source of "inspection," come from the same verb. The connection with Gk *eidos* = form or idea, is a semantic and not an etymological one.)

**species concept(s)**
*See* **species**.

**specific mate-recognition system**
*See* **recognition species concept**.

**speciose**
A taxonomic hypothesis that emphasizes a larger rather than a smaller number of species. A researcher who favors speciose taxonomies is called a splitter (*which see*). (syn. taxic.) *See also* **splitter**.

**speech**
*See* **spoken language**.

**speleothem**
The inclusive term for a range of rock types (e.g., flowstones, stalactites, and stalagmites) that are formed when water containing dissolved calcium carbonate reaches an air-filled cave and precipitates. Flowstones, which are formed when water flows on the floor or walls of a cave, are approximately horizontal sheets that lie against the walls or floor of the cave or between blocks of sediments. Stalactites are icicle-shaped mineral deposits that grow down from cave roofs, whereas stalagmites are conical mineral deposits fed from drip waters above that grow up from cave floors. (L. *speleon* = cave and *them* = deposit.) *See also* **calcium carbonate; uranium-series dating**.

**spheroid**
*See* **Developed Oldowan**.

**sp. indet.**
When dealing with fragmentary specimens it is sometimes possible to identify the genus a specimen belongs to but it may not be possible to be sure which species it belongs to. In that case the taxonomic allocation researchers should use is "sp. indet." (e.g., the mandible KNM-ER 1506 has been referred to *Homo* sp. indet.). (etym. sp. indet. is an abbreviation of "species indeterminate.")

**splitter**
A label for a researcher who favors dividing the fossil record into a larger number of exclusive taxa (i.e., with more stringent membership criteria) rather than a smaller number of more inclusive taxa (i.e., with more relaxed membership criteria). Some researchers who are trying to reconstruct aspects of the biology of extinct taxa such as life history have suggested that splitting taxonomies are to be preferred because lumping taxonomies potentially jumble characteristics from very different subgroups, producing chimeric life histories. Others claim that splitting makes the hominin fossil record unnecessarily confusing. (OD *splitten* = to divide.) *See also* **alpha taxonomy; lumper**.

## spoken language

Spoken language, or speech, is ubiquitous among living modern human populations. It involves the use of a learned repertoire of sound units assembled in a distinctive sequence in rapid succession to convey information and meaning in an energetically efficient way. Among the advantages of spoken language are that the individuals involved in its transmission and reception need not be in visual contact with one another, and spoken language is also an exceptionally efficient way of communicating with more than one person at a time. Modern human languages use around 100 acoustically distinctive sound units, or phonemes. Phonemes are broken up into the consonants and vowels that make up syllables. Vowels involve phonation (modification of laryngeal air flow by the vocal folds, tongue, and lips) whereas consonants involve the blockage and subsequent release of air flow. Compared with nonhuman primates, modern human spoken language involves sound sequences that are an order of magnitude more rapid and are more than twice as long as those seen in nonhuman primates. The supralaryngeal vocal tract of modern humans is distinctive in that the horizontal (aka $SVT_H$) and vertical (aka $SVT_V$) components are nearly equal in length; in nonhuman primates and in other animals the $SVT_H$ is usually longer than the $SVT_V$. It is this 1:1 $SVT_H/SVT_V$ ratio and the approximately 90° angle between the $SVT_H$ and the $SVT_V$ that makes it possible for modern humans to produce vowels such as [a], [i], and [u]. Researchers interested in language origins have been searching for a hard-tissue proxy for the shape and proportions of the supralaryngeal part of the vocal tract. They thought they had found it in the form of the cranial base angle, but flexion of the basicranium and laryngeal descent are not well enough related for the cranial base angle to be used a proxy for the height of $SVT_V$. (ME *speche* = speech and *language* from OF *langage* from L. *lingua* = tongue or speech.)

## Spy

Two adult presumed male partial skeletons (Spy 1 and 2) were found in 1885 in the Grotte de Spy (aka Béche-aux-Roches cave), a limestone cave about 15 km/9 miles east of Namur, Belgium. Subsequently two teeth and a tibia of a juvenile (Spy 3) were located and, more recently, researchers identified 24 new fragments, including part of the mandible and some of the mandibular dentition of an infant (the Spy VI child). All have been assigned to *Homo neanderthalensis* and all are c.36 ka. (Location 50°28′21.64″N, 04°40′50.52″E, Belgium.)

## square-cube law
See **body size**.

## Sr/Ca ratios
See **strontium/calcium ratios**.

## SR-μCT
Abbreviation of **synchrotron radiation micro-computed tomography** (*which see*).

## stabilizing selection
See **natural selection**.

### stable isotope biogeochemistry
The science that involves determining the stable isotope composition of biological, atmospheric, and geologic samples using isotope ratio mass spectrometers, and then interpreting the ecological, chemical, or geological processes driving stable isotope ratios in sediments and in plant or animal tissues. Stable isotope biogeochemistry is a systems science in the sense that it aims to determine how nutrients are cycled at scales ranging from the global down to an individual organism.

### stable isotopes
An isotope is one of two or more atomic forms of the same element that have the same atomic number (i.e., same number of protons) but different atomic masses (i.e., different numbers of neutrons). Unlike the isotopes used in isotopic dating, which necessarily undergo decay, the proportions of stable isotopes do not change with the passage of time. Instead their proportions are determined by processes such as photosynthesis (e.g., $^{13}C/^{12}C$), or climate-related variables and/or diet ($^{2}H/^{1}H$, $^{13}C/^{12}C$, $^{15}N/^{14}N$, $^{18}O/^{16}O$). Thus, the stable isotope ratios in collagen recovered from fossils can be used for diet reconstruction and to reconstruct paleoclimate. Stable isotope ratios are frequently referred to using delta (or δ) notation, where the isotope ratio in a sample is measured relative to an internationally agreed standard. If there is more of one of the two isotopes, the system is said to be enriched with respect to that particular isotope; when there is less of one of the two isotopes, the system is said to be depleted with respect to that particular isotope. Combining data from two different stable isotope pairs is frequently more meaningful than only using one isotope system (e.g., in regions where prehistoric modern human populations had access to both marine foods and $C_4$ foods such as maize, which confound the use of $\delta^{13}C$ values, bone collagen $\delta^{15}N$ values can help identify dependence on marine foods). The $^{87}Sr/^{86}Sr$ ratios of animal apatite (bone and teeth) is a characteristic of the rocks from which the strontium derives and researchers have made ingenious use of these ratios to investigate the use of the landscape by *Australopithecus africanus* and *Paranthropus robustus*. See also **$C_3$ and $C_4$**; **$^{13}C/^{12}C$**; **$^{2}H/^{1}H$**; **$^{15}N/^{14}N$**; **$^{18}O/^{16}O$**.

### St. Acheul
See **Saint-Acheul**; **Acheulean**.

### stalactite
See **calcium carbonate**; **speleothem**.

### stalagmite
See **calcium carbonate**; **speleothem**.

### stance phase
See **walking cycle**.

### Standardized African Site Enumeration System
(or SASES) A system developed in the 1960s for recording the location of African archeological sites. The SASES divides the African continent into squares that measure 6° of longitude or latitude; these are identified by capital letters. Then each square is subdivided into units that

correspond to 15′ of longitude or latitude; these are assigned lower-case letters. If there is more than one site in a square, the sites are assigned an Arabic numeral based on the order in which they were recorded. For example, the SASES designation of the KBS site at Koobi Fora, east of Lake Turkana in northern Kenya, is FxJj1. The first two letters represent the latitude, the second two the longitude, and the Arabic numeral tells the user that this was the first site recorded at that location.

## stapes
One of the three auditory ossicles in the middle ear. *See* **auditory ossicles**; **middle ear**.

## starch grains
One of several kinds of plant microfossil (the others are phytoliths and pollen). The morphologically most diagnostic forms of starch grains are found within the areas of the plant that are designed for long-term energy storage (e.g., seeds, fruits, and underground storage organs) and fortunately this is also where they are most abundant. Starch grains are best preserved in protected archeological environments, and have been recovered from stone tools, ceramics, and dental calculus. Cooking and other types of food processing causes predictable damage to starch grains, and thus the presence of such damage can be used as evidence that food has been cooked, or otherwise processed. (ME *starche* = substance used to stiffen cloth, from *sterchen* = to stiffen, from OE *stercan*.)

## stasis
Randomly directed change is constantly occurring within living species, but the term stasis is used for a period of time (it could be the temporal span of a species) during which there is no evidence of any long-term directional change (aka trend) in morphology. For example, there are very few temporal trends evident within the time spans of *Paranthropus boisei sensu stricto* or *Australopithecus afarensis* and the lack of any significant trend between the first and last appearance dates of those taxa make them examples of stasis within the hominin fossil record. (Gk *stasis* = standstill.) *See also* **punctuated equilibrium**.

## stature
Stature (aka standing height) varies within and between populations and has been used in modern humans as an indicator of diet, environmental stress, and gene flow. Stature is also one of several ways of summarizing the overall size of an individual hominin and it has been used as a scaling factor. Standing height has increased over the course of hominin evolution. The most complete *Australopithecus afarensis* skeleton recovered, A.L. 288-1, with an estimated stature of 1.05 m contrasts with the estimated stature of 1.6 m for the Nariokotome *Homo erectus* skeleton, KNM-WT 15000. There is no evidence of modern human-like statures prior to *H. erectus*. Among modern humans, mean stature varies among populations and to varying degrees between males and females within populations. When only skeletal elements are available, stature is most commonly estimated from linear dimensions of long bones. It is important to note that all stature estimates (like all mathematical predictions based on regression analysis), especially those for fossil hominins for which there are no reference populations, are subject to varying amounts of error and therefore may not be accurate. (L. *statura* = height.) *See also* **stature estimation**.

## stature estimation

Stature is usually estimated by relating skeletal dimensions, often the lengths of the long bones of the lower limb, to living stature. Most standard contemporary methods of stature estimation regress the lengths of the femur, the tibia, or both bones against known statures to yield predictive equations for that reference sample. Relatively few modern human reference samples exist for which both living stature and directly observed limb-bone lengths are available, and obviously there are no directly relevant reference samples that can be used to estimate the stature of extinct hominin taxa. Skeletal elements that contribute a greater amount to total stature (e.g., the femur vs talus) should yield smaller estimation errors and therefore more accurate predictions of stature, but stature estimations from long bones assume that lengths of the skeletal elements are proportionally similar to total stature in the reference and estimated samples. This has far-reaching implications for stature estimations of extinct hominins, many of which had overall body and internal limb proportions that differ from those of modern humans. In anatomical stature estimation methods the constituent skeletal elements that directly comprise stature [i.e., the length of the lower limb (including tarsal height), the length of the vertebral column, and the height of the cranium] are measured and added together, but few early hominin associated skeletons are complete enough to justify the use of anatomical techniques. (L. *statura* = height.)

## Steinheim

Only one hominin specimen, a cranium, is known from the gravel deposits at the Steinheim site in Germany. Due to its combination of archaic and modern features and plastic deformation, the c.225 ka Steinheim cranium has proved to be somewhat difficult to classify. The Steinheim cranium is one of a group of specimens (e.g., Reilingen, Sima de los Huesos, Swanscombe) referred to *Homo neanderthalensis*. Researchers who interpret that taxon as an evolving lineage would interpret the Steinheim cranium as an example of "Stage 2"; a stage that has also been referred to as pre-Neanderthal. (Location 48°58′N, 09°17′E, Germany.)

## stem group

The term used for a clade (aka total group) minus its crown group. For example, for hominins it would be the hominin clade minus the *Homo sapiens*/*Homo neanderthalensis* subclade. (OE *stemn* = supporting stalk.) See also **clade**; **crown group**; **total group**.

## stenotope

A species adapted to a restricted or narrow range of environmental conditions, or to a narrow ecological niche (e.g., an ant-eater). (Gk *steno* = narrow and *topos* = place; syn. specialist.) See also **stenotopy**.

## stenotopy

The condition of being ecologically specialized. A stenotopic (aka specialist) species is one that can use or consume only a narrow range of ecological resources, and that can live in only a limited set of habitats. The term was introduced to paleoanthropology in a series of hypotheses seeking to explain how environmental change influences evolutionary patterns in the fossil record, but several studies have inferred that most early hominin species, particularly with respect to diet, were not ecological specialists but rather were generalists. (Gk *steno* = narrow and *topos* = place.) *See also* **effect hypothesis**; **eurytope**; **habitat theory hypothesis**; **turnover-pulse hypothesis**.

## stenotypic
*See* **stenotope**; **stenotopy**.

## Sterkfontein
The umbrella term for a system of breccia-filled caves [e.g., Extension site, Jakovec Cavern, Silberberg Grotto, Tourist Cave(s), Type site, etc.] formed from solution cavities within Precambrian dolomite in the Blauuwbank Valley near Krugersdorp in Gauteng Province, South Africa. The first hominin to be recovered from Sterkfontein was TM 1511 in 1936. This fossil cranium was the holotype of *Plesianthropus transvaalensis*, but has since been transferred to *Australopithecus africanus*. Most hominins recovered from Sterkfontein have come from Member 4 and the majority of these have been assigned to *A. africanus*. Some researchers think there is more than one species belonging to the genus *Australopithecus* represented at the site, including fossils from Member 4 breccia exposed in the main cave (e.g., Sts 71, Stw 252, Stw 505), and from Member 2 breccia exposed in the Silberberg Grotto (e.g., Stw 573) and in the Jakovec Cavern, but opinions are divided about exactly which specimens do not belong to *Au. africanus*, and what any second taxon should be called. Other taxa possibly represented in the Sterkfontein hominin fossil record are *Homo* sp. and *Paranthropus robustus* from Member 5a. Archeological evidence includes stone artifacts (Oldowan from Member 5; Acheulean from the Extension site; MSA from the Lincoln Cave) and digging sticks fashioned from bone (Member 5). The vast majority, if not all of the australopith-bearing deposits [i.e., the Jakovec Cavern, Silberberg Grotto (Member 2), and Member 4] likely date to between 2.6 and 2.0 Ma. The Stw 573 fossil in Member 2 is below a 2.3–2.2 Ma uranium-series dated flowstone. Member 5 has traditionally been separated into the "StW 53" (aka Member 5a), "Oldowan" (aka Member 5b), and "Acheulian" (aka Member 5c) infills. The StW 53 infill is dated to between 1.8 and 1.5 Ma, the Oldowan infill to between 1.4 and 1.2 Ma, and the Acheulean infill to between 1.3 and 1.1 Ma. (Location 25°58′08″S, 27°45′21″E, South Africa.)

## stone cache hypothesis
A hypothesis that suggests Oldowan sites were processing areas where artifacts and manuports were deposited at various points in the foraging range of hominins. These caches were either established consciously or they developed as an unconscious byproduct of hominin discard behavior, and then they became secondary sources of raw material. According to this view, archeological sites served not as home bases but as recurrently visited stockpiles of stone.

## strain
A measure of the relative deformation that an object undergoes in response to an applied stress. The different types of strain are linear, volume, and shear; these involve linear, volumetric, and angular distortion, respectively. In biological anthropology the most commonly discussed type of strain is linear (aka tensile) strain. Linear strain is technically defined as the difference between the loaded and the unloaded (i.e., the original) length relative to the original length of

a structure ($\varepsilon = \Delta L/L$). Any change in length of a bone is usually very small, so strain is normally expressed as microstrain ($\mu\varepsilon$), indicating a change in length on the order of a thousandth of the original length. Experimental evidence has shown that during normal activities (e.g., running, biting hard objects, etc.) most mammalian bones experience a peak of about 2000–3000 $\mu\varepsilon$ (which indicates a 0.2–0.3% change in length). Mammalian bones typically break at around 6000 $\mu\varepsilon$, indicating that most bones are overbuilt relative to their peak normal strains and thus they have a substantial safety factor of between 2 and 3. Some amount of dynamic strain is required for normal bone growth, but excessive strains can result in bone modeling and bone remodeling. *See also* **stress**.

### strain gauge
A device that measures strain because the material from which it is made changes its electrical resistance when it is stretched. A rosette strain gauge, which consists of three elements at 45° to each other, makes it possible to calculate the linear strain in all directions within the plane of the gauge (e.g., on the external surface of the body of the mandible) and to determine the directions in which the maximum and minimum principal strain lie. Multiple gauges around the circumference of a long bone can be used to calculate (using certain assumptions) the strain distribution throughout the bone's cross-section. Strain gauges are also used to calculate forces in force plates. Strain gauges placed on experimental animals provided evidence to refute the hypothesis that brow ridge development in higher primates, including fossil hominins, is structurally related to countering stresses resulting from biting and chewing. *See also* **chewing cycle**.

### strata
More than one layer of sedimentary rock. (Gk *stratos* = to cover or spread.)

### stratigraphic criterion
A controversial criterion for deciding which state of a character is primitive and which is derived (aka character polarity). When employing the stratigraphic criterion, the character state that occurs in the oldest deposit is the one that is deemed to be primitive. *See also* **cladistic analysis**.

### stratigraphy
The study of sedimentary rocks, especially their formation, composition, sequence, and correlation. It also refers to the work of identifying and tracing layers of sedimentary rock. The "stratigraphy" of a site is the formal description of the sequence of strata found at that site. (Gk *stratos* = to cover or spread and *graphos* = to draw or write.)

### stratum
A single layer of sedimentary rock. (Gk *stratos* = to cover or spread.)

### strength
In material science the strength of a material (e.g., bone, enamel) is its ability to withstand an applied stress without failure. There are two common types of failure: yielding and fracture. Yielding, which is defined by the material's yield strength, occurs when the material

undergoes permanent deformation. Fracture is defined by the ultimate strength of the material, which is the largest absolute value of applied stress the material can withstand without fracturing. The type of loading (e.g., tensile, compressive, or shear) also influences the likelihood of failure.

## stress

A measure of the intensity of the internal forces within a loaded object. It can be quantified as the force per unit area that results in strain. If the stress acts perpendicular to a plane, it is called a normal stress component; a negative normal stress implies a compressive stress and a positive normal stress implies the normal stress is tensile. Stress that acts tangential to a plane is called shear. Shear stresses tend to distort the material without changing its volume, whereas normal stresses tend to elongate or compress the material in the directions in which they act. *See also* **strain**.

## striae of Retzius

Long-period incremental lines in enamel that represent the position of the developing enamel front at a point in time. They appear on the lateral surface of teeth as perikymata (aka imbrication lines). Although the exact cause of striae of Retzius is not known, they result from a regular slowing of enamel matrix secretion that follows an intrinsic and consistent temporal rhythm. The temporal repeat interval of striae of Retzius (aka periodicity) is assessed by counting the number of daily cross-striations between consecutive striae. In dentine, the long-period incremental markings that are analogous to striae of Retzius are called Andresen lines. Researchers have examined the curvature, length, and the angle of intersection between the striae of Retzius and the enamel–dentine junction in attempts to discriminate among primate taxa. In reflected light, striae look blue but in transmitted light they look brown (hence the original description of "brown striae"). (L. *stria* = a furrow-like phenomenon described by Anders Retzius in 1837; syn. brown striae, Retzius' lines.) *See also* **enamel development**.

## *striae parietalis*

Descriptive term for the fine ridges and grooves on the external (aka ectocranial) surface of the parietal bones that run posterosuperiorly away from the border with the squamous part of the temporal bone. This distinctive morphology, which is probably related to the unusual degree of overlap at the temporoparietal suture, is best seen in *Paranthropus boisei* and *Paranthropus aethiopicus*. (L. *striae* = scores or grooves and *parietalis* = wall.)

## striate cortex

The striate cortex (aka primary visual cortex, V1, Brodmann's area 17) is located at the posterior, or caudal, pole of the occipital lobe of the cerebral hemisphere of the brain. Most of it is on the medial aspect of the occipital lobe and much of it is buried within the walls of the calcarine sulcus. The name striate derives from the prominent "striped" appearance of the stria of Gennari in layer IVB (which is made up of myelinated axons) of primates. Because of this distinctive myelination pattern, the striate cortex is readily seen in sections of the brain, even with the naked eye. The striate cortex in each hemisphere contains a complete retinotopic map of the contralateral (i.e., the opposite side) visual field, with anthropoid primates having a magnified representation of the central portion of the visual field. Neurons in the striate cortex are tuned to relatively simple features of color and spatial frequency. The volume of the striate cortex in modern humans is considerably smaller than that predicted by the observed allometry for a generalized primate of the same brain size. This relative reduction of primary visual cortex in modern humans has been argued to reflect a reorganization and reallocation of cortical area, with a greater proportion of neocortical tissue devoted to the adjacent posterior parietal association cortex. Histological features in the striate cortex of modern humans suggest an enhanced ability to process motion information, possibly in relation to tracking the movement of lips during the perception of spoken language. (L. *striatus* = furrowed, striped, ridged and *cortex* = husk, shell.)

## stride
See **walking cycle**.

## striking platform
A morphologically distinctive part of a stone artifact, it is the place on the core that is struck during the act of percussion (freehand or otherwise) to remove a flake.

## strontium/calcium ratios
(or Sr/Ca ratios) A system of stable isotopes used for diet reconstruction that can indicate whether a hominin consumed more meat or plants, and what kinds of plants or animals they consumed. Herbivores discriminate against dietary strontium, so their bones and tooth enamel have lower Sr/Ca ratios than the plants they consume. The carnivores that consume those herbivores also discriminate against dietary strontium, so their bones and tooth enamel have lower Sr/Ca ratios than the herbivores they consume. But leaves of broad-leaved plants have less Sr than grass, so grazing herbivores have a higher Sr/Ca ratio than browsing herbivores. Dietary items such as insects and underground storage organs are particularly rich in strontium, but grazers and insectivores are also rich in barium (Ba) so relatively high Ba/Ca ratios should discriminate between high Sr/Ca ratios due to grazing or insectivory from high Sr/Ca ratios due to the consumption of, say, underground storage organs. Enamel apatite is best for Sr/Ca analysis; the chemical changes that occur during the process of fossilization are more of a problem with respect to bone than with enamel. *See also* **stable isotope biogeochemistry**.

## strontium isotope ratios
(or $^{87}Sr/^{86}Sr$ ratios) Many types of bedrock have distinctive ratios of stable strontium isotopes and these ratios are passed, unchanged, up the food chain. By analyzing the strontium isotope ratios in the teeth of fossil hominins, and comparing these values to those from the local geology, it is possible to reconstruct where, and at what point in their life history, hominins used different parts of the landscape. One such study has indicated that male and female *Australopithecus africanus* and *Paranthropus robustus* had different landscape use patterns, and they also suggested that females moved away from their childhood residential groups. *See also* **stable isotope biogeochemistry**.

## structure–function relationship

The relationship between the structure of an object or body part and its function. It is relatively easy to infer function from structure when a trait seen in an extinct taxon is also seen in the same context in an extant taxon in which structure–function relationships can be observed directly. Inferring function is much more difficult when the trait is found only in extinct taxa. In such cases it has been suggested that the trait should be compared with structures that are well suited to their function. For example, a pestle and mortar is well-suited to fracturing relatively small pieces of brittle material, so a pestle and mortar would be a "structural paradigm" for that particular function. The fossil trait (e.g., the shape of postcanine tooth crowns) is then compared to various relevant structural paradigms, and its function is taken to be closest to the structural paradigm it most closely resembles.

## Sts 5

A well-preserved subadult hominin cranium that lacks all of the tooth crowns, some of the alveolar process, and some details of the cranial base morphology. This 2.2–2.0 Ma cranium from Member 4C at Sterkfontein, South Africa, is considered to be a female *Australopithecus africanus*. Although not the type specimen of *Au. africanus* (that is TM 1511), Sts 5 has come to represent the modal cranial morphology of that taxon. This is unfortunate because it is one of the more prognathic, if not the most prognathic, specimen in the *Au. africanus* hypodigm.

## Stw 573

This unusually well-preserved adult associated skeleton was found in and around a flowstone in Member 2 in the Silberberg Grotto at Sterkfontein, South Africa. It was previously estimated to be *c*.3.0–2.6 Ma, but a combination of uranium-series dating, electron spin resonance dating, and magnetostratigraphy suggests its age is between 2.6 and 2.2 Ma. It most likely belongs to *Australopithecus africanus*, but some claim it is evidence of a second *Australopithecus* taxon at Sterkfontein that is more *Paranthropus*-like.

## subchron

An abbreviation of the term "polarity subchronozone." It refers to a relatively short period of consistent geomagnetic polarity. Subchrons are subdivisions of chrons, and like chrons they are numbered (NB: the magnetic direction of a subchron is identified by a lower-case n for normal and r for reversed; thus the first reversed subchron in the normal chron 3An is 3A.1r). Numbered subchrons have replaced the named events (e.g., Jaramillo, Olduvai, etc.) that were used in the older geomagnetic polarity time scales. (Gk *kronos* = time.) *See also* **geomagnetic polarity time scale**.

## subjective synonym

*See* **synonym**.

## subnasal prognathism

*See* **prognathic**.

### substitution
Used in genetics to refer to one allele replacing another or in molecular evolution to refer to a single nucleotide or base change in DNA. (L. *substituere* = in place of.) *See also* **mutation**; **single nucleotide polymorphism**.

### suid
Informal term for taxa within the family Suidae. (syn. pig, swine.) *See also* **Suidae**.

### Suidae
The formal name for the mammalian family that includes the pigs. Pigs are omnivorous artiodactyls with simple nonruminating stomachs. In addition to the ubiquitous domestic pig, *Sus scrofa*, the suid family contains wild species including three indigenous extant African wild hogs: *Phacochoerus aethiopicus*, the warthog; *Hylochoerus meinertzageni*, the giant forest hog; and *Potamochoerus porcus*, the bush pig. While paleoanthropologists are interested in all members of the ecological communities in which hominins evolved, there at least two reasons why the Suidae have a special bearing on human evolution. First, the Suidae underwent an adaptive radiation during the Pliocene and Pleistocene of Africa. This radiation is characterized by relatively rapid speciation, resulting in the evolution of taxa with a larger body size than the ancestral forms and with taller tooth crowns (i.e., hypsodonty). The speed at which this radiation occurred has made the Suidae a useful biochronological tool for dating hominin sites (e.g., Koobi Fora, Sterkfontein, etc.). Second, because they are found alongside hominins at fossil sites it is assumed that there ecological preferences are similar to those of hominins (i.e., they are also large-bodied, terrestrial, omnivores). Thus, they are a useful comparative tool for examining our own evolution. (L. *sus* = swine.)

### sulcus
A deep furrow. In the central nervous system the crest of the folds in the cerebral cortex is called a gyrus and the deepest part of a fold is a sulcus. (L. *sulcus* = furrow.) *See also* **lunate sulcus**.

### sulcus supratoralis
*See* **supratoral sulcus**.

### Sunda
The term Sunda is used to refer to both the Sunda Shelf (or Sundaland) and to the Sunda Islands. Sundaland is the name given to the part of the continental landmass that includes mainland Southeast Asia and the Greater (Borneo, Sumatra, Java) and Lesser (Lombok, Sumbawa, Flores, Sumba, Timor) Sunda Islands. During particularly cold periods when substantial volumes of water were locked up in ice accumulations at high latitudes, sea levels would have fallen low enough for much of Sundaland to have been one continuous landmass. However, the water channels between the Greater and Lesser Sunda Islands, and between Timor and the smaller Lesser Sunda Islands (i.e., Flores, Sumba, etc.), are deep enough to constitute a substantial barrier to migration. The same goes for the deep channel between Timor and Sahul (i.e., Greater Australasia). (etym. in Hindu lore, variants of the word Sunda were used as names by ancient creatures.) *See also* **Sahul**.

## superciliary arch

A bilateral bony arch (aka eminence) that runs laterally and superiorly from the medial end of the superior orbital margin in some pre-modern *Homo* crania and in most *Homo sapiens* crania. In such specimens the supraorbital region takes the form of a composite structure that consists of the superciliary arch medially and the supraorbital trigon laterally. [L. *cilium* = eyelid and *supercilium* = eyebrow; NB: the term "ciliary" refers to anything (e.g., lids, muscles, nerves, vessels) that pertains to the eye.]

## superciliary eminence

See **superciliary arch**.

## superior nuchal line

A roughened line on the occipital bone produced by the attachment of the fascia that covers one of the biggest nuchal muscles, the semispinalis capitis. If the muscle and the fascia are well developed the line may take the form of a crest of bone called the nuchal crest.

## supernumerary tooth

When the number of teeth exceeds the typical, or modal, number for that tooth class in each quadrant of a jaw (e.g., for Catarrhines, including hominoids, incisors = 2, canine = 1, deciduous molars = 2, premolars = 2, molars = 3). Supernumerary teeth may resemble neighboring teeth or show less morphological complexity than the other teeth in the respective tooth class. There are many documented cases of fourth molars among hominoid primates, and in mammals supernumerary teeth are found in higher frequency in hybrids, possibly due to the breakdown of normal genetic regulation in these individuals. (L. *super* = above and *numerus* = number.)

## superposition, law of

Principle, or law, that within a sequence of strata the strata at the top are younger than those at the bottom. (L. *super* = above.)

## suprainiac fossa

A small depression, or fossa, above inion (the most prominent projection of the occipital bone at the lower back part of the skull). It has been proposed as a unique, derived trait (autapomorphy) of *Homo neanderthalensis*, but others have suggested that its expression is not restricted to that taxon and that it is also seen in fossil crania belonging to *Homo sapiens*. However, the suprainiac fossae of *H. neanderthalensis* and the supranuchal fossae of *H. sapiens* (the feature usually confused with a true suprainiac fossa) differ in their ontogeny as well as in their adult morphology, suggesting that the suprainiac fossa is a reliable trait for distinguishing Neanderthal crania.

## supralaryngeal vocal tract

See **spoken language**.

## supramastoid crest

If the posterior-most fibers of the temporalis muscle are especially well developed, then the edge of that muscle's attachment in the region of the mastoid on the cranium may be raised up to form a supramastoid crest. The supramastoid crest and the mastoid crest are well developed in *Paranthropus* and are usually separated by a supramastoid sulcus. (L. *supra* = above, Gk *mastos* = breast, *oeides* = shape, and L. *crista* = crest.)

### supramastoid sulcus
See **mastoid process**; **supramastoid crest**.

### supranuchal fossa
See **suprainiac fossa**.

### supraorbital bar
The name given to a strut of bone that extends across the frontal bone above the orbits and the nose. It is seen in taxa that manifest substantial postorbital constriction (e.g., chimpanzees, *Australopithecus afarensis*). In these cases, that constriction effectively separates the supraorbital bar from the rest of the cranial vault. A supraorbital bar has a midline (aka glabellar) component and two lateral components; in hyper-megadont archaic hominin crania such as OH 5 the glabellar component projects ahead of the lateral components. (L. *supra* = above and *orbis* = wheel or hoop.)

### supraorbital region
The region of the cranial vault above the superior orbital margins. (L. *supra* = above and *orbis* = wheel or hoop.)

### supraorbital torus
A relatively straight and continuous morphological bar of bone that extends across the frontal bone above the orbits and the nose. It has a midline, or glabellar, component and two lateral components. In taxa that manifest substantial postorbital constriction, or postorbital constriction (e.g., chimpanzees, *Australopithecus afarensis*), the analogous structure is referred to as a supraorbital bar. (L. *supra* = above, *orbis* = wheel or hoop, and *torus* = bulge.) *See also* **supraorbital bar**.

### supraorbital trigone
The lateral of the two components of the compound structure that forms the supraorbital margin in most *Homo sapiens* crania. (L. *supra* = above, *orbis* = wheel or hoop, and *trigonum* = triangle.)

### supratoral sulcus
A hollowed area in the midline between lambda and inion on the occipital bone of many *Homo erectus* crania.

### susceptibility
Biology The state of being at risk from a disease or condition. It is the probability that an individual will suffer from a condition because of environmental exposure to an infective agent or from a condition that is influenced by both genotype and environment (i.e., genetic susceptibility to a disease such as type II diabetes or malaria). Scenarios for human evolution seldom take into account the possibility that disease may have played an important role in our evolutionary history, especially when hominins migrated to regions where they will have had no chance to

develop immunity to infectious diseases. Earth sciences Used to describe rocks that retain enough of a magnetic signal that they can be used for magnetostratigraphy.

## suture

A fibrous joint between two bones of the cranial vault. The fibrous tissue in these joints gradually disappears as the bone fuse together. (L. *sutura* = a seam.)

## Swanscombe

A gravel pit (aka Barnfield Pit) in the 30 m terrace just inland from the south bank of the estuary of the River Thames, in England. The site is known for its Clactonian stone artifacts recovered *in situ* from the middle gravel beds and for a partial hominin calvaria (Swanscombe 1) represented by three cranial vault fragments. The fauna from the site is consistent with the interglacial that preceded the Anglian glaciation (i.e., Marine Isotope Stage 11, *c.*400 ka). The Swanscombe calvaria is one of a group of specimens (e.g., Reilingen, Sima de los Huesos, Steinheim) referred to *Homo neanderthalensis*. Researchers who interpret that taxon as an evolving lineage would interpret the Swanscombe 1 calvaria as an example of "Stage 2," a stage also referred to as pre-Neanderthal. *See* also **pre-Neanderthal hypothesis**.

## Swartkrans

This cave formed from a series of solution cavities within Precambrian dolomite in the Blauuwbank Valley near Krugersdorp in Gauteng Province, South Africa, and had been mined for lime in the 1930s. The first hominins (SK 2–SK 6) were recovered in 1948. They were initially assigned to *Paranthropus crassidens*, a taxon that most researchers now subsume into *Paranthropus robustus*. Researchers now recognize five Members (1–5). Member 1 (the old "pink breccia") is the most extensive and is divided into the Hanging Remnant (HR) and the "lower bank" (LB) components. Most of the Swartkrans *P. robustus* hypodigm comes from Member 1, but both Member 2 (this includes the old "brown breccia") and Member 3 also contain fossil evidence of *P. robustus*. There are two newly recognized sedimentary units, the Talus Cone Deposit (TCD) and the Lower Bank East Extension. The former contains *P. robustus* fossils and the latter, which is presently the lowest sedimentary unit in the cave, is equivalent to the "lower bank" of Member 1. Approximate ages for Members 1–3 are *c.*1.8–1.0 Ma with Member 1 *c.*1.8 Ma, Member 2 *c.*1.5 Ma, and Member 3 *c.*1.0 Ma. In addition, fossils assigned to *Homo* sp. have been recovered from Members 1 and 2, and some researchers have suggested that one, or more, hominin postcranial fossils from Member 3 may belong to *Homo*. Remains attributed to *Homo* make up approximately 5% of the hominins in Member 1, and around 20% of the hominins in Member 2. In addition to SK 15 (the holotype of *Telanthropus capensis*) these presumed *Homo* fossils include the composite cranium, SK 847, the mandible fragment, SK 45, and several isolated teeth. Stone artifacts, including cores and choppers resembling the Developed Oldowan from East Africa, have been recovered from Members 1–3, as have bones that appear to have been used as digging sticks. Assemblages from Member 1 (Lower Bank) and Members 2 and 3 have been interpreted as evidence for systematic butchery by early hominins, and it has been suggested that burned bone recovered in substantial quantities from Member 3 is evidence of the deliberate use of fire. (Location 25°58′08″S, 27°45′21″E, South Africa.)

## swing phase
See **walking cycle**.

## symbol
One of three categories used by communication systems to refer to things in the world (icons and indexes are the others). Symbols are arbitrarily related to their referent, but the relationship between a given symbol and its referent is derived by an explicit and learned social convention. While indexical and iconic signs are common in animal communication systems, symbols are not. Archeologists have emphasized the use of symbols (e.g., the curation of ochre, some types of which are presumably used to color the skin) as an important element of what makes the behavior of modern humans different from that of earlier extinct hominins.

## symbolic
An adjective that describes anything that serves to effectively reference a socially shared idea, object, concept, or phenomenon. The presence of symbolic behavior is a central component of many debates on the origin of modern human behavior. Among extant populations symbolism takes many forms, including material representations as well as social performances such as dance, myth, folklore, song, gesture, and speech. An object, image, or other item of material culture is not inherently symbolic, but rather its symbolic nature is socially ascribed or constructed. It is through collective agreement on the meaning of the symbol that it comes to be symbolic and the symbol itself may or may not bear a direct resemblance to that which it symbolizes. This renders interpretation of symbolism in the archeological record difficult, as we lack anyone to inform us of the cultural context necessary to interpret such items. Consider wedding rings. For many in the USA, Canada, and the UK a person wearing a ring on the fourth finger of the left hand is probably married. However, from the perspective of the archeologist, there is nothing about the physical characteristics of (most) wedding rings that would link them to marriage; rather, it is the understanding of their meaning shared by members of a culture or community that lend them that significance. Early artifacts (e.g., pierced shells from Grotte des Pigeons) that may have served as pendants may carry similarly complex meanings, but their interpretation presents a unique challenge to archeologists. (L. *symbolus* = a token or sign.)

## sympatric
Organisms with significantly overlapping geographic ranges. It is possible that several hominin species (e.g., *Paranthropus boisei* and *Homo habilis*) were sympatric, but co-occurrence in the fossil record does not necessarily imply the animals lived at precisely the same time or in the same habitat. (Gk *sym* from *sumbion* = to live together and *patris* = fatherland.) See also **synecology**; **taphonomy**.

## sympatric speciation
A mode of speciation in which new species are formed when two or more populations within a single species diverge ecologically within the same geographic range. The ecological divergence results in disruptive selection that drives the populations apart anatomically and/or behaviorally, resulting eventually in genetic divergence and reproductive isolation between the

groups. There is a considerable debate within evolutionary biology as to how frequently sympatric speciation occurs, if at all.

## sympatry
The state when two or more organisms are sympatric. *See also* **sympatric**.

## symphyses
Pl. of **symphysis** (*which see*). *See also* **joint**.

## symphysis
A midline secondary cartilaginous joint such as that between the bodies of the right and left sides of the developing mandible and between the paired pubic components of the pelvic bones. (Gk *sym* from *sumbion* = to live together and *phyein* = to grow; pl. symphyses.) *See also* **joint**.

## symplesiomorphic
*See* **symplesiomorphy**.

## symplesiomorphy
A term used in cladistics to refer to the primitive condition of a character (i.e., the character state of the hypothetical common ancestor of a clade, or of the outgroup). For example, a small canine crown is probably a symplesiomorphy, or symplesiomorphic, for the hominin clade, and a slender mandibular corpus is probably a symplesiomorphy for the *Homo* clade. (Gk *sym* from *sumbion* = to live together, *plesios* = near to, and *morphe* = form; syn. plesiomorphy.)

## synapomorphy
A term used in cladistics for a character state that is neither symplesiomorphic (primitive or shared by many taxa) nor autapomorphic (confined to one taxon). Synapomorphies are character states shared by at least two taxa. For example, extreme postcanine megadontia is probably a synapomorphy of *Paranthropus boisei* and *Paranthropus aethiopicus*. Extreme postcanine megadontia is also seen in *Australopithecus garhi*, but it may be a homoplasy in that taxon and not a synapomorphy shared with the aforementioned *Paranthropus* taxa. (Gk *syn* from *sumbion* = to live together, *apo* = different from, and *morphe* = form; syn. apomorphy.)

## synchondroses
*See* **joint**.

## synchronic
Term implying that events took place together, or that fossils are of the same geological age. But "together" and the "same" are being used in the sense of geological time. The events could have been hundreds or even a few thousands of years apart, and the organisms represented by the fossils could have been living at times that were hundreds or a few thousands of years apart. (Gk *sunkhronos* = contemporaneous.)

### synchrotron

A machine that sends accelerated particles (electrons) through a cyclical course determined by magnetic fields. When the electrons are deviated by the magnets, synchrotron light is emitted, which can range from radio frequencies to high-energy X-rays. Synchrotron X-rays have properties that make them superior to laboratory or medical computed tomography for imaging. *See also* **synchrotron radiation micro-computed tomography**.

### synchrotron radiation micro-computed tomography

A synchrotron imaging technique (abbreviated to SR-μCT) that reveals structures invisible with conventional computed tomography (or CT). The small size and focused high energy of the source beam means that, compared to CT, SR-μCT can deliver an advantageous signal-to-noise ratio on even highly mineralized bone and tooth samples. It has enabled researchers to image the incremental development of intact teeth, reveal the fine details of the enamel–dentine junction in heavily mineralized fossil teeth, and determine the age at death in juvenile fossil hominins from dental microstructure.

### syncline

In structural geology a syncline is a type of fold where the youngest rocks occupy the center of the structure and the rocks become progressively older towards the margins. The simplest form of syncline is a symmetrical U-shaped fold. Such folds are important in geological and paleontological fieldwork, as they will determine the direction in which successively older or younger strata are encountered and will cause a repetition of the outcrop pattern on either side of the fold's axis. (Gk *syn* = together or with and *klinein* = to slope.)

### syndesmoses

*See* **joint**.

### synecology

The branch of ecology concerned with the interactions of communities of organisms. There is great potential for synecological studies in paleoanthropology. Some hominins may have interacted or lived alongside other hominin species (e.g., at Koobi Fora as many as four contemporaneous hominin species have been identified: *Homo ergaster*, *Homo habilis*, *Homo rudolfensis*, and *Paranthropus boisei*). Although co-occurrence in the fossil record does not necessarily imply that the animals lived at precisely the same time or in precisely the same place, the fact that these species are found together in the same geological unit and in the same region suggests they may well have been sympatric. Although several arboreal primate species live in mixed-species groups, analogy with modern terrestrial primates suggests that sympatric hominins may have avoided competition by (a) having different home ranges, (b) exploiting different foodstuffs, or (c) interacting only where their ranges met. (Gk *syn* = together, *auto* = self, and *logos* = knowledge.) *See also* **community ecology**; **taphonomy**.

## synonym

Any available formal name that denotes the same taxon as another formal name. The earliest available name is the senior synonym; any other name denoting the same taxon is a junior synonym. For example, the names *Paranthropus robustus* Broom, 1938 and *Paranthropus crassidens* Broom, 1949 are considered to denote the same taxon, so are deemed to be synonyms. Because *P. robustus* has 11 years' priority it is the senior synonym and *P. crassidens* is a junior synonym. The same logic means that *Pithecanthropus* is a junior synonym of *Homo*. If it is someone's opinion that two or more named taxa actually denote the same taxon, they are subjective synonyms. Occasionally, however, two or more taxa are established on the same type specimen. In this case, the names given to them are called objective synonyms. For example, when *Anthropopithecus erectus* Dubois, 1893 was described, with the Trinil calotte as its type specimen, several other scientists gave their own names to the same taxon (e.g., *Homo javanensis primigenius* Houzé, 1896; *Homo pithecanthropus* Manouvrier, 1896; *Hylobates giganteus* Bumuller, 1899; *Pithecanthropus duboisii* Morselli, 1901; *Hylobates gigas* Krause, 1909; and *Homo trinilis* Alsberg, 1922). They are all examples of objective synonyms. Synonymy is the condition that exists when more than one name has been given to the same taxon. (Gk *sun* = with and *onuma* = name.)

## synonymy

See **synonym**.

## syntactic rules

See **syntax**.

## syntax

A property of language involving the meaningful order of words. Language is not just about using signs (e.g., sounds or gestures) with specific meanings to refer to things in the world; languages are also characterized by syntax. Consider the phrase *the boy hit the girl* versus *the girl hit the boy*. While the words in each sentence are identical, their meaning is radically different; what governs this shift in meaning is syntax. Experiments suggest that there are cognitive constraints that limit syntactic processing in nonhuman primates. These constraints may explain the universality of modern human languages and the failure of the various ape language projects. *See also* **language**.

## syntype

All of the specimens of a particular species that are known when that species is first described. If a holotype was not designated at the time of the original description of a taxon then one of the syntypes (i.e., all the specimens known to the authors at the time of the initial publication) can subsequently be designated as the lectotype (the alternative to a holotype). For example, when the taxon *Pithecanthropus rudolfensis* was introduced by Alexeev he did not formally designate a holotype. Subsequently KNM-ER 1470, one of the syntypes of the original taxon, was formally designated as the lectotype of *Homo rudolfensis*. (Gk *syn* = together and *typus* = image.)

**systematics**
An inclusive term that subsumes all of the activities involved in the study of the diversity and origins of living and extinct organisms. Thus systematics includes (a) identification (i.e., identifying individual living, or fossil, organisms, and assembling them into groups), (b) classification (i.e., the formalization of those groups as taxa, giving formal names to the taxa, allocating the taxa to taxonomic categories, and then assembling the taxonomic categories into a hierarchical scheme), and (c) phylogeny reconstruction (i.e., generating hypotheses about the branching pattern of the tree of life). Taxonomy is the study of the principles and theory of classification. Nomenclature, which combines the principles used in the allocation of formal names to taxa and the rules (e.g., priority, synonymy, etc.) governing how those names may be used, is a subcomponent of classification, and thus of taxonomy. (Gk *systema* = a whole made of several parts.)

# T

## Tabarin
A locality within the lowest part of the Chemeron Formation, which is in the Tugen Hills sequence to the west of Lake Baringo, Kenya. It has produced abundant fossil fauna and one hominin, KNM-TH 13150, a partial adult mandible. At the time of its discovery in 1984 the estimated age of *c.*4.42 Ma for KNM-TH 13150 made it the oldest known hominin. The KNM-TH 13150 mandible may belong to *Ardipithecus*. (Location 00°75′N, 35°86′E, Kenya.) See also **KNM-TH 13150**.

## Tabon Cave
This cave site on the western coast of Palawan Island is one of the few Southeast Asian island sites that document early colonization by modern humans. Several *c.*34 ka hominin fossils were recovered including a *Homo sapiens* frontal bone, plus two fragmentary mandibles and several cranial and postcranial specimens. (Location 09°16′N, 117°58′E, Palawan Island, The Philippines.)

## Tabun
Several levels (B, C, and E) at this cave site on the western slope of Mount Carmel, Israel, have yielded the remains of up to 14 individuals belonging to *Homo neanderthalensis*, plus Mousterian and Levallois artifacts. The fossils and artifacts were originally estimated to be *c.*40 ka but electron spin resonance spectroscopy dating suggests that specimens such as the Tabun 1 cranium are *c.*122 ka. This means that some of the remains of *H. neanderthalensis* in the Near East antedate evidence of *Homo sapiens* (e.g., Qafzeh and Skuhl) in that region. (Location 32°40′N, 34°58′E, Israel.)

## *Taenia*
See **taeniid tapeworms**.

## taeniid tapeworms
Modern humans are the definitive host for three different species of taeniid tapeworm (*Taenia saginata*, *Taenia asiatica*, and *Taenia solium*). Domesticated cattle are the intermediate hosts for *T. saginata*, whereas domestic swine are the intermediate hosts for *T. asiatica* and *T. solium*. For a long time it was assumed that modern humans had acquired their taeniid parasites when cattle and swine were domesticated, but evidence from molecular biology suggests that it was

the other way round. Molecular evidence suggests that the three taeniid tapeworms that infest modern humans belong to two different subclades (*T. solium* (*T. saginata*, *T. asiatica*)); this means there must have been at least two, and probably three, separate episodes of infestation. There is also evidence that the definitive hosts of the immediate precursors of modern human taeniid tapeworms were carnivores (canids, felids, or hyaenids) and their intermediate hosts were bovids. This is consistent with the hypothesis that host-switching, the passage of a parasite or infectious agent from one species to another, occurred in sub-Saharan Africa prior to the domestication of ungulates, and that it most likely occurred when early hominins were competing with carnivores for meat. This research on tapeworms is an example of how parasitological data can be used to investigate the behaviour of extinct hominins.

## talon
The distal (aka heel) component of a maxillary (upper) postcanine tooth crown. In hominin maxillary molars it comprises the hypocone distolingually, plus any accessory cusps or cuspules associated with the hypocone. (L. *talus* = heel.)

## talonid
The distal (aka heel) component of a mandibular (lower, hence the postfix "-id") postcanine tooth crown. In hominin lower molars it comprises the hypoconid buccally, the entoconid lingually, plus the hypoconulid (if present) and any C6s. Premolars with a well-developed talonid (e.g., *Paranthropus boisei*) are referred to as "molarized" or "molariform." Modern human lower premolars do not normally have a talonid. (L. *talus* = heel.) *See also* **molarized**.

## talus cone
A cone-shaped heap of debris that forms in a cave when soil is washed in via an opening in the roof. As new material is washed into the cave it adds to the height of the talus cone, but most runs down the sides thus increasing the area of the base. Talus cones are one of the reasons why some of the sediments in the southern African early hominin cave sites (e.g., Silberberg Grotto, Sterkfontein, Swartkrans) violate the law of superposition. (OF *talu* = the sloping sides of an earthwork.)

## tandemly repeated sequences
An inclusive term for DNA sequences, two or more bases long, that are repeated and adjacent (e.g., microsatellites, minisatellites, satellite DNA).

## tandem repeat array
*See* **tandemly repeated sequences**.

## tandem repeats
*See* **tandemly repeated sequences**.

## tang
The narrow elongated region at the base of a point-like stone tool that is designed to facilitate hafting. Tangs are frequently seen on arrowheads and tanged points are considered diagnostic of Aterian sites in North Africa. (etym. ME of Scandinavian origin, similar to the Norse *tangi*, or point of land.)

## Tangier
See **Mugharet el ʿAliya**.

## Tangshan Huludong
(汤山葫芦洞) The Tangshan locality consists of the karstic cave site of Huludong (Hulu Cave) near Nanjing in China. Uranium-series dating of the overlying stalagmite and flowstone provides a possible age of c.580 ka for a hominin partial cranium and calotte, which have been assigned to *Homo erectus*. (Location 32°03′24.94″N, 119°02′40.41″E, eastern China.)

## tapeworms
See **taeniid tapeworms**.

## taphonomy
The factors involved in the death, decay, preservation, and fossilization of organisms. In other words, taphonomy is the study of the transition of organisms from being part of the biosphere to potentially being preserved in the geological record. Taphonomy includes the consideration of why some organisms are more likely to fossilize than others (aka differential preservation, or differential survivorship). By understanding the factors that bias the fossil record, allowances can be made for those biases when researchers try to reconstruct the structure of a fossil paleocommunity or when they use the archeological record to interpret hominin behavioral patterns. Some general biases are well known. For example, bones with more dense cortical bone (e.g., the mandible) tend to survive better than bones that are mostly made up of more fragile cancellous bone (e.g., vertebrae). Small animals tend to be systematically under-represented and large ones over-represented in the mammalian fossil record. However, size is not the sole determinant in all groups. For example, vole teeth, which are small but dense and durable, are well represented in the fossil record. Taphonomy is obviously important to paleontologists but it is also of critical importance to archeologists (e.g., zooarcheologists, paleoethnobotanists) who focus on the organic component of the archeological record. In zooarcheological contexts, taphonomy encompasses all the processes involved from the death of an animal (e.g., its butchery, transport, and discard), to the deposition, preservation, recovery, and analysis of that animal in the form of fossilized faunal remains. (Gk *taphos* = grave and L. *nomos* = law.)

## Taung
(formerly Taungs; aka the Buxton-Norlim Limeworks) A cave that was exposed in an abandoned limeworks in North-West Province, South Africa. Taung is used to refer to the deposits associated with the Taung 1 child's skull, but there are at least 17 sites that span the Pliocene to Holocene within the Buxton-Norlim Limeworks site complex. The exact location of where the Taung child's skull was found is a matter of debate because the find spot was partly if not entirely destroyed by mining. It has been suggested that the Taung 1 deposits are 2.8–2.6 Ma, but others consider this is likely a minimum age. The only hominin recovered from the site, Taung 1, is the holotype of *Australopithecus africanus*. (Location 27°32′S, 24°48′E, South Africa.) See also **Taung 1**.

### Taung 1
A mostly undistorted skull of a juvenile hominin that preserves most of the right half of a natural endocast, the face, the upper jaw, part of the cranial base, and the mandible. The natural endocast preserves some endocranial morphology in exquisite detail, but frustratingly the area where the lunate sulcus is most likely located is poorly preserved (see below). The dentition of Taung 1 includes the maxillary and mandibular permanent molars, all of the deciduous teeth on at least one side, and the germs of the deciduous premolars, canines, and incisors. Its endocranial volume is about 400 cm$^3$. The Taung 1 skull is iconic for many reasons. It was the first evidence of an early hominin to be recovered from Africa, it is the holotype of *Australopithecus africanus*, it was the first hominin skull to be imaged using computed tomography (or CT), and its natural endocast has been the focus of one of the longest-running disputes in paleoanthropology, namely at what stage in hominin evolution the parietal association cortex expanded. When the discovery of the Taung 1 skull was announced in 1925 its combination of a human-like dentition and a small brain were the opposite of what was then conventional wisdom of what a human ancestor would look like (i.e., the large skull and ape-like jaw of Piltdown). This meant that it was some time before Taung 1 was accepted as a possible human ancestor. Damage to the Taung 1 cranium is consistent with the hypothesis that the Taung child was taken to the cave by a large raptor. *See also* **lunate sulcus**.

### taurodont
Molar teeth that have tall pulp chambers relative to the overall height of the tooth. High incidences of taurodontism are claimed for the molars of *Homo neanderthalensis* and for African modern humans. (Gk *tauros* = bull and *dont* = teeth; ant. cynodont.) *See also* **pulp chamber**.

### Tautavel Cave
*See* **Caune de l'Arago**.

### taxa
Pl. of **taxon** (*which see*).

### taxic
An approach to paleontology in general, and to paleoanthropology in particular, that emphasizes the origin and identification of species as opposed to the recognition and identification of lineages. (Gk *taxis* = to arrange or "put in order.") *See also* **species**.

### taxon
A group recognized at any level, or category, in the Linnaean hierarchy (e.g, the tribe Hominini, the genus *Paranthropus*, and the species *Paranthropus boisei* could each be an example of a taxon). (Gk *taxis* = to arrange or "put in order"; pl. taxa.)

### taxonomy
The principles involved in assembling individual organisms into groups, formalizing those groups as taxa, giving those taxa formal names, allocating the taxa to ranks, or taxonomic categories, and then assembling the taxonomic categories into a hierarchical classification.

The term is also sometimes used to refer to the taxonomic hypothesis that results when Linnaean taxonomic principles are applied to a particular group of organisms (e.g., a taxonomy of the hominin clade). Alpha taxonomy refers to the part of taxonomy involved in diagnosing, defining, and naming species. (Gk *taxis* = arrange or "put in order.") *See also* **alpha taxonomy**.

## TD6
*See* **Gran Dolina**.

## technostratigraphy
The use of archeological evidence (e.g., Mousterian, Oldowan) to provide a relative or correlated date for a layer or site.

## teeth
In modern humans and hominins teeth are used for processing food, or as tools for holding, softening, or otherwise processing materials (e.g., hide-softening or basket-making). Teeth are occasionally involved in culture and ritual (e.g., tooth jewelry, tooth sharpening, deliberate removal). In the closest living relatives of modern humans, the canine teeth are used as a means of social interaction, either directly as weapons in interpersonal conflict or indirectly as part of threat displays. The shape of a tooth is related to how it is used. Teeth with sharp ridges, such as modern human incisors, primate lophodont (aka sharply ridged) molars, and the steep-sided carnassial molars of carnivores, are used for cutting tough materials such as leaves and meat. Teeth with low, rounded (aka bunodont) cusps, such as the molars of orangutans and megadont and hyper-megadont fossil hominins, are used for processing hard material like nuts and seeds.

Teeth consist of a crown and one, or more, roots. The outer part of the crown is made up of an enamel cap that covers a core of dentine; roots are made of a core of dentine outside of which is a thin layer of cementum. Dental tissues (i.e., enamel, dentine, and cementum) are heavily mineralized, and this gives teeth the high density that contributes to their prominence in the fossil and archeological records. Hominins and the great apes, like all primates, have two sets of teeth, deciduous and permanent. Deciduous incisors and canines are replaced by equivalent permanent teeth, but deciduous molars are replaced by permanent premolars; permanent molars have no precursors. Teeth are described using the letter of the tooth type in lower or upper case to indicate deciduous or permanent teeth, respectively (incisor = i or I; canine = c or C; premolars = P; and molars = m or M), plus a number that reflects their place in a sequence of the same tooth type. The number is a superscript for teeth in the upper jaw and a subscript for teeth in the lower jaw. In each quadrant of the upper and lower jaw, catarrhines (i.e., humans, apes, and Old World monkeys) normally have five deciduous teeth [mandible: $di_1$, $di_2$, $d\bar{c}$, $dm_1$, $dm_2$ (or $dp_1$ and $dp_2$); maxilla: $di^1$, $di^2$, $d\underline{c}$, $dm^1$, $dm^2$ (or $dp^1$ and $dp^2$)] and eight permanent teeth (mandible: $I_1$, $I_2$, $\bar{C}$, $P_3$, $P_4$, $M_1$, $M_2$, $M_3$; maxilla: $I^1$, $I^2$, $\underline{C}$, $P^3$, $P^4$, $M^1$, $M^2$, $M^3$). Platyrrhines (i.e., New World monkeys) have an additional premolar in each quadrant ($P^2$ and $P_2$). Note that, because catarrhines retain only the distal two of the four premolars seen in primitive mammals, the two premolars are numbered P3 and P4 (platyrrhines retain three of the four premolars). (OE *toth* = tooth.) *See also* **deciduous dentition**; **permanent dentition**.

## tektites
Approximately spherical natural glass rocks, usually black or olive-green, up to a few centimeters in size. Showers of tektites are formed when large meteorites hit the Earth's surface. The area of the Earth's surface a particular tektite shower covers is called its strewn-field. The presence of distinctive tektites is used to correlate some Asian fossil sites. (Gk *tektos* = molten.)

## *Telanthropus*
Genus established in 1949 to accommodate the species *Telanthropus capensis* Broom and Robinson, 1949. In 1961 it was formally sunk into *Homo*. *See also* **Telanthropus capensis**.

## *Telanthropus capensis* Broom and Robinson, 1949
Hominin species established to accommodate SK 15, an adult mandible recovered from Member 2 at Swartkrans, South Africa. It is now widely regarded to be a junior synonym of *Homo erectus*. (Gk *tele* = distant, *anthropos* = human being, and *cape* = reference to land that includes the Cape of Good Hope.) *See also* **SK 15**; **Swartkrans**.

## telencephalon
*See* **forebrain**.

## temperate
Refers to the regions (and the organisms that occupy them) between the Tropic of Cancer (approximately 23°N) and the northern polar circle, and between the Tropic of Capricorn (approximately 23°S) and the southern polar circle (i.e., between latitude 23° and 66° both north and south). These regions experience seasons that usually have significant seasonal variation in both sunlight (and therefore temperature) and rainfall. In general biodiversity is lower in temperate regions than in the tropics (aka Rapoport's Rule). Very few contemporary primates live in temperate regions but in the past, when the Earth's climate was warmer, the latitudinal range of primates was greater. Anatomically modern humans and their recent extinct relatives used cultural and behavioral adaptations to continue to exploit temperate regions despite climate trends that have made those regions colder than they were in the past. (L. *temperatus* = restrained or regulated.)

## tempo
In connection with evolution, tempo refers to the time it takes for evolution to occur (i.e., the rate at which evolution occurs). The other main variable is the mode, or the pattern, of evolution. (L. *tempus* = time.) *See also* **mode**.

## temporalis
*See* **mastication, muscles of**.

## temporal lobe
One of the four main subdivisions of the cerebral cortex of each cerebral hemisphere. The temporal lobe, which is located ventral to (i.e., beneath) the Sylvian fissure, has the primary auditory cortex along the posterior part of its superior surface. Other parts, or areas, within the temporal lobe are involved in language, higher-order auditory processing, and the processing

of visual information important for object perception and recognition. The medial part of the temporal lobe contains the hippocampus and the amygdala.

## temporomandibular joint
(or TMJ) A modified synovial joint between the condyle of the mandible and the mandibular (aka glenoid) fossa of the temporal bone of the cranium. The shape of the various parts of the TMJ varies among hominins. In modern humans the well-defined and deeply concave TMJ is compressed between well-defined bony structures. This contrasts with the much more "open" morphology of the mandibular fossa of the great apes; the fossa is shallow with poorly defined bony margins. Mandibular condyles are rare in the hominin fossil record but the mandibular fossa is relatively well represented, so it is possible to reconstruct jaw movements based on the shape of the fossa. The earliest hominins (*Sahelanthropus*, *Ardipithecus*, *Australopithecus anamensis*) have shallow, ape-like TMJs. The fossae of *Australopithecus afarensis*, *Australopithecus africanus*, and *Kenyanthropus platyops* are a little deeper but they are still more ape-like than modern human-like. The *Paranthropus robustus* mandibular fossa is similar to that of *Au. africanus* except for its larger size and steeper joint surface. The fossa of *Paranthropus aethiopicus*, which generally resembles the TMJ of *Gorilla*, contrasts with the derived mandibular fossa of *Paranthropus boisei*. The mandibular fossa of early *Homo* is generally narrow and deeper than that of *Paranthropus* or earlier hominins. In *Homo erectus sensu lato* the form of the mandibular fossa varies considerably; the African *H. erectus* specimens are more similar to early *Homo*, whereas in Asian *H. erectus* the mandibular fossae are more derived in size and shape (i.e., even narrower and deeper). The form of the mandibular fossa in *Homo heidelbergensis sensu lato* shows regional variation, with African Middle Pleistocene specimens (e.g., Kabwe) having a steeper joint surface than is found in European specimens (e.g., Sima de los Huesos). The less inclined anterior joint surface of the European specimens is similar to the morphology seen in *Homo neanderthalensis*.

## tephra
A collective term for rock fragments that have been ejected explosively from a volcanic source (aka pyroclastic material) that form unconsolidated (i.e., loosely bonded) deposits. Tephra range in size from fine ash (about 0.063 mm in diameter) to larger "bombs" (more than 64 mm in diameter). A consolidated (i.e., firm or hardened) tephra deposit is called a tuff. (Gk *tephra* = ash.) *See also* **tuff**.

## tephrochronology
The use of dated tuffs to provide a regional chronology. (Gk *tephra* = ash and *kronos* = time.)

## tephrostratigraphy
The use of chemically or lithologically distinctive tuffs to correlate strata from one locality to another or from one site to another. (Gk *tephra* = ash, *stratos* = to cover or spread, and *graphos* = to draw or write.)

## termites
Termites are an important source of protein and lipids for some chimpanzee communities and they may have been part of the diet of some early hominins. Chimpanzees use sticks or grass, sometimes extensively modified, to remove termites from the conspicuous mounds that are

abundant on many African landscapes (a process also known as extractive foraging). Potential evidence for the consumption of termites by early hominins includes the use-wear patterns observed on bone tools from the sites Swartkrans and Drimolen in southern Africa. Actualistic studies showed that microscopic wear on the bone tools resembled the fine and highly oriented wear seen when experimental bone tools were used to dig termite mounds. Other potential support for termite consumption comes from evidence that some archaic hominins consumed $^{13}$C-rich $C_4$ foods and many termites, including the mound-building termites belonging to the genus *Macrotermes*, consume $C_4$ vegetation.

## Ternifine

The old name for the site of Tighenif (or Tighennif). See also **Tighenif**.

## Terra Amata

A controversial site located on an ancient beach about 20 m above the current sea level near Nice, France. Some claim it is *c*.380 ka evidence of shelters and fires on a series of living floors; others consider it to be more recent (*c*.230 ka) and suggest that the stones were accumulated naturally. (Location 43°41′51″N, 07°17′20″E, France.)

## terrestrial

An animal that lives on the ground. The vast majority of living primate taxa are dependent on trees for living, sleeping, and foraging, but a few primate species spend a significant proportion of their time on the ground. The most terrestrial living nonhuman primate is *Theropithecus gelada* but many other taxa, including ring-tailed lemurs, common baboons, patas monkeys, vervets, mandrills, rhesus macaques, and gorillas, also spend significant amounts of time on the ground. Terrestriality is an important adaptation in areas where tree cover is broken, which could partially explain why the ancestors of hominins eventually became habitually terrestrial. Increased susceptibility to predation is an important cost of terrestriality and terrestrial primates therefore tend to live in relatively large groups and seek refuge from predators in trees, which are often used as sleeping sites. Terrestrial and bipedal locomotion are usually treated synonymously in debates about early hominins, but at least one researcher has argued that bipedalism might have evolved in part as a postural rather than a locomotor adaptation. Terrestrial primates tend to have relatively large home and day ranges. (L. *terrestris* = earthly.) See also **seed-eating hypothesis; postural feeding hypothesis**.

## terrestriality

The tendency to live partially, or wholly, on the ground. See also **terrestrial**.

## Tertiary

The term is a remnant of the old classification of geologic strata (and time) into four units: Primary, Secondary, Tertiary, and Quaternary. It was decided to split up the Tertiary into the Paleogene and Neogene periods in order for the Cenozoic era to more closely match the duration of the Mesozoic and Paleozoic eras. Although it is no longer recognized as a formal geochronologic or chronostratigraphic unit, many still use the Tertiary to refer to the pre-Pleistocene part of the Cenozoic era. (L. *tertius* = the third in a series.) See also **Neogene**.

## tertiary dentine
See **dentine**.

## Teshik-Tash
A cave site in the Baisin Tau Mountains of Uzbekistan that includes the skeleton of a c.150–100 ka *Homo neanderthalensis* child between the ages of 8 and 11 years. Some scholars have questioned whether the remains can be unequivocally linked to *H. neanderthalensis*, but mitochondrial DNA from the left femur confirmed that this specimen is a Neanderthal as did a geometric morphometric analysis of the frontal bone. The lithic assemblage has been characterized as Mousterian, but most tools are sidescrapers and transverse flakes; blades and Levallois flakes are rare. (Location approximately 38°17′24″N, 67°02′46″E, Baisan Tau Mountains, near Pas-Machai in Uzbekistan.)

## The High Cave
See **Mugharet el 'Aliya**.

## theory of mind
The ability to attribute perceptions, knowledge, intentions, goals, and beliefs to oneself and others. Theory of mind tests may be divided into two categories: tests of true beliefs and tests of false beliefs. Tests of true beliefs assess second-order intentionality; that is, an individual's ability to attribute beliefs, intentions, goals, etc. to another individual. Tests of false beliefs assess third-order intentionality, which involves beliefs about (second-order) beliefs. At present there is no consensus as to whether or not nonhuman primates understand either true beliefs or false beliefs. Some authors have argued that chimpanzees and rhesus monkeys understand true beliefs such as "seeing" and "hearing"; others argue that whereas chimpanzees understand true beliefs, they do not understand false beliefs.

## thermal ionization mass spectrometry
See **uranium-series dating**.

## thermoluminescence dating
See **luminescence dating**.

## *Theropithecus* Geoffroy, 1843
A genus of Old World monkey with only one extant species, the highly terrestrial *Theropithecus gelada* whose habitat is presently restricted to the Ethiopian highlands. Although *Theropithecus* is only distantly related to the hominin clade it is relevant to hominin evolution because (a) the genus underwent an adaptive radiation in the Plio-Pleistocene at nearly the same time as the hominin clade, (b) at most African early hominin sites *Theropithecus* is usually the most closely related primate, (c) any success at reconstructing the habitats of fossil theropiths might help in the reconstruction of early hominin habitats, and (d) the unique status of fossil *Theropithecus* as large-bodied, open habitat primates has led to their use as an analogue for hominin behavioral evolution (e.g., the seed-eating hypothesis). (Gk *ther* = wild beast and *pithekos* = ape.) See also **seed-eating hypothesis**.

## Thomas Quarry

One of several sites discovered as a result of sandstone quarrying activities during harbor construction near Casablanca, Morocco. The Thomas Quarry site was initially divided into three quarries: I, II, and III. Caves in Thomas Quarry I and Thomas Quarry III have yielded hominin remains, but the Thomas III (also known as Oulad Hamida 1) cave has subsequently been destroyed. Four hominin teeth and two mandibles have been recovered from the GH cave of Thomas Quarry I. Along with the fossils from Salé and Sidi Abderrahman, the Thomas Quarry fossils that are dated to 700–400 ka represent a distinctive north African group of hominins that falls morphologically somewhere between *Homo erectus* and *Homo sapiens*. The lithic artifacts from Thomas Quarry include Acheulean handaxes, cleavers, cores, flakes, and other lithic debris associated with a modest faunal assemblage. (Location 33°34′N, 07°42′W, Morocco.)

## thoracic cage, comparative anatomy and evolution

The thoracic cage consists of the ribs and the thoracic vertebrae to which they are attached, along with the sternum and costal cartilages. In modern humans the ventral ends of the first six ribs articulate with the sternum via independent costal cartilages and ribs 7–10 share a joint costal cartilage; the caudal two ribs do not articulate with the sternum, nor do they have costotransverse articulations. The thoracic cages of the extant great apes are narrow at the top and wide at the bottom, and the long lower ribs sit adjacent to the iliac crest due to their elongated pelves and short, stiff lumbar spines. Monkeys, gibbons, and modern humans have longer lumbar regions, and their lower rib-cage dimensions do not have to match the pelvis as tightly. Modern humans resemble gibbons in having a barrel-shaped thoracic cage, whereas the thoracic cage of the great apes is cone-shaped. Modern humans also have their thoracic vertebral column more centrally located in the body relative to the rest of the rib cage, so that the neck of the rib and thoracic vertebral transverse processes flare dorsally relative to the vertebral body, before swinging laterally and anteriorly. The initial reconstruction of the thoracic cage of *Australopithecus* suggested it was cone-shaped, like that of apes, but the new *Australopithecus afarensis* skeleton (KSD-VP-1/1) from the Woranso-Mille study area in Ethiopia has a nearly complete second rib that is similar in curvature to those of modern humans and recent pre-modern *Homo*. Other *Au. afarensis* thoracic vertebrae reflect a modern human-like pattern and *Australopithecus sediba* also appears to have had a modern human-like thoracic cage. This calls into question reconstructions of a large ape-like gut made under the assumption of a caudally flaring, cone-shaped rib cage that would have contrasted with a more modern human-like smaller gut; however, reconstructing gut size from fossil evidence is necessarily highly speculative. Early *Homo* appears to have had a thoracic cage most like that of modern humans. The thoracic cage of *Homo neanderthalensis*, which is similar overall to modern humans, is barrel-shaped (i.e., it is expanded both anteriorly and laterally), probably reflecting their stocky, presumably cold-adapted, body form. *See also* **expensive tissue hypothesis**.

## three-dimensional morphometrics

*See* **geometric morphometrics**; **morphology**.

## Tighenif
(alternate spellings: Ternifine, Tighennif) This Middle Pleistocene site is a commercial sandpit 17 km/10 miles southeast of Mascara, Algeria, in the foothills of the Atlas Mountains. The c.780 ka hominin fossils, which include three mandibles (Tighenif 1–3) and a parietal fragment (Tighenif 4), were initially assigned to their own genus and species as *Atlanthropus mauritanicus*, a taxon that was later subsumed within *Homo erectus*. However, the most recent detailed taxonomic assessment of the Tighenif mandibles concluded that the remains should be assigned to *Homo heidelbergensis*. (Location 35°30′N, 00°20′E, Algeria.)

## time-averaging
Any collection of fossils that is thought to represent a single species will consist of multiple individuals that lived during different time intervals within what may be a very long (e.g., 1 Ma) time frame. There may have been random phenotypic changes within the species during this long time frame, so cumulative descriptions of the morphology of a fossil hominin taxon are considered to be "time-averaged." Time-averaging complicates attempts to use museum collections of extant taxa, which were obviously accumulated over a much shorter period of time, as comparative analogues for determining whether multiple taxa may be represented within a fossil sample.

## time scale
Framework for assigning ages to geological deposits or to events in the geological record. For example, the astronomical time scale is based on Earth/sun orbital systems and the geomagnetic polarity time scale uses calibrated changes in the direction of the Earth's magnetic field.

## timing
*See* **heterochrony**; **tempo**.

## TIMS
Acronym for thermal ionization mass spectrometry. *See* **uranium-series dating**.

## TL
Abbreviation of thermoluminescence dating. *See* **luminescence dating**.

## TM 1517
This c.2.0–1.5 Ma associated skeleton, made up of a skull, isolated teeth, the distal end of the right humerus, the proximal end of the right ulna, hand bones, a right talus, and other foot bones, from Member 3 at Kromdraai B in South Africa, is the holotype, or type specimen, of *Paranthropus robustus*. It is one of the few associated skeletons of that taxon. *See also* **Paranthropus robustus**.

## TM 266-01-060-1
This c.7 Ma nearly complete cranium with many of the roots and some of the crowns of the maxillary dentition is from the Anthracotheriid unit at locality TM 266 at Toros-Menalla, Chad, Central Africa. It is the holotype, or type specimen, of *Sahelanthropus tchadensis*. *See also* **Sahelanthropus tchadensis**; **Toros-Menalla**.

## TMJ
Abbreviation of **temporomandibular joint** (*which see*).

## toe-off
See **push-off**.

## ToM
Abbreviation of **theory of mind** (*which see*).

## tool
Any material object (modified or unmodified) used to accomplish a task. Evidence from Gona, Ethiopia (dated to *c*.2.55 Ma), which consists of flakes and the cores from which they were struck, includes the oldest stone tools currently known in the archeological record. However, if the cutmarks from the Dikika study area in Ethiopia (*c*.3.39 Ma) are anthropogenic, then they would represent the first use of stone flakes, manufactured or natural, as tools. The Gona first appearance datum almost certainly underestimates the onset of habitual tool use, but because evidence of scattered unmodified stones or perishable items of the sort used by other primates is difficult to detect we may never know when habitual tool use began. (OE *tool* = to "prepare for use.")

## tool-use hypothesis
One of several hypotheses that attempt to explain the origin of upright posture and bipedal locomotion within the hominin clade. First proposed by Charles Darwin in *The Descent of Man, and Selection in Relation to Sex* (1871), the basic premise of the tool-use hypothesis is that bipedalism freed the hands from locomotion, thereby allowing them to be employed for tool use, which in turn stimulated the evolution of many other derived characteristics of later hominins. Later versions of the hypothesis proposed a sophisticated system of selective feedback mechanisms in which bipedalism allowed and/or was caused by tool use, which in turn favored canine reduction, hunting, and the evolution of a large brain. This brain enlargement and the concomitant increase in intelligence led to the evolution of ever more sophisticated tools, which re-energized the feedback loop (i.e., better tools led to enhanced bipedalism and a larger brain, etc.). Ultimately, increased brain size and hunting behavior led to language and complex human social behavior. The tool-use hypothesis was effectively falsified when it became clear that the appearance of undoubted hominins (*c*.4.2 Ma) and inferred bipedal locomotion substantially pre-dated the earliest archeologically visible stone tools (*c*.2.6 Ma). See also **bipedal**.

## tooth crypt
See **alveolus**.

## tooth emergence
See **eruption**.

## tooth eruption
See **eruption**.

## tooth formation

Process of crown and root development that begins with soft tissue differentiation inside tooth crypts in jaws and ends with apical closure of the root(s). It is divided into discrete categories or stages that can be assessed using dental radiographs. *See also* **dentine; enamel development; eruption**.

## tooth germ

An immature tooth bud within the mandible or maxilla. Initially tooth germs are small and comprise only soft tissues, but as mineralization begins tooth germs enlarge to fill the crypts in the jaws. Because they are so fragile tooth germs are rare in the hominin fossil record. *See also* **dentine; enamel development**.

## tooth marks

*See* **carnivore modification; cutmarks**.

## tooth nomenclature

*See* **teeth**.

## tooth size

When researchers write and talk about hominin tooth size, unless they specify to the contrary, they are almost certainly referring to the maximum size of a tooth crown in a plane at right angles to its long axis. For the incisors and canines this is usually the part of the tooth crown closest to the occlusal plane; for the postcanine teeth it is the size of the junction between the crown and the root. Tooth crown size can be expressed in terms of one or more linear dimensions or as an area measurement. When a single linear dimension is used as a proxy for tooth size it is almost always the mesiodistal length of the crown (e.g., premolar and molar chords, see below), but mesiodistal measurements need to take into account loss of enamel due to the interproximal wear that occurs between adjacent teeth when they move against each other during chewing. Advances in methods for digital image capture that allow for three-dimensional (3D) measurements, as well as advances in 3D analytical methods, are enabling researchers to investigate the absolute and relative sizes of the component cusps of multicuspid tooth crowns and for taxonomic purposes these studies are beginning to replace more simplistic assessments of overall tooth size. For the great apes, possible primitive hominins, and early hominins the operative measure of canine size is the height of the crown.

## tooth wear

The process of tooth wear (the loss of enamel volume, and then later the loss of dentine volume) begins as soon as a tooth loses its gingival covering and is exposed in the mouth. Most research on tooth wear in hominins is concentrated on the wear that occurs on the functional, or occlusal, surface of a tooth (aka occlusal wear). But wear also takes place on the non-occlusal surfaces of the teeth (due to the action of acid) and interproximal (aka approximal) wear occurs between tooth crowns when adjacent teeth move slightly against each other in the mouth during chewing. Occlusal wear can be divided into tooth–tooth wear (aka attrition), the wear that occurs from teeth processing food (tooth–food wear is also called abrasion), and the wear that results from chemical erosion. However, these processes also interact in a complex fashion. Occlusal tooth wear that can be seen with the naked eye is called gross wear (aka dental macrowear or dental mesowear), while tooth wear that can only be seen with a microscope is called dental microwear. Observations about gross dental wear focus on the development of wear facets, including measurements of their size and orientation. Studies of dental microwear focus on the size, number, and orientation of microscopic scratches, pits, etc. on the enamel, or on overall complexity and isotropy of the enamel surface. Dental macrowear is a measure of the abrasiveness of the diet in the long term, whereas dental microwear indicates whether the food ingested in the days or weeks before death contained hard or abrasive material.

## Toros-Menalla

A fossiliferous area in the Chad Basin of the Djurab Desert of northern Chad where fossil evidence of *Sahelanthropus tchadensis* has been recovered. Three localities, TM 247, TM 266, and TM 292, have yielded hominin remains, and all three are within the Anthracotheriid unit. That unit has been dated using biostratigraphy to between 7 and 6 Ma, and by $^{10}Be/^9Be$ cosmogenic nuclide dating to between 7.2 and 6.8 Ma. The hominins recovered include TM 266-01-060-1, the holotype of *S. tchadensis*, plus other cranial remains. It is also understood there are as yet unpublished postcranial remains. There is no archeological evidence. (Location 16°14–15′N, 17°28–30′E, Chad.) *See also* **Sahelanthropus tchadensis**; **TM 266-01-060-1**.

## torque

*See* **petalia**.

## Torralba

This *c.*240 ka site in Spain, which is significantly older than the nearby site of Ambrona, was best known for being interpreted as evidence of hunting, but now researchers are less sure that the preserved fauna were the result of organized hunting. Torralba also has at least two pieces of elephant bones that have been intentionally flaked and shaped into points; these are among the oldest evidence of flaked bone tools. (Location 41°08′19″N, 02°29′51″W, Spain.)

## torus

An area of projecting bone with a rounded profile (e.g., the occipital and frontal tori on the cranium and the transverse tori on the mandible). (L. *torus* = bulge.)

## total group

A clade (aka monophyletic group) that includes all of the taxa that are more closely related to the living taxon in that clade than they are to any other living taxon (e.g., the hominin and panin clades are both total groups). A total group is made up of a crown group plus the stem group. (L. *totus* = whole.) *See also* **clade**; **crown group**; **stem group**.

## total morphological pattern

A concept that emphasizes the importance of taking into account all of the available morphological evidence (e.g., not focusing on just one region of the body) when generating systematic hypotheses.

## "Toumaï"
The nickname given to TM 266-01-060-1, the cranium that is the holotype of *Sahelanthropus tchadensis*. In the Goran language it means "hope of life." *See also* **Sahelanthropus tchadensis**; **TM 266-01-060-1**.

## trabecular bone
*See* **bone**.

## trace element
A chemical (e.g., aluminum, zinc) that is only required in small (or trace) amounts in the diet. Trace elements are not nutrients, but they may be essential for nutrients to be effective.

## trace fossil
Fossils that provide direct evidence that an organism has been at a particular place at a particular time, but no remnant of the organism itself, either in the form of hard (i.e., bones and teeth) or soft (i.e., skin, muscle, etc.) tissues, is preserved. There are two categories of trace fossils. In the first, sediments have functioned like casting material and they have retained details of structures and/or behaviors long after any evidence of the individual responsible for them has disappeared. Examples include footprints (e.g., Laetoli and Koobi Fora hominin footprints) that preserve the impressions made by external movements of the foot, and natural endocranial casts that faithfully reproduce the inner (i.e., endocranial) surface of the brain case. The second much smaller category comprises fossilized solid excreta in the form of coprolites (i.e., fossilized feces). Trace fossils (aka ichnofossils) make up a small, but important, fraction of the fossil evidence for human evolution. (L. *tractus* = a track and *fossile* = something that is dug up.) *See also* **coprolite**; **endocranial cast**; **track**; **trackway**.

## trachyte
A family of fine-grained volcanic rocks mostly rich in phenocrysts of alkali feldspar and other minerals. When it was available trachyte appears to have been used frequently for stone tool production (e.g., the blades at the archeological sites in the Kapthurin Formation in the Tugen Hills of Kenya). (Gk *trachys* = rough.)

## track
A general term that refers to trace fossils created by the interaction between locomotor anatomy and the substrate over which an animal was moving. For bipedal hominins the term "footprint" is frequently used, but "track" is more common in reference to the trace fossils created by animals other than hominins. A sequence of tracks is called a trackway. *See also* **hominin footprints**; **trackway**.

## trackway
A general term for a series of tracks. Because they preserve direct evidence of gait cycles, trackways provide a unique form of paleontological data that can be used to reconstruct the locomotion of extinct taxa. *See also* **hominin footprints**; **track**.

## trade-off

A mutually exclusive interaction between two life history traits (i.e., traits that concern growth, maintenance, and/or reproduction). For each pair, you can have the benefit of one, or the benefit of the other, but you cannot have the benefit of both. Examples of trade-offs include those that occur between the following pairs of variables: extended life span versus early reproduction; egg number versus egg size; reproductive effort versus energy storage. *See also* **life history**.

## trait

Any measurable feature or nonmetrical component of the phenotype (e.g., the characters used in a phylogenetic analysis are referred to as traits, as are the features used in a species diagnosis). (L. *tractus* = drawing. It refers to the distinctive pencil or brush strokes in a drawing or painting.) *See also* **nonmetrical trait**.

## transcription

The process whereby the information encoded by DNA is transferred (aka copied) into messenger RNA (or mRNA): the RNA product of this process is called a transcript. The information encoded in mRNA then takes part in a second process, called translation, to form a polypeptide chain. DNA transcription is initiated or regulated by proteins called transcription factors (e.g., in modern humans the *FOXP2* gene codes for the FOXP2 protein, which is a transcription factor that influences a number of target genes involved with the cognitive and motor aspects of language). (L. *trans* = across and *scribere* = to write, thus to transcribe is to "write across", or to copy.)

## transcription factor

A protein complex that initiates or regulates the process of DNA transcription (e.g., in modern humans the *FOXP2* gene codes for the FOXP2 protein which influences a number of target genes involved with the cognitive and the motor aspects of language). Transcription factors may bind directly to the promoter or other regulatory regions of genes to influence transcription, or they may bind to other transcription factors to affect their activity. Transcription factors are one of many variables that control gene expression.

## transformation grid

A device used in the pre-computer era for visualizing shape differences between two objects whereby one (the reference shape) is typically mapped onto a two-dimensional squared-paper grid and the other (the target shape) is represented as a warping of that grid. The method is the basis of the thin-plate splines that are used in landmark-based geometric morphometrics. Differences between two sets of corresponding (presumably homologous) landmarks are visualized by warping the square grid of one set to fit the differences in the second set; the spline function bends and stretches the grid lines. The warped lines provide an intuitive and appealing model of how the target shape differs from the reference. (L. *transformare* = to change shape, and grid is a back-formation of gridiron, which is of obscure origin, perhaps from *gredire* which likely derives from OF *gredil* = griddle plus ME *ire* = iron.) *See also* **geometric morphometrics**.

## transgenic mice
See **knockout**.

## translocation
The transfer of a section of one chromosome to another, usually nonhomologous, chromosome. Robertsonian translocation is where the long arms of two acrocentric chromosomes fuse at the centromere (acrocentric describes chromosomes for which the centromere is close to one end). A Robertsonian translocation may have been responsible for producing the modern human chromosome 2 from two great ape chromosomes, thus reducing the diploid number of chromosomes from 48 to 46. (L. *trans* = to move across and *locus* = place, so to move from one place to another.) See also **chromosome**.

## transmutation
See **Lamarckism**.

## transverse arch
See **foot**.

## transverse-sigmoid system
See **cranial venous drainage**.

## transverse sinus
See **cranial venous drainage**.

## transverse tori
See **mandible**.

## travertine
See **calcium carbonate**.

## tree
In systematics, a tree is a diagram that depicts the descent relationships among a group of taxa. The term tree is used interchangeably with the term cladogram (syn. cladogram).

## tree ring dating
See **radiocarbon dating**.

## tree topology
The branching pattern of a phylogenetic tree or cladogram. (Gk *topos* = a place and *logos* = word.)

## tribe
The category in the Linnaean hierarchy between a genus and a family. It is the category that some researchers prefer for the *Pan* (Panini) and *Homo* (Hominini) clades. There is no formal definition of a tribe.

## tribosphenic
The name given to the triangular configuration of the three main cusps that forms the basic structure of the upper and lower postcanine tooth crowns. Hence the etymology "wedge-shaped structures that rub together." (Gk *tribein* = to rub and *sphen* = wedge.) *See also* **cusp morphology**.

## trigon
The basic triangular three-cusped (aka tribosphenic or triconodont) structure of a maxillary (upper) postcanine tooth crown in many higher mammals. The three cusps are the protocone at the apex of the triangle on the lingual aspect, and on the buccal aspect the paracone mesially and the metacone distally. (Gk *trigonon* = triangle.)

## trigonid
The basic triangular three-cusped (aka tribosphenic or triconodont) structure of a mandibular (lower) postcanine tooth crown in many higher mammals. The three cusps are the protoconid at the apex of the triangle on the buccal aspect, the paraconid mesially (this cusp has been lost in all higher primates including hominins), and the metaconid distally on the lingual aspect. (Gk *trigonon* = triangle.)

## trigonid crest
A structure on a mandibular molar tooth crown that some call the distal trigonid crest. The incidence of the trigonid crest varies qualitatively and quantitatively among hominin taxa (e.g., it is higher in *Homo neanderthalensis* than in modern humans). The trigonid crest is present on both the outer enamel surface and at the enamel–dentine junction and comparative analysis suggests that a trigonid crest is likely to be the primitive condition for the hominin clade.

## *trigonum frontale*
*See* **frontal trigon**.

## Trinchera del Ferrocarril
One of several sediment-filled cave systems in the Sierra de Atapuerca east of Burgos in northern Spain. The fossiliferous cave/fissure complexes within this system include the sites of Galería, Gran Dolina, and Sima del Elefante. *See also* **Atapuerca; Galería; Gran Dolina; Sima del Elefante**.

## Trinchera Dolina
*See* **Gran Dolina**.

## Trinil
This site on the banks of the Solo River in Indonesia includes exposures of the Bapang (formerly Kabuh) Formation and the underlying Sangiran (formerly Pucangan) Formation; these hominin fossil-bearing sediments are between 1 Ma and 700 ka. Trinil was excavated by Eugène Dubois from 1891 to 1894, and his field assistants continued systematic excavations until 1900. Subsequently, Selenka excavated there from 1906 to 1908 and a joint Indonesian–Japanese

team worked there in 1976–7. The vertebrate fossils (e.g., mammals, crocodiles, and turtles) excavated by Dubois from the site where Trinil 2 (see below) was recovered are known as the Trinil H.K. Fauna (H.K. is an abbreviation of *Hauptknochenschicht* meaning "main bone layer" in German). The fauna suggests that the paleoenvironment of Trinil was diverse and included open savannas, densely covered river valleys, and upland forests. Nine hominin specimens (Trinil 1–9) were recovered by Dubois and his collectors between 1891 and 1900. A calotte found in October 1891, Trinil 2, is the holotype of *Pithecanthropus erectus*. There is no archeological evidence found at this site. (Location 07°22′S, 110°34′E, Indonesia.)

## Trinil 2
This calotte with an endocranial volume of about 940 cm$^3$ was recovered from the Trinil Beds of the Bapang Formation in Indonesia in October 1891. Dubois initially assigned it to *Anthropopithecus erectus* but he subsequently made it the holotype of *Pithecanthropus erectus*; that taxon was sunk into *Homo*, as *Homo erectus*, in 1940. Trinil 2 was the first good evidence for an early hominin outside of Europe. See also **Pithecanthropus erectus**.

## Trinil 3
This complete adult hominin left femur was recovered in August 1892 from the Trinil Beds of the Bapang Formation in Indonesia. The resemblance between Trinil 3 and the femora of upright-walking modern humans was the reason Dubois used the species name *Pithecanthropus erectus*, but doubts have been expressed about Trinil 3's contemporaneity with Trinil 2, the type specimen of *Pithecanthropus erectus* (now *Homo erectus*).

## Trinil fauna
The inclusive name for the fauna found at Trinil and at Trinil-aged sites elsewhere in Java, Indonesia. Not to be confused with the more exclusive Trinil H.K. fauna found in the same horizon as Trinil 2 at Trinil proper.

## Trinil femora
The relatively complete and well-preserved Trinil 3 femur (also referred to as "Femur I") was recovered in August 1892. Four more femora were excavated in 1900, but Dubois did not recognize them as hominins until 1932. A sixth femur (Femur VI in Dubois' scheme) was attributed to *Pithecanthropus erectus*, but subsequent analysis has confirmed that it is not a hominin femur. It is likely that the Trinil femora belong to *Homo sapiens* and not to *Homo erectus*.

## Trinil H.K. fauna
See **Trinil**; **Trinil fauna**.

## triplet codon
See **genetic code**.

## trophic
Any behaviors involved with the acquisition and the processing of food are examples of trophic behavior. (Gk *trephein* = to nourish.)

## trophic level

The position an organism occupies in a food chain. Primary producers, typically photosynthetic plants, use sunlight as their energy source. Herbivores obtain their energy by consuming live plants, and carnivores, called primary consumers, obtain their energy by eating herbivores. Substantial amounts of energy and/or biomass are lost at each stage of the food chain as (a) waste (e.g., feces and urine) and the energy expended on waste excretion, (b) energy used for locomotion, mastication, etc., and (c) energy consumed to heat and cool the body (especially by warm-blooded creatures). It is unlikely that any member of the hominin clade was either an obligate herbivore or an obligate carnivore; they were most likely eclectic feeders (e.g., omnivores). (Gk *trephein* = to nourish.)

## tropical

The region (and the organisms occupying that region) between the Tropics of Cancer (approximately 23° latitude in the northern hemisphere) and Capricorn (approximately 23° latitude in the southern hemisphere) where at some point in the year the sun is directly overhead. Day length in the tropics is less variable than in temperate regions and climate tends to be determined by rainfall rather than temperature. The tropics have high levels of biodiversity, with the latitudinal range of each species generally being less than in the species found in temperate areas (aka Rapoport's Rule). Primates are primarily a tropical order and the hominin clade almost certainly emerged in the tropics. (Gk *tropikos* = pertaining to the solstice.)

## true chin

To be a true chin the mental protuberance must be separated from the alveolar process of the mandible by a hollowed area called an *incurvatio mandibulae*. When that is the case, the mental protuberance is referred to as a *mentum osseum*. See also **mandible**.

## true fossils

Remnants of an organism itself in the form of either hard (i.e., bones and teeth) or soft (i.e., skin, muscle, etc.) tissues. True hard-tissue fossils (i.e., bones and teeth) make up the vast majority of the fossil evidence for human evolution. (OE *tríewe* = steadfast, characterized by good faith and L. *fossile* = something that is dug up.)

## *t* test

A statistical test used to determine whether the mean of a sample of measurements differs significantly from some hypothetical value (usually zero), or whether the means of two different samples of measurements differ significantly from each other (e.g., whether mean cranial capacity size differs significantly between a sample of modern humans and a sample of Neanderthals). A paired *t* test determines whether two sets of observations on the same sample differ significantly from each other (e.g., whether mean endocranial volume is significantly different than mean brain volume within a sample of modern human cadavers). All *t* tests assume that data are normally distributed, both populations have equal variances, and the data were sampled independently; if not, a nonparametric statistical test such as the Mann–Whitney U test or the Wilcoxon signed rank test should be used.

## tuber
See **underground storage organ**.

## tubercle
A small rounded projection on a bone (e.g., scaphoid tubercle) or tooth. (L. *tuber* = lump.)

## tuberculum
An accessory cusp on a tooth crown [e.g., tuberculum intermedium (aka C7) and tuberculum sextum (aka C6)]. (L. *tuberculum* = a small rounded projection, diminutive of L. *tuber* = lump.)

## tufa
A type of speleothem, a tufa is a sedimentary rock (predominantly composed of calcium carbonate) made up of sediments deposited from water percolating through a porous rock in a spring or a lake. The Taung cave was formed as a solution cavity within a flow made up of tufa. (OIt *tufo* = volcanic ash.) See also **calcium carbonate**; **speleothem**.

## tuff
A rock made of small-sized (less than about 4 mm) volcanic debris often in the form of an ash. Most fragments contain enough glass crystals for their age to be determined using the argon-argon or potassium-argon dating methods (e.g., KBS Tuff in the Koobi Fora Formation). (OIt *tufo* = volcanic ash.) See also **tephra**.

## tuffaceous
Any sediment that has a significant tephra (or tuff) component (e.g., reworked volcanic ash deposits). See also **tephra**; **tuff**.

## Turkana Basin
See **Omo-Turkana Basin**.

## "Turkana boy"
See **KNM-WT 15000**.

## turnover-pulse hypothesis
A hypothesis about how environmental change influences evolutionary patterns in the fossil record. The turnover-pulse hypothesis states that most of the turnover within lineages (i.e., the extinction of existing species and the appearance of new ones) is driven by environmental change, and that if environmental changes are strong enough to induce turnover within one lineage then they should be strong enough to induce turnover in other lineages. It proposes that specialist (aka stenotopic) species are more affected by environmental change than generalist (aka eurytopic) ones, and it predicts there will be a temporal ordering of extinction and speciation effects, with stenotopes being affected first and eurytopes later. It was suggested that the diversification of mammal lineages, including hominins, that occurred around when African habitats became cooler and dryer at c.2.5 Ma was an example of a "turnover pulse." However, a detailed investigation of fossil mammals within the Turkana Basin failed to find conclusive evidence of such a pulse at the predicted time.

## type
A specimen to which the name of a taxon is irrevocably attached. If it is designated in the original description it is called the holotype, but if it is designated subsequently it is called the lectotype. (L. *typus* = image.) *See also* **type specimen**.

## Type I error
The type of error that occurs if a result is accepted as significant when it is actually not significant. For example, when the difference in cranial capacity between two samples is accepted as significant when it actually is not, or when two variables are thought to be significantly associated when there is not actually a significant relationship.

## Type II error
The type of error that occurs when a result is judged to be nonsignificant when it actually is significant. For example, when the cranial capacities in two samples are thought to be equal when they are actually different, or when two variables are thought to be independent of each other when there is actually a significant relationship between them.

## type locality
The locality at which the type specimen of a species was collected.

## type section
The geological section specified in the first formal description of a stratum.

## type site
In archeology, a type site is considered to be the model of a particular technocomplex (e.g., Saint-Acheul is the type site of the Acheulean). In geology, a type site is considered to be typical of a particular rock formation.

## type species
A species of a given genus that has been chosen (by the describer of the genus, or, failing that, by a subsequent reviser) to represent the genus. The generic name is thereafter irrevocably attached to the type species, just as a species name is irrevocably attached to its type specimen (e.g., *Homo sapiens* is the type species of the genus *Homo*).

## type specimen
The specimen to which a species' name is irrevocably attached (e.g., Neanderthal 1 with *Homo neanderthalensis*; OH 7 with *Homo habilis*; Trinil 2 with *Homo erectus*). Whatever species the specimen belongs to, it carries its Linnaean binominal with it, and if its binominal has priority (i.e., it has the earliest publication date) then it provides the name of the species. Thus, if someone decided that the type specimen of *H. habilis* really belonged to *Homo rudolfensis*, but that all other members of the *H. habilis* hypodigm were still distinct, then *H. rudolfensis* would be sunk into *H. habilis* (*H. habilis*, 1964 has priority over *H. rudolfensis*, 1986). A new Linnaean binominal would then have to be found for the "old" *H. habilis* hypodigm (minus the transferred type specimen); in this case the appropriate binominal is *Homo microcranous*

Ferguson, 1995. If the type specimen (holotype or lectotype) is destroyed then a new type specimen, called a neotype, can be designated from one of the syntypes. *See also* **holotype; lectotype; paratype**.

## typological

A typological definition of a species is one that allows for relatively little variation. All the individual members of the species would closely conform to the specifications set out in the definition. It is an outmoded (but alas still extant) way of thinking about species.

# U

### UA 31
This c.1 Ma almost complete cranium from Buia in Eritrea is notable for its location and morphology. The site is close to the prehistoric landbridge between Africa and Asia and although in many ways it resembles *Homo erectus*, including its endocranial capacity of around 950 cm$^3$, it lacks an occipital torus and it possesses some features (e.g., a high position of maximum parietal breath) that are seen in later humans.

### 'Ubeidiya
This site, just south of the Sea of Galilee in Israel, is important because of its location and age. Biostratigraphic dating suggests a range of c.1.5–1.4 Ma and is consistent with the magnetostratigraphy that indicates a sequence between c.1.53 and 1.2 Ma. A right $I_2$ is the only *in situ* Lower Pleistocene hominin from the site (NB: fragments of two hominin parietals and one temporal and two teeth are most likely intrusive). The archeological evidence most closely resembles the Developed Oldowan B and the Early Achulean from Olduvai Gorge. (Location 32°41'N, 35°33'E, Israel.)

### uncalibrated radiocarbon age
*See* **radiocarbon dating**.

### unconformity
A break in the stratigraphic record due to either a period of nondeposition or a period of active erosion. In some cases there is an angular difference in the amount and/or direction of dip in the rocks on either side of the stratigraphic break. When there is no such angular difference an unconformity may be very difficult to spot within the stratigraphic record yet within a given sequence unconformities can account for substantial amounts of time.

### underground storage organ
(or USO) The belowground or below-water part of a plant that has evolved to store energy and/or water. Depending on the part of the plant (e.g., root, stem, leaf base) that has been modified these storage bodies are called tubers, corms, bulbs, or rhizomes. USOs can be cryptic and are sometimes deeply buried and therefore may be difficult to acquire. Because of their high caloric value USOs are an important food of modern-day and historic hunter–gatherer groups (e.g., the Hadza of Tanzania). Several researchers have suggested USOs may have

played an important role in hominin evolutionary history, perhaps as fallback foods for the hyper-megadont archaic hominins. This is at least in part because many USOs use the $C_4$ photosynthetic pathway and their consumption by hominins could explain the elevated $\delta^{13}C$ levels seen in some hominins (e.g., *Paranthropus boisei*). They are harder and/or tougher than foods eaten by extant great apes and their consumption could therefore possibly explain the derived craniofacial morphology of *Paranthropus*. USOs have also been proposed as a key dietary component that allowed the evolution of the modern human-like body plan and life history of *Homo erectus* (syn. geophyte). *See also* **grandmother hypothesis**.

## ungual tuft
*See* **apical tuft**.

## ungulate
The informal term for the group of mammals that possesses hoofs. The group contains horses, sheep, goats, and cows, as well as their wild relatives contained within the orders Perissodactyla and Artiodactyla. Ungulates are generally large-bodied, are relatively long-lived, give birth to single (or rarely, multiple) live young, and they are either herbivores or omnivores. Recent genetic studies suggest that ungulates are a polyphyletic group; nonetheless the use of the term to describe the ubiquitous hoofed animals remains a useful, even if cladistically inaccurate, shorthand. (L. *ungulātus* = hoof, from *unguis* = nail.)

## unifacial
*See* **unifacially worked**.

## unifacially worked
A term used in the analysis of stone tools that refers to a piece that is worked (aka shaped) on one surface (aka face). It contrasts with a piece that is worked on two surfaces; such pieces are said to be bifacially worked. (L. *unus* = one.)

## uniformitarianism
A fundamental assumption of all sciences is that natural laws do not change across time. It is an example of inductive logic in that it assumes that modern processes operated in the past. Examples of the assumption of uniformitarianism in human evolutionary studies include (a) reconstructing the biomechanics of early hominin chewing using data derived from observational and experimental studies of living species, (b) dating tuffaceous rocks using the assumption that decay rates of potassium have been invariant through time, (c) estimating body mass of past species using relationships between a proxy (e.g., femoral head diameter) and body mass deduced for living species, and (d) reconstructing aspects of paleoclimate or diet using the stable isotopic behavior of oxygen or carbon as measured in living systems.

## unit
In earth science, an interval of rock strata. Units are the basic components by which a stratigraphic sequence is described and measured in the field. They are commonly bounded by natural breaks or discontinuities, but the scale of the chosen units is determined primarily

by the field geologist according to the scale of analysis and type of investigation being undertaken.

## Upper Cave
See **Zhoukoudian**.

## Upper Paleolithic
A term used to describe the stage of the European and Near Eastern Paleolithic that is defined by the use of Mode 4 stone tool technology (i.e., blade and burin tools). It is characterized by an increase in number and variety of tool types made from stone and organic materials and the proliferation of artwork and symbolic behavior including personal ornaments (beads, jewelry), portable art (figurines), and parietal art (cave paintings and engravings). It appears in eastern Europe around 40 ka (e.g., at Istállóskő and Bacho Kiro) and spreads westward over the course of 5,000–10,000 years and, depending on the area, it is replaced by Mode 5 microlithic technologies between 20 and 10 ka. It is usually linked with *Homo sapiens*, but some early Upper Paleolithic or so-called transitional technocomplexes (e.g., the Châtelperronian) may have been made by *Homo neanderthalensis*. The equivalent term for African sites with similar technologies (but not necessarily similar ages) is Later Stone Age.

## upstream
A single strand of DNA or RNA has a direction determined by the numbered carbon molecules of the sugars that make up its backbone. The term upstream refers to any relative position in a DNA or RNA molecule that is toward the 5′ end of that molecule.

## UR 501
See **HCRP UR 501**.

## Uraha
This site is located in the southernmost area of the Chiwondo Beds close to Lake Malawi. It is significant for its intermediate location between the regional concentrations of early hominin sites in East and southern Africa, for its age of c.2.5–2.3 Ma (according to biostratigraphy), and for the fact that a well-preserved fragment of an adult hominin mandibular corpus (HCRP UR 501) was recovered at the site. The mandible has been assigned to *Homo rudolfensis* or to *Homo habilis sensu lato*. (Location 10°21′0.6″S, 34°09′23.3″E, Malawi.)

## uranium-series dating
An isotopic dating method that measures the daughter products resulting from the decay of uranium, and which is widely used for dating Quaternary deposits and the hominins therein. Uranium-thorium dating (aka $^{230}$Th/$^{234}$U or U-Th) can only be applied to materials from the past c.0.5 Ma because over this length of time all detectable amounts of $^{230}$Th will have decayed. Suitable material older than this can, however, be dated with uranium-lead dating (aka $^{206}$Pb/$^{238}$U or U-Pb). This latter technique, which is relatively new, has been used recently to provide direct radiometric ages for the southern African hominin-bearing cave deposits. The advantages of uranium-series dating are its precision (e.g., errors are typically 0.1–1% for U-Th and 1–10% for U-Pb) and its independence from ambient conditions such as temperature.

Inductively coupled plasma mass spectrometry (ICP-MS) and thermal ionization mass spectrometry (TIMS) are sensitive new counting methods that are up to 10 times more precise than traditional alpha counters. For very high-precision work TIMS is the best, although ICP-MS has the advantage of simpler sample preparation and a higher sample throughput. The relationship between uranium-series dating and ICP-MS is much the same as that of accelerator mass spectrometry (or AMS) to regular $^{14}$C estimations. For U-Th, a single analysis of the U and Th isotopes can produce an age, whereas U-Pb dating requires multiple analyses to construct isochrons from which the age is calculated. The most suitable materials for uranium-series dating are inorganic carbonates, both marine (coral) and terrestrial (stalagmites, flowstones, tufa). Organic material (e.g., bones, teeth, and ostrich egg shell) can be dated by U-Th, but these are open systems, meaning that the uranium can enter or leave the fossil, and the researcher must use models that make assumptions about the uptake or loss of uranium. Simple "early uptake" or "late-uptake" models can be employed, and these assume that the uranium entered the fossil soon after burial, or much more recently. However, most bone and mollusc shell samples have a much more complex uranium-uptake history. Mollusc shells are notorious for giving misleadingly young ages. The speleothem material best suited to uranium-series dating is often not directly associated with fossil hominins or archeological remains; thus it is important to have precise and accurate site stratigraphy.

### ursid
The informal name for the Ursidae (bears), one of the caniform families whose members are found at some hominin sites. *See also* **Carnivora**.

### U-series dating
*See* **uranium-series dating**.

### use wear
Polishes, pits, and striations on the surface of a stone, bone, or metal tool that are the result of the use of that tool. Archeologists are able to examine use wear and develop hypotheses regarding the tool motion and contact material on the basis of actualistic studies. *See also* **use-wear analysis**.

### use-wear analysis
The study of polishes, pits, fractures, and striations on the surface of a stone, bone, or metal tool to understand how the tool was likely used. Although several blind tests using both high-power (high magnification) and low-power (low-magnification) analysis have shown that the method is robust, many researchers in the archeological community are still wary of use-wear analysis, because of early problems with accuracy and reliability. But after years of methodological improvements use-wear analyses are increasingly being used to answer questions about past life and behavior.

## Usno Formation
A group of outcrops immediately to the west of the Omo River in southern Ethiopia and approximately 30 km/19 miles north of the northernmost extent of the Shungura Formation. The oldest sediments of the Usno Formation are c.3.6 Ma and the youngest are c.2.7 Ma;

it is equivalent to Member A and most of Member B of the Shungura Formation. It is exposed at two localities, known as Brown Sands and White Sands; the only hominins recovered are isolated teeth that are similar to those from Hadar and elsewhere that have been assigned to *Australopithecus afarensis*.

## USO
Acronym for **underground storage organ** (*which see*).

# V

## valgus
A joint or an angle within a bone (e.g., the neck-shaft angle of the femur) that is "bent" so that the limb segment beyond the joint, or the shaft of a long bone beyond the angulation, are inclined further away from the midline than normal. Thus, when children have knees that are abnormally close together because the long axis of the lower leg appears to be inclined away from the midline, that "knock-knee'd" deformity is called genu valgum. But valgus and its antonym varus are so often misunderstood their use should be avoided. (L. *valgus* = bent out; ant. varus.) See also **varus**.

## valid
Among what may be several available names for a taxon, only one, the senior synonym, is valid in the sense that it is the correct name according to the rules and recommendations of the International Code of Zoological Nomenclature. For example, *Australopithecus africanus* is the senior synonym for the taxon for which Taung 1 is the holotype. *Plesianthropus transvaalensis* is one of several junior synonyms of *Au. africanus*, so although it is an available taxon name it is not the valid name because it is not the senior synonym. (L. *validus* = strong.) See also **International Code of Zoological Nomenclature**.

## Vallesian
See **European mammal neogene**.

## variability
The tendency for developmental systems to exhibit variation in the face of particular combinations of genetic and environmental variance. The terms variation and variability are often used interchangeably in the literature, but this is not strictly correct. Variability may be thought of as referring to how different a new observation from a population is likely to be compared to earlier observations, whereas variation refers to the amount of difference in a set of observations. Like solubility, variability is a dispositional concept and refers not to the observation of variation but rather to the underlying property of having the potential to exhibit variation. Variability cannot be observed directly, and it can only be inferred from observations about variation in samples of the parent population. (L. *varius* = various.)

## variability selection

A mode of natural selection that is hypothesized to occur when environmental conditions are highly unstable (i.e., they are prone to both large- and small-scale unpredictable changes that may occur over geologically short periods of time). The variability selection hypothesis suggests that under such conditions alleles that confer behavioral flexibility would be favored over alleles that experience strong positive selection during a given climatic extreme (e.g., dry habitats) but negative selection during the opposite extreme (wet habitats). Many modern human adaptations (e.g., intelligence, language, manual dexterity, tool use) that confer behavioral flexibility evolved during the Pleistocene epoch when oscillations in global climates increased in both amplitude and frequency. It has been suggested that variability selection may explain the evolution of these traits, but whereas variability selection is an intuitively appealing hypothesis it is not immediately apparent how it can be tested in the context of human evolution. *See also* **natural selection**.

## variable

When used as a noun, variable refers to any trait, feature, characteristic, or property of interest that can be measured, described, or categorized (e.g., the mesiodistal length of mandibular first molars or the shape of the orbit), which differs in some way among the members of a population in a sample of that population. The variables in a statistical study can be organized by their properties [e.g., continuous, discrete, nominal (aka categorical), ordinal, dependent, independent, etc.]. Variables can be computed (derived) from other variables. Because variables must take on more than one value, sample statistics are typically used to describe the values for a set of observations of a variable for a sample drawn from a population (e.g., the mean, range, or standard deviation for continuous variables; frequencies for categorical variables, etc.). When used as an adjective, variable refers to how much observations within a sample differ from each other. Compared to other samples of the same taxon, or to a sample of a different taxon, the sample in question can be characterized as showing little, average, or substantial variation. (L. *varius* = various.) *See also* **variability**.

## variance

A quantity that expresses the variability (aka dispersion) in a sample. To compute the variance, the deviation of each value from the sample mean is squared. The variance of the population is equal to the sum of these values (aka "sum of squares") divided by $N$, the total number of samples. But extreme values tend to be underrepresented in samples so the sample variance is a biased estimate of the population variance. This bias is corrected by dividing the sample sum of squares by $N-1$ rather than $N$. The bias-corrected sample variance is also equivalent to the mean of the squared deviations of a variable's observed values from their mean, which is also known as the "mean squared error." The square root of the variance is the standard deviation. (L. *varius* = various.)

## variation

The observation that the values for a variable or trait differ from each other among the individuals in a sample. The quality of being subject to variation is known as variability. Variation within a population can be estimated and quantified with sample statistics such as the standard deviation, variance, or range. A homogeneous sample exhibits little variation; a heterogeneous sample exhibits substantial amounts of variation. An emphasis on the study of variation is a key feature of evolutionary biology. (L. *varius* = various.) *See also* **variability**.

## varus
A joint, or an angle within a bone (e.g., the neck-shaft angle of the femur) that is "bent" so that the limb segment beyond the joint, or the shaft of a long bone beyond the angulation, are inclined further towards the midline than is normal. In some children with an unusually obtuse (i.e., large) femoral neck-shaft angle, the knees are so far apart that the lower leg appears to be inclined towards the midline. Thus, that deformity is called genu varum. But varus and its antonym valgus are terms that are often misunderstood so their use should be avoided. (L. *varus* = bent or turned in; ant. valgus). *See also* **valgus**.

## vault
*See* **cranial vault**.

## Venta Micena
*See* **Orce region**.

## Venus figurines
*See* **Gravettian**; **Hohle Fels Venus**.

## vertebral number
The count of vertebrae is recorded as the total for the whole vertebral column [i.e., cervical (Cx), thoracic (T), lumbar (L), sacral (S), and caudal (Ca)]. It is usually written out as the modal numbers for each of the above five regions, in the order given above (e.g., 7, 13, 4, 6, 3 for *Pan troglodytes*), or as the modal number within just one of the five regions of the vertebral column [e.g., for *P. troglodytes* lumbar (L) = 4 (range 2–5)]. Total vertebral number is precisely related to the number of somites (one of the repeated segments of the body that are important during development). All but the most rostral (aka anterior) somites that are incorporated into the cranium give rise to sclerotomes and the sclerotomes give rise to vertebrae and ribs. Each vertebra receives a contribution from two sclerotomes: its own and the one caudal to it (e.g., the C4 vertebra is formed from the caudal part of the C4 sclerotome and the cranial part of the C5 sclerotome). Regional modal numbers of vertebrae can be combined as species-specific vertebral formulae or patterns (e.g., 7, 12, 5, 5, 4 for *Homo sapiens*). Regional vertebral number varies within as well as between species, especially in hominoids. In addition to different patterns, the total number of vertebrae varies within species, usually with a range of 4 or 5. In hominoids, as in almost all mammals, cervical number is effectively invariant at 7, with vertebral number variation tending to increase as you progress caudally along the vertebral column. Of both phylogenetic and functional interest are the total numbers of thoracic plus lumbar vertebrae (documenting trunk length and degree of fore- and hindlimb separation) and their relative contributions to trunk length. For *H. sapiens*, the dominant modal pattern for thoracic and lumbar vertebrae is 12:5, whereas in the African apes it is 13:4. Reasonably complete vertebral columns are rare in the early hominin fossil record, and they are only seen in useful numbers in *Homo neanderthalensis*, whose vertebral column resembles that of *H. sapiens*.

There is no consensus about whether the one *Homo erectus* and two *Australopithecus africanus* individuals that preserve at least some evidence of the vertebral column had five or six lumbar vertebrae. Prior to *c.*3 Ma the hominin fossil record is mute about the organization of the vertebral column. (L. *verter* = to turn.) *See also* **lumbar vertebral column**.

## Vértesszöllős

A travertine quarry west of Budapest that has a range of ages by uranium-series dating of >350 ka to *c.*225–185 ka, equivalent to Marine Isotope Stages 6–11 (424–123 ka). A squamous part of an adult occipital that dates from Marine Isotope Stages 6–7 (243–123 ka) is one of a group of specimens (e.g., Reilingen, Sima de los Huesos, Steinheim, and Swanscombe) that some refer to as *Homo neanderthalensis*. Researchers who interpret *H. neanderthalensis* as an evolving lineage would interpret the Vértesszöllős occipital as an example of "Stage 2" (or perhaps "Stage 1") in this lineage. These stages are also referred to as pre-Neanderthal and "early pre-Neanderthal," respectively. (Location 47°41′N, 18°20′E, Hungary.)

## vicariance

The subdivision of a population or species range that results in two descendant populations separated by a biogeographic barrier (e.g., the split of the aboriginal *Pan* population into *Pan paniscus* and *Pan troglodytes* by the Congo River). Vicariance, which may lead to speciation or extinction in one or more of the descendant populations, is caused by fragmentation of the environment, either because a geographic barrier has developed (e.g., an enlarged river or a recently formed mountain chain) or because of a change in climate. The study of the role played by vicariance events in evolution is called vicariance biogeography. (L. *vicarious* = a substitute, from *vicis* = to change.) *See also* **allopatry**; **vicariance biogeography**.

## vicariance biogeography

A mode of biogeographic analysis that uses cladistic methods to determine whether or not vicariance best explains the distributions of closely related taxa within multiple clades. The principle underlying the method is that because a single ecological/geological change might be expected to affect several different clades in similar ways, vicariance due to environmental or geophysical factors is more parsimonious than dispersal as an explanation of the same geographic distribution applying to several taxa. This is because in the latter case researchers must invoke multiple independent dispersal events to explain the distribution of taxa. (L. *vicarious* = a substitute, from *vicis* = to change, and Gk *bios* = life, *geo* = earth, and *graphein* = to write.)

## vicariance event

*See* **vicariance**.

## Villafranchian

*See* **European mammal neogene**.

## Vindija

More than 60 hominin specimens associated with Mousterian artifacts and identified as *Homo neanderthalensis* have been recovered from the *c.*40 ka levels G3 and G1 at this rock-shelter near Ravna Gora. One of the Vindija hominins, Vi-80 (33.16) from level G3, was an important source of information about the genome of *H. neanderthalensis* because its mitochondrial and nuclear DNA were exceptionally well preserved. (Location 46°17′N, 16°06′E, Croatia.)

### virtual reconstruction.
See **reassembly**.

### viscerocranium
One scheme for subdividing the cranium draws a distinction between the neurocranium and the viscerocranium. The latter is more or less equivalent to the face (aka the part of the cranium that covers the anterior aspect of the brain). (L. *viscus* = organ and Gk *kranion* = brain case.) See **face**.

### visor
See **facial visor**.

### visual cortex
See **striate cortex**.

### vitric tuff
A deposit of pyroclastic material erupted from a volcano in which the majority of the particles are the noncrystalline glass form of ash. Vitric tuffs are central to the geochemical correlation of individual volcanic eruptions because the glass phase of the ash preserves a unique subsample of magma chemistry at the time of an eruption. The correlation of vitric tuffs has been used to relate the products of the same eruption in regional volcanic-rich deposits (e.g., within the Omo- Turkana Basin). (L. *vitrium* = glass.)

### viverrid
Informal name for the Viverridae (aka civets and their relatives), one of the carnivore families whose members are found at some hominin sites. See also **Carnivora**.

### vocal imitation
See **social learning; theory of mind**.

### volar
The volar surface refers to the inferior surface (aka sole) of the foot and the anterior surface (aka palm) of the hand. A movement in the volar direction is a movement that in modern humans takes a finger or thumb towards the palm of the hand. (L. *vola* = the sole of the foot or the palm of the hand).

### volcanic ash correlation
See **tephrochronology; tephrostratigraphy**.

### von Ebner's lines
See **dentine**.

**von Economo neurons**
(or VEN) A specialized class of neurons found at high densities within the anterior cingulate cortex and frontoinsular cerebral cortex of great apes and modern humans. VENs have also been found in the anterior insula of the macaque. These brain areas are involved in monitoring social information and feedback regarding internal body states, and therefore von Economo neurons might play a role in the evolution of self-awareness. Because these neurons were previously only identified in large-brained, social animals such as apes, elephants, and whales, they were initially thought to be part of neural circuits that evolved through convergent evolution to facilitate complex social cognition. However, von Economo neurons were recently located in the hippopotamus, zebra, manatee, and walrus. The interpretation of VENs has since changed from having a purely social role to being a component of independently evolved, taxon-specific networks whose function depends on the cortical distribution of the VENs. (etym. named after Constantin von Economo, the neuroanatomist who described them in 1926.)

**voxel**
*See* **computed tomography**.

# W

## Wadjak
See **Wajak**.

## Wajak
Two modern human-like crania and various fragments of a hominin postcranial skeleton were discovered in 1889–90 at this site in East Java, Indonesia. Uranium-series dating suggests this site has a minimum age of c.37–28 ka. It is one of the few sites in Southeast Asia that has evidence of modern human fossils. (Location 08°06′S, 112°02′E, Indonesia.)

## walking
See **foot function**; **gait**; **walking cycle**.

## walking cycle
A walking cycle describes the movements, distance covered, and time elapsed during one stride of a walking gait [i.e., from the initial contact of one foot with the substrate through to the next time the same (aka ipsilateral) foot touches the substrate]. The beginning of the human walking cycle (conventionally this is 0%) is initiated with heel strike after which that foot is in contact with the ground during the stance phase (this lasts for approximately 60% of the cycle). Toe-off (aka push-off) initiates the swing phase that comprises the remaining 40% of the walking cycle. During the stance phase the center of gravity starts low, rises to its highest point in the middle of stance phase, and falls again at toe-off. For two periods during a single walking cycle the opposite (aka contralateral) foot is in stance phase at the same time as the ipsilateral foot; this is called the double stance phase. The first of these periods occurs from 0% to approximately 10% of the walking cycle. The second occurs prior to toe-off when heel contact has already been made by the contralateral foot. This second double stance phase lasts for approximately 50–60% of the walking cycle. *See also* **foot arches**; **push-off**; **windlass effect**.

## Wallacea
The region in Southeast Asia/Australasia where Alfred Russel Wallace showed that the fauna is neither Southeast Asian nor Australian. Its western boundary is Wallace's Line, which runs between Bali and Lombok in the south then northwards between the eastern

---

*Wiley Blackwell Student Dictionary of Human Evolution*, First Edition. Edited by Bernard Wood.
© 2015 John Wiley & Sons, Ltd. Published 2015 by John Wiley & Sons, Ltd.

edge of the Sunda continental land mass (i.e., Borneo) to the west and Kalimantan and Sulawesi to the east. The eastern boundary of Wallacea (aka Lydekker's Line) is the northwestern edge of Sahul (i.e., Wallacea excludes the New Guinea and the Aru Islands). Wallacea includes islands such as Sulawesi and the Lesser Sunda Islands (e.g., Flores, Lombok).

## Wallace's Line
The western boundary of Wallacea. *See also* **Wallacea**.

## warp
A warp is a representation of the differences in the coordinates of landmarks or semilandmarks between two objects of interest. (ME *warpen* = to twist, so to twist or bend out of shape.) *See also* **transformation grid**.

## wear facets
*See* **tooth wear**.

## Weimar-Ehringsdorf
*See* **Ehringsdorf**.

## Western Margin
*See* **Middle Awash study area**.

## West Turkana
A substantial area of fossil-rich exposures on the western shore of Lake Turkana in Kenya. It consists of an approximately 10 km-/6 mile-wide strip of Plio-Pleistocene sediments between Lake Turkana to the east and the Labur and Murua Rith hills to the west. The sediments exposed in West Turkana, which belong to the Nachukui Formation that ranges in age from 4.3 to 0.7 Ma, include many of the tuffs recognized in the Koobi Fora Formation to the east and the Shungura Formation exposed to the north in the lower reaches of the Omo River. The Nachukui Formation comprises eight members; from oldest to youngest they are the Lonyumun, Kataboi, Lomekwi, Lokalalei, Kalochoro, Kaitio, Natoo, and Nariokotome. Fossil hominins recovered from West Turkana have been assigned to *Australopithecus afarensis* (e.g., KNM-WT 8556), *Kenyanthropus platyops* (e.g., KNM-WT 40000, the holotype), *Paranthropus aethiopicus* (e.g., KNM-WT 16002, KNM-WT 17000), *Paranthropus boisei* (e.g., KNM-WT 17400), and *Homo erectus* (e.g., KNM-WT 15000, aka "Turkana boy"). More than 60 Oldowan and Acheulean sites have been recorded in West Turkana, grouped into eight major complexes and ranging from 2.3 to 0.7 Ma. The KS4 assemblage at Kokiselei pushed the first appearance date for the Acheulean back to 1.76 Ma and it is also an example of the co-occurrence of the Oldowan and the Acheulean. (Location 03°35'–04°30'N, 35°40–55'E, Kenya.)

## Wilcoxon signed rank test
*See* **nonparametric statistics**.

## Willandra Lakes

A series of dried-up lake basins in New South Wales, Australia. The earliest hominin remains to be recovered from what is now referred to as the Willandra Lakes were discovered at Lake Mungo. When fossils began to be discovered at other lakes in the same complex of dried-up lakes it was decided that the name should be broadened from "Lake Mungo" to "Willandra Lakes." Thus the former Mungo 1 skeleton is now called WLH 1, etc. The Willandra Lakes Hominid 50 (WLH 50) partial cranium was discovered in a deflating lake-shore dune between the dry lakes of Garnpung and Leaghur, and hominin footprints ($n=124$) were discovered about 10 km/6 miles north of the WLH 50 find site. When all of the lines of dating evidence are taken into account the likely age of the WLH 50 hominin is $c.32$–$12$ka. It is the focus of an ongoing debate regarding its relevance to modern human origins. One group of researchers claims that WLH 50's robusticity and morphology suggest descent from archaic Southeast Asian hominins such as Ngandong, thereby supporting an evolutionary explanation in line with the multiregional hypothesis. Others claim these same features can be explained by pathology and/or intentional cranial-deformation practices. The modern human footprints were made by adults, adolescents, and children and may indicate collective activities, such as the hunting of water birds at ephemeral water sources. (Location 33°S, 143°E, Australia.)

## Wilson bands

Abnormal accentuated markings in enamel that follow the contour or alignment of the normal, regular, striae of Retzius. The disruptive physiological or pathological events that cause accentuated striae may not coincide exactly with the formation of a normal stria of Retzius, so clusters, or groups, of irregularly spaced Wilson bands are superimposed onto the regular underlying pattern of long-period striae of Retzius. Counts and measurements of Wilson bands are an important way of assessing the degree of stress experienced by an individual during development. The equivalent of Wilson bands in dentine are called Owen's lines. *See also* **striae of Retzius**.

## windlass effect

The mechanism that converts the foot into a solid lever during the push-off phase of walking. The plantar aponeurosis, which attaches posteriorly to the calcaneus, runs across the plantar aspect of the foot, wraps around the metatarsal heads to insert into the base of the proximal phalanges. During the push-off phase of walking the metatarsophalangeal joints dorsiflex, tightening the plantar aponeurosis and raising the longitudinal arch while shortening the foot. At the same time the hindfoot inverts relative to the forefoot, locks the calcaneocuboid joint and thus converts the midfoot into a stiff lever that facilitates efficient propulsion during bipedal locomotion. *See also* **foot arches**; **push-off**; **walking cycle**.

## winnowing

The removal by flowing water of the smallest subset of flaked stone artifacts or bone from an archeological or paleontological site. The effects of winnowing are typically recognized in archeological contexts by comparing the size distribution of artifacts in an assemblage to those from experimentally derived lithic assemblages. The degree to which an assemblage has been winnowed is important to consider when making behavioral inferences since winnowing can alter the spatial integrity of an archeological site and remove certain classes of evidence. (OE *windwain* = the use of wind to separate the lighter chaff from the denser, heavier, seed.)

## WLH 50
See **Willandra Lakes**.

## Wonderwerk Cave
This large (approximately 2400 m$^2$) solution cavity in dolomitic limestone in modern day South Africa is unusual in that it preserves an archeological sequence that spans the Early, Middle, and Later Stone Ages, and this sequence is characterized by excellent organic preservation. Magnetostratigraphy suggests the lower-most strata date to the Olduvai subchron, a hypothesis supported by a single cosmogenic nuclide dating estimate of *c.*2 Ma. Uranium-series dating on speleothems, a single result from optically stimulated luminescence dating, and radiocarbon dating on charcoal constrain the ages of the overlying Middle Stone Age levels to between >220 and *c.*70 ka. The site's preservation of multiple Early Stone Age strata in a cave setting is unusual. Furthermore the cave preserves some of the earliest evidence for fire at *c.*1 Ma. (Location 27°50'45"S, 29°33'19"E, Northern Cape Province, South Africa.)

## woodland
The most widespread biome in tropical Africa, woodlands are areas of clumped trees that do not have a continuous canopy and thus they have a grassy understory (NB: forests do *not* have a grass understory). In general, woodlands are 8–20 m in height and the canopy covers at least 40% of the surface; the density of coverage is related to annual precipitation. Like other biomes, woodlands are a continuum of ecosystems that run from more open to more closed. Most trees are deciduous or semideciduous, but many woodlands contain some evergreen species. Woodland trees tend to be smaller than forest trees. Woodlands of different types (e.g., *Acacia*, miombo) are a significant component of modern sub-Saharan African landscapes and were likely to have been so throughout the whole of hominin evolutionary history. Early hominin sites (e.g., Hadar) are described as mosaic habitats that included open and closed woodlands, shrublands, edaphic grasslands, and gallery forests. While *Australopithecus afarensis* lived in several different types of biome, the most common of these were woodland habitats. Based on faunal evidence from Olduvai Gorge Bed I, it is likely that some African early Pleistocene woodland habitats were more species rich than modern woodlands in the same region, suggesting there were differences between past and present woodland ecosystems.

## Woranso-Mille study area
The Woranso-Mille study area consists of four collection areas (Am-Ado, Aralee Issie, Mesgid Dora, and Makah Mera) that extend along the north bank of the Mille River, Ethiopia, in the northwest part of the study area, plus other collection areas (e.g., Korsi Dora and Burtele) in other parts of the study area. The hominin fossils recovered include 26 dentognathic hominin specimens, plus an associated skeleton (KSD-VP-1/1) and the skeleton of a forefoot (BRT-VP-2/73). Most of the hominin fossils are bracketed by an overlying tuff (KT = *c.*3.6 Ma) and an underlying basalt (*c.*3.82 Ma). Other hominins are just above, within, or just below the *c.*3.7 Ma Arala Issie Tuff (AT). The mandibular and dental morphology and the dental metrics

are mostly in the zone of overlap between the Allia Bay part of the hypodigm of *Australopithecus anamensis* and the Laetoli part of the hypodigm of *Australopithecus afarensis*. (Location 11°30′N, 40°30′E, Central Afar region, Ethiopia.)

**working memory**
*See* **memory**.

**working side**
*See* **chewing**.

**woven bone**
*See* **bone**.

**Würm**
*See* **glacial cycles**.

### xeric
An environment or habitat that is dry. (Gk *xeros* = dry.)

### Xujiayao
(许家窑) This open-air site in Yanggao County, Shanxi Province, China, which consists of two localities, 74093 and 73113, is one of the most productive and important (but least-discussed) paleoanthropological sites in mainland East Asia. A recent magnetostratigraphic study places it in the early Brunhes (chron 1n, *c.*750 ka), indicating that Xujiayao is coeval with Zhoukoudian Locality 1. The Xujiayao hominin collection consists of at least 11 individuals. More than 13,500 stone artifacts have been recovered, including polyhedral cores, scrapers, points, anvils, gravers, choppers, spheroids (potential bolas balls), and other multifunctional implements. Most tools are small and it has been hypothesized that Xujiayao may constitute one of the principal forerunners of the North Chinese microlithic tradition. A taphonomic study of percussion and tooth-mark frequency and cutmark patterns indicated that the Xujiayao hominins had regular primary access to intact artiodactyls and equid long bones. (Location 40°06′N, 113°59′E, northern China.)

# Y

### Y chromosome
One of the sex chromosomes in modern humans, the great apes, and in most mammals. Normal males possess one Y chromosome, whereas normal females have no Y chromosomes. The Y chromosome is much smaller than the X chromosome and only the tips of the Y chromosome (aka the pseudoautosomal regions) undergo recombination with the X chromosome during meiosis. The nonrecombining portion of the Y chromosome has been used to investigate modern human population history, thus providing a male-based scenario that complements reconstructions using female-based (e.g., mitochondrial DNA) data.

### Younger Dryas
A cold interval in the North Atlantic between 12.8 and 11.5 ka that briefly reversed the warming that took place after the Last Glacial Maximum. Evidence of ice rafting in the North Atlantic qualifies it as the most recent Heinrich event (a cold period during which sea ice extends into the ocean). Cold sea surface temperatures in the North Atlantic and the brief resumption of a glacial-type climate led to a dryer climate in Africa at this time. (etym. named after *Dryas octopetala*, a cold-tolerant alpine/tundra wildflower plant species found in Scandinavia.) *See also* **Heinrich events**.

### Yuanmou
Two hominin incisors attributed to the genus *Homo* were recovered from this site in the Yuanmou Formation, China. Recent work shows that the hominin teeth occurred in reverse polarity sediments above the upper boundary of the Olduvai subchron; this *c*.1.7 Ma age is consistent with the occurrence of early Pleistocene mammalian species in the fauna. It is among the earliest evidence for hominins in mainland Asia. (Location 25°40′N, 101°53′E, China.)

### Yunxian Quyuanhekou
(郧县曲远河口) An open-air site located in Yunxian County, Hubei Province, China, that has yielded two large, presumably male, badly crushed adult hominin crania and more than 300 artifacts. They include large cores, uni- and bifacial tools worked on pebbles and large cobbles (e.g., choppers), flakes, and flake fragments. Electron spin resonance spectroscopy dating and magnetostratigraphy provide conflicting evidence about the age of the site, but it almost certainly dates to the middle of the Middle Pleistocene (i.e., *c*.400 ka). (Location 32°50′23.97″N, 110°34′42.35″E, central China.)

# Z

## Zafarraya

This karstic cave, which is located in the northeast of the Málaga province, Spain, includes a typical Mousterian lithic assemblage and hominin fossils. It has been securely dated to $c.30$ ka and this means that the *Homo neanderthalensis* remains recovered from the cave are among the most recent evidence of that taxon. (Location 36°57′05″N, 04°07′36°W, Spain.)

## Zhoukoudian

(or Choukoutien) This cave complex, which produced both the "Peking Man" (originally attributed to *Sinanthropus pekinensis* and now referred to *Homo erectus*) and the "Upper Cave" *Homo sapiens* fossils, is located approximately 42 km/26 miles south of Beijing, China. It was first investigated in 1921 for its quartz stone tools. Two years later, in 1923, two isolated teeth were recognized as being potentially hominin and in 1926 a hominin lower first molar was distinctive enough for a new taxon, *S. pekinensis*, to be proposed to accommodate it. Nearly 200 *H. erectus* fossils representing up to 45 individuals were recovered from Zhoukoudian between 1921 and 1966 as well as stone artifacts and fossils representing at least 98 nonhuman mammalian species. All of the *H. erectus* fossils derive from Zhoukoudian Locality 1 in the Lower Cave, an approximately 40 m-thick depositional sequence that has been divided into layers numbered 1–17. Layer 1 is the highest and the youngest stratum and Layer 17 the deepest and the oldest; most of the hominin fossils derive from Layers 8–9 and 3–4. Thermal ionization mass spectrometry uranium-series dating of Layer 3, which yielded Skull V, are $c.500$–400 ka; the dates for Layers 8–11, which yielded Skulls II, III, X, XI, and XII, are >600 ka; and a recent cosmogenic nuclide dating study suggests an age of $770 \pm 80$ ka for Layers 7–10. Together with climatic correlations, the hominin-bearing layers have been bracketed between 780 and 400 ka [i.e., Hexian (440–390 ka) and Tangshan Huludong ($c.620$ ka) are coeval with the later Zhoukoudian hominins]. Most of the stone tools from Locality 1 are choppers, scrapers, points, or burins. Initial interpretations suggested that Locality 1 was a place where hunted animals were consumed, tools were made, and fire was controlled, but a more recent and taphonomically informed study suggests that giant cave hyenas were the primary residents of the Lower Cave and that *H. erectus* could be best characterized as a transient scavenger who used (but did not necessarily control) fire during their sporadic and ephemeral occupations of the site.

The Upper Cave at Zhoukoudian was discovered in 1930 and excavations from 1933 to 1934 produced three skulls; the remains of at least eight individuals have been attributed to modern

*Wiley Blackwell Student Dictionary of Human Evolution*, First Edition. Edited by Bernard Wood.
© 2015 John Wiley & Sons, Ltd. Published 2015 by John Wiley & Sons, Ltd.

*Homo sapiens* and date to the Upper Pleistocene. The archeological evidence from the Upper Cave is dominated by small stone flake tools and evidence of what are assumed to be personal adornments (e.g., perforated shell, bone, teeth, and stone). Some of these perforated artifacts were found near the necks of some of the interred humans and they are most likely necklaces. A perforated bone needle suggests sewn clothing, and several bones and artifacts were covered with red hematite. Most of the artifacts are associated with the modern human burials from Layer 4, but some were excavated from the upper layers. The presence of sea shells (*Areca* sp.) attests to either a regional trade network or to long-range mobility. (Location 39°41'18.77"N, 115°55'32.50"E, northern China.)

## Zhoukoudian Locality 1
See **Zhoukoudian; Zhoukoudian Locality 1 hominins**.

## Zhoukoudian Locality 1 hominins
There are several well-preserved calvariae (e.g., Skulls II, III, X, XI, and XII), isolated cranial bones, maxillary and mandibular fragments (some with teeth) (e.g., Maxillae III, V, and VI; Mandibles B I, G I), isolated teeth, and some postcranial fossils (e.g., Femora I and IV and Humerus II) from Zhoukoudian Locality 1. The endocranial volumes of the calvariae range from 915 cm$^3$ (Skull III) to 1225 cm$^3$ (Skull V). Both adults and juveniles are represented in the collection. There has never been any serious suggestion that the hominins recovered from Locality 1 sample a taxon other than *Homo erectus* and as such they comprise the single largest site collection of that taxon.

## Zhoukoudian Upper Cave
See **Zhoukoudian; Zhoukoudian Upper Cave hominins**.

## Zhoukoudian Upper Cave hominins
The hominin fossils, which represent at least eight individuals, include three relatively intact crania (UC 101, 102, 103), four mandibles, dozens of loose teeth, and an assortment of postcranial fragments. Some have suggested that the Upper Cave hominins provide evidence of morphological continuity between *Homo erectus* from Zhoukoudian Locality 1 and modern Chinese. However, the results of multivariate analyses suggest that none of the Upper Cave crania have particularly strong phenetic ties to modern-day Chinese.

## *Zinjanthropus* Leakey, 1959
Genus established to accommodate fossil hominins recovered in 1955 and 1959 in Bed I, Olduvai Gorge, Tanzania. Contemporary researchers regard *Zinjanthropus* as a junior synonym of either *Australopithecus* or *Paranthropus*. (Swa. *Zinj* = East Africa and Gk *anthropos* = human being.)

## *Zinjanthropus boisei* Leakey, 1959
Hominin species established in 1959 to accommodate fossil hominins recovered in 1955 and 1959 in Bed I, Olduvai Gorge, Tanzania. Contemporary researchers regard *Zinjanthropus boisei* as a junior synonym of either *Australopithecus* (as *Australopithecus boisei*) or of *Paranthropus* (as *Paranthropus boisei*). It is arguably the most distinctive early hominin taxon. The skull

resembles an exaggerated version of *Paranthropus robustus*, with ectocranial crests, a massive, flat and broad face, large and robust mandibular corpora, small anterior teeth, molarized premolar crowns and roots, and large molar tooth crowns. There is also substantial cranial sexual dimorphism. Until the discovery of *Homo habilis* it was assumed that *Z. boisei* was the manufacturer of the "Oldowan pre-Chelles-Acheul culture." (etym. Swa. *Zinj* = East Africa, Gk *anthropos* = human being, and *boisei* to recognize the substantial help provided to Louis and Mary Leakey by Charles Boise.) *See also* **OH 5**; ***Paranthropus boisei***.

## zooarcheology

Zooarcheology is a subdiscipline of archeology concerned with the study of the animal remains from archeological sites, including bone, teeth, antlers, and shells. Zooarcheological data are used to reconstruct hominin diets and subsistence strategies, past environments, and the interactions between people and animal communities. As a result, zooarcheology provides information critical to understanding both hominin evolution and changes to environments through time. (Gk *zoio* = a living being and *arkhaiologia* = the study of antiquity.) *See also* **diet reconstruction**.

## Zuttiyeh

This >200 ka cave site northwest of the Sea of Galilee, Israel, is best known for a nearly complete frontal with part of the left zygomatic that researchers suggest might sample the population that may have given rise to the hominins sampled at sites such as Skhul. The archeological evidence found at the site was originally regarded as being Mousterian, but it is now recognized as being more similar to the much older Acheulo-Yabrudian. (Location 32°51′N, 35°30′E, Israel.)

## zygomatic prominence

A bulbous protuberance at the anterior end of the zygomatic arch, which is seen in *Australopithecus africanus* (e.g., Sts 5, Sts 71), *Paranthropus robustus*, and KNM-WT 17000. The zygomatic prominence marks the more-or-less abrupt transition from the forward-facing facial surface of the zygomatics to the laterally facing surface of the temporal process. When the zygomatic prominences project anteriorly they contribute to the "dishing" of the face seen in some archaic hominins.